Microbial Consortium and Biotransformation for Pollution Decontamination

Advances in Pollution Research

Microbial Consortium and Biotransformation for Pollution Decontamination

Edited by

GOWHAR HAMID DAR

Department of Environmental Science, Sri Pratap College, Cluster University Srinagar, Jammu and Kashmir, India

ROUF AHMAD BHAT

Department of School Education, Government of Jammu and Kashmir, Jammu & Kashmir, India

HUMAIRA QADRI

Department of Environment and Water Management, Sri Pratap College, Cluster University Srinagar, Jammu and Kashmir, India

KHALID REHMAN HAKEEM

Department of Biological Sciences, Faculty of Science, King Abdulaziz University, Jeddah, Saudi Arabia

Series Editor

MUHAMMAD ZAFFAR HASHMI

ELSEVIER

Elsevier
Radarweg 29, PO Box 211, 1000 AE Amsterdam, Netherlands
The Boulevard, Langford Lane, Kidlington, Oxford OX5 1GB, United Kingdom
50 Hampshire Street, 5th Floor, Cambridge, MA 02139, United States

Notices

Knowledge and best practice in this field are constantly changing. As new research and experience broaden our understanding, changes in research methods, professional practices, or medical treatment may become necessary.

Practitioners and researchers must always rely on their own experience and knowledge in evaluating and using any information, methods, compounds, or experiments described herein. In using such information or methods they should be mindful of their own safety and the safety of others, including parties for whom they have a professional responsibility.

To the fullest extent of the law, neither the Publisher nor the authors, contributors, or editors, assume any liability for any injury and/or damage to persons or property as a matter of products liability, negligence or otherwise, or from any use or operation of any methods, products, instructions, or ideas contained in the material herein.

ISBN: 978-0-323-91893-0

Publisher: Candice Janco
Acquisitions Editor: Jessica Mack
Editorial Project Manager: Joshua Mearns
Production Project Manager: Kumar Anbazhagan
Cover Designer: Greg Harris

Typeset by MPS Limited, Chennai, India

Dedication

Dedicated to all those who lost their life due to COVID-19 pandemic.

Contents

List of contributors

Shackira A.M.
Department of Botany, Sir Syed College, Taliparamba, Kannur, Kerala, India

Shahid Adeel
Department of Chemistry, Government College University Faisalabad, Faisalabad, Pakistan

Muhammad Rahil Afzal
Faculty of Life Sciences, Institute of Environmental and Agricultural Sciences, University of Okara, Okara, Pakistan

Tawseef Ahmad
Department of Biotechnology, Punjabi university, Patiala, India

Zeeshan Ahmed
Institute of Industrial Biotechnology, Government College University, Lahore, Pakistan

Fatima Akram
Institute of Industrial Biotechnology, Government College University, Lahore, Pakistan

Nadia Akram
Department of Chemistry, Government College University Faisalabad, Faisalabad, Pakistan

Zunaira Amjad
Department of Chemistry, Government College University Faisalabad, Faisalabad, Pakistan

Sapna Anam
Department of Chemistry, Government College University Faisalabad, Faisalabad, Pakistan

Yesenia Ithaí Ángeles-López
Department of Entomology and Plant Pathology, North Carolina State University, Raleigh North Carolina, United States

Muhammad Anwar
Guangdong Key Laboratory of Plant Epigenetics, Guangdong Technology Research Center for Marine Algal Bioengineering, College of Life Sciences and Oceanography, Shenzhen, P.R. China

Amna Aqeel
Institute of Industrial Biotechnology, Government College University, Lahore, Pakistan

Priyanka Basera
The Energy and Resources Institute, New Delhi, India

Fatima Batool
Department of Botany, Division of Science and Technology, University of Education Lahore, Lahore, Pakistan

Ashish Bedi
School of Life and Environmental Sciences, Deakin University, Geelong, VIC, Australia

Sashidhar Rao Beedu
Department of Biochemistry, University College of Science, Osmania University, Hyderabad, Telangana, India

Manish Biyani
Department of Bioscience and Biotechnology, Japan Advanced Institute of Science and Technology, Nomi, Japan

Tanveer Hussain Bokhari
Department of Chemistry, Government College University Faisalabad, Faisalabad, Pakistan

Candy Carranza-Álvarez
Facultad de Estudios Profesionales Zona Huasteca, Universidad Autónoma de San Luis Potosí, Ciudad Valles San Luis Potosí, México

Juan Antonio Cervantes-Montelongo
Departamento de Ingeniería Bioquímica, Tecnológico Nacional de México, Instituto Tecnológico de Celaya, Celaya Guanajuato, México

Gowhar Hamid Dar
Department of Environmental Science, Sri Pratap College, Cluster University Srinagar, Srinagar, Jammu & Kashmir, India

Shoaib Ali Dar
Department of Zoology, University of Kashmir, Srinagar, Jammu and Kashmir, India

Sumit Kumar Dubey
Department of Biotechnology, Govt. E Raghavendra Rao Postgraduate Science College, Bilaspur, Chhattisgarh, India

Xiaorong Fan
State Key Laboratory of Crop Genetics and Germplasm Enhancement, Nanjing Agricultural University, Nanjing, P.R. China; Key Laboratory of Plant Nutrition and Fertilization in Lower-Middle Reaches of the Yangtze River, Ministry of Agriculture, Nanjing Agricultural University, Nanjing, P.R. China; Institute of Environment and Ecology, School of The Environment and Safety Engineering, Jiangsu University, Jiangsu Province, P.R. China

Bashir Ahmad Ganai
Centre of Research for Development (CORD), University of Kashmir, Srinagar, Jammu & Kashmir, India

Luis Fernando García-Ortega
Departamento de Ingeniería Genética, Unidad Irapuato, Centro de Investigación y de Estudios Avanzados del Instituto Politécnico Nacional, Irapuato Guanajuato, México

Bashir Ahmad Sheer Gojree
Department of Environmental Science/Centre of Research for Development (CORD), University of Kashmir, Srinagar, Jammu & Kashmir, India

Mir Zahoor Gul
Department of Biochemistry, University College of Science, Osmania University, Hyderabad, Telangana, India

Ikram ul Haq
Institute of Industrial Biotechnology, Government College University, Lahore, Pakistan

Akanksha Jaiswar
The Energy and Resources Institute, New Delhi, India

Ishrat Jan
Department of Zoology, University of Kashmir, Srinagar, Jammu and Kashmir, India

Muhammad Jafar Jaskani
Faculty of Agriculture, Institute of Horticultural Sciences, University of Agriculture Faisalabad, Faisalabad, Pakistan

Rahul Singh Jasrotia
Department of Biological Science, Florida State University, Tallahassee, FL, United States

Arun Kumar Kashyap
Department of Biotechnology, Govt. E Raghavendra Rao Postgraduate Science College, Bilaspur, Chhattisgarh, India

Manmeet Kaur
Department of Microbiology, Punjab Agricultural University, Ludhiana, Punjab, India

Johra Khan
Department of Medical Laboratory Sciences, College of Applied Medical Sciences, Majmaah University, Al Majmaah, Saudi Arabia

Shumaila Kiran
Department of Applied Chemistry, Government College University Faisalabad, Faisalabad, Pakistan

Nalinee Kumari
Department of Zoology, DPG Degree College, Gurugram, India

Tarun Kumar Kumawat
Department of Biotechnology, Biyani Girls College, University of Rajasthan, Jaipur, India

Varsha Kumawat
Naturilk Organic and Dairy Foods Pvt. Ltd., Jaipur, India

Claudia Geraldine León-Ramírez
Departamento de Ingeniería Genética, Unidad Irapuato, Centro de Investigación y de Estudios Avanzados del Instituto Politécnico Nacional, Irapuato Guanajuato, México

Nazir Ahmad Malik
Department of Botany, Dolphin PG college of Science and Agriculture, Chunni Kalan, Punjab, India

Domingo Martínez-Soto
Department of Biochemistry and Molecular Biology, University of Massachusetts Amherst, Amherst Massachusetts, United States; Ingeniería en Innovación Agrícola Sustentable, Instituto Tecnológico Superior de Los Reyes, Tecnológico Nacional de México, Los Reyes Michocán, México

Masarat Nabi
Department of Environmental Science/Centre of Research for Development (CORD), University of Kashmir, Srinagar, Jammu & Kashmir, India

Misbah Naz
Institute of Environment and Ecology, School of The Environment and Safety Engineering, Jiangsu University, Jiangsu Province, P.R. China

Saba Naz
Department of Applied Chemistry, Government College University Faisalabad, Faisalabad, Pakistan

Mohamed Nouri
Faculty of Polydisciplinary, Sultan Moulay Slimane University, Beni Mellal, Morocco

Lucila Ortiz-Castellanos
Departamento de Ingeniería Genética, Unidad Irapuato, Centro de Investigación y de Estudios Avanzados del Instituto Politécnico Nacional, Irapuato Guanajuato, México

Muhammad Ammar Raza
Key Laboratory of Fruits and Vegetables Postharvest and Processing Technology Research of Zhejiang Province, College of Food Science and Biotechnology, Zhejiang Gongshang University, Hangzhou, P.R. China

Rosario Razo-Belman
Departamento de Ingeniería Genética, Unidad Irapuato, Centro de Investigación y de Estudios Avanzados del Instituto Politécnico Nacional, Irapuato Guanajuato, México

Muhammad Riaz
School of Agriculture and Biology, Shanghai Jiao Tong University, Shanghai, P.R. China

Karuna Rupula
Department of Biochemistry, University College of Science, Osmania University, Hyderabad, Telangana, India

Neha Sharma
The Energy and Resources Institute, New Delhi, India

Shivani Sharma
Department of Microbiology, Punjab Agricultural University, Ludhiana, Punjab, India

Vishnu Sharma
Department of Biotechnology, Biyani Girls College, University of Rajasthan, Jaipur, India

Vandana Singh
Department of Microbiology, School of Allied Health Science, Sharda University, Greater Noida, Uttar Pradesh, India

Bhat Mohd Skinder
Department of Environmental Science/Centre of Research for Development (CORD), University of Kashmir, Srinagar, Jammu & Kashmir, India

H.S. Sodhi
Department of Microbiology, Punjab Agricultural University, Ludhiana, Punjab, India

Younas Rasheed Tantary
Department of Botany, Government Degree College, Tangdhar, Jammu and Kashmir, India

Muhammad Tariq
Department of Pharmacology, Lahore Pharmacy Collage, Lahore, Pakistan

Muhammad Usman
Department of Chemistry, Government College University Faisalabad, Faisalabad, Pakistan

John Martin Velez-Haro
Departamento de Ingeniería Bioquímica, Tecnológico Nacional de México, Instituto Tecnológico de Celaya, Celaya Guanajuato, México

Mohammad Saleem Wani
Department of Agriculture, Govt. Mohindra College, Patiala, India

Naseema Aqbar Wani
Department of Botany, Punjabi university, Patiala, India

Javeria Zafar
Institute of Industrial Biotechnology, Government College University, Lahore, Pakistan

About the editors

Dr. Gowhar Hamid Dar (PhD) is currently working as an assistant professor in Environmental Science, Sri Pratap College, Cluster University Srinagar, Department of Higher Education (J&K). He has a PhD in Environmental Science with specialization in Environmental Microbiology (Fish Microbiology, Fish Pathology, Industrial Microbiology, Taxonomy and Limnology). He has been teaching postgraduate and graduate students for the past many years at Post Graduate Department of Environmental Science, Sri Pratap College, Cluster University Srinagar. He has more than 50 papers (h-index 12; i-index 13; total citation >360) in international and national journals of repute and a number of books with international publishers (*Springer, Elsevier, CRC Press Taylor and Francis, Apple Academic Press, John Wiley, IGI Global*) to his credit. Moreover, he is supervising a number of students for the completion of degrees. He has been working on the isolation, identification, and characterization of microbes, their pathogenic behavior, and impact of pollution on development of diseases in fish fauna for the last several years. He has received many awards and appreciations for his services toward the science and development. Besides, he also acts as a member of various research and academic committees.

Dr. Rouf Ahmad Bhat (PhD) has pursued his doctorate at Sher-e-Kashmir University of Agricultural Sciences and Technology Kashmir (Division of Environmental Science) and presently working in the Department of School Education, Government of Jammu and Kashmir. Dr. Bhat has been teaching graduate and postgraduate students of environmental sciences for the past 3 years. He is an author of more than 50 research articles (h-index 14; i-index 19; total citation >591) and 40 book chapters, and has published more than 28 books with international publishers (*Springer, Elsevier, CRC Press Taylor and Francis, Apple Academic*

Press, John Wiley, IGI Global). He has his specialization in Limnology, Toxicology, Phytochemistry, and Phytoremediation. Dr. Bhat has presented and participated in numerous state, national, and international conferences, seminars, workshops, and symposium. Besides, he has worked as an associate environmental expert in World Bank-funded Flood Recovery Project and also the environmental support staff in the Asian Development Bank (ADB)—funded development projects. He has received many awards, appreciations, and recognition for his services to the science of water testing, air, and noise analysis. He has served as an editorial board member and a reviewer of reputed international journals. Dr. Bhat is still writing and experimenting with diverse capacities of plants for use in aquatic pollution remediation.

Dr. Humaira Qadri (PhD) is actively involved in teaching the postgraduate students of environmental science for the past 10 years in Sri Pratap College Campus, Cluster University Srinagar, Jammu and Kashmir, India. She is also heading the Department of Environment and Water Management. A gold medalist at her master's level, she has bagged a number of awards and certificates of merit. Her specialization is in Limnology, Nutrient Dynamics, and Phytoremediation. She has published scores of papers in international journals and has more than 10 books with national and international publishers. She is also the reviewer of various international journals and is the principal investigator of some major projects on phytoremediation. She is guiding a number of research students for PhD program and has supervised more than 60 master's dissertations. She also has been on the scientific board of various international conferences and holds life memberships of various international organizations. With a number of national scientific events to her credit, she is an active participant in national and international scientific events and has organized a number of conferences as well as seminars of national and international repute.

Dr. Khalid Rehman Hakeem (PhD) is a professor at King Abdulaziz University, Jeddah, Saudi Arabia. After completing his doctorate (Botany; specialization in Plant Eco-physiology and Molecular Biology) from Jamia Hamdard, New Delhi, India, in 2011, he worked as a lecturer at the University of Kashmir, Srinagar, for a short period. Later, he joined Universiti Putra Malaysia, Selangor, Malaysia, and worked there as a postdoctorate fellow in 2012 and fellow researcher (associate professor) from 2013 to 2016. Dr. Hakeem has more than 10 years of teaching and research experience in plant eco-physiology, biotechnology, and molecular biology, medicinal plant research, plant–microbe–soil interactions, as well as in environmental studies. He is the recipient of several fellowships at both national and international levels; also, he has served as the visiting scientist at Jinan University, Guangzhou, China. Currently, he is involved with a number of international research projects with different government organizations.

So far, Dr. Hakeem has authored and edited more than 65 books with international publishers, including *Springer Nature, Academic Press (Elsevier), and CRC Press*. He also has to his credit more than 135 research publications in peer-reviewed international journals and 62 book chapters in edited volumes with international publishers.

At present, Dr. Hakeem serves as an editorial board member and a reviewer of several high-impact international scientific journals from Elsevier, Springer Nature, Taylor and Francis, Cambridge, and John Wiley publishers. He is included in the advisory board of Cambridge Scholars Publishing, UK. Prof. Hakeem has recently been elected as the fellow of the Royal Society of Biology, London, UK. He is also a fellow of Plantae group of the American Society of Plant Biologists, member of the World Academy of Sciences, member of the International Society for Development and Sustainability, Japan, and member of Asian Federation of Biotechnology, Korea. Dr. Hakeem has been listed in Marquis Who's Who in the World, from 2014 to 2019. Currently, Dr. Hakeem is engaged in studying the plant processes at eco-physiological as well as molecular levels.

Foreword

The book "*Microbial Consortium and Biotransformation for Pollution Decontamination*" to be published by Elsevier Academic Press (AP), having Gowhar Hamid Dar, PhD, Rouf Ahmad Bhat, PhD, Humaira Qadri, PhD, and Khalid Rehman Hakeem, PhD, as editors—four important scientific personalities—with achievements in related fields: environmental contamination, microbial consortia, biotransformation, enzymatic bioremediation, microbial genome, will be a landmark for those interested in this information.

Microorganisms can be considered chemical machines, which transform various substances both inside and outside. This is due to its enzymatic activity, which accelerates the rates of specific chemical reactions. Microorganisms are found in virtually every environment on Earth, familiar or not to higher life forms. The application of molecular biology techniques in the field of microbial ecology in this book will make it possible to explore existing microbial biodiversity.

According to the information in this book, it will even be possible to identify microorganisms in their natural microhabitats, that is, in situ. This is especially important in the study of extremophilic microorganisms, whose optimal growth conditions are complex to simulate in the laboratory. On the other hand, recombinant DNA technology with the use of genetically modified microorganisms has allowed the elimination of environmental pollutants in bioremediation processes.

The book "*Microbial Consortium and Biotransformation for Pollution Decontamination*" combines microbiology with ecological theory, by studying the diversity of microbial functional roles in their natural environment.

Microorganisms are the largest biomass on Earth, so it is not surprising that their ecological functions or roles affect the ecological history of ecosystems, and there is so much information that explains their activity. An example of this influence is the appearance of aerobic life forms due to the accumulation of oxygen (O_2) in the primitive atmosphere, generated by the photosynthetic activity of cyanobacteria.

Microbial ecology is transversal to all other disciplines of microbiology and as evidenced in its content: microbial diversity and its evolutionary history; interactions between microorganisms in a population and between populations in a community; interactions between microorganisms and plants; phytopathogens (bacterial, fungal, and viral); interactions between microorganisms and animals, microbial communities, their composition and succession processes; microbial adaptations to environmental conditions, but also some types of microbial habitats (atmosphere-ecosphere, hydro-ecosphere, litho-ecosphere, and extreme habitats).

Decontamination studies the application of biological agents (microorganisms and/or enzymes and their plants) in the recovery processes of soils and water contaminated with substances dangerous to human health and the environment. Many of the current environmental problems can be solved with the use of the microbial component of the global ecosystem.

Research areas for bioremediation in this book will be highlighted by the microbial metabolic capacities applicable in environmental sanitation processes and microbial interactions with inorganic and xenobiotic pollutants (toxic synthetic products, which are not generated by natural biosynthetic processes).

Among the most studied xenobiotic compounds are halocarbons, nitroaromatics, polychlorinated biphenyls, dioxins, alkylbenzyl sulfonates, petroleum hydrocarbons, and pesticides. Among the most studied inorganic elements are heavy metals and finally the biodegradability of environmental pollutants in situ and in the laboratory.

This book will be a useful reference for the readers to access information about the discovery of new microbial metabolic pathways with potential applications in commercial value processes; reconstruction of microbial phylogenetic relationships; analysis of aquifers and public drinking water supply; dissolution or leaching (biolixation) of metals in the environment, for their recovery; biohydrometallurgy or biomining of heavy metals, in the bioremediation processes of contaminated areas; biocontrol of microorganisms involved in the biocorrosion of radioactive waste containers dissolved in underground aquifers; the construction of primitive terrestrial history, the paleoenvironment, and primitive life forms; building useful models in search of fossilized life on other planets, and sanitizing areas contaminated with xenobiotic or inorganic substances, among many other applications.

Monica Butnariu
Chemistry and Biochemistry Discipline, Banat's University of Agricultural Sciences and Veterinary Medicine "King Michael I of Romania", Timisoara, Romania

Preface

While the global population continues to rise at an astonishing rate, the intensive agricultural and industrial systems needed to support such a large number of people inevitably cause a huge rise of soil, water, and air pollution. Estimates have attributed pollution to 62 million deaths each year, which is 40% of the global total. While the scientists all over the world are busy looking for sustainable solutions for pollution prevention and control, bioremediation is gaining pace as the most viable technology.

Microorganisms are well known for their ability to break down a huge range of organic compounds and absorb inorganic substances and are finding vast applications in pollution treatment processes. However, microbial remediation can be effective only in places where environmental conditions permit microbial growth and activity. Where the conditions are not favorable for their growth, manipulation of environmental parameters is carried out to allow microbial growth and degradation at a faster rate.

In this backdrop, this present book has been compiled as an extension of the bioremediation technologies in context of the microbial applications. This book has been divided into 16 chapters.

Chapter 1 concerns the rising reuse of treated effluent in the agricultural sector for irrigation. It explicitly explains the benefits of the wastewater irrigation which will minimize purification and expense of fertilization, since soil and crops act as biofilters, whereas sewage water holds large quantities of nutrients.

In order to address the concerns regarding potential adverse health impacts for the general population, Chapter 2 details the complete epidemiological literature and representative case reports connected to emerging contaminants and humans including perfluorinated chemicals, gasoline additives, man-made nanomaterials, human and veterinary medications, and ultraviolet filters.

Furthermore, in Chapter 3 an effort has been made to summarize different types of wastewater, its effects on freshwater ecosystem, and remedies to reduce the effect of these effluents on freshwater ecosystem and on humans.

At a time when numerous strategies have been developed for the treatment of wastewater, conventional detection methods have become expensive and inefficient in tracing various wastewater pollutants.

Biosensors have drawn considerable attention in this field because of their highly sensitive, selective, cost-effective, and on-site detection capability. Chapter 4 provides a comprehensive overview of various classes of biosensors that have been used for monitoring organic, physical, and chemical pollutants. Various limitations and future prospects of the strategies involving biosensing have also been discussed.

In Chapter 5 microbiological systems and their impact on animal, plants, changes in atmosphere, and sustainable development have been discussed in detail. Microbial consortia research, on the other hand, is still in its infancy, with numerous unknown factors and hurdles in cell signaling and the establishment of robust and regulated microbial consortia processes. In this context Chapter six highlights some of the success stories of defined microbial consortia in the area of environmental bioremediation. Chapter 6 reviews various biological transformation methods for resource pollution control naturally providing a critical review of the extensive work done by various research groups throughout the last 5 years to address this issue followed by a detailed account of the decontamination techniques involving biological transformation in Chapter 7.

Enhanced biological phosphorous removal (EBPR) is a biological process for phosphorous removal by using certain anoxic, anaerobic, and aerobic steps, and optimized pH, temperature, and concentrations of polyphosphate-accumulating organisms (PAOs) may play an important role in regulation of this process. Chapter 8 explains the basic concepts, mechanisms, and factors affecting EBPR process. Furthermore, future technologies for P-removal using biological processes and PAOs are effectively explained.

Environmental challenges may be solved via genetic engineering. The genetically engineered bacteria can break down harmful contaminants in the environment. Chapter 9 describes the advancements and potential of genetically engineered bacteria for constructing a sustainable environment for human happiness. Keeping in view the extensive use of microbes in the industries, Chapter 10 elaborates the fungal decolorization systems and their particular function in decolorization of textile dye color effluents followed by an elaborate account of *endophytic and growth-promoting bacteria in the bioremediation of contaminants in Chapter 11.*

Chapter 12 deals with an updated discussion of including their biological mechanisms as well as symbiotic/nonsymbiotic relationships between fungi with other microorganisms and plants for the use in the bioremediation of contaminated sites. Chapter 13 is mainly emphasized on the classification of biosurfactant, its microbial origin, and its applications as bioremediator of hydrocarbons, oil, and petroleum.

Traditional trickling filter systems and activated sludge biofilm wastewater treatment systems are applicable for removal of pollutants from the wastewater without using harmful chemicals. Resistance of biofilm to harsh environmental conditions adds benefit to microbial communities to deal with toxicity on biofilm-based treatment systems. Additionally biofilm-based wastewater treatment processes are stable and facilitate physiologically and biochemically active pathway and optimized processes. Chapter 14 deals with the biofilm formation, molecular biology of biofilm formation and mechanism, current status, and future prospects.

The possible use of the microbes, conditions, and their mode of action for treatment of effluent are detailed in Chapter 15. This chapter will be useful for the global community who are nowadays using advanced methods for the treatment of water to make effluent water recycled for use. In Chapter 16 a comprehensive overview of various physical methods is given pointing out their merits and demerits. It also includes their positive economic and environmental aspects.

In general, this book is covering all the aspects of microbial technology that will lead to a better pollution-free tomorrow where development and preservation of environment shall go hand in hand.

Gowhar Hamid Dar
Rouf Ahmad Bhat
Humaira Qadri
Khalid Rehman Hakeem

Acknowledgments

Editors would like to thank and appreciate the efforts of all the authors who contributed for this volume on "Microbial Consortium and Biotransformation for Pollution Decontamination". Acknowledgment is also extended to the Principal of Sri Pratap College, Higher Education Department (J&K) and Cluster University Srinagar for the encouragement and support in bringing this volume detailing the effectiveness of microorganisms against the rising pollution levels. Editors would also like to take this opportunity to thank the faculty and nonteaching staff of PG Department of Environmental Science and Department of Environment and Water Management, Sri Pratap College, Cluster University Srinagar, for their kind guidance and support. Editors would also like to acknowledge the help of all those friends and colleagues who encouraged us to edit this volume. Special thanks to Mr. Asif Ahmad Bhat and Mr. Shahid Bashir Haji for their lovable and adorable support while conscripting and editing this book, and it would not have been possible without their positive encouragement throughout.

About the book

This book entitled "Microbial Consortium and Biotransformation for Pollution Decontamination" picturized a detailed insight into the various essential aspects of the microbiological consortia and their role in the cleanup of different environmental pollutants with the help of suitable practically viable case studies. Besides, the main objective by bringing this volume was to provide a detailed concept pertaining to present time's demand of eco-friendly technology and cost-effective alternative techniques to depollute the contaminated environs.

The book consists of 16 chapters addressing linked themes that reconnoiter various features of microbial-based techniques and mechanisms to deal with different types of environmental contaminations. These chapters are summarized comprehensibly and provide a broad view about the applicability, sustainability, and eco-friendly nature of microorganisms to remediate environmental pollution. Biosensors for monitoring contamination, polyphosphate-accumulating organisms in enhanced biological phosphorous removal, genetically engineered bacteria: a novel technique for environment decontamination, degradation of Azo dyes, endophytic and growth-promoting bacteria in the bioremediation of contaminants, biosurfactants for the recovery and remediation of oil and petroleum waste, and microbial biotechnology have been given an important space in this book. In general, the book provides an outline about the vitality of potential microorganisms to reestablish polluted environs.

CHAPTER 1

Threats and consequences of untreated wastewater on freshwater environments

Ishrat Jan[1], Tawseef Ahmad[2], Mohammad Saleem Wani[3], Shoaib Ali Dar[1], Naseema Aqbar Wani[4], Nazir Ahmad Malik[5] and Younas Rasheed Tantary[6]

[1]Department of Zoology, University of Kashmir, Srinagar, Jammu and Kashmir, India
[2]Department of Biotechnology, Punjabi university, Patiala, India
[3]Department of Agriculture, Govt. Mohindra College, Patiala, India
[4]Department of Botany, Punjabi university, Patiala, India
[5]Department of Botany, Dolphin PG college of Science and Agriculture, Chunni Kalan, Punjab, India
[6]Department of Botany, Government Degree College, Tangdhar, Jammu and Kashmir, India

1.1 Introduction

By 2050 the human population is predicted to exceed 9.3 billion, causing a high demand for drinking water and resulting in a significant rise in domestic and industrial wastewater production. An unexpected release of waste into the marine ecosystem and the unusual inability of wastewater treatment plants (WWTPs) to eradicate certain toxic compounds have increased concerns about the imminent quality of water in lakes, rivers, estuaries, and oceans. For many species, these water bodies are significant, as they fill in as nursing grounds and as important fish and bird food resources and supplements (Allanson & Baird, 1999; Bhat et al., 2021). Many natural and synthetic chemicals, such as pharmaceutical materials, endocrine disruptor metals, NH_3, pesticides, and polycyclic aromatic hydrocarbons, have been detected in diluted domestic wastewater or liquid sewage and also in purified domestic sewage (Chambers et al., 1997; Daughton & Ternes, 1999; Kummerer, 2001). A large number of these chemicals are also reported in the communities which are consuming the sewage either directly or indirectly (Kolpin et al., 2002).

The life-threatening effects were observed in aquatic organisms (fish and molluscs) which are subjected to checked and unchecked urban wastewater These effects include increased metallothionein and blended-function oxidase activity, damage to DNA, induction of vitellogenin, proliferation of lymphocytes, and decline in phagocytic activity (Diniz, Peres, & Pihan, 2005; Gagne

Microbial Consortium and Biotransformation for Pollution Decontamination
DOI: https://doi.org/10.1016/B978-0-323-91893-0.00009-2

et al., 2002; Hoeger, Van den Heuvel, Hitzfeld, & Dietrich, 2004). Waste from populated neighborhoods, factories, and industrial plants flow into water bodies, including rivers, which also function as drainage systems. Elevated concentrations of organic waste from these sites lead to an elevated demand for oxygen because of consequent upsurge in decomposition processes. The consumption of dissolved oxygen (DO) and ammonia from the decomposition of organic nitrogen compounds results in increased degrading processes.

Typically, industrial discharges are believed to be the main industrial effluents involving acids, alkalis, suspended solids, and various other organic and inorganic compounds (Behera & Misra, 1982). The industries producing phosphate compost also produce a specific type of extremely acidic contaminant (pH-2) that comprises substantial quantities of fluoride and phosphate (Altintas, Ayok, & Gozmen, 1987; Grzmil & Wronkowski, 2006). The significant amount of fluoride and phosphate chemicals and, from the fertilizer industry, the huge daily flow rate of phosphoric acid wastewater as well as the output of triple superphosphate result in several thousand tons of phosphate being removed annually and more (Altintas et al., 1987; Grzmil & Wronkowski, 2006). In industrial runoff, elevated fluoride can present many health issues (Agarwal, Gupta, & Gupta, 1999; Hichour, Persin, Sandeaux, & Gavach, 2000; Islam & Patel, 2007). The ruining of water resources with these phosphate chemicals poses a significant problem of water contamination, particularly "eutrophication" of lotic water (Ruixia, Jinlong, & Hongxiao, 2002). In addition, sustainability is increasingly imitated as a critical ecological concern in government policy, business strategy, and international treaties (Morse, Lester, & Perry, 1993). The removal procedures to set for fluoride and phosphate impurities will permit the reprocessing of these two impurities as viable products for use either in agriculture or as raw materials for application in industries (Altintas et al., 1987; Battistoni, De Angelis, Prisciandaro, Boccadoro, & Bolzonella, 2002; Ben Ayed & Bouaziz, 2008; Kara & Stevens, 2002). This is worth highlighting that for the growth and development of plants, phosphorus is an important supplement. Phosphorous has a crucial role in protein activation, energy transfer, and the regulation of metabolism (Priya & Sahi, 2009). Though phosphorus abounds in many soils, it is one of the least accessible mineral nutrients in the plants, as a large volume of total soil P (50–80%) is present in organic forms (Dalai, 1977). Plant growth is highly influenced by soil phosphorus availability (Van den Berg, Gomassen, Roelofs, & Bokkink, 2005) and affects its configuration. It adds to the enormous need for phosphate composts using numerous methods (Garg & Bahl, 2008). In soil, fluorides are rarely leached as they are very immobile. Some metal ions may influence the bioavailability of fluoride to plants because of precipitation, pH changes, complexation, etc. (Cronin, Manoharan, Hedley, & Loganathan, 2000). Fluoride absorption is highly based on pH of soil. Natural soil has strong affinity for fluoride and holds this element sturdily. The uptake of excess fluoride by plants (such as leaching loss) from soil is usually low. However, in acidic soils, the fluoride solubility is high, allowing high absorption by plants (Daines, Leone, & Brennane, 1952). Fluoride absorption by plants is influenced by type of soil, organic matter, pH, and amount of P and Ca (Hall & Cain, 1972; Treshow, 1971). Through mixing huge amount of soluble fluoride into acidic soils (pH 5), plant absorption of fluoride increases, but this absorption remains limited for well-limed soils (Dar, Bandh, & Kamili, 2011; Singh et al., 1995). The growing reuse of treated waste in agriculture helps to endorse sustainable agriculture, improve the quality of the ecosystem, and control scarcity of water supplies. In addition, watering with wastewater will minimize fertilization and cleaning costs, since crops and soil are used as biofilters, whereas there is a huge amount of various nutrients in wastewater (Haruvy, 1997). Discharge from phosphate compost industry harbors several beneficial features for farming, since it includes phosphorus (P), magnesium (Mg), potassium (K), and minimal quantities of heavy metals. Therefore consideration was specified to authorizing sustainable treatment strategies for using treated sewage water in farming (Hu, Lo, Kuan, & Lee, 2005). Hence, the runoff which is rich in phosphate and poor in fluoride quantities must be produced to endorse the use of treated sewage water in farming. Water phytotoxic impacts should be overcome because they are the result of a mixture of numerous variables (Zucconi, Pera, Forte, & De Bertoldi, 1981), such as low-molecular-weight fatty acids (Zucconi et al., 1981), various heavy metals (Wollan, Davis, & Jenner, 1978), and salts (Adriano, Chang, Pratt, & Sharpless, 1973) that are liable for inhibitory impacts (Zucconi, Monaco, Forte, & De Bertoldi, 1985). The important tools for measuring putative ecological menaces are the bioassays (Gopalan, 1999). They are consistent, fast, simple, and financially savvy (Gustavson, Sonsthagen, Crunkilton, & Harkin, 2000). The utilization of plants provides a favorable position over different life forms since they can be increasingly delicate to ecological pressures (Dutka, 1989). It is easy to monitor and store plants, as they are cost-effective and also provide strong correlation (Fiskesjo, 1993; Valerio, García, & Peinado, 2007).

Untreated impurities from industrial area are released in local water bodies ensuing in stink, discoloration, and slimy nature of these water bodies (Akaninwor, Anosike, & Egwim, 2007). Much of the farming land had moderate rates of wastewater treatment in about 70 countries, exposing 885 million urban customers as well as agronomists and food suppliers to grave health issues. The bulk of the agricultural land is occupied by five countries—Iran, China, India, Pakistan, and Mexico. These figures give a sense of the crucial role wastewater plays in meeting the need of water and food around the biosphere, as well as the need to put resources into both treating wastewater and more stable practices to ensure ecological and general well-being. Wastewater, if appropriately treated, is an incredible chance; however, we have to discover methods of utilizing it without risking food quality. Wastewater is a benefit if the correct arrangements, advances, and money-related motivators are set up. And yet, awareness needs to be raised among those managing wastewater reuse to avert illness and boost benefits. Wastewater release vigorously impacts both organisms and our water bodies. We estimate that extreme pathogenic contamination—essentially from an absence of sanitation treatment—can be found in up to 33% of streams in developing nations, prompting direct health effects.

Vital sources of freshwater such as rivers also function as the finest sinks for discharge of household and industrial pollutants (Das & Acharya, 2003; Tukura, Kagbu, & Gimba, 2009). Sewage is the major contributor of water contamination issue. Several developing countries generate wastewater on an average of 30–70 mm^3 per annum per person. Due to the absence or inadequate treatment systems for wastewater, it along with its dregs are frequently drawn into open water bodies, which act as reservoirs for local and industrial waste causing contamination of ecosystem. Low-quality wastewater effluent is the source for the deterioration of open water bodies. Proper methods must be followed for treating the effluents of the wastewater and prevent risks to the well-being of the people using such water resources. The ejection of inappropriately treated sewage into various water pools has several consequences on the ecosystem and the human comfort. Accordingly, there should be valid water and environmental laws for the protection of the power of both rural and municipal population.

1.2 What is sewage?

The discarded liquids and solids frequently carried away by drains are called sewage. As per the "International Journal of Environmental Research and

Public Health," "wastewater is defined as any storm water runoff, as well as industrial, domestic or commercial sewage or any combination thereof carried by water." Various kinds of wastewater are as follows:

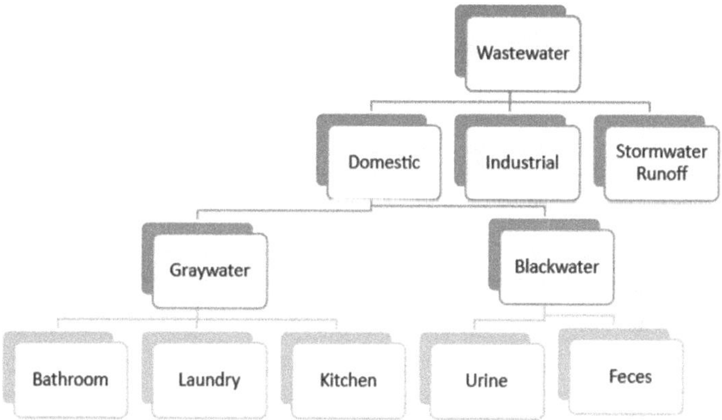

The four fundamental kinds of wastewater are industrial, domestic, urban, and rural. Local wastewater comprises fecal matter of humans and animals and wastes from other activities such as washing, bathing, cooking, and gardening. Industrial wastewater contains agricultural waste, such as mud, plastic, petrochemical overflow, salts, acids, and organic chemicals. Agrarian sewage originates from farming operations, polluted soil, and agricultural practices, particularly those related to pesticides and fertilizers. Municipal sewage is characterized as a mixture of industrial and household wastewater and sewage invasion and downpour water.

1.3 Contaminant sources of emerging concerns

1.3.1 Wastewater

One of the principle sources of contaminants of emerging concern (CEC) is the municipal sewage such as industries, point and nonpoint sources, tempest water, water treatment facilities, and wastewater from family units into the ecosystem. Once these contaminants of emerging concern are stored in sewer lines and devoured by populace, CEC stream to municipal wastewater treatment plants (MWWTPs), are generally not intended to evacuate those (Lofrano, Pedrazzani, Libralato, & Carotenuto, 2017). Wu et al. (2016) in Shanghai checked 10 antibiotics in fourWWTPs detailing their presence in most emanating trials. WWTPs serve as antibiotic hotspots for the open water bodies. Comparable outcomes were accounted for, which investigated the doom of 19 antibiotics in one stabilization pond (SP), one constructed wetland (CW), one micropower biofilm (MP), and one activated sludge (AS) (Dong, Yuan, Wang, & Qiang, 2016). Despite the fact that target antibiotics had lower mean profluent concentrations than the dominant ones, their elimination was typically insufficient. The AS and CW beat the MP and SP processes. Both the AS and CW forms displayed advanced evacuation competences in summer rather in winter, demonstrating that natural degradation could undertake a significant part in antibiotic squalor regardless of whether their total expulsion stood unaccomplished. Clinic wastewater is found to contain high amount of pharmaceuticals. For instance, ciprofloxacin was found in the concentrations of somewhere in the range of 0.7 and 124.5 mg/L in clinic profluent. Similarly, ampicillin was found in concentrations of somewhere in the range of 20 and 80 mg/L in the profluent. Antibiotic concentrations determined and estimated in hospital effluents are of the similar significant degree as base inhibitory amounts for defenseless disease-causing microorganisms. The weakening of clinical waste by metropolitan wastewater will bring down the quantity of antibiotics only moderately on the grounds that city wastewater additionally holds disinfectants and antibiotic substances from veterinary sources, households, and to a minor degree from domesticated animals (Kummerer, Al-Ahmad, & Mersch-Sundermann, 2000). Triclosan (TCS) and triclocarban (TCC) are among top 10 most regularly recognized biological wastewater composites for recurrence and concentration. Triclosan has been recognized in WWTPs gushing in amount more noteworthy than 10 mg/L (Brausch & Rand, 2011). Triclocarban has been distinguished in large amounts and often in WWTP profluent and surface waters (SWs) than triclosan or methyl-TCS in the course of the last 5 years (Brausch & Rand, 2011). PBDEs (polybrominated diphenyl ethers) enter WWTPs by the removal of domestic wastewater because of the release of water from industrial materials containing PBDEs and washing of mats, covers, or dim water and leachate from landfilled PBDE-containing items. Engineered NPs (nanoparticles) are discharged in the ecosystem in a roundabout way from items during their production and their undeniably normal use through wastewater release.

1.3.2 Sewage sludge

It is the by-product arising from the treatment processes that occur in WWTPs. It might encompass microbes isolated from the fluid stage, just as an extensive range of natural and inorganic matters, together with CEC (Martin-Pozo et al., 2019). Attributable to the physical, chemical, and natural procedures engaged with the WWTPS, the slop will in general ponder heavy metals and ineffectively decomposable trace organic compounds just as conceivably disease-causing vectors (protozoans, bacteria, viruses, and so on.) which occur in wastewaters. Slime is rich in supplements, for example, nitrogen, fluoride, phosphorous, and important organic matter that is valuable if soils are subjected to erosion. Amid a few removal courses for sewage slush (burning, landfill, fertilizing the soil, and so forth), its deposition on rural land speaks to a far-reaching technique because of its little economic expense and natural benefits to soil. Organic matter and nutrients are the two principle components of wastewater on land, which aid in the dispersal of manure or act as an appropriate natural soil reformer. The production of sewage—muck was of 9.4 million tons from urban MWWTPs in the EU in 2015, from which 39.6% was reused for agricultural purposes, though the occurrence and likely dangers of developing contaminants in water and the rest 60.4% were utilized as fertilizers or for other uses. Since 2010 these figures have marginally diminished with total slush output hitting 9.7 million tons in the EU, 45.2% of which was applied to land. In any case, this method could be related to perilous results in the ecosystem and public well-being particularly as to CECs' value. Ground water (GW) sullying due to draining or land spillovers could prove hazardous to people's well-being through the food chain because of consuming stale water or utilization of polluted harvests (Healy et al., 2017). Sewage slime reuse is supported by the European Directives (Council of the European Communities, Directive 91/271/EEC of May 21, 1991) to forestall further amphibian condition contamination and energize a productive administration of the slush. To this end, EU (Council of the European Communities, Directive 86/278/EEC of 12 June, 1986) and Spanish legislations have set limits for the concentration of metals and supplements found in the slime (Real Decreto 1310/1990, Ministerio de Agricultura, 1990). A few emerging contaminants have been distinguished and estimated in processed sewage slop for quite a long time. As of now, no guidelines are set for the existence of these pollutants in wastewater, though a continuous working archive containing limit esteems for concentrations of seven gatherings of natural pollutants (11 polycyclic aromatic hydrocarbons, bis(2-ethylhexyl) phthalate, linear alkylbenzene sulfonates, nonylphenolethoxylates, halogenated organic compounds (AOX), and 7 polychlorinated biphenyls just as dibenzofurans

(PCDD/F) and polychlorinated dibenzodioxins, that is, chlorinated furans and dioxins) at present is in audit by the European Commission (Martin-Pozo et al., 2019). Investigating CEC in slime is a troublesome assignment, as slop is a complex network in which the pollutants are regularly seen in low amounts. Creating a proficient pretreatment technique in order to remove the objective pollutants is yet a challenge.

1.3.3 Urban solid waste

In spite of the fact that wastewater has been focused in depth as the significant CEC source for nature, moderately slight examination has been dedicated to landfill leachate from municipal solid waste (MSW) as likely wellspring of these elements. MSW is a kind of waste that contains daily by-products, for example, packaging materials, plastic boxes, and food wastes. These days, this solid waste is essentially arranged in landfills in view of the financial focal points. Items regularly disposed of to landfills include paints, batteries, pharmaceuticals, oils, and electrical items. Residential route by urban WWTPs is perceived as the principle course of synthetic clinical materials to enter the aquatic environs. A watery by-product of landfilling, landfill leachate is created by the precipitation permeation through biochemical procedures in waste cells and the characteristic water substance of wastes themselves (Qi et al., 2018). The capacity of landfill leachate created at every landfill relies upon its geography, provincial climatic conditions, and surface zone, for example, precipitation, and the cover type. The sort and convergence of synthetic substances present in landfill leachate will be decided by the actual structure of waste existing in landfills combined with the amount of leachate produced. It might hold an extensive assortment of CEC devised from landfilled MSW themselves (Qi et al., 2018). For example, the normal quantity of medicines (sulfamethoxazole, oxytetracycline, and tetracycline) in MSW from Xi'an city (capital of Shaanxi Province in China) was 100.9 ± 141.81, 63.8 ± 37.7, and 47.9 ± 8.1 mg/kg, respectively (Song et al., 2016). In that capacity, landfill leachate is viewed as repository of CEC and wellspring of CEC to the local surroundings. Poisonousness measures inspected by various experimental species have additionally affirmed the likely damage of landfill leachates and the need to treat it in order to fulfill the criteria for release in open water bodies. For more established landfills and those without any liners, leachate is once in a while gathered and inclined to permeate into the GW bit by bit. For recently fabricated landfills and those with liners, the leachate is gathered and cured nearby with customary management (organic and synthetic/physical treatment), film treatment, or moved off-site to an sewage treatment plants. Anyways, most CEC are not organically tainted and might be gathered in the biosolids or discharged into the earth through sewage emission.

1.4 Fate of contaminants

This is a fundamental piece of progressive exploration to investigate the source, events, ecological destinies, and contribution of these pollutants from sources to receptors. The multifaceted nature of these pollutants and their physicochemical properties in water, air, and soil thwart their exact data list. A variety of critical endeavors have been developed by specialists for the availability of these pollutants from sources to a receptor to guarantee its potential impact. In the event that the CEC are not corrupted or dispensed with, while wastewater treatment, in soil and in other natural sections, they will arrive at GW and SW, and possibly water for consumption. When the leftover contaminants of WWTP are released into SWs, CEC stream to water receptors and may invade to GW or stay adsorbed on soil and dregs, contingent in its association with the natural framework and upon their physicochemical properties where it has been moved. Release of untreated sewage into SWs, inappropriate damp of expired medicine/beauty products in latrines or landfills, might likewise be a basis for CEC to GW and SW. When an EC or its metabolite or transformation item arrives at an ecological section, it might experience different marvels, for example, adsorption, ingestion, hydrolysis, dilution, photolysis, volatilization, biodegradation, complexation, or oxidation (Starling, Amorim, & Leão, 2019). These phenomenon prompts corruption, transformation, or tirelessness of the compound in the ecosystem. (Kummerer, Al-Ahmad, & Mersch-Sundermann, 2000) revealed that only a couple of pharmaceuticals were somewhat biodegraded under experimental settings in aquatic frameworks. Many of which are relentless. The genotoxicity of elements, for example, metronidazole or quinolones was not evacuated. Quinolones, for instance, adsorb unequivocally onto slime, residue, and soils and were not biodegraded in experiments with dregs. Under 1% of sarafloxacin, a fluoroquinolone affirmed for the counteraction of poultry illnesses and was wiped out from various soils in 80 days, likely in view of its great capacity to tie to soil. Virginiamycin, an antitoxin diet added substance directed orally as a development advertiser in livestock, was found to biodegrade in various soils, however just having extended half-life. Cyclosporin A appeared to vitiate simply after a long time in experiments of wet nursery soil, in spite of the way only a few degrading varieties have been disengaged from soil. Loffler, Rombke, Meller, and Ternes (2005) sorted 10 pharmaceuticals and pharmaceutical metabolites into low-, moderate-, and high-diligence composites as indicated by their dissemination time (DT50) in dregs/water tests. Ibuprofen, paracetamol, CBZ-diol, and 2-hydroxyibuprofen were categorized as indicating little persistence (DT50 ¼ 3.1e7 days). Ivermectin, oxazepam, and iopromide were esteemed to be tolerably tireless (DT50 ¼ 15e54 days), while diazepam, carbamazepine, and clofibric acid were appraised for being profoundly persistent (DT50 ¼ 119e328 days). A later report exhibited the anxiolytic medication (oxazepam) to show broadened ingenuity in freshwater lakes because of previous data and growing municipal populace (Klaminder et al., 2015). An USGS study observing 95 compounds in SW all over the United States found that TCS as one of the most regularly recognized matter with SW quantities as great as 2.3 mg/L (Kolpin et al., 2002). For every published data directed to date, TCS has been distinguished in 56.8% of SW tests with a middle grouping of around 50 ng/L. Since 1957 TCC has been utilized in personal care products and has been seen in SW at an amount of up to 6.75 ng/L (Halden & Paull, 2005). PBDEs quickly get absorbed to natural and particulate material, being bound to be found and segmented into the slop in WWTPs. Along these lines, PBDEs are brought in ecosystem through land application of biosolids. PBDEs can likewise diffuse to aquatic environments through runoffs from polluted soils. Be that as it may, sewage slush and biosolids are perplexing grids in which the degrees of natural synthetic compounds, for example, PBDEs may fluctuate contingent upon various topographical wellsprings of wastewater discharge.

1.5 Ecological risk and health assessment of emerging contaminant in untreated water

In spite of the fact that CEC in SW, GW, and soils are present in low quantities and they pose a grave danger to the integrity of the environment. Inferable from the development of ecological ecotoxicology in the previous years, CEC have been connected to feminization (Kidd et al., 2007) and social changes in fish, (Kidd et al., 2007) imposex (Cacciatore et al., 2018; Titley-O'Neal, Munkittrick, & MacDonald, 2011), immunological, conceptive, and neurological disorder in organisms (Fowler et al., 2012), malignant growths (Bolong, Ismail, Salim, & Matsuura, 2009), improvement of antibiotic-resistant genes (ARGs) in microbes (Pruden, Pei, Storteboom, & Carlson, 2006), heftiness, heart maladies, and advancement of diabetes type 2 in people (Vrijheid, Casas, Gascon, Valvi, & Nieuwenhuijsen, 2016) and decrease of decent variety of small-scale invertebrate in streams (Malaj et al., 2014) among other ecotoxicological impacts. Notwithstanding, for a large portion of them concise controls for environmental hazard evaluation are by and large absent. It is the condition of a few PPCPs recognized in low quantities in the freshwater ecosystem; a large number of PPCPs and their metabolites are dynamic in nature and can affect nontarget water animals. Most significant worry concerning the harmful ramifications of medicines is about their explicit plan to amplify their organic action at meager portions and to focus on certain enzymatic, metabolic, or cell-flagging instruments. The geological preservation of the target compounds in an organism conceivably builds the likelihood that these medicines will be medicologically dynamic in nontarget life forms. A few examinations have analyzed the impact of PPCPs on nontarget living beings particularly fish. Alternative examination by Vernouillet et al. (2010) uncovered that in algae *Raphidocelis subcapitata* and the shellfish *Thamnocephalus platyurus*, there is bioaccumulation of the antiepileptic medicine carbamazepine (CBZ). Moreover, there is a detailed account on the take-up and depuration of medicines in recycled water by mosquito fish (*Gambusia holbrooki*) (Wang & Gardinali, 2013). The bioaccumulation factor estimated for different medicines accumulated in aquatic environment, such as carbamazepine, ibuprofen, caffeine, diphenhydramine, and diltiazem, was 1.4, 28, 2.0, 16, and 16, respectively. Brodin et al. (2014) found that in Eurasian roost fish, oxazepam was notably at a high concentration with a bioaccumulation factor of 12. In addition, the bioaccumulation of fluoxetine with a factor of 3000 was found in snails (Du, Haddad, Scott, Chambliss, & Brooks, 2015) and the wild and confined mussels were found to have accumulated 145 PPCPs from the Grand River, Ontario (de Solla et al., 2016). While examining the mussel tissue, about 43 medicinal chemicals from various classes were distinguished, with bioaccumulation factors extending from 0.66 to 32,022 for metformin and sertraline respectively. UV channels have been recognized in different aquatic creatures, for example, marine mussels (*Mytilus edulis*) up to 256 ng/g (EHMC) and 7112 ng/g (OC) dry weight (dw) along the French Atlantic and Mediterranean coasts (Bachelot et al., 2012) and in brown trout (*Salmo trutta fario*) up to 1800 ng/g (4-methylbenzylidene camphor, 4-MBC) and 2400 ng/g (OC) lipid weight (lw) in Swiss waterways (Buser, Balmer, Schmid, & Kohler, 2006). In addition, Fent, Zenker, and Rapp (2010) recommended natural pecking order aggregation of EHMC, announcing its concentrations in cormorants and fish from six Swiss streams up to 701 and 337 ng/g lw, respectively. The collection of the harmful chemicals in creatures is a worry since natural ultraviolet channels and their metabolites were found to meddle with endocrine system by serving as ecological hormones both in vitro and in vivo. Besides, these chemicals appeared to instigate blanching in coral reefs by stimulating infectious diseases (Danovaro et al., 2008). Stevens, Kim, Adhikari, Vadapalli, and Venables (2009) have found that in aquatic plants, there is no bioaccumulation of TCS in spite of the fact that its methyl group in methyl triclosan (M-TCS) gets bioaccumulated within 28 days in *Sesbania herbacea*. Based on the facts, different investigations have negated these discoveries showing TCS bioaccumulates to an a lot more noteworthy.

1.6 Untreated wastewater as a cause of antibiotic resistance

During 1990s the introduction of antibiotics into human medication has been proved to be the most significant medications for the treatment of infectious illnesses. Be that as it may, due to the excessive use of antibiotics, microorganisms have developed resistance against them and, in response to selection pressure, these have spread infection among human and animal communities worldwide. For example, some antibiotic-resistant microbes such as, carbapenem-resistant *Enterobacteriaceae* (Patrice, Thierry, & Laurent, 2011) and methicillin-resistant *Staphylococcus aureus* (Grundmann, Aires-De-Sousa, Boyce, & Tiemersma, 2006) present noteworthy effects on human well-being. Sanatorium sewage is exceptionally perilous on the grounds that it is not just irresistible and harmful, but at the same time, it is a significant source of antimicrobial-resistant bacteria (ARB) and ARGs. Lacking appropriate management, ARB or ARGs from medical sources may be scattered and may flourish under conditions, quickening the improvement of MARB (multiple antibiotic-resistant bacteria). Due to the expanding worries about antibiotic contamination of water systems, a few examinations have evaluated the availability of ARBs, ARGs, and antibiotics in clinical effluents (Le et al., 2016; Rodríguez-Mozaz et al., 2015; Timraz, Xiong, Al Qarni, & Hong, 2017; Varela, Nunes, & Manaia, 2016), and different investigations have analyzed the connection between ARBs and antibiotics (Varela, André, Nunes, & Manaia, 2014) or between ARGs and antibiotics (Li et al., 2016). There have not been many investigations of clinical aquatic environments than on different situations, for example, city sewage. Any investigations concentrated on the pollutants from medical clinic treatment plants, where the decent variety and conveyance of ARBs, ARGs, and antibiotics were changed in reaction to the discerning force drawn by stretched haul of antibiotic presentation. Hardly any investigations were drawn on to contemplate the event of antimicrobial obstruction in the crude medical clinic sewage water, where ARBs and ARGs are influenced by the antibiotics. These sewage waters mirror the incorporated releases of people, particularly when ill people rush is alarming, and hence mirror the first ARBs and ARGs in human medical trials. The microflora of humans, particularly intestinal microorganisms, which is a store of assorted ARGs. These microorganisms trade the qualities of opposition among themselves; however, these can likewise communicate with the microscopic organisms that go through the body, making those microbes obtain and spread ARGs (Sommer, Church, & Dantas, 2010; Van den Braak et al., 1998). Salyers, Gupta, and Wang (2004) described that these ARBs and ARGs which are discharged from the humans may again come back to unique living spaces (e.g., food, water, and soil) and represent an expanded ecological hazard. In this manner, examination on resistance of antibiotics in crude medical clinic sewage water is useful for knowing how ecological ARBs are influenced by the resistance of clinical antibiotics.

1.7 Impact of wastewater on cities

Principally, in low-pay regions of urban communities and towns inside developing nations, an enormous extent of wastewater is released straightforwardly in the nearest stream or casual seepage channel, at some point without or with little treatment. Notwithstanding family profluent and domestic waste, city hospitals and industries, for example, little mining and engine garages regularly landfill exceptionally poisonous synthetic concoctions and clinical leftover into the sewage framework. Indeed, in urban communities where sewage water is gathered and treated, effectiveness of treatment might differ as indicated by the framework utilized. In any case, water utilized by metropolitan expert for inundating green spaces or cleaning avenues shouldn't be managed to receive a consumable norm. The development of municipal interest for water will need novel ways to deal with sewage water assortment and the treatment. In fact, recycled sewage water might aid to address different difficulties such as production of food and different industrial procedures.

1.8 Impact of wastewater on industry

Cultural and natural weights over recent years have prompted a growing development for industry to decrease its wastewater and to treat it before release. Wastewater is currently observed as a possible asset and its utilization, or reusing after appropriate treatment, can give economic and monetary advantages. Wastewater can be utilized inside the business itself or between a few organizations through "industrial symbiosis." Modern water utilization is responsible for 22% of worldwide water use. In 2009 in Europe and North America, water utilization by enterprises was half when contrasted with 4%–12% in developing nations. It is normal that in quickly industrializing nations, this extent could increment by a factor of 5 in the following 10–20 years. Subsequently, there is a solid motivating force to utilize wastewater in-house and privately, in view of cost investment funds alone. Organizations can legitimately utilize some wastewater, giving it a reason to fit; for example, utilizing process water for cooling or warming, or water from rooftop assortment or solid covers for latrine flushing, water system, or vehicle washing.

1.9 Impact of wastewater on agriculture

Water system utilization of cured sewage and slop should be addressed thoughtfully in arranging effluent removal plot. There is thorough rationale in restoring the hard waste back to the soil and recycling the sewage water at whatever point plausible. The raised degrees of persistent poisons, for example, metals and somewhat insecticides in removal of cured sewage water and slop for water systems and revision of farming areas might form in greater quantities in land because of aggregation. There is negative effect of heavy metals on crop yield, hindering development in many ways. Nonetheless, pH is one of the most significant variables of the soils. Soluble soil pH would confine the preparation of the metals in soil lattice and thus the uptake of metals by crop plants would be limited, clearly diminishing the danger of toxicity of metals. Usage of chemical fertilizers and pesticides has increased in past few years both in industrial and small agricultural sectors, making farming a possible source of ecological contamination. Contamination of water bodies by farming and use of crude or insufficiently cured sewage water is a significant problem in various developing countries where sewage plants are established. Developed sewage water management can mend the health of laborers, mainly in farming, by decreasing the microbe exposure threat. At whatever time applied steadily, sewage water is an important source of water and supplements. We all are surrounded by sewage water. It is a by-product of our cutting-edge way of life, from the spillover which originates from the wet streets to the water running down our shower pipe. The water we drink and shower in is sieved and treated to remove any pollutants such as sewage or synthetic chemicals, leading to advanced wastewater treatment innovation.

1.10 Impact of wastewater on natural bodies of water

Everyday both freshwater and saltwater bodies are indiscriminately contaminated by sewage water. According to the US Ecological Protection Agency (EPA), about 3.28 billion gallons of sewage from domestic and industrial sources are dumped into the water bodies of the country, or about 1.2 trillion gallons of sewage water per year. This not only affects creatures in water but proves hazardous to the humans as well. These waste materials will ultimately wind up in our home if appropriate measures are not taken on time. This shows the significance of WWTP and framework. Drainage and sewage drain into the water bodies from various ways range from surface runoff and septic structures to sewage treatment facilities for drainage and tempest channel outfalls. About 3.5 million American people get chronically ill from outdoor activities such as swimming and drifting due to the fact that the water is tainted. Most people do not use the water they used during their illness. The effect of contaminated water on marine ecosystem goes a long way beyond human ailment, be that as it may.

1.11 Impact of untreated wastewater on microbial diversity

Soil pollution is of incredible worry in croplands in light of expected impacts on sanitation and soil biological systems, particularly microflora (Chen et al., 2015; Dar et al., 2012; Muhammad, Xu, Li, Wang, & Yao, 2005; Mekki, Awali, Aloui, Loukil, & Sayadi, 2017). The wealth, assembly, and action of soil microbial networks extraordinarily influence the strength and development of plants and animals. The unsettling influence of organisms in soil can decrease soil fertility and hence can severely influence the nature of crop yield (Dar, Bandh, Kamili, Nazir, & Bhat, 2013). Contaminants in cropland soils for the most part begin from barometrical testimony, sewage water irrigation system, fertilization, and pesticide showering (Luo, Ma, Zhang, Wei, & Zhu, 2009; Xie et al., 2009). Cheng (2003) and Wu et al. (2010) evaluated that almost about one-fifth of the total farmland in China has been contaminated with heavy metals. In order to improve the crop yield, wastewater is used throughout the world by using accessible water which is premixed with certain macro and micronutrients. There are certain poisonous compounds in wastewater which pollute the groundwater and land. Consequently, a large amount of toxins get introduced into the natural food chain due to long-haul irrigation of wastewater and influence soil microflora network. From 1927 to 2002 the Hunpu territory in one examination was watered by wastewater and the source of the water was the stream and dregs of Xihe River which was highly contaminated. Previous examinations have shown that the significant source of land contamination was inundated water. Likewise, the silt is the major cause of expected poisonousness for land and groundwater close to waterway.

1.12 Impact of wastewater in aquatic environments

Household drainage contains contaminations extending from natural perils and microplastic elements to cleansers and flabs. Rural sewage contains natural dangers, composts, pesticides, and salts. City sewage incorporates local and industrial drainage water yet in addition holds overflow from squall channels. Squall channels convey contaminations from local areas, parks, boulevards, and parking garages (fats, fuel, filth, and garbage). Wastewater from factories holds an extensive scope of synthetic substances which incorporate acids, petrochemicals, radioactive materials, salts, and different synthetics. Earlier discoveries have shown that an assortment of medications likewise pollute waterways. As per the 2018 report, the University of Michigan takes note that as per the US EPA, "53% of waterway and stream miles, 71% of lake sections of land, 79% of estuarine square miles, and 98% of Great Lakes shoreline miles that have been surveyed are named as impaired (unsatisfactory for in any event one assigned use)."

1.13 Biologic hazards in aquatic environments

Natural perils found in sewage incorporate microscopic organisms such as fungi, bacteria, protozoa, and viruses; for example, bacterial sicknesses from *Escherichia coli* and fever, typhoid, salmonellosis, shigellosis, and cholera by others. Parasites incorporate roundworms, giardia, and cryptosporidium into the host that causes severe illness. Infections such as hepatitis A can likewise be caused due to contamination of river water by sewage. Every year about 3.5 million American citizens suffer from various medical issues due to wastewater contamination. As estimated, 50% of untreated sewage water enters the Mediterranean. Natural waste from ranches, houses, stops, and sea shores mess well-being up that sway more than people.

Microscopic organisms and different creatures in lakes and rivers use high quantities of oxygen to metabolize the waste they go with. These microbes can cause hypoxic (oxygen-drained) dead zones while separating the wastewater. There is a high demand of oxygen in these dead zones that the creatures in these water bodies need to survive. In marine situations, microorganisms present in digestive tract of humans can contaminate coral reefs and causes blenching of coral reefs and its destruction. At the point when coral lose their regular microscopic organisms and green growth, they collapse, bringing about area in which the coral biological system, from microorganisms to fish populaces, perishes.

1.14 Major threats

Wastewater can have genuine outcomes which are chiefly brought about by industrialization and urbanization. Water contamination and degradation of habitat are significant danger to freshwater environments (Dirzo et al., 2014; Dudgeon, Arthington, & Gessner, 2006), along these lines undermining the assorted variety of existence (Moss, 2008; Vorosmarty et al., 2010). Streams which are a source of drinking water get stress-causing compounds which have anthropogenic basis, comprising organic and inorganic supplements (Moss, 1998), and toxins from factory-developed sewage (Meybeck, 2004; Sabater, Barcel, & Castro-Catal, 2016). This can have genuine outcomes in the fish diversity. Sewage runoffs are liable for the dilapidation of water bodies, for example, streams, rivers, and lakes. It relies upon volume of the release, compound and microbiological concentration, and structure of wastes. Degradation likewise relies upon sort of the release whether it is measure of organic substance or organochlorines, heavy metals, perilous toxins, and suspended solids (Owuli, 2003). Introduction to poisons produced by these living beings can cause different maladies, for example, liver damage, gastroenteritis, skin disturbance, sensory system disability, and malignant growth of liver in animals (EPA, 2000; Eynard, Mez, & Walther, 2000; WHO, 2006).

There has been a long and written history confirming the effects of freshwater contamination, particularly organic pollution (Hynes, 1963; Mason, 2001); sewage water is a genuine and rising issue in most of the developing nations (Girija, Mahanta, & Chandramouli, 2007; Haque, 2017; Omosa, Wang, Cheng, & Li, 2012). It is principally in light of the fact that industrial, rural, civil, and sewage release cause weakening of accepting the quality of water (Dudgeon, 1992; Paul, 2017; Vaseem & Banerjee, 2013). Inland water ecosystems become unfit for consumption of living organisms of aquatic and amphibious flora and fauna (Karaouzas, Smeti, & Vourka, 2018). Medicinal wastes can bring about hormonal disturbance in fish, diminished bounty, and extinction of certain sensitive aquatic species (Dar et al., 2016; Dar, Kamili, Dar, Chishti, & Ahmad, 2015; Sanchez, Sremski, & Piccini, 2011).

Wastewater effluents can have harmful effects on degradation of the accepting water bodies; natural effects, for example, diminished degrees of DO, and well-being impacts, for example, infections brought about by microscopic organisms, protozoa and viruses, pose serious dangers to well-being related with untreated recreational and drinking waters.

1.15 Why should wastewater be treated?

Untreated wastewater makes significant harm to the earth and to human well-being. Quite often, subsequently, wastewater ought to be treated so as to reduce the transmission of excreta-related maladies and to decrease water contamination and the ensuing harm to aquatic biota.

1.16 Challenges and opportunities

The important basic factor for health of people, public and economic turn of events, and the biological system is the good quality of water. Be that as it may, as communities grow and natural ecosystems become polluted, it is becoming increasingly difficult to guarantee sufficient and secure water sources for all. A vital part of the agreement is to create low contamination, and to enhance the way wastewater is tracked. An economy that is becoming increasingly circular and thus progressively supportable requires us to value wastewater for its latent prospective, rather than disposing or ignoring it. Safe wastewater management will help protect our biological systems beyond an elective source of water and provide us nutrients, energy, and other recoverable resources. Water must be managed sensibly in all aspects of the water cycle from pretreatment, posttreatment, freshwater reflection, dissemination, assortment and use to the utilization of cured sewage and its definitive comeback in ecosystem to be fit to begin the cycle once more. Because of populace development, quickened urbanization, and financial turn of events, the amount of wastewater produced and its general contamination load are expanding internationally.

Accessibility of harmless and sufficient sources of water is inseparably related to how wastewater is handled. Prolonged untreated wastewater initiatives together with farming spillover and industrial release have impaired the quality of water and tainted its assets around the globe. All inclusive, 80% of sewage water drains again in the biological system without being dealt with or recycled, leading to a situation in which about 1.8 billion humans use drinking water source polluted with dung, putting the people into the danger of diseases such as typhoid, cholera, polio, and dysentery contracting. Apart from being anything to neglect or dispose, sewage should play an important part in meeting the increasing demand for pure water in quickly developing urban areas, improving energy production and industrial growth, and promoting manageable agribusiness.

1.17 Conclusion

The general effects of wastewater effluents found in water bodies need to be cured and reduced. Sewage needs to be cured prior to its release. It could be accomplished by using fitting treatment forms/techniques that may assist in limiting the dangers to general well-being as well as the ecosystem. To accomplish untainted sewage release in open water bodies, cautious planning, satisfactory and reasonable treatment, normal checking, and proper enactment are required.

References

Adriano, D. C., Chang, A. C., Pratt, P. F., & Sharpless, R. (1973). Effect of soil application of dairy manure on germination and emergence of some selected crops. *Journal of Environmental Quality*, *2*, 396–399.

Agarwal, K. C., Gupta, S. K., & Gupta, A. B. (1999). Development of new low cost defluoridation technology (krass). *Water Science and Technology: a Journal of the International Association on Water Pollution Research*, *40*, 167–173.
Akaninwor, J. O., Anosike, E. O., & Egwim, O. (2007). Effect of Indomie industrial effluent discharge on Microbial properties of new Calabar River. *Journal of Science Research and Essay*, *2*, 001–005.
Allanson, B. R., & Baird, D. (1999). *Estuarines of South Africa*. Cambridge: Cambridge University Press.
Altintas, N., Ayok, T., & Gozmen, T. (1987). Fluorine pollution in the Marmara region, Turkey. Production methods for fluorine removal and recovery. *Water Research*, *21*, 781–788.
Bachelot, M., Li, Z., Munaron, D., Le Gall, P., Casellas, C., Fenet, H., & Gomez, E. (2012). Organic UV filter concentrations in marine mussels from French coastal regions. *Science of the Total Environment*, *420*, 273–279.
Battistoni, P., De Angelis, A., Prisciandaro, M., Boccadoro, R., & Bolzonella, D. (2002). P removal from anaerobic supernatants by struvite crystallization: long term validation and process modelling. *Water Research*, *36*, 1927–1938.
Behera, B. K., & Misra, B. N. (1982). Analysis of the effect of industrial effluent on growth and development of rice seedlings. *Environmental Research*, *28*, 10–20.
Ben Ayed, F., & Bouaziz, J. (2008). Sintering of tricalcium phosphate–fluorapatite composites by addition of alumina. *Ceramics International.*, *34*, 1885–1892.
Bhat, R. A., Singh, D. V., Qadri, H., Dar, G. H., Dervash, M. A., Bhat, S. A., … Yousaf, B. (2021). Vulnerability of municipal solid waste: An emerging threat to aquatic ecosystems. *Chemosphere*, *287*(2022), 132223. Available from https://doi.org/10.1016/j.chemosphere.2021.132223.
Bolong, N., Ismail, A. F., Salim, M. R., & Matsuura, T. (2009). A review of the effects of emerging contaminants in wastewater and options for their removal. *Desalination*, *239*, 229–246.
Brausch, J. M., & Rand, G. M. (2011). A review of personal care products in the aquatic environment: Environmental concentrations and toxicity. *Chemosphere*, *82*, 1518–1532.
Brodin, T., Piovano, S., Fick, J., Klaminder, J., Heynen, M., & Jonsson, M. (2014). Ecological effects of pharmaceuticals in aquatic systems impacts through behavioural alterations. *Philosophical Transactions of the Royal Society of London. Series B, Biological Sciences*, *369*.
Buser, H. R., Balmer, M. E., Schmid, P., & Kohler, M. (2006). Occurrence of UV filters 4-methylbenzylidene camphor and octocrylene in fish from various Swiss rivers with inputs from wastewater treatment plants. *Environmental Science & Technology*, *40*, 1427–1431.
Cacciatore, F., Noventa, S., Antonini, C., Formalewicz, M., Gion, C., Berto, D., … Brusà, R. B. (2018). Imposex in *Nassarius nitidus* (Jeffreys, 1867) as apossible investigative tool to monitor butyltin contamination according to the Water Framework Directive: A case study in the Venice Lagoon (Italy). *Ecotoxicology and Environmental Safety*, *148*, 1078–1089.
Chambers, P. A., Allard, M., Walker, S. L., Marsalek, J., Lawrence, J., Servos, M., … Adare, K. (1997). Impacts of municipal wastewater effluents on Canadian waters: A review. *Water Quality Research Journal of Canada*, *32*, 659–713.
Chen, M., Xu, P., Zeng, G., Yang, C., Huang, D., & Zhang, J. (2015). Bioremediation of soils contaminated with polycyclic aromatic hydrocarbons, petroleum, pesticides, chlorophenols and heavy metals by composting: applications, microbes and future research needs. *Biotechnology Advances*, *33*, 745–755.
Cheng, S. (2003). Heavy metal pollution in China: Origin, pattern and control. *Environmental Science and Pollution Research*, *10*, 192–198.
Council of the European Communities, Directive 86/278/EEC of 12 June. (1986). on the protection of the environment, and in particular of the soil, when sewage sludge is used in agriculture. *Official Journal of the European Communities*, *L181*(1986), 6.

Council of the European Communities, Directive 91/271/EEC of May 21. (1991). concerning urban wastewater treatment. *Official Journal of the European Communities, L135*(1991), 40.

Cronin, S. J., Manoharan, V., Hedley, M. J., & Loganathan, P. (2000). Fluoride: A review of its fate, bioavailability, and risks of fluorosis in grazed-pasture systems in New Zealand. *New Zealand Journal of Agricultural Research, 43*, 295–321.

Daines, R. M., Leone, I., & Brennane, E. (1952). *The effect of fluoride on plants as determined by soil*. New York: Nutrition and Fumigation StudiesMcGraw-Hill Book Co.

Dalai, R. C. (1977). Soil organic phosphorus. *Advances in Agronomy, 29*, 83–117.

Danovaro, R., Bongiorni, L., Corinaldesi, C., Giovannelli, D., Damiani, E., Astolfi, P., Greci, L., & Pusceddu, A. (2008). Sunscreens cause coral bleaching by promoting viral infections. *Environmental Health Perspectives, 116*, 441–447.

Dar, G. H., Bandh, S. A., Kamili, A. N., Nazir, R., & Bhat, R. A. (2013). Comparative analysis of different types of bacterial colonies from the soils of Yusmarg Forest, Kashmir valley India. *Ecologia Balkanica, 5*(1), 31–35.

Dar, G. H., Kamili, A. N., Chishti, M. Z., Dar, S. A., Tantry, T. A., & Ahmad, F. (2016). Characterization of *Aeromonas sobria* Isolated from Fish Rohu (*Labeo rohita*) collected from polluted pond. *Journal of Bacteriology and Parasitology*, 7(3), 1–5. Available from https://doi.org/10.4172/2155-9597.1000273.

Dar, G. H., Kamili, A. N., Dar, S. A., Chishti, M. Z., & Ahmad, F. (2015). Detection and characterization of potentially pathogenic *Aeromonas sobria* isolated from fish *Hypophthalmichthys molitrix* (Cypriniformes: Cyprinidae). *Microbial Pathogenesis, 2015*. Available from https://doi.org/10.1016/j.micpath.2015.10.017.

Dar, G. H., Bandh, S. A., & Kamili, A. N. (2011). Bacteriological analysis of soil from Yusmarg Health Resort of Kashmir Valley. *International Journal of Current Research., 3* (10), 53–56, 2011.

Dar, G. H., Kamili., Azra, N., Nazir, R., Bandh, S. A., & Bhat, R. A. (2012). A preliminary study of colony forming units of bacteria from the soils of Yusmarg forest, Kashmir valley, India. *International Journal of Current Research, 4*(12), 467–472.

Das, J., & Acharya, B. C. (2003). Hydrology and assessment of lotic water quality in Cuttack City, India. *Water, Air, and Soil Pollution, 150*, 163–175.

Daughton, C. G., & Ternes, T. A. (1999). Pharmaceuticals and personal care products in the environment: Agents of subtle change? *Environmental Health Perspectives, 107*, 907–938.

de Solla, S. R., Gilroy, E. A. M., Klinck, J. S., King, L. E., Mcinnis, R., Struger, J., ... Gillis, P. L. (2016). Bioaccumulation of pharmaceuticals and personal care products in the unionid mussel *Lasmigona costata* in a river receiving wastewater effluent. *Chemosphere, 146*, 486–496.

Diniz, M. S., Peres, I., & Pihan, J. C. (2005). Comparative study of the estrogenic responses of mirror carp (*Cyprinus carpio*) exposed to treated municipal sewage effluent (Lisbon) during two periods in different seasons. *Science of the Total Environment, 349*, 129–139.

Dirzo, R., Young, H. S., Galetti, M., Ceballos, G., Isaac, N. J. B., & Collen, B. (2014). Defaunation in the anthropocene. *Science (New York, N.Y.), 345*, 401–406.

Dong, H., Yuan, X., Wang, W., & Qiang, Z. (2016). Occurrence and removal of antibiotics in ecological and conventional wastewater treatment processes: A field study. *Journal of Environmental Management, 178*, 11–19.

Du, B., Haddad, S. P., Scott, W. C., Chambliss, C. K., & Brooks, B. W. (2015). Pharmaceutical bioaccumulation by periphyton and snails in an effluent dependent stream during an extreme drought. *Chemosphere, 119*, 927–934.

Dudgeon, D. (1992). Endangered ecosystems: a review of the conservation status of tropical Asian rivers. *Hydrobiologia, 248*, 167–191.

Dudgeon, D., Arthington, A. H., Gessner, M. O., et al. (2006). Freshwater biodiversity: Importance, threats, status and conservation challenges. *Biological Reviews of the Cambridge Philosophical Society, 81*, 163–182.

Dutka, B. J. (1989). *Methods for microbiological and toxicological analysis of waters, wastewaters and sediments*. Canada, Burlington: National Water Research Institute (NWRI).
EPA (2000). *Nutrient criteria technical guidance manual-rivers and streams*. EPA-822-B-00–002. Washington DC.
Eynard, F., Mez, K., & Walther, J. L. (2000). Risk of cyanobacterial toxins in Riga waters (LATVIA). *Water Research*, *30*, 2979–2988.
Fent, K., Zenker, A., & Rapp, M. (2010). Widespread occurrence of estrogenic UV filters in aquatic ecosystems in Switzerland. *Environmental Pollution*, *158*, 1817–1824.
Fiskesjo, G. (1993). The allium test in wastewater monitoring. *Environmental Toxicology and Water Quality*, *8*, 291–298.
Fowler, P. A., Bellingham, M., Sinclair, K. D., Evans, N. P., Pocar, P., Fischer, B., ... O'Shaughnessy, P. J. (2012). Impact of endocrine-disrupting compounds (EDCs) on female reproductive health. *Molecular and Cellular Endocrinology*, *355*, 231–239.
Gagne, F., Blaise, C., Aoyama, I., Luo, R., Gagnon, C., Couillard, Y., ... Salazar, M. (2002). Biomarker study of a municipal effluent dispersion plume in two species of freshwater mussels. *Environmental Toxicology*, *17*, 149–159.
Garg, S., & Bahl, G. S. (2008). Phosphorus availability to maize as influenced by organic manures and fertilizer P associated phosphatase activity in soils. *Bioresource Technology*, *99*, 5773–5777.
Girija, T. R., Mahanta, C., & Chandramouli, V. (2007). Water quality assessment of an untreated effluent impacted urban stream: The Bharalu Tributary of the Brahmaputra River, India. *Environmental Monitoring and Assessment*, *130*, 221–236.
Gopalan, H. N. B. (1999). Ecosystem health and human well being: the mission of the international programme plant bioassays. *Mutation Research*, *426*, 99–102.
Grundmann, H., Aires-De-Sousa, M., Boyce, J., & Tiemersma, E. (2006). Emergence and resurgence of meticillin-resistant *Staphylococcus aureus* as a public-health threat. *Lancet*, *368*, 874. Available from https://doi.org/10.1016/s0140-6736(06)68853-3.
Grzmil, B., & Wronkowski, J. (2006). Removal of phosphates and fluorides from industrial wastewater. *Desalination*, *189*, 261–268.
Gustavson, K. E., Sonsthagen, S. A., Crunkilton, R. A., & Harkin, J. M. (2000). Groundwater toxicity assessment using bioassay, chemical, and toxicity identification evaluation analysis. *Environmental Toxicology*, *15*, 421–430.
Halden, R. U., & Paull, D. H. (2005). Co-occurrence of triclocarban and triclosan in US water resources. *Environmental Science & Technology*, *39*, 1420–1426.
Hall, R. J., & Cain, R. G. (1972). Organic fluorine in tropical soils. *The New Phytologist*, *71*, 839–853.
Haque, N. (2017). Exploratory analysis of fines for water pollution in Bangladesh. *Water Resources and Industry*, *18*, 1–8.
Haruvy, N. (1997). Agricultural reuse of wastewater: Nation-wide cost–benefit analysis. *Agriculture, Ecosystems & Environment*, *66*, 113–119.
Healy, M. G., Fenton, O., Cormican, M., Peyton, D. P., Ordsmith, N., Kimber, K., & Morrison, L. (2017). Antimicrobial compounds (triclosan and triclocarban) in sewage sludges, and their presence in runoff following land application. *Ecotoxicology and Environmental Safety*, *142*, 448–453.
Hichour, M., Persin, F., Sandeaux, J., & Gavach, C. (2000). Fluroide removal from waters by Donnan dialysis. *Separation and Purification Technology*, *18*, 1–11.
Hoeger, B., Van den Heuvel, M. R., Hitzfeld, B. C., & Dietrich, D. R. (2004). Effects of treated sewage on immune function in rainbow trout (*Oncorhynchus mykiss*). *Aquatic Toxicology (Amsterdam, Netherlands)*, *70*, 345–355.
Hu, C. Y., Lo, S. L., Kuan, W. H., & Lee, Y. D. (2005). Removal of fluoride from semiconductor wastewater by electrocoagulation-flotation. *Water Research*, *39*, 895–901.
Hynes, H. B. N. (1963). *The biology of polluted waters*. Liverpool: Liverpool University Press.

Islam, M., & Patel, R. K. (2007). Evaluation of removal efficiency of fluoride from aqueous solution using quick lime. *Journal of Hazardous Materials, 143*, 303–310.
Kara, A., & Stevens, R. (2002). Characterisation of biscuit fired bone china body microstructure. Part I: XRD and SEM of crystalline phases. *Journal of the European Ceramic Society, 22*, 731–736.
Karaouzas, I., Smeti, E., Vourka, A., et al. (2018). Assessing the ecological effects of water stress and pollution in a temporary river – Implications for water management. *The Science of the Total Environment, 618*, 1591–1604.
Kidd, K. A., Blanchfield, P. J., Mills, K. H., Palace, V. P., Evans, R. E., Lazorchak, J. M., & Flick, R. W. (2007). Collapse of a fish population after exposure to a synthetic estrogen. *Proceedings of the National Academy of Sciences of the United States of America, 104*, 8897–8901.
Klaminder, J., Brodin, T., Sundelin, A., Anderson, N. J., Fahlman, J., Jonsson, M., & Fick, J. (2015). Long-term persistence of an anxiolytic drug (Oxazepam) in a large freshwater lake. *Environmental Science & Technology, 49*, 10406–10412.
Kolpin, D. W., Furlong, E. T., Meyer, M. T., Thurman, E. M., Zaugg, S. D., Barber, L. B., & Buxton, H. T. (2002). Pharmaceuticals, hormones and other organic wastewater contaminants in US streams, 1999–2000: a national reconnaissance. *Environmental Science & Technology, 36*, 1202–1211.
Kummerer, K. (2001). Drugs in the environment: emission of drugs, diagnostic aids and did infectants into wastewater by hospitals in relation to other sources—A review. *Chemosphere, 45*, 957–969.
Kummerer, K., Al-Ahmad, A., & Mersch-Sundermann, V. (2000). Biodegradability of some antibiotics, elimination of the genotoxicity and affection of wastewater bacteria in a simple test. *Chemosphere, 40*, 701–710.
Le, T. H., Ng, C., Chen, H., Yi, X. Z., Koh, T. H., Barkham, T. M. S., et al. (2016). Occurrences and characterization of antibiotic-resistant bacteria and genetic determinants of hospital wastewater in a tropical country. *Antimicrobial Agents and Chemotherapy, 60*, 7449–7456.
Li, C., Lu, J., Liu, J., Zhang, G., Tong, Y., & Ma, N. (2016). Exploring the correlations between antibiotics and antibiotic resistance genes in the wastewater treatment plants of hospitals in Xinjiang, China. *Environmental Science and Pollution Research, 23*, 15111–15121.
Loffler, D., Rombke, J., Meller, M., & Ternes, T. A. (2005). Environmental fate of pharmaceuticals in water/sediment systems. *Environmental Science & Technology, 39*, 5209–5218.
Lofrano, G., Pedrazzani, R., Libralato, G., & Carotenuto, M. (2017). Advanced oxidation processes for antibiotics removal: A review. *Current Organic Chemistry, 621*, 1–14.
Luo, L., Ma, Y., Zhang, S., Wei, D., & Zhu, Y. G. (2009). An inventory of trace element inputs to agricultural soils in China. *Journal of Environmental Management, 90*, 2524–2530.
Malaj, E., von der Ohe, P. C., Grote, M., Kuhne, R., Mondy, C. P., Usseglio- Polatera, P., . . . Schäfer, R. B. (2014). Organic chemicals jeopardize the health of freshwater ecosystems on the continental scale. *Proceedings of the National Academy of Sciences of the United States of America, 111*, 9549–9554.
Martin-Pozo, L., de Alarcon-Gomeza, B., Rodriguez-Gomeza, R., Garcia-Corcolesa, M. T., Cipa, M., & Zafra-Gomeza, A. (2019). Analytical methods for the determination of emerging contaminants in sewage sludge samples. A review. *Talanta, 192*, 508–533.
Mason, C. F. (2001). *Biology of freshwater pollution* (4th (ed.), p. 400). New Jersey: Prentice Hall.
Mekki, A., Awali, A., Aloui, F., Loukil, S., & Sayadi, S. (2017). Characterization and toxicity assessment of wastewater from rock phosphate processing in Tunisia. *Mine Water and the Environment, 36*, 502–507.

Meybeck, M. (2004). The global change of continental aquatic systems: dominant impacts of human activities. *Water Science and Technology: A Journal of the International Association on Water Pollution Research*, *49*, 73–83.

Morse, G. K., Lester, J. N., & Perry, R. (1993). *The economic and environmental impact of phosphorus removal from wastewater in the European Community*. London: Selper Publications.

Moss, B. (1998). *Ecology of fresh waters: man and medium, past to future* (p. 572) Oxford: Blackwell Science.

Moss, B. (2008). Water pollution by agriculture. *Philosophical Transactions of the Royal Society B*, *363*, 659–666.

Muhammad, A., Xu, J., Li, Z., Wang, H., & Yao, H. (2005). Effects of lead and cadmium nitrate on biomass and substrate utilization pattern of soil microbial communities. *Chemosphere*, *60*, 508–514.

Omosa, I. B., Wang, H., Cheng, S., & Li, F. (2012). Sustainable tertiary wastewater treatment is required for water resources pollution control in Africa. *Environmental Science & Technology*, *46*, 7065–7066.

Owuli, M. A. (2003). *Assessment of impact of sewage effluents on coastal water quality in Hafnarfjordur*, Iceland. The United Nations Fishery Training Program, Final Report.

Patrice, N., Thierry, N., & Laurent, P. (2011). Global spread of carbapenemase-producing Enterobacteriaceae. *Emerging Infectious Diseases*, *17*, 1791. Available from https://doi.org/10.1128/microbe.3.516.1.

Paul, D. (2017). Research on heavy metal pollution of river Ganga: A review. *Annals of Agrarian Science*, *15*, 278–286.

Priya, P., & Sahi, S. V. (2009). Influence of phosphorus nutrition on growth and metabolism of Duo grass (*Duo festulolium*). *Plant Physiology and Biochemistry*, *47*, 31–36.

Pruden, A., Pei, R., Storteboom, H., & Carlson, K. H. (2006). Antibiotic resistance genes as emerging contaminants: studies in Northern Colorado. *Environmental Science & Technology*, *40*, 7445–7450.

Qi, C., Huang, J., Wang, B., Deng, S., Wang, Y., & Yu, G. (2018). Contaminants of emerging concern in landfill leachate in China: A review. *Emerging Contaminants*, *4*, 1–10.

Real Decreto 1310/1990, Ministerio de Agricultura (1990). Pesca y Alimentacion, BoletinOficial del Estado (BOE). State Official Newsletter, 32339. BOE-A-1990-26490, 262.

Rodríguez-Mozaz, S., Chamorro, S., Marti, E., Huerta, B., Gros, M., Sànchez-Melsió, A., et al. (2015). Occurrence of antibiotics and antibiotic resistance genes in hospital and urban wastewaters and their impact on the receiving river. *Water Research*, *69*, 234–242.

Ruixia, L., Jinlong, G., & Hongxiao, T. (2002). Adsorption of fluoride, phosphate, and arsenate ions on a new type of ion exchange fiber. *Journal of Colloid and Interface Science*, *248*, 268–274.

Sabater, S., Barcel, D., Castro-Catal, N. D., et al. (2016). Shared effects of organic microcontaminants and environmental stressors on biofilms and invertebrates in impaired rivers. *Environmental Pollution (Barking, Essex: 1987)*, *210*, 303–314.

Salyers, A. A., Gupta, A., & Wang, Y. (2004). Human intestinal bacteria as reservoirs for antibiotic resistance genes. *Trends in Microbiology*, *12*, 412–416.

Sanchez, W., Sremski, W., Piccini, B., et al. (2011). Adverse effects in wild fish living downstream from pharmaceutical manufacture discharges. *Environment International*, *37*, 1342–1348.

Singh, V., Gupta, M. K., Rajwanshi, P., Mishra, S., Srivastava, S., Srivastava, R., . . . Dass, S. (1995). Plant uptake of fluoride in irrigation water by ladyfinger (*Abelmorchusesculentus*). *Food and Chemical Toxicology: An International Journal Published for the British Industrial Biological Research Association*, *33*, 399–402.

Sommer, M. O., Church, G. M., & Dantas, G. (2010). The human microbiome harbors a diverse reservoir of antibiotic resistance genes. *Virulence*, *1*, 299–303.

Song, L., Li, L., Yang, S., Lan, J., He, H., McElmurry, S. P., & Zhao, Y. (2016). Sulfamethoxazole, tetracycline and oxytetracycline and related antibiotic resistance genes in a large-scale landfill, China. *Science of the Total Environment*, *551*, 9–15.

Starling, M. C. V., Amorim, C. C., & Leão, M. M. D. (2019). Occurrence, control and fate of contaminants of emerging concern in environmental compartments in Brazil. *Journal of Hazardous Materials, 372*, 17–36.

Stevens, K. J., Kim, S. Y., Adhikari, S., Vadapalli, V., & Venables, B. J. (2009). Effects of triclosan on seed germination and seedling development of three wetland plants: *Sesbania herbacea, Eclipta prostrata*, and *Bidens frondosa. Environmental Toxicology and Chemistry/SETAC, 28*, 2598–2609.

Timraz, K., Xiong, Y., Al Qarni, H., & Hong, P. Y. (2017). Removal of bacterial cells, antibiotic resistance genes and integrase genes by on-site hospital wastewater treatment plants: surveillance of treated hospital effluent quality. *Environmental Science: Water Research & Technology, 3*, 293–303.

Titley-O'Neal, C. P., Munkittrick, K. R., & MacDonald, B. A. (2011). The effects of organotin on female gastropods. *Journal of Environmental Monitoring: JEM, 13*, 2360–2388.

Treshow, M. (1971). Fluorides as air pollutants affecting plants. *Annual Review of Phytopathology, 9*, 21–44.

Tukura, B. W., Kagbu, J. A., & Gimba, C. E. (2009). Effects of pH and seasonal variations on dissolved and suspended heavy metals in dam surface water. *Chemistry Class The Journal, 6*, 27–30.

Valerio, M. E., García, J. F., & Peinado, F. M. (2007). Determination of phytotoxicity of soluble elements in soils, based on a bioassay with lettuce (*Lactuca sativa* L.). *The Science of the Total Environment, 378*, 63–66.

Van den Berg, L. J. L., Gomassen, H. B. M., Roelofs, J. G. M., & Bokkink, R. (2005). Effects of nitrogen enrichment on coastal dune grassland: A mesocosm study. *Environmental Pollution (Barking, Essex: 1987), 138*, 77–85.

Van den Braak, N., van Belkum, A., van Keulen, M., Vliegenthart, J., Verbrugh, H. A., & Endtz, H. P. (1998). Molecular characterization of vancomycin-resistant enterococci from hospitalized patients and poultry products in The Netherlands. *Journal of Clinical Microbiology, 36*, 1927–1932.

Varela, A. R., André, S., Nunes, O. C., & Manaia, C. M. (2014). Insights into the relationship between antimicrobial residues and bacterial populations in a hospital-urban wastewater treatment plant system. *Water Research, 54*, 327–336.

Varela, A. R., Nunes, O. C., & Manaia, C. M. (2016). Quinolone resistant Aeromonas spp. as carriers and potential tracers of acquired antibiotic resistance in hospital and municipal wastewater. *The Science of the Total Environment, 542*, 665–671.

Vaseem, H., & Banerjee, T. K. (2013). Contamination of the River Ganga and its toxic implication in the blood parameters of the major carp *Labeo rohita* (Ham). *Environmental Science and Pollution Research, 20*, 5673–5681.

Vernouillet, G., Eullaffroy, P., Lajeunesse, A., Blaise, C., Gagne, F., & Juneau, P. (2010). Toxic effects and bioaccumulation of carbamazepine evaluated by biomarkers measured in organisms of different trophic levels. *Chemosphere, 80*, 1062–1068.

Vorosmarty, C. J., Mcintyre, P. B., Gessner, M. O., Dudgeon, D., Prusevich, A., Green, P., et al. (2010). Global threats to human water security and river biodiversity. *Nature, 467*, 555–561.

Vrijheid, M., Casas, M., Gascon, M., Valvi, D., & Nieuwenhuijsen, M. (2016). Environmental pollutants and child health—A review of recent concerns. *International Journal of Hygiene and Environmental Health, 219*, 331–342.

Wang, J., & Gardinali, P. R. (2013). Uptake and depuration of pharmaceuticals in reclaimed water by mosquito fish (*Gambusia holbrooki*): a worst case, multiple-exposure scenario. *Environmental Toxicology and Chemistry / SETAC, 32*, 1752–1758.

WHO. (2006). *Guidelines for the safe use of wastewater, excreta and greater.* Geneva, Switzerland: World Health Organisation Press, vol. 3.

Wollan, E., Davis, R. D., & Jenner, S. (1978). Effects of sewage sludge on seed germination. *Environmental Pollution (Barking, Essex: 1987)*, *17*, 195–205.

Wu, G., Kang, H., Zhang, X., Shao, H., Chu, L., & Ruan, C. (2010). A critical review on the bioremoval of hazardous heavy metals from contaminated soils: Issues, progress, ecoenvironmental concerns and opportunities. *Journal of Hazardous Materials*, *174*, 1–8.

Wu, M. H., Que, C. J., Xu, G., Sun, Y. F., Ma, J., Xu, H., … Tang, L. (2016). Occurrence, fate and interrelation of selected antibiotics in sewage treatment plants and their receiving surface water. *Ecotoxicology and Environmental Safety*, *132*, 132–139.

Xie, W., Zhou, J., Wang, H., Chen, X., Lu, Z., Yu, J., et al. (2009). Short-term effects of copper, cadmium and cypermethrin on dehydrogenase activity and microbial functional diversity in soils after long-term mineral or organic fertilization. *Agriculture, Ecosystems & Environment*, *129*, 450–456.

Zucconi, F., Monaco, A., Forte, M., & De Bertoldi, M. (1985). Phytotoxins during the stabilization of organic matter. In J. K. R. Gasser (Ed.), *Composting of Agricultural and Other Wastes* (pp. 73–86). London, UK: Elsevier.

Zucconi, F., Pera, A., Forte, M., & De Bertoldi, M. (1981). Evaluating toxicity of immature compost. *Biocycle*, *22*, 54–57.

CHAPTER 2

Unraveling a correlation between environmental contaminants and human health

Misbah Naz[1], Muhammad Ammar Raza[2], Muhammad Tariq[3], Shackira A.M.[4] and Xiaorong Fan[1,5,6]

[1]Institute of Environment and Ecology, School of The Environment and Safety Engineering, Jiangsu University, Jiangsu Province, P.R. China
[2]Key Laboratory of Fruits and Vegetables Postharvest and Processing Technology Research of Zhejiang Province, College of Food Science and Biotechnology, Zhejiang Gongshang University, Hangzhou, P.R. China
[3]Department of Pharmacology, Lahore Pharmacy Collage, Lahore, Pakistan
[4]Department of Botany, Sir Syed College, Taliparamba, Kannur, Kerala, India
[5]Key Laboratory of Plant Nutrition and Fertilization in Lower-Middle Reaches of the Yangtze River, Ministry of Agriculture, Nanjing Agricultural University, Nanjing, P.R. China
[6]State Key Laboratory of Crop Genetics and Germplasm Enhancement, Nanjing Agricultural University, Nanjing, P.R. China

2.1 Introduction

The natural environment presents many challenges to human health and well-being, many of which we can reasonably predict and which we cannot. Furthermore, socioeconomic factors mean that different social classes are affected in different ways to different degrees (Altomonte et al., 2020). The natural environment makes an important contribution to people's health through the quality of the air we breathe and the food and water we consume. On the one hand, it offers economic and recreational opportunities that enhance health; on the other, it is influenced by transportation, industrial processes, and agricultural and waste management practices (Southwick, Bonanno, Masten, Panter-Brick, & Yehuda, 2014). Environmental pollutants and potentially pathogenic organisms can pass through a series of complex transportation and exposure route. In general, the natural environment can promote health. Several medicines, for example, are derived from natural sources and plants, giving a compelling rationale for environmental protection (Kraemer, Ramachandran, & Perron, 2019). Furthermore, exposure to nature might become directly good to one's health; if that's the fact, the field of environmental health must go above toxicity concerns. To understand the interplay between

Microbial Consortium and Biotransformation for Pollution Decontamination
DOI: https://doi.org/10.1016/B978-0-323-91893-0.00005-5

human health and the natural environment at all levels, from the most intimate (how an individual interacts with the world around them) to the global (i.e., how countries have agreed to decrease pollution, practice appropriate agricultural and land use practices, and nourish their population) (Flandroy et al., 2018). Biodiversity can also be utilized as a number of metrics of environmental change, such as studies on endocrine-disrupting chemicals in wild animals and the global fall and extinction of amphibian species. According to Lovejoy, biodiversity becomes the highest integrators of all things affecting biological systems. Some other strategy to connect biodiversity to public health is to reduce the environmental diseases. An active area is the use of biological processes (mostly microbes) to perceive a lack of waste from the environment, including polychlorinated biphenyls (PCBs), oil spills, and heavy metals (Holmes, Nelson, Coops, & Wulder, 2013). Individuals are becoming more interested in biological systems as a technique of cleansing water sources by repairing catchment regions. A healthy environment is crucial for human health, according to good judgment. Yet, science and technology has only recently equipped us with a technique to assess the relationship between a healthy environment and a healthy body. The natural environment, in which we live, as well as the countries and international organizations that must discuss protection of the environment add one local and global perspective to environmental health (Bohensky et al., 2011). Even though there is a strong relationship between public health and the environment, conference respondents stated that we have scarce resources to analyze health concerns, as well as to adopt policies. Numerous workshop participants stated that in order to face these difficulties, we ought to build a more systematic and disciplined approach to environmental health, which should include concerns of human biology and ecological health, as well as a deeper understanding of these interconnections The systemic perspective of ecologists must be accepted as a new approach for environmental health experts (Gray, 2018) (Fig. 2.1).

2.2 Environmental toxicology and related human health risks

Abovementioned threats to human health and environment mediated by various contaminants highlight the significance of a safe and healthy environment, which has emerged as a topmost issue in recent research (Berglund et al., 1992). People have become more aware of the effects of

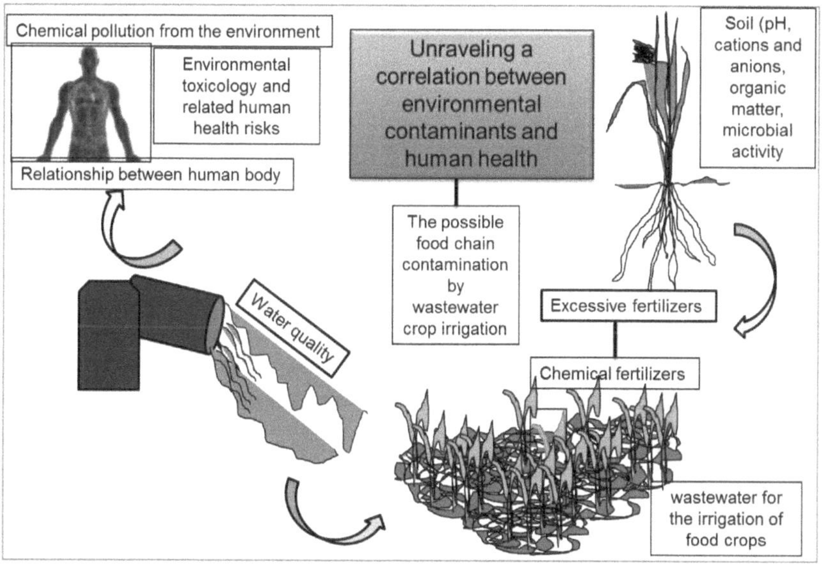

Figure 2.1 Environmental toxic elements and its effects on the ecosystem.

pollution on humans and the environment, and they are demanding fairer and ecologically sound progress. As a result, environmental conservation and environmental sustainability are likewise high priorities in the globalist agenda, and they are inextricably linked (Alshuwaikhat & Abubakar, 2008). As a result, it is critical to stimulate the growth of ecologically responsible technologies for assessing, monitoring, and managing dangerous contaminants As a result, healthy collaboration involving researchers, scientists, citizens, and lawmakers is critical for the long-term and extensive development of the communities (Blokhuis, Jones, Geers, Miele, & Veissier, 2003). Human-caused ecosystem toxicity is a pervasive and significant issue confronting society, scientists, and regulatory authorities all around the world currently, the world is seeing rapid and imbalanced industrial and technical advancement, as well as the unjust and unsustainable exploitation of nonrenewable resources (Jain, Ranjan, Dasgupta, & Ramalingam, 2018). As a result, several products and by-products have just become contaminants, threatening environmental health and stability. Numerous chemical compounds will damage our water, land, and air, negatively impacting both the ecosystem and our health. The majority of contaminants get up in the environment from industrial and commercial activities (Schwarzenbach, Egli, Hofstetter, Von Gunten, & Wehrli, 2010); for example, spills of petroleum and chemicals, as well as nonpoint sources (NPSs) such as roadways, parking lots, and storm drains, as well as wastewater treatment plants and sewage treatment systems (Bhat et al., 2021). For decades, many hazardous material dumps and industrial operations have damaged the environment and are continuing to do so (Jeong, Choi, Lee, Lim, & Ra, 2020).

2.2.1 Air pollution

The fundamental worry of the modern civilized world is air pollution, which has major deleterious impacts on human health and environment. There are several emission types, but motor vehicles and industrial production activities account for the majority of air pollution (Zhang & Smith, 2003). Particulate pollution, ground-level ozone, carbon monoxide, sulfur oxides, nitrogen oxides, and lead are the six main air pollutants. According to the World Health Organization, long-term and short-term contact to airborne toxins can cause a variety of toxicological effects in humans, involving respiratory and cardiovascular disease, neuropsychiatric problems, eye irritation, skin illness, and long-term chronic disorders (such as cancer) (Ghorani-Azam, Riahi-Zanjani, & Balali-Mood, 2016). Various studies have found a direct correlation between poor air quality and increased morbidity and death that are primarily caused by cardiovascular and respiratory disorders. Air pollution is regarded as the primary environmental risk factor for the emergence and progression of many diseases, including asthma, lung cancer, ventricular hypertrophy, Alzheimer's and Parkinson's disease, psychological difficulties, autism, retinopathy, and fetal harm growth and low birth weight. The purpose of this review article is to describe the toxicology of important air pollutants, their causes of emission, and their impact on human health (Tyagi, Garg, & Paudel, 2014). We also offered practical strategy to minimize air pollution in Iran. Air pollution has a direct impact on public health, causing and triggering a variety of diseases that result in high morbidity and death, particularly in developing nations such as Iran. As a result, controlling air pollution is critical and must be prioritized by the government. These countries' policymakers and lawmakers should reform all air pollution standards and policies (Mills et al., 2009). A good protection of the environment organization must lead the coordination of the various agencies engaged in air pollution. An efficient environmental safety organization ought to have an adequate fund for environmental management, research, development, surveillance, and overall environmental control, covering air pollution (Staniskis & Stasiskiene, 2006).

2.2.2 Hazard effect on health

Based on health risks, an air toxin is any aberrant persistent substance in the air that interferes with the normal operation of human organs. As per existing data, the main hazardous effects of air pollution exposure are mostly on the respiratory system, cardiovascular system, ophthalmic system, dermatology, neuropsychiatry, hematological, immunology, and reproductive system (Dada et al., 2021). Furthermore, molecular and cytotoxicity might, in the long term, cause a range of cancers. Besides this, even a small portion of airborne poisons have been shown to be hazardous to vulnerable populations, such as children and elderly as a patient with respiratory and cardiovascular disorders (Buzea, Pacheco, & Robbie, 2007).

2.2.3 Nonpoint source pollution

When pollution occurs from a source that cannot be traced assigned to a particular site, we refer to it as "nonpoint source pollution." Such pollution happens when septic tanks leak or rainwater runoff penetrates into streams and rivers, carrying sediments, fertilizers, pet waste, or excrement into estuaries and coastal water (Xepapadeas, 2011). NOAA and the Environmental Protection Agency collaborated to administer the coastal NPS pollution control plan in order to alleviate the contaminated runoff problem. All states and areas with authorized coastal management plans are obligated to establish and implement coastal plans to limit the quantity of NPS pollution entering our waterways, according to the plan (Ongley, Xiaolan, & Tao, 2010). Excess fertilizers, herbicides, and pesticides from farming and residential areas are the examples of NPS pollution. Oil, grease, and harmful substances are released into the environment as a result of urban runoff and power generation. The sediments from poorly managed construction sites, agriculture and forests, and eroded river banks. NPS pollution is pollution for which a specific source or magnitude of emissions cannot be accurately seen or identified (Talabi & Kayode, 2019). Environmental levels of pollutants associated to individual unobserved emission are typically detected in NPS pollution. The primary source of water pollution, eutrophication, and hypoxia is NPS pollution generated by agricultural runoff. Traditional environmental policy mechanisms (such as emissions taxes or tradable permits) are extremely hard to set up to limit NPS pollution problems to knowledge asymmetries and unpredictable effects (Stephenson & Shabman, 2017). The chapter examines concerns linked to NPS pollution monitoring and their link with theoretically presented instruments, as well as the main theoretical techniques of NPS pollution supervision thus far-input-based schemes, environmental schemes, and endogenous surveillance.

2.2.4 Chemical pollution from the environment

Environmental chemical infection is a serious food standard concern around the world, posing a substantial hazard to human health. Metals/metalloids, polycyclic aromatic hydrocarbons, persistent organic pollutants, perfluorinated compounds, pharmaceuticals, and personal-care products, radioactive elements, electronic waste, plastics, and nanoparticles are among the many kinds of chemicals (Rather, Koh, Paek, & Lim, 2017) (Fig. 2.2). Some are found naturally in the environment, while some are created by man-made means they have the potential to pollute our food (crops, animals, and fisheries) and drinking water, as well as harm our health (Sardar et al., 2013). It's really critical to assess the associated possible risks. Assessing pollution levels, developing standard precautions, including remedial measures, and taking sociopolitical effects into account are all key components in supplying safer food worldwide (Lu et al., 2015). In certain circumstances, the

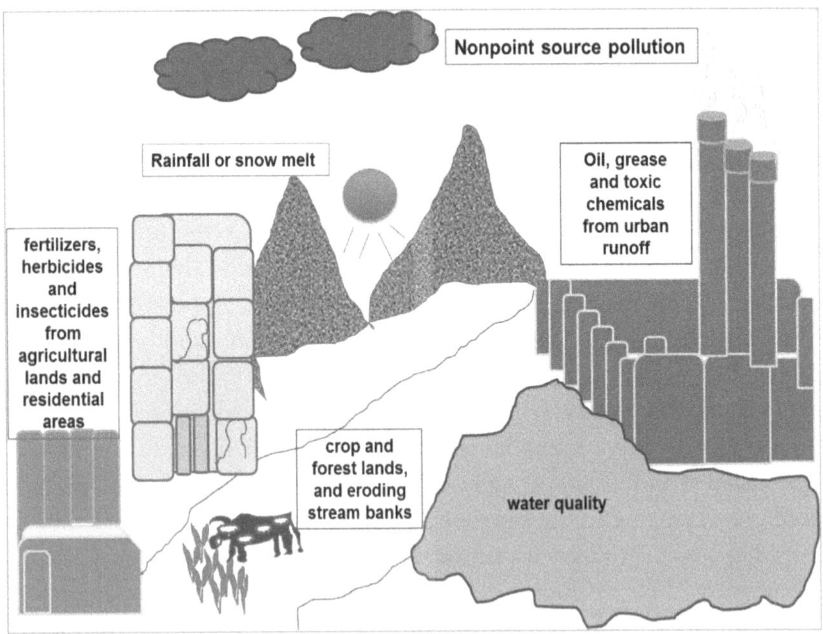

Figure 2.2 Nonpoint source pollution and its effects.

environment might just be the source of contaminants. That's also true for metals such as lead and mercury, as well as dioxins and PCBs. Pesticide being used in agriculture might result in food spoilage. Consequently, medications used in humans and animals may poison streams and endanger consumers' health. Furthermore, food packaging methods could be a source of pollution, as "migrants" leached from packing materials. These contaminants can have both acute and chronic harmful effects (Motarjemi & Käferstein, 1999). Toxicities could be related to the method of contact and intake, as well as human variables, including age and health state.

2.3 The environmental impact of chemical fertilizers and excessive fertilizers on water quality

Manure and fertilizer nutrients reach lakes and streams via runoff and soil erosion. In general, as the nitrogen (N) and phosphorus (P) levels in the soil test rise, so will the nitrogen and phosphorus accessible to a wide range of plants (Nangia, de Fraiture, & Turral, 2008). Runoff from farms with substantial soil and nitrogen and phosphorus concentration could include huge levels of these dissolved nutrients, potentially damaging streams, marshes, and lakes (King et al., 2015). Furthermore, erosion would transport small particles in the soil that is rich in nutrients (Dar, Bandh, & Kamili, 2011; Dar et al., 2012; Dar, Bandh, Kamili, Nazir, & Bhat, 2013). Eroded soil particles with connected nutrients would aggregate as sediments in water supplies and act as a basis of accessible nutrients for a long period of time (Barling & Moore, 1994).

2.3.1 Oxygen consumption

When fertilizers or commercial fertilizers reach surface water, the nutrients released drive the development of microorganisms (Khan, Mobin, Abbas, & Alamri, 2018). Microorganism development and reproduction decrease the dissolved oxygen level of the water bodies. Fish and other aquatic life could suffocate and die when there is insufficient dissolved oxygen in the groundwater (Dar, Kamili, Dar, Chishti, & Ahmad, 2015; Dar et al., 2016). The dead fish and other sea creatures that arise would affect water quality and generate foul smells (Rowe et al., 2016).

2.3.2 Weed growth and algae bloom

As the intake of energy (particularly nitrogen and phosphorus) increases, so does the quantity of plants and algae in lakes, ponds, and other bodies of water (Wurtsbaugh, Paerl, & Dodds, 2019). Manure contains enough nitrogen and phosphate to be utilized as a fertilizer for plant growth and has some nutrient benefit. The impact on algae and aquatic plants is comparable, and nutrient usage, including crop production, is critical to the growth of aquatic plants and algae (Carpenter et al., 1998). The availability of nutrients with least quantity of growth would limit the aquatic system's productivity. Even tiny amounts of restricted nutrients introduced into crops or aquatic systems can significantly enhance yield. That's also beneficial to crops (Snyder, Bruulsema, Jensen, & Fixen, 2009). Stimulating the yield of aquatic plants and algae, on the other hand, is harmful to water supplies. "Eutrophication" is a term that describes natural or expedited process wherein water bodies get overgrown with aquatic plants and have low levels of oxygen. Microorganisms feed on organic debris as these aquatic plants die. The microbes replicate and develop a new, consuming the oxygen from the water (Paerl, 2006). Rise in aquatic plant growth would ultimately result in a drop in the oxygen in the water level of the water body, suffocating fish and other aquatic organisms (Camargo & Alonso, 2006) (Table 2.1).

2.4 Method to reveal the relationship between human body, environment, and emotion data

Environmental model and data aggregation platforms have made great advances in recent years, thanks to advances in sensor and mobile technologies. Yet, fewer individuals have focused on measuring the link between external alterations and their effects on our bodies in the actual world thus far (Blais, 2004). We have researched a data-driven strategy obtained by direct and continual sensor information to evaluate the surroundings and the effects of physiological changes and feelings in the chapter (Na, Choi, Kim, & Kim, 2019). Therefore our purpose here is to investigate the feasibility of combining human physiological signals, environmental sensory data, and based on the person emotion indicators to (1) collect the required information and (2) mimic the short-term effects of the environment on the body. In order to attain this objective, we performed "field" serious research on the human body and mobile sensors using human sensor and environment data. The information was gathered from respondents while they strolled across Nottingham's city center, with the goal of constructing analytic and prediction models. Despite controlling for potential confounding variables, regression analysis revealed a substantial

Table 2.1 Sources of environmental stress and effects on environment and human health, its causes, and how the ecosystem responds to changes in its intensity.

Environmental stressors	Effects on environment and human health	References
Physical stress	Disorderly events as a tempest or tornado, a seismic sea wave (tsunami), the blast of a volcanic eruption, an explosion, or trampling by heavy machinery, instruments, or hikers.	Negro-Vilar (1993)
Wildfire	Includes the unrestrained burning of the biomass of an environment.	Kalabokidis (2000)
Chemical pollution	Pesticides, gases such as ozone and sulfur dioxide, and toxic elements such as arsenic and mercury. Too much nutrients can also cause pollution, which can distort productivity and other ecological functions.	Schwarzenbach et al. (2010)
Thermal pollution	It is caused by the release of heat (thermal energy) into the environment, and because species have different tolerance to extreme temperatures, this leads to ecological pressure.	Najjar (2011)
Radiation stress	This is caused by excessive exposure to ionizing energy. The radiation may be from nuclear waste or explosions, or it may be diagnostic X-rays or solar ultraviolet energy.	Frumkin (2001)
Climatic stress	It is related to insufficient or excessive temperature, humidity, solar radiation, wind, or a combination of these factors alone.	McMichael, Woodruff, and Hales (2006)
Biological pollution	This may involve introducing invasive species and changing natural habitats, or it may be the release of pathogens into the environment through the discharge of raw sewage.	Khan and Ghouri (2011)
Biological stressors	It is related to the interactions that occur between organisms, such as competition, herbivore, predation, parasitism, and disease. For example, individuals of the same or different species may compete for basic resources with limited supply. Herbivory, predation, parasitism, and disease are nutritional interactions, one species uses another species.	Estok, Freedman, and Boyle (1989)

relationship among noise exposure and heart rate (Mandryk, Atkins, & Inkpen, 2006). Similarly, UV radiation and environmental noise have been shown to have a profound impact on variations in electrical skin activity. The largest impact of the observed variations in body temperature and movement is provided by barometric pressure. Furthermore, it was shown that there is a substantial relationship between air pressure and heart rate (Mandryk et al., 2006). Ultimately, decision-making fusion is applied to the categorization result from several modalities. This is, as far as we understand, the first try to merge and model data from environmental and physiological inputs acquired from sensors in a real-world context. The world's population densities and water quality are unequally distributed. Farmers are forced to utilize wastewater to irrigate food crops. Due to this practice, both positive and negative effects have been observed in agricultural use, with pollution and toxicity being the negative effects. Although wastewater is an important source of nutrients for plants, the presence of harmful chemicals and bacteria in some wastewater, as well as environmental, sanitation and health concerns, are related to the use of wastewater for agricultural irrigation (Qadir, Boers, Schubert, Ghafoor, & Murtaza, 2003). This chapter discusses the negative and positive consequences of wastewater irrigation on the physical, biological, and chemical aspects of soil (pH, cations and anions, organic matter, and microbial activity). We evaluate the possible toxic elements that store in soil and, as a response, are transferred to plants and humans. The potential human health consequences of using industrial sewage for agriculture irrigation are also foreseen and addressed. We analyzed the level of agricultural wastewater reuse in developing and developed nations, as well as associated environmental and health challenges. In addition, taking into account the global and regional context, as well as the current status of wastewater use in crop production, sanitation and planning difficulties, restoration techniques, civil society concerns, and government and related interests (Khalid et al., 2018).

2.5 Conclusion

Humans and the entire universe are subjected to a multitude of developing contaminants via numerous known and undiscovered methods. According to our findings, these contaminants pose great new and serious threats to soil, water, air, and ecosystems (especially human health) Furthermore, newer chemical compounds grow and frequently outstrip the capacities of safety remedies, threat assessment methods, monitoring technologies, and cutting-edge technologies. For primary prevention, it is important to analyze the influence of pollutants on human health under current conditions, and for that, the key interaction considerations including pollutant content, type and nature of pollutants, scale and amount of pollution, severity of harm to health, and accessible resources must be taken. The removal of rapidly developing contaminants from the environment and the human body is indicative of future technology and study to build comprehensive remediation theories and methodologies.

References

Alshuwaikhat, H. M., & Abubakar, I. (2008). An integrated approach to achieving campus sustainability: assessment of the current campus environmental management practices. *Journal of Cleaner Production, 16*(16), 1777–1785.

Altomonte, S., Allen, J., Bluyssen, P. M., Brager, G., Heschong, L., Loder, A., ... Wargocki, P. (2020). Ten questions concerning well-being in the built environment. *Building and Environment, 180*, 106949.

Barling, R. D., & Moore, I. D. (1994). Role of buffer strips in management of waterway pollution: A review. *Environmental Management, 18*(4), 543–558.

Berglund, B., Brunekreef, B., Knöppe, H., Lindvall, T., Maroni, M., Mølhave, L., ... Skov, P. (1992). Effects of indoor air pollution on human health. *Indoor Air, 2*(1), 2–25.

Bhat, R. A., Singh, D. V., Qadri, H., Dar, G. H., Dervash, M. A., Bhat, S. A., ... Yousaf, B. (2021). Vulnerability of municipal solid waste: An emerging threat to aquatic ecosystems. *Chemosphere, 287*(2022), 132223. Available from https://doi.org/10.1016/j.chemosphere.2021.132223.

Blais, F. (2004). Review of 20 years of range sensor development. *Journal of Electronic Imaging, 13*(1), 231–243.

Blokhuis, H., Jones, R., Geers, R., Miele, M., & Veissier, I. (2003). Measuring and monitoring animal welfare: Transparency in the food product quality chain. *Animal welfare-potters bar then wheathampstead, 12*(4), 445–456.

Bohensky, E., Butler, J. R., Costanza, R., Bohnet, I., Delisle, A., Fabricius, K., & Pert, P. (2011). Future makers or future takers? A scenario analysis of climate change and the Great Barrier Reef. *Global Environmental Change, 21*(3), 876–893.

Buzea, C., Pacheco, I. I., & Robbie, K. (2007). Nanomaterials and nanoparticles: Sources and toxicity. *Biointerphases, 2*(4), MR17–MR71.

Camargo, J. A., & Alonso, A. (2006). Ecological and toxicological effects of inorganic nitrogen pollution in aquatic ecosystems: A global assessment. *Environment International, 32*(6), 831–849.

Carpenter, S. R., Caraco, N. F., Correll, D. L., Howarth, R. W., Sharpley, A. N., & Smith, V. H. (1998). Nonpoint pollution of surface waters with phosphorus and nitrogen. *Ecological Applications, 8*(3), 559–568.

Dada, O. T., Odufuwa, B. O., Badiora, A. I., Agbabiaka, H. I., Ogunseye, N. O., & Samuel, O. S. (2021). Environmental hazard and health risks associated with slaughterhouses in Ibadan, Nigeria. *Environmental Hazards, 20*(2), 146–162.

Dar, G. H., Bandh, S. A., Kamili, A. N., Nazir, R., & Bhat, R. A. (2013). Comparative analysis of different types of bacterial colonies from the soils of Yusmarg Forest, Kashmir valley India. *Ecologia Balkanica, 5*(1), 31–35.

Dar, G. H., Kamili, A. N., Chishti, M. Z., Dar, S. A., Tantry, T. A., & Ahmad, F. (2016). Characterization of *Aeromonas sobria* isolated from Fish Rohu (*Labeo rohita*) collected from polluted pond. *Journal of Bacteriology and Parasitology*, 7(3), 1–5. Available from https://doi.org/10.4172/2155-9597.1000273.

Dar, G. H., Kamili, A. N., Dar, S. A., Chishti, M. Z., & Ahmad, F. (2015). Detection and characterization of potentially pathogenic *Aeromonas sobria* isolated from fish *Hypophthalmichthys molitrix* (Cypriniformes: Cyprinidae). *Microbial Pathogenesis*, *91*, 136–140. Available from https://doi.org/10.1016/j.micpath.2015.10.017.

Dar, G. H., Bandh, S. A., & Kamili, A. N. (2011). Bacteriological analysis of soil from Yusmarg Health Resort of Kashmir Valley. *International Journal of Current Research*, *3* (10), 53–56, 2011.

Dar, G. H., Kamili, A. N., Nazir, R., Bandh, S. A., & Bhat, R. A. (2012). A preliminary study of colony forming units of bacteria from the soils of Yusmarg Forest, Kashmir Valley, India. *International Journal of Current Research*, *4*(12), 467–472.

Estok, D., Freedman, B., & Boyle, D. (1989). Effects of the herbicides 2, 4-D, glyphosate, hexazinone, and triclopyr on the growth of three species of ectomycorrhizal fungi. *Bulletin of Environmental Contamination and Toxicology*, *42*(6), 835–839.

Flandroy, L., Poutahidis, T., Berg, G., Clarke, G., Dao, M.-C., Decaestecker, E., & Plovier, H. (2018). The impact of human activities and lifestyles on the interlinked microbiota and health of humans and of ecosystems. *Science of the Total Environment*, *627*, 1018–1038.

Frumkin, H. (2001). Beyond toxicity: Human health and the natural environment. *American Journal of Preventive Medicine*, *20*(3), 234–240.

Ghorani-Azam, A., Riahi-Zanjani, B., & Balali-Mood, M. (2016). Effects of air pollution on human health and practical measures for prevention in Iran. *Journal of Research in Medical Sciences: The Official Journal of Isfahan University of Medical Sciences*, *21*.

Gray, K. M. (2018). From content knowledge to community change: A review of representations of environmental health literacy. *International Journal of Environmental Research and Public Health*, *15*(3), 466.

Holmes, K. R., Nelson, T. A., Coops, N. C., & Wulder, M. A. (2013). Biodiversity indicators show climate change will alter vegetation in parks and protected areas. *Diversity*, *5*(2), 352–373.

Jain, A., Ranjan, S., Dasgupta, N., & Ramalingam, C. (2018). Nanomaterials in food and agriculture: An overview on their safety concerns and regulatory issues. *Critical Reviews in Food Science and Nutrition*, *58*(2), 297–317.

Jeong, H., Choi, J. Y., Lee, J., Lim, J., & Ra, K. (2020). Heavy metal pollution by road-deposited sediments and its contribution to total suspended solids in rainfall runoff from intensive industrial areas. *Environmental Pollution*, *265*, 115028.

Kalabokidis, K. D. (2000). Effects of wildfire suppression chemicals on people and the environment—A review. *Global Nest: The International Journal*, *2*(2), 129–137.

Khalid, S., Shahid, M., Bibi, I., Sarwar, T., Shah, A. H., & Niazi, N. K. (2018). A review of environmental contamination and health risk assessment of wastewater use for crop irrigation with a focus on low and high-income countries. *International Journal of Environmental Research and Public Health*, *15*(5), 895.

Khan, M. A., & Ghouri, A. M. (2011). Environmental pollution: Its effects on life and its remedies. *Researcher World: Journal of Arts, Science & Commerce*, *2*(2), 276–285.

Khan, M., Mobin, M., Abbas, Z., & Alamri, S. (2018). Fertilizers and their contaminants in soils, surface and groundwater. *Encyclopedia of the Anthropocene*, *5*, 225–240.

King, K. W., Williams, M. R., Macrae, M. L., Fausey, N. R., Frankenberger, J., Smith, D. R., & Brown, L. C. (2015). Phosphorus transport in agricultural subsurface drainage: A review. *Journal of Environmental Quality*, *44*(2), 467–485.

Kraemer, S. A., Ramachandran, A., & Perron, G. G. (2019). Antibiotic pollution in the environment: From microbial ecology to public policy. *Microorganisms*, 7(6), 180.

Lu, Y., Song, S., Wang, R., Liu, Z., Meng, J., Sweetman, A. J., & Luo, W. (2015). Impacts of soil and water pollution on food safety and health risks in China. *Environment International*, 77, 5–15.

Mandryk, R. L., Atkins, M. S., & Inkpen, K. M. (2006). A continuous and objective evaluation of emotional experience with interactive play environments. Paper presented at the Proceedings of the SIGCHI conference on Human Factors in computing systems.

McMichael, A. J., Woodruff, R. E., & Hales, S. (2006). Climate change and human health: present and future risks. *The Lancet*, *367*(9513), 859–869.

Mills, N. L., Donaldson, K., Hadoke, P. W., Boon, N. A., MacNee, W., Cassee, F. R., & Newby, D. E. (2009). Adverse cardiovascular effects of air pollution. *Nature Clinical Practice. Cardiovascular Medicine*, *6*(1), 36–44.

Motarjemi, Y., & Käferstein, F. (1999). Food safety, Hazard Analysis and Critical Control Point and the increase in foodborne diseases: A paradox? *Food Control*, *10*(4–5), 325–333.

Na, H., Choi, J.-H., Kim, H., & Kim, T. (2019). Development of a human metabolic rate prediction model based on the use of Kinect-camera generated visual data-driven approaches. *Building and Environment*, *160*, 106216.

Najjar, Y. S. (2011). Gaseous pollutants formation and their harmful effects on health and environment. *Innovative Energy Policies*, *1*, 1–9.

Nangia, V., de Fraiture, C., & Turral, H. (2008). Water quality implications of raising crop water productivity. *Agricultural Water Management*, *95*(7), 825–835.

Negro-Vilar, A. (1993). Stress and other environmental factors affecting fertility in men and women: overview. *Environmental Health Perspectives*, *101*(Suppl 2), 59–64.

Ongley, E. D., Xiaolan, Z., & Tao, Y. (2010). Current status of agricultural and rural non-point source pollution assessment in China. *Environmental Pollution*, *158*(5), 1159–1168.

Paerl, H. W. (2006). Assessing and managing nutrient-enhanced eutrophication in estuarine and coastal waters: Interactive effects of human and climatic perturbations. *Ecological Engineering*, *26*(1), 40–54.

Qadir, M., Boers, T. M., Schubert, S., Ghafoor, A., & Murtaza, G. (2003). Agricultural water management in water-starved countries: Challenges and opportunities. *Agricultural Water Management*, *62*(3), 165–185.

Rather, I. A., Koh, W. Y., Paek, W. K., & Lim, J. (2017). The sources of chemical contaminants in food and their health implications. *Frontiers in pharmacology*, *8*, 830.

Rowe, H., Withers, P. J., Baas, P., Chan, N. I., Doody, D., Holiman, J., & McDowell, R. (2016). Integrating legacy soil phosphorus into sustainable nutrient management strategies for future food, bioenergy and water security. *Nutrient Cycling in Agroecosystems*, *104*(3), 393–412.

Sardar, K., Ali, S., Hameed, S., Afzal, S., Fatima, S., Shakoor, M. B., & Tauqeer, H. M. (2013). Heavy metals contamination and what are the impacts on living organisms. *Greener Journal of Environmental Management and Public Safety*, *2*(4), 172–179.

Schwarzenbach, R. P., Egli, T., Hofstetter, T. B., Von Gunten, U., & Wehrli, B. (2010). Global water pollution and human health. *Annual Review of Environment and Resources*, *35*, 109–136.

Snyder, C. S., Bruulsema, T. W., Jensen, T. L., & Fixen, P. E. (2009). Review of greenhouse gas emissions from crop production systems and fertilizer management effects. *Agriculture, Ecosystems & Environment*, *133*(3–4), 247–266.

Southwick, S. M., Bonanno, G. A., Masten, A. S., Panter-Brick, C., & Yehuda, R. (2014). Resilience definitions, theory, and challenges: Interdisciplinary perspectives. *European Journal of Psychotraumatology*, *5*(1), 25338.

Staniskis, J. K., & Stasiskiene, Z. (2006). Environmental management accounting in Lithuania: exploratory study of current practices, opportunities and strategic intents. *Journal of Cleaner Production*, *14*(14), 1252–1261.

Stephenson, K., & Shabman, L. (2017). Can water quality trading fix the agricultural non-point source problem? *Annual Review of Resource Economics*, *9*, 95–116.

Talabi, A. O., & Kayode, T. J. (2019). Groundwater pollution and remediation. *Journal of Water Resource and Protection*, *11*(01), 1.

Tyagi, S., Garg, N., & Paudel, R. (2014). Environmental degradation: Causes and consequences. *European Researcher*, *81*(8-2), 1491.

Wurtsbaugh, W. A., Paerl, H. W., & Dodds, W. K. (2019). Nutrients, eutrophication and harmful algal blooms along the freshwater to marine continuum. *Wiley Interdisciplinary Reviews: Water*, *6*(5), e1373.

Xepapadeas, A. (2011). The economics of non-point-source pollution. *Annual Review of Resource Economics*, *3*(1), 355–373.

Zhang, J., & Smith, K. R. (2003). Indoor air pollution: A global health concern. *British Medical Bulletin*, *68*(1), 209–225.

CHAPTER 3

Effect of wastewater from industries on freshwater ecosystem: threats and remedies

Johra Khan
Department of Medical Laboratory Sciences, College of Applied Medical Sciences, Majmaah University, Al Majmaah, Saudi Arabia

3.1 Introduction

"Industrialization" is a term used to represent the revolution took place in manufacturing industries around the late 17th century. This revolution not only brought a dramatic change in human development but also in the environment around us in a global scenario (Carley & Spapens, 2017). With industrialization, the other development in human civilization was the development of big cities around industries. Industrialization and urbanization helped us in development, but at the same time, they caused a lot of damage to our environment, especially to freshwater, the availability of which is just 0.5% of the total water on the Earth (Elliott, Day, Ramachandran, & Wolanski, 2019).

Freshwater ecosystem is one of the most diverse ecosystems on our planet. The damage caused to this ecosystem is creating a significant danger to plants, animals, and humans. Around the world, we see freshwater habitat consisting of freshwater streams, lakes, rivers, reservoirs, ponds, and connective channels as well as their sources (glaciers, aquifers, or rainfall) (Elliott et al., 2019). Freshwater ecosystem includes a diverse variety of organisms which either live permanently in water or a large portion of their life cycle passes in freshwater. The freshwater ecosystem of Central Asia and Europe is considered as extremely diverse. Based on different species of water plants, fishes, and other water animals freshwater ecosystem is divided into 60 ecoregions (Ribeiro et al., 2017b). In comparison to terrestrial and marine ecosystem, the freshwater ecosystem is at high risk of degradation due to which the quality of freshwater and quantity of abundant species found in freshwater are constantly declining (Vollmer et al., 2018).

Microbial Consortium and Biotransformation for Pollution Decontamination
DOI: https://doi.org/10.1016/B978-0-323-91893-0.00010-9

Freshwater ecosystem is the home of nearly 10% of all species found on the Earth and 30% of all vertebrate. According to Canadian Natural Resources Limited (CNRL), 35% of freshwater amphibians are either facing threat or extinct, and 46% of mammals and 38% of turtles are about to get extinct. In this chapter, we are focusing on different industrial wastewater, their threat on freshwater ecosystem, and remedies or methods to reduce their impact on fauna and flora of freshwater ecosystem.

3.2 Saline wastewater: its impact and treatment

Saline wastewater is the outcome of many industries such as agro-food, petroleum, textile, and leather. To deal with the salification, European Union Directive 2000/60/EC established a policy measure to monitor the salinity of freshwater ecosystem and biological community (Dar, Bandh, & Kamili, 2011; Dar, Kamili, Nazir, Bandh, & Bhat, 2012; Farmaki & Tranoulidis, 2018). The plants present in freshwater such as macrophytes, epiphytic algae, and riparian vegetation were found to be very sensitive to increased salinity (Dar, Bandh, Kamili, Nazir, & Bhat, 2013; Tibby, Adamson, & Kershaw, 2020) as they grow best in freshwater; hence, their strength to survive reduces as the salinity increases. Many countries around the world have documented the maximum concentration of salinity such as Australia and the United States specifies that electrical conductivities of 2 mS/cm can be acceptable for drinking and irrigation (Cañedo-Argüelles et al., 2019). One of the important determinations that was recorded that salinity of 4000 mg/L seems to be considered as the upper tolerance limit by widespread freshwater macrophytes. (Steinmuller & Chambers, 2018).

3.2.1 Effect of salinity on freshwater ecosystem

To study the effect of salinity, it is important to understand the process affected by salinity. Osmoregulation is one of these processes. Osmoregulation is a very important phenomenon for the survival of freshwater animals (Baldisserotto, 2019). A study done on mayfly species concerning salinity causes substantial mortality, if the osmolality decreases in comparison to their internal fluid. High salinity can causes pH irregularity or Na poisoning. High salinity can have drastic effect on fitness (Chamberlain et al., 2019) and survival of freshwater organisms (Bonga, 1972).

Increased salinity in freshwater species causes ion toxicity as it exceeds the threshold limit for them (Kefford, Dalton, Palmer, & Nugegoda, 2004).

The other effects of high salt concentration observed are oxidative stress (Cañedo-Argüelles et al., 2016), delayed growth (Sala et al., 2016), malformation (Hopkins, French, & Brodie, 2017), and reduction in feeding rate (Arribas et al., 2019). Increase in total salinity, pH, and base cation concentration develops freshwater salinization syndrome (Kaushal et al., 2019), commonly found in river and lakes of North America and Europe (Cañedo-Argüelles, Kefford, & Schäfer, 2019). A meta-analysis study on lethal salinity concentration (Coldsnow, Mattes, Hintz, & Relyea, 2017) found that salinity most affects microinvertebrates and amphibians (Iglesias, 2020). A literature review of the Germany on freshwater salinity (Schulz & Cañedo-Argüelles, 2019) reported the salinity sensitivity of freshwater fauna and gave the sequence of sensitivity from more to less as fish > macroinvertebrates > phytoplankton > macrophytes > diatoms (Hopkins et al., 2017). The saline toxicity is very clear from a lot of research on this issue but the response of organisms to salinity is due to different ions as total salt concentration or certain ion concentration needs more research (Cañedo-Argüelles et al., 2017) (Fig. 3.1).

3.3 Food-processing industry wastewater

Food-processing industry produces two main types of wastewater: (1) saline wastewater and (2) slaughterhouse wastewater (SWW) (Jensen, Pinches, Desai, & Street, 2017). The agro-food industry uses salt for food preservative. Salt in high amount is used for canning fish, meat, pickled vegetables, cheese, and dairy products. Salt is used to remove water so that the activity of microbes can be reduced or limited for long-time preservation of products. The saline effluents produced are produced due to the use of brine in canning fish and pickles. Other than canned food industry, olive oil and seafood industries produce large quantity of saline wastewater (Dar, Kamili, Dar, Chishti, & Ahmad, 2015; Dar et al., 2016; Liang et al., 2017). Fishery industry produces not only saline water but also protein nitrogen and organic matter (Bhat et al., 2021; Letelier-Gordo, Aalto, Suurnäkki, & Pedersen, 2020). SWW is categorized as most harmful wastewater to freshwater ecosystem by the United States Environmental Protection Agency (US EPA). The meat-processing industry is the major consumer of freshwater among food- and beverage-processing sectors. Meat-processing industry uses up to 24% of the total freshwater (Orssatto et al., 2017). Wastewater in slaughterhouses is more harmful as it is a mixture of complex fats, proteins, and fibers. This wastewater has high blood, mucosa, and tallow which increase its biochemical oxygen

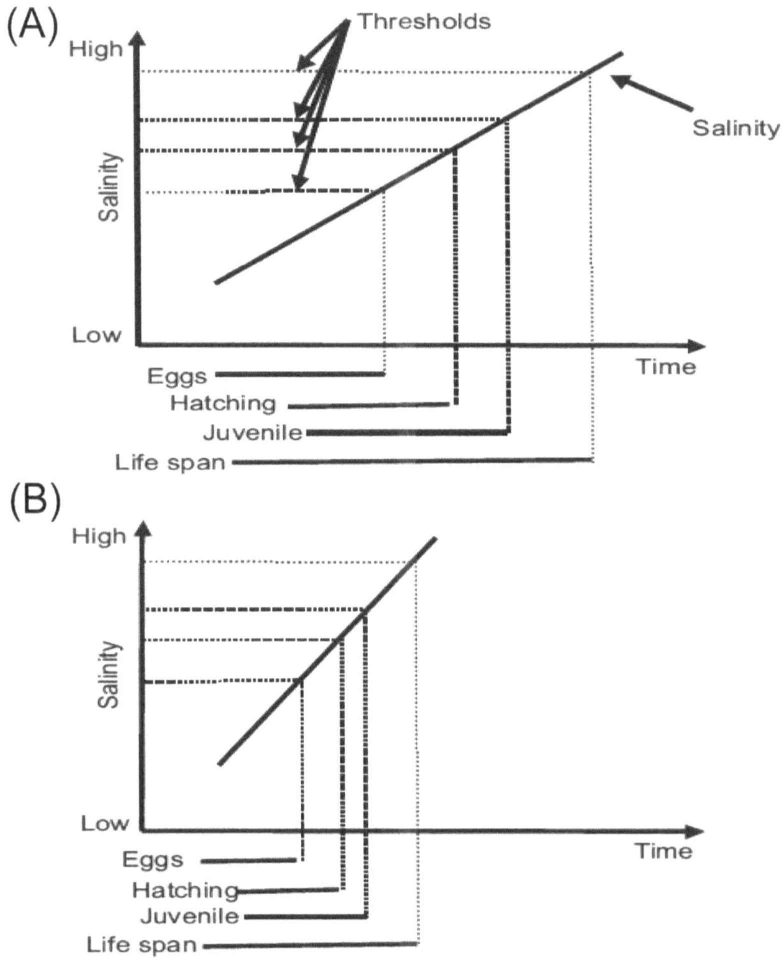

Figure 3.1 Effect of salinity on life history traits. (A) Natural delivery. (B) Increase salinity effect in comparison to time (Nielsen, Brock, Rees, & Baldwin, 2003).

demand and chemical oxygen demand. Due to high value of total suspended solids, SWW can cause deoxygenation of rivers and contamination of groundwater (Bustillo-Lecompte & Mehrvar, 2017).

3.4 Leather industry wastewater

Leather industry belongs to one of the most polluting industries due to the tanning process of animal skin into leather (Apruzzese, Bhardwaj, Degrassi, & Bhardwaj, 2017). These industry uses huge amount of salt and freshwater.

Some processes such as chromium tanning and soak liquor, in which skin and hides that are soaked in high salt concentration can generate highly saline wastewater with that leather industry also produce colored compounds, sodium chloride, sulfate, various organic and inorganic substances, toxic metallic compounds, and many different types of tanning containing oxidizable materials, and large quantities of putrefying suspended matter (Ali, 2017). The effluents after the completion of tanning process contain sulfides and salts of ammonium and calcium, making wastewater slightly alkaline. The total chemicals that are used only 20% of them are absorbed by leather (Ma, Ding, Xin, Ma, & Xia, 2019). The study conducted around the world shows that the wastewater effluents contain around 300–400 million tons of heavy metals, toxic sludge, and many solvents (Salih, 2018). The wastewater from leather industries contains high load of suspended solids and organic nitrogen and ammonia with high pH.

3.5 Effluents from petroleum industry

Petrochemical industry produces wastewater full of hydrocarbons, which are complex organic compound toxic in nature (Jain, Sharma, Dureja, Sarma, & Lal, 2017). Wastewater discharge from these industries affects freshwater ecosystem and also groundwater sources (Flörke, Bärlund, van Vliet, Bouwman, & Wada, 2019). Petroleum industry effluents are produced primarily from the industries involved in crude oil refining and fuel manufacturing, lubricants, and other petrochemical-based products. The research data shows that effluents produced from the oil industry will continue to be produced and discharged into the main water bodies. Pollutants that are found in the effluent are known to cause serious toxic hazards to the environment. Petroleum refinery effluents vary depending upon the oil type being processed, the plant's configuration, and procedures followed during the operation (Noor & Mutamim, 2017).

Wastewater produced by oilfield industries contains high amount of artificial surfactants used in crude oil extraction (Rehman, Imran, Amin, & Afzal, 2018). The wastewater produced during crude oil extraction contains polymers, surfactants, benzenes, and many other heavy mineral oils (Dickhout et al., 2017).

3.6 Plastic industries and micro- and nanoplastic in freshwater ecosystem

Plastic industries use different polymer-based materials to develop a variety of products in combination to different compounds. With development in technology, newly developed nanoscale elements are used in different products for weight loss and nanosilver parts and nanochip are used in instruments for medical application and also in developing optical fibers and many more. Due to huge expansion of nanoplastic use, it is estimated that in 2020, in total share percent of nanocomposites in the United States will reach up to 7% of total plastic use. Recent studies show huge variety of micro- and nanoplastics found in freshwater. A report on Rhine River (Germany) showed that average concentration of microplastic was around 892,777 particles/km^2 to a maximum of 3.9 million particles/km^2. In Asia, Africa, and Europe, this concentration was found very high because separation and identification method used for studying are based on visual observation method and many of them are excluded due to their submicron size. To understand the severity of microplastic pollution and its effect, it is important to understand their transportation from industries to freshwater bodies with industrial wastewater (Fig. 3.2).

3.6.1 Effect of microplastic on freshwater ecosystem

The data of 19,000 samples about the presence of microplastic in freshwater taken from Bayreuth found the presence of 4335 as plastic particles out of which more than 99% of these plastic particles are found to be

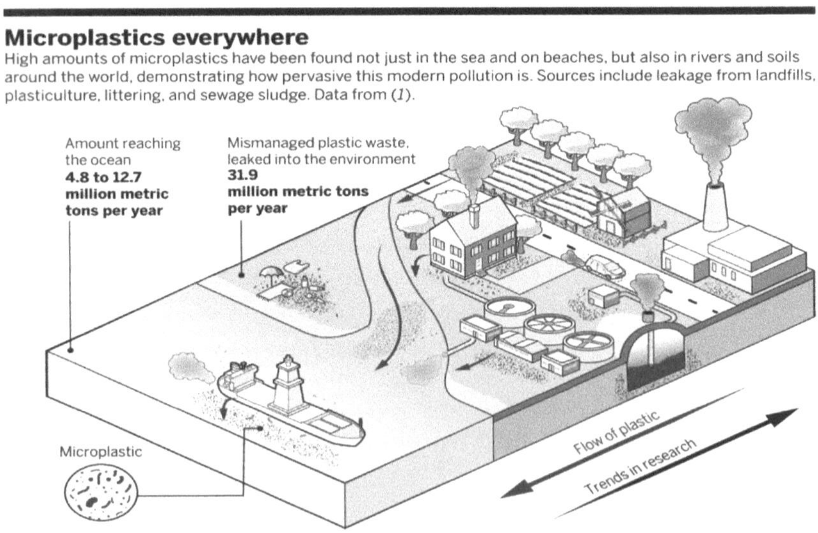

Figure 3.2 Flow of microplastic. *From online.*

microplastics with a size smaller than 5 mm. The tiny microplastic size ranges between a diameter of 0.02 and 0.3 mm (20–300 μm).

These microplastic particles enter the body of fishes and zooplankton through ingestion or dermal uptake from gills. A study on *Bosmina coregoni*, *Daphnia cucullata* in combination with algae when exposed to different size of microplastic particles between 2 and 6 μm, demonstrated that *Daphnia magna* ingests nano (20 nm) and micro (1 μm) particles but 20 nm were retained in their body in greater extent. The exposer to microplastic causes physical stress in freshwater animals and plants but the extent of stress depends on the uptake of particle size.

Some studies showed that these particles cause blockage of gills and digestive tract, whereas the expose to particle size of 2.6 μm causes increase in mortality rate, interbrood period, and decreased reproduction.

The effect of microplastics as vector for exotic species and pathogen is a cause of concern because some studies showed the presence of microorganisms developing biofilm on microplastic particles. Many members of *Vibro* genus that are opportunistic human pathogens are also plastisphere members developing biofilm around microplastic particles found in lakes and rivers. So, these microplastic particles can act as vectors for waterborne pathogens for both human and freshwater fishes.

3.7 Effect of different wastewater from industries on freshwater organisms

The organisms living in freshwater from microscopic diatoms to big fishes all are affected by wastewater pollutants. Wastewater from chemical and fertilizer industries was reported to alter the activity of many metabolic enzymes in different species of fishers (Chowdhary, Bharagava, Mishra, & Khan, 2020). The enzyme lactate dehydrogenase which catalyzes reversible reaction of reduction of pyruvate into lactate (Kumar, Nagarajan, & Uchil, 2018) was found to be inhibited due to alteration in active site of enzyme in freshwater fishes found in river and lakes receiving wastewater from fertilizer industries (Álvarez-García, Urrestarazu, Guil-Guerrero, & Jiménez-Becker, 2019). The degree of alternations in cellular enzyme activity depends on magnitude, type, and time of exposure (Tasneem & Yasmeen, 2018). The toxic effect of effluents of fertilizer industry on *Channa striatus*, (Singh, Tiwari, & Pandey, 2019) a teleost fish commonly found to inhabit the freshwater ponds and rivers of India, shows lethal effect on lactate dehydrogenase (LDH) in different tissues when exposed to wastewater for short (96 h) and long (15 days) treatment duration (Bakshi & Panigrahi, 2018).

Chromium, which is one of the ubiquitous pollutants (JC & Sekar, 2020), commonly found as the effluents of industries such as textiles, tanneries, electroplating workshops, ore mining, dyeing, printing-photographic, and pharmaceuticals (Sajjad & Arif, 2018), are observed to cause mutagenic alterations when exposed to a combination of trivalent chromium and macromolecules (Sajjad & Arif, 2018). In comparison to trivalent chromium, hexavalent chromium is more toxic due to its easy permeability across cell membrane (Ahmad, Shukla, Swami, & Shukla, 2017). It was reported that exposure to chromium even at low concentration in *Clarias gariepinus* (African freshwater catfish) causes mortality of embryo and problem in hatching (Mustapha, 2017). In *Labeo rohita* chromium hexavalent exposure causes decline in glycogen content in gills, muscles, and liver (Jayakumar, Senthilmurugan, Vijayan, & Tamizhazhagan, 2018). In many fishes, chromium intoxication affects biochemical mechanisms (Sharma, 2019).

The wastewater from pharmaceutical industries release huge amount of chemicals in freshwater bodies which affect its ecosystem by disrupting its food chain and worst effect is seen on the lowest in position but biggest in number on zooplanktons. Zooplanktons are considered as the biological indicators of water quality of rivers and receiving water bodies such as lakes and ponds (Balogh, 1988). Pharmaceutical effluents create more problem to freshwater ecosystem due to their nontarget nature (Sarma, Garcia-Garcia, Nandini, & Saucedo-Campos, 2017). Some drugs such as metformin remain in wastewater in huge amount even after treatment (Yan et al., 2019). The concentration of metformin in wastewater was found from 1 to 140 μg/L in Mexico City (Poursat et al., 2019). Very low concentration of metformin can cause toxicity to zooplankton (Fonseca & Reis-Santos, 2019). Easy solubility of metformin and other pharma drugs is an important factor for acute or chronic toxicity in fishes and zooplankton (Garcia-Garcia, Reyes-Carrillo, Sarma, & Nandini, 2017). Other than zooplankton, many phytoplankton species such as *Selenastrum*, *Pseudokirchneriella*, and *Scenedesmus* (green algae species) (Tetorou, Makhatova, & Poulopoulos, 2019) subject to stress from antihyperglycemic pharmaceutical effluents, causing reduced or complete inhibition of growth (Bai & Acharya, 2017) and cellular dysfunctions (Bai & Acharya, 2017; de la Cruz Quiroz, Barceló, Iqbal, & Parra-Saldivar, 2019).

Use of silver nanoparticles in different industries such as from food processing to medical equipment to textiles and automobiles becomes an important advancement in technology (Lasek et al., 2019). Silver ions when released in freshwater cause biocidal effect (Kumar, Farrukh, & Faisal, 2020). Silver nanoparticles on entering freshwater affect permeability of cell membrane and enter the food chain by accumulation in bacterioplankton and phytoplankton (Norman et al., 2019). These phytoplankton and zooplankton cause accumulation of silver ions in food chain causing metabolic and genetic disorders in birds and human (Norman et al., 2019). Many studies suggest that silver nanoparticles produce toxic effects on algae (Domingo, Bracale, & Vannini, 2019), invertebrates (Ribeiro et al., 2017a), bacteria (Jin, Zhang, Shi, Zhang, & Wang, 2017), fish (Ale et al., 2018), and freshwater plants (Ale et al., 2018).

3.8 Remedies to reduce industrial effluents

The presence of toxic compound in industrial effluents and their effect on freshwater ecosystem are big problems and concerns to all the countries around the world (Kumar, Othman, & Asharuddin, 2017). Different techniques and remedies such as electrochemical oxidation for the removal of organic pollutants from pharmaceuticals (Xia & Dai, 2018; Flores et al., 2017) and pesticides (Liu et al., 2017; Ren et al., 2019; Daij, Bellebia, & Bengharez, 2017), removal of azo dyes (Hamad, Bassyouni, El-Ashtoukhy, Amin, & El-Latif, 2018; Bedolla-Guzman, Feria-Reyes, Gutierrez-Granados, & Peralta-Hernández, 2017) and carboxylic acids (Barhoumi et al., 2017). Many other techniques are also in use, but the technique that is most accepted and eco-friendly now is the use of biotechnology in bioremediation (Card et al., 2017). Bioremediation is a conventional, less invasive phytochemical technique developed by scientists using environmental biotechnological approach (Kidd, Alvarez-Lopez, Becerra-Castro, Cabello-Conejo, & Prieto-Fernandez, 2017).

The process of bioremediation is successfully used for heavy metals and organic compounds for one time (Deng et al., 2020). Use of microorganisms like actinobacteria (Alvarez et al., 2017) belonging to Gram-positive (Neuenschwander, Ghai, Pernthaler, & Salcher, 2018), facultative anaerobes or anaerobes (Vick, Greenfield, Tetu, Midgley, & Paulsen, 2019), which have a rigid cell wall that contains muramic acid (Alvarez et al., 2017) is done for many contaminations such as oil (Chaturvedi & Khurana, 2019), rubber, plastics, pesticides, and heavy metals (Analía, Maria, Sebastián, & Marta, 2019) from last 15 years. Other than actinobacteria, microalgae, yeast, and fungi are used for these processes.

Microalgae is very efficient to grow under stress conditions such as heavy metal toxicity, salinity, pH, and temperature, making it most deserving candidate for bioremediation in phycoremediation of heavy metal for its high binding affinity and large surface area (Agüera, Plaza-Bolaños, & Fernández, 2020).

Microalgae can use heavy metals such as boron (B), cobalt (Co), copper (Cu), iron (Fe), molybdenum (Mo), manganese (Mn), and zinc (Zn) as trace elements for its enzymatic process and cell metabolism (Chiellini, Guglielminetti, Pistelli, & Ciurli, 2020). Microalgae shows hormesis phenomenon (Agathokleous, Feng, Iavicoli, & Calabrese, 2019), due to which low-toxic metals such as As, Cd, Cr, Pb, and Hg stimulate the growth and metabolism of microalgae (Xu, Ge, Li, Wan, & Xiao, 2019).

Mercury toxicity in freshwater ecosystem is a big threat; using bioreactor, this toxic metal can be reduced (Amin, Sarwar, Saleem, Latif, & Opella, 2019). The bioremediation technique depends on many factors such as pH (Aryal, 2020), microbial biomass surface area (Ashokkumar, Loashini, & Bhavya, 2017), and pre-treatment of wastewater (Ayangbenro & Babalola, 2017).

Fungi as we know are responsible for decomposition and help in continuation of nutrition cycle (Abdi & Kazemi, 2015); at the same time, these is known for their survival under extreme stress continents (Archana & Jaitly, 2015). Use of fungi in bioremediation is also known as mycoremdiation (Hamba & Tamiru, 2016). Mycoremdiation is a process in which fungi, living or dead, are used to remove toxic contaminants such as heavy metal from wastewater (Ogbo & Okhuoya, 2011). Use of fungi is more preferable in comparison to other microbes in bioremediation as per the research due to the ability to accumulate heavy metals in their fruit bodies (Ogbo & Okhuoya, 2011) in an efficient manner, making them unavailable or decreasing their concentration (Ogbo & Okhuoya, 2011; Srivastava & Thakur, 2006). The species of fungi such as *Saccharomyces cerevisiae* is stated to remove up to 65%–79% of Cd from industrial effluent (Demirbaş, 2002). Similarly, *Aspergillus* sp. were reported to remove up to 65% of chromium from tannery wastewater (Singh, 2006).

3.9 Conclusion

Today's world is facing problem of increasing population and pollution at the same time. Due to daily increasing population, the need for higher production of food and other necessities is increasing and to fulfill them more industries are also coming up. But the increased number of industries is increasing pollution of land, air, and water. We have limited natural resources and we not only need to use them but we have to save them for the survival of different life forms on the Earth.

Industrial wastewater entering freshwater ecosystem is contaminating it with heavy metals, oil, rubber, chemicals, and many more. Once it enters freshwater ecosystem, it destroys the balance and many living species come under threat. Accumulation of these pollutants enter our food chain and finally our body, causing different health issues.

We can reduce the effect by taking some measures such as treatment of industrial wastewater using bioremediation, which is natural method to get rid of these pollutants without further damage to the Earth.

References

Abdi, O., & Kazemi, M. (2015). A review study of biosorption of heavy metals and comparison between different biosorbents. *Journal of Materials and Environmental Science, 6* (5), 1386–1399.

Agathokleous, E., Feng, Z., Iavicoli, I., & Calabrese, E. J. (2019). The two faces of nanomaterials: A quantification of hormesis in algae and plants. *Environment International, 131*, 105044.

Agüera, A., Plaza-Bolaños, P., & Fernández, F. A. (2020). *Removal of contaminants of emerging concern by microalgae-based wastewater treatments and related analytical techniques. Current Developments in Biotechnology and Bioengineering* (pp. 503–525). Elsevier.

Ahmad, S., Shukla, S., Swami, V. P., & Shukla, S. (2017). Ethological response of catfish *Heteropneustes* fossils after acute and sub-acute exposure of cadmium. *Imperial Journal of Interdisciplinary Research, 3*, 506–511.

Ale, A., Bacchetta, C., Rossi, A. S., Galdopórpora, J., Desimone, M. F., Fernando, R., … Cazenave, J. (2018). Nanosilver toxicity in gills of a neotropical fish: Metal accumulation, oxidative stress, histopathology and other physiological effects. *Ecotoxicology and Environmental Safety, 148*, 976–984.

Ali, F. (2017). *A study on occupational health and safety for tannery workers at Khartoum State.* Sudan University of Science & Technology.

Álvarez-García, M., Urrestarazu, M., Guil-Guerrero, J. L., & Jiménez-Becker, S. (2019). Effect of fertigation using fish production wastewater on Pelargonium x Zonale growth and nutrient content. *Agricultural Water Management, 223*, 105726.

Alvarez, A., Saez, J. M., Costa, J. S. D., Colin, V. L., Fuentes, M. S., Cuozzo, S. A., … Amoroso, M. J. (2017). Actinobacteria: current research and perspectives for bioremediation of pesticides and heavy metals. *Chemosphere, 166*, 41–62.

Amin, A., Sarwar, A., Saleem, M. A., Latif, Z., & Opella, S. J. (2019). Expression and purification of transmembrane protein MerE from mercury-resistant *Bacillus cereus. Journal of Microbiology and Biotechnology, 29*(2), 274–282.

Analía, A., Maria, S. J., Sebastián, D. C. J., & Marta, P. (2019). *A discussion about different strategies using actinobacteria. Bioremediation of Agricultural Soils.* CRC Press, 130.

Apruzzese, I., Bhardwaj, V., Degrassi, G., & Bhardwaj, R. K. (2017). Biogas production from organic wastes of paper and leather industries. *IOSR Journal of Biotechnology and Biochemistry*, *3*(4), 8–14.
Archana, A., & Jaitly, A. (2015). Mycoremediation: utilization of fungi for reclamation of heavy metals at their optimum remediation conditions. *Biolife*, *3*(1), 77–106.
Arribas, P., Gutiérrez-Cánovas, C., Botella-Cruz, M., Cañedo-Argüelles, M., Antonio Carbonell, J., Millán, A., ... Sánchez-Fernández, D. (2019). Insect communities in saline waters consist of realized but not fundamental niche specialists. *Philosophical Transactions of the Royal Society B*, *374*(1764), 20180008.
Aryal, M. (2020). A comprehensive study on the bacterial biosorption of heavy metals: materials, performances, mechanisms, and mathematical modellings. *Reviews in Chemical Engineering*, *1*. (ahead-of-print).
Ashokkumar, P., Loashini, V. M., & Bhavya, V. (2017). Effect of pH, temperature and biomass on biosorption of heavy metals by *Sphaerotilus natans*. *International Journal of Microbiology and Mycology*, *6*(1), 32–38.
Ayangbenro, A. S., & Babalola, O. O. (2017). A new strategy for heavy metal polluted environments: A review of microbial biosorbents. *International Journal of Environmental Research and Public Health*, *14*(1), 94.
Bai, X., & Acharya, K. (2017). Algae-mediated removal of selected pharmaceutical and personal care products (PPCPs) from Lake Mead water. *Science of the Total Environment*, *581*, 734–740.
Bakshi, A., & Panigrahi, A. (2018). A comprehensive review on chromium induced alterations in fresh water fishes. *Toxicology Reports*, *5*, 440–447.
Baldisserotto, B. (2019). *Fish osmoregulation*. CRC Press.
Balogh, K. V. (1988). Comparison of mussels and crustacean plankton to monitor heavy metal pollution. *Water, Air, and Soil Pollution*, *37*(3–4), 281–292.
Barhoumi, N., Olvera-Vargas, H., Oturan, N., Huguenot, D., Gadri, A., Ammar, S., ... Oturan, M. A. (2017). Kinetics of oxidative degradation/mineralization pathways of the antibiotic tetracycline by the novel heterogeneous electro-Fenton process with solid catalyst chalcopyrite. *Applied Catalysis B: Environmental*, *209*, 637–647.
Bedolla-Guzman, A., Feria-Reyes, R., Gutierrez-Granados, S., & Peralta-Hernández, J. M. (2017). Decolorization and degradation of reactive yellow HF aqueous solutions by electrochemical advanced oxidation processes. *Environmental Science and Pollution Research*, *24*(14), 12506–12514.
Bhat, R. A., Singh, D. V., Qadri, H., Dar, G. H., Dervash, M. A., Bhat, S. A., ... Yousaf, B. (2021). Vulnerability of municipal solid waste: An emerging threat to aquatic ecosystems. *Chemosphere*, *287*(2022), 132223. Available from https://doi.org/10.1016/j.chemosphere.2021.132223.
Bonga, S. E. W. (1972). Neuroendocrine involvement in osmoregulation in a freshwater mollusc, *Lymnaea stagnalis*. *General and Comparative Endocrinology*, *3*, 308–316.
Bustillo-Lecompte, C. F., & Mehrvar, M. (2017). Treatment of actual slaughterhouse wastewater by combined anaerobic–aerobic processes for biogas generation and removal of organics and nutrients: An optimization study towards a cleaner production in the meat processing industry. *Journal of Cleaner Production*, *141*, 278–289.
Card, M. L., Gomez-Alvarez, V., Lee, W.-H., Lynch, D. G., Orentas, N. S., Lee, M. T., ... Boethling, R. S. (2017). History of EPI Suite™ and future perspectives on chemical property estimation in US Toxic Substances Control Act new chemical risk assessments. *Environmental Science: Processes & Impacts*, *19*(3), 203–212.
Carley, M., & Spapens, P. (2017). *Sharing the world: sustainable living and global equity in the 21st century*. Routledge.

Cañedo-Argüelles, M., Brucet, S., Carrasco, S., Flor-Arnau, N., Ordeix, M., Ponsá, S., & Coring, E. (2017). Effects of potash mining on river ecosystems: An experimental study. *Environmental Pollution, 224*, 759–770.

Cañedo-Argüelles, M., Kefford, B., & Schäfer, R. (2019). Salt in freshwaters: causes, effects and prospects-introduction to the theme issue. *Philosophical Transactions of the Royal Society B, 374*(1764), 20180002.

Cañedo-Argüelles, M., Sala, M., Peixoto, G., Prat, N., Faria, M., Soares, A. M., ... Kefford, B. (2016). Can salinity trigger cascade effects on streams? A mesocosm approach. *Science of the Total Environment, 540*, 3–10.

Chamberlain, S. D., Hemes, K. S., Eichelmann, E., Szutu, D. J., Verfaillie, J. G., & Baldocchi, D. D. (2019). Effect of drought-induced salinization on wetland methane emissions, gross ecosystem productivity, and their interactions. *Ecosystems*, 1–14.

Chaturvedi, S., & Khurana, S. P. (2019). *Importance of actinobacteria for bioremediation. Plant Biotechnology: Progress in Genomic Era* (pp. 277–307). Springer.

Chiellini, C., Guglielminetti, L., Pistelli, L., & Ciurli, A. (2020). Screening of trace metal elements for pollution tolerance of freshwater and marine microalgal strains: Overview and perspectives. Algal. *Research; a Journal of Science and its Applications, 45*, 101751.

Chowdhary, P., Bharagava, R. N., Mishra, S., & Khan, N. (2020). *Role of industries in water scarcity and its adverse effects on environment and human health. Environmental Concerns and Sustainable Development* (pp. 235–256). Springer.

Coldsnow, K. D., Mattes, B. M., Hintz, W. D., & Relyea, R. A. (2017). Rapid evolution of tolerance to road salt in zooplankton. *Environmental Pollution, 222*, 367–373.

Daij, K. B., Bellebia, S., & Bengharez, Z. (2017). Comparative experimental study on the COD removal in aqueous solution of pesticides by the electrocoagulation process using monopolar iron electrodes. *Chemistry International, 3*(4), 420–427.

Dar, G. H., Bandh, S. A., & Kamili, A. N. (2011). Bacteriological analysis of soil from Yusmarg health resort of Kashmir Valley. *International Journal of Current Research, 3*(10), 53–56, 2011.

Dar, G. H., Bandh, S. A., Kamili, A. N., Nazir, R., & Bhat, R. A. (2013). Comparative analysis of different types of bacterial colonies from the soils of Yusmarg Forest. *Kashmir valley India. Ecologia Balkanica, 5*(1), 31–35.

Dar, G. H., Kamili, A. N., Chishti, M. Z., Dar, S. A., Tantry, T. A., & Ahmad, F. (2016). Characterization of *Aeromonas sobria* isolated from fish rohu (*Labeo rohita*) collected from polluted pond. *Journal of Bacteriology and Parasitology*, 7(3), 1–5. Available from https://doi.org/10.4172/2155-9597.1000273.

Dar, G. H., Kamili, A. N., Dar, S. A., Chishti, M. Z., & Ahmad, F. (2015). Detection and characterization of potentially pathogenic *Aeromonas sobria* isolated from fish *Hypophthalmichthys molitrix* (Cypriniformes: Cyprinidae). *Microbial Pathogenesis*. Available from https://doi.org/10.1016/j.micpath.2015.100.017. (2015).

Dar, G. H., Kamili, A. N., Nazir, R., Bandh, S. A., & Bhat, R. A. (2012). A preliminary study of colony forming units of bacteria from the soils of Yusmarg forest, Kashmir valley, India. *International Journal of Current Research, 4*(12), 467–472.

de la Cruz Quiroz, R., Barceló, D., Iqbal, H., & Parra-Saldivar, R. (2019). Algal-based removal strategies for hazardous contaminants from the environment—A review. *The Science of the Total Environment, 665*, 358–366.

Demirbaş, A. (2002). Metal ion uptake by mushrooms from natural and artificially enriched soils. *Food Chemistry, 78*(1), 89–93.

Deng, J., Fu, D., Hu, W., Lu, X., Wu, Y., & Bryan, H. (2020). Physiological responses and accumulation ability of *Microcystis aeruginosa* to zinc and cadmium: Implications for bioremediation of heavy metal pollution. *Bioresource Technology, 303*, 122963.

Dickhout, J. M., Moreno, J., Biesheuvel, P., Boels, L., Lammertink, R. G., & de Vos, W. M. (2017). Produced water treatment by membranes: A review from a colloidal perspective. *Journal of Colloid and Interface Science*, *487*, 523–534.

Domingo, G., Bracale, M., & Vannini, C. (2019). *Phytotoxicity of silver nanoparticles to aquatic plants, algae, and microorganisms. Nanomaterials in Plants, Algae and Microorganisms* (pp. 143–168). Elsevier.

Elliott, M., Day, J. W., Ramachandran, R., & Wolanski, E. (2019). *A synthesis: what is the future for coasts, estuaries, deltas and other transitional habitats in 2050 and beyond? Coasts and Estuaries* (pp. 1–28). Elsevier.

Farmaki, P, & Tranoulidis, A. (2018). Water policy in Greece: Management and pricing under the provisions of the European Water Framework Directive 2000/60/Ec. In *CBU international conference proceedings*. pp. 107–112.

Flores, N., Sirés, I., Rodríguez, R. M., Centellas, F., Cabot, P. L., Garrido, J. A., & Brillas, E. (2017). Removal of 4-hydroxyphenylacetic acid from aqueous medium by electrochemical oxidation with a BDD anode: Mineralization, kinetics and oxidation products. *Journal of Electroanalytical Chemistry*, *793*, 58–65.

Flörke, M., Bärlund, I., van Vliet, M. T., Bouwman, A. F., & Wada, Y. (2019). Analysing trade-offs between SDGs related to water quality using salinity as a marker. *Current Opinion in Environmental Sustainability*, *36*, 96–104.

Fonseca, V. F., & Reis-Santos, P. (2019). *Chapter 7 Ecotoxicology of pharmaceuticals in coastal and marine organisms. Ecotoxicology of Marine Organisms* (pp. 158–184). CRC Press.

Garcia-Garcia, G., Reyes-Carrillo, G., Sarma, S., & Nandini, S. (2017). Population level responses of rotifers (*Brachionus calyciflorus* and *Plationus patulus*) to the anti-diabetic drug, metformin. *Journal of Environmental Biology*, *38*(6), 1213–1219.

Hamad, H., Bassyouni, D., El-Ashtoukhy, E.-S., Amin, N., & El-Latif, M. A. (2018). Electrocatalytic degradation and minimization of specific energy consumption of synthetic azo dye from wastewater by anodic oxidation process with an emphasis on enhancing economic efficiency and reaction mechanism. *Ecotoxicology and Environmental Safety*, *148*, 501–512.

Hamba, Y., & Tamiru, M. (2016). Mycoremediation of heavy metals and hydrocarbons contaminated environment. *Asian Journal of Natural & Applied Sciences*, *5*(2).

Hopkins, G. R., French, S. S., & Brodie, E. D., Jr (2017). Interacting stressors and the potential for adaptation in a changing world: Responses of populations and individuals. *Royal Society open science*, *4*(6), 161057.

Iglesias, M. C.-A. (2020). A review of recent advances and future challenges in freshwater salinization. *Limnetica*, *39*(1), 185–211.

Jain, P., Sharma, M., Dureja, P., Sarma, P. M., & Lal, B. (2017). Bioelectrochemical approaches for removal of sulfate, hydrocarbon and salinity from produced water. *Chemosphere*, *166*, 96–108.

Jayakumar, V., Senthilmurugan, S., Vijayan, P., & Tamizhazhagan, V. (2018). Indian major carp *Labeo rohita* (Hamilton, 1882) toxicology approaches on overview. *International Journal of Biology Research*, *3*(1), 332–338.

JC, N., & Sekar, R. R. J. (2020). Acute and chronic effects of chromium metal toxicity on the growth and reproduction of zebra fish, *Danio rerio*. *Studies in Indian Place Names*, *40*(70), 3400–3409.

Jensen, P., Pinches, D., Desai, D., & Street, W. (2017). *Robust membrane systems for enhanced primary treatment and energy recovery of abattoir waste water*.

Jin, J., Zhang, L., Shi, M., Zhang, Y., & Wang, Q. (2017). Ti-GO-Ag nanocomposite: The effect of content level on the antimicrobial activity and cytotoxicity. *International Journal of Nanomedicine*, *12*, 4209.

Kaushal, S. S., Likens, G. E., Pace, M. L., Haq, S., Wood, K. L., Galella, J. G., ... Kortelainen, P. (2019). Novel 'chemical cocktails' in inland waters are a consequence of the freshwater salinization syndrome. *Philosophical Transactions of the Royal Society B, 374*(1764), 20180017.

Kefford, B. J., Dalton, A., Palmer, C. G., & Nugegoda, D. (2004). The salinity tolerance of eggs and hatchlings of selected aquatic macroinvertebrates in south-east Australia and South Africa. *Hydrobiologia, 517*(1–3), 179–192.

Kidd, P. S., Alvarez-Lopez, V., Becerra-Castro, C., Cabello-Conejo, M., & Prieto-Fernandez, A. (2017). *Potential role of plant-associated bacteria in plant metal uptake and implications in phytotechnologies*. *Advances in botanical research* (vol 83, pp. 87–126). Elsevier.

Kumar, D., Farrukh, M., & Faisal, N. (2020). *Nanocomposites in the food packaging industry: Recent trends and applications. Diverse Applications of Organic-Inorganic Nanocomposites: Emerging Research and Opportunities* (pp. 103–135). IGI Global.

Kumar, P., Nagarajan, A., & Uchil, P. D. (2018). Analysis of cell viability by the lactate dehydrogenase assay. *Cold Spring Harbor Protocols, 2018*(6), pdb. prot095497.

Kumar, V., Othman, N., & Asharuddin, S. (2017). Applications of natural coagulants to treat wastewater – A review. In *MATEC web of conferences*. EDP Sciences, p. 06016.

Lasek, F., Leitner, N. K. V., Rauwel, G., Blanchier, L., Castel, O., Ayraud-Thevenot, S., & Deborde, M. (2019). Discharge of biocidal products from healthcare activities into a sewage system—A case study at a French university hospital. *Environmental Science and Pollution Research, 26*(5), 4938–4951.

Letelier-Gordo, C. O., Aalto, S. L., Suurnäkki, S., & Pedersen, P. B. (2020). Increased sulfate availability in saline water promotes hydrogen sulfide production in fish organic waste. *Aquacultural Engineering, 89*, 102062.

Liang, Y., Zhu, H., Bañuelos, G., Yan, B., Zhou, Q., Yu, X., & Cheng, X. (2017). Constructed wetlands for saline wastewater treatment: A review. *Ecological Engineering, 98*, 275–285.

Liu, S., Wang, Y., Zhou, X., Han, W., Li, J., Sun, X., ... Wang, L. (2017). Improved degradation of the aqueous flutriafol using a nanostructure macroporous PbO2 as reactive electrochemical membrane. *Electrochimica Acta, 253*, 357–367.

Ma, Y., Ding, S., Xin, D., Ma, F., & Xia, R. (2019). *Optimization of Hot-pressing Technique for Genuine Full-grain Leather Made of Natural Rubber Latex*. *Annales de Chimie-Science des Matériaux* (vol 3, pp. 165–172).

Mustapha, M. K. (2017). Dynamics of hexavalent chromium in four types of aquaculture ponds and its effects on the morphology and behavior of cultured *Clarias gariepinus* (Burchell 1822). *Toxicological research, 33*(2), 119–124.

Neuenschwander, S. M., Ghai, R., Pernthaler, J., & Salcher, M. M. (2018). Microdiversification in genome-streamlined ubiquitous freshwater Actinobacteria. The. *ISME journal, 12*(1), 185–198.

Nielsen, D., Brock, M., Rees, G., & Baldwin, D. S. (2003). Effects of increasing salinity on freshwater ecosystems in Australia. *Australian Journal of Botany, 51*(6), 655–665.

Noor, Z. Z., & Mutamim, N. S. A. (2017). *Removal of micro-pollutants from wastewater through MBR technologies: A case study on spent caustic wastewater. Sustainable Water Treatment* (pp. 81–94). CRC Press.

Norman, B. C., Frost, P. C., Blakelock, G. C., Higgins, S. N., Hoque, M. E., Vincent, J. L., ... Xenopoulos, M. A. (2019). Muted responses to Ag accumulation by plankton to chronic and pulse exposure to silver nanoparticles in a boreal lake. *FACETS, 4*(1), 566–583.

Ogbo, E., & Okhuoya, J. (2011). Bio-absorption of some heavy metals by *Pleurotus tuber-regium* Fr. Singer (an edible mushroom) from crude oil polluted soils amended with fertilizers and cellulosic wastes. *International Journal of Soil science, 6*, 34–48.

Orssatto, F., Ferreira Tavares, M. H., Manente da Silva, F., Eyng, E., Farias Biassi, B., & Fleck, L. (2017). Optimization of the pretreatment of wastewater from a

slaughterhouse and packing plant through electrocoagulation in a batch reactor. *Environmental Technology*, *38*(19), 2465–2475.
Poursat, B. A., van Spanning, R. J., Braster, M., Helmus, R., de Voogt, P., & Parsons, J. R. (2019). Biodegradation of metformin and its transformation product, guanylurea, by natural and exposed microbial communities. *Ecotoxicology and Environmental Safety*, *182*, 109414.
Rehman, K., Imran, A., Amin, I., & Afzal, M. (2018). Inoculation with bacteria in floating treatment wetlands positively modulates the phytoremediation of oil field wastewater. *Journal of Hazardous Materials*, *349*, 242–251.
Ren, Q., Yin, C., Chen, Z., Cheng, M., Ren, Y., Xie, X., … Yang, H. (2019). Efficient sonoelectrochemical decomposition of chlorpyrifos in aqueous solution. *Microchemical Journal*, *145*, 146–153.
Ribeiro, F., Van Gestel, C. A., Pavlaki, M. D., Azevedo, S., Soares, A. M., & Loureiro, S. (2017a). Bioaccumulation of silver in *Daphnia magna*: Waterborne and dietary exposure to nanoparticles and dissolved silver. *Science of The Total Environment*, *574*, 1633–1639.
Ribeiro, V. R., da Silva, P. R. L., Gubiani, É. A., Faria, L., Daga, V. S., & Vitule, J. R. S. (2017b). Imminent threat of the predator fish invasion *Salminus brasiliensis* in a Neotropical ecoregion: Eco-vandalism masked as an environmental project. *Perspectives in Ecology and Conservation*, *15*(2), 132–135.
Sajjad, H., & Arif, N. (2018). Prevalence of hexavalent chromium (VI) induced alterations in fish: A review. *Journal of Analytical Toxicology*, *1*(2), 5.
Sala, M., Faria, M., Sarasúa, I., Barata, C., Bonada, N., Brucet, S., … Soares, A. M. (2016). Chloride and sulphate toxicity to *Hydropsyche exocellata* (Trichoptera, Hydropsychidae): exploring intraspecific variation and sub-lethal endpoints. *Science of the Total Environment*, *566*, 1032–1041.
Salih, A. M. (2018). *The purification of industrial wastewater to remove heavy metals and investigation into the use of zeolite as a remediation tool.*
Sarma, S., Garcia-Garcia, G., Nandini, S., & Saucedo-Campos, A. (2017). Effects of antidiabetic pharmaceuticals to non-target species in freshwater ecosystems: A review. *Journal of Environmental Biology*, *38*(6), 1249–1254.
Schulz, C.-J., & Cañedo-Argüelles, M. (2019). Lost in translation: the German literature on freshwater salinization. *Philosophical Transactions of the Royal Society B*, *374*(1764), 20180007.
Sharma, M. (2019). Behavioural responses in effect to chemical stress in fish: A review. *International Journal of Fisheries and Aquatic Studies*, 7, 1–5.
Singh, H. (2006). *Mycoremediation: Fungal bioremediation*. John Wiley & Sons.
Singh, U., Tiwari, R. K., & Pandey, R. S. (2019). Physicochemical characteristics of fertilizer industry effluent and its toxicological impact on the activity of acetylcholinesterase (AChE) in freshwater teleosts *Heteropneustes* fossils and *Labeo rohita*. *Croatian Journal of Fisheries*, 77(2), 77–86.
Srivastava, S., & Thakur, I. S. (2006). Isolation and process parameter optimization of *Aspergillus* sp. for removal of chromium from tannery effluent. *Bioresource Technology*, *97*(10), 1167–1173.
Steinmuller, H. E., & Chambers, L. G. (2018). Can saltwater intrusion accelerate nutrient export from freshwater wetland soils? An experimental approach. *Soil Science Society of America Journal*, *82*(1), 283–292.
Tasneem, S., & Yasmeen, R. (2018). Biochemical changes in carbohydrate metabolism of the fish—*Cyprinus carpio* during sub-lethal exposure to biopesticide—Derisom. *Iranian Journal of Fisheries Sciences*.
Tetorou, A., Makhatova, A., & Poulopoulos, S. G. (2019). Photochemical mineralization of Ibuprofen medicinal product by means of UV, hydrogen peroxide, titanium dioxide and iron. *Water Science and Technology*, *80*(11), 2200–2205.

Tibby, J., Adamson, K., & Kershaw, A. (2020). An 1800-year water-quality and vegetation record from Junction Park Billabong, Murray River, Australia: An assessment of European impacts and sensitivity to climate. *Journal of Paleolimnology*, 1–17.
Vick, S. H., Greenfield, P., Tetu, S. G., Midgley, D. J., & Paulsen, I. T. (2019). Genomic and phenotypic insights point to diverse ecological strategies by facultative anaerobes obtained from subsurface coal seams. *Scientific Reports*, *9*(1), 1–13.
Vollmer, D., Shaad, K., Souter, N. J., Farrell, T., Dudgeon, D., Sullivan, C. A., ... Power, A. G. (2018). Integrating the social, hydrological and ecological dimensions of freshwater health: The Freshwater Health Index. *Science of the Total Environment*, *627*, 304–313.
Xia, Y., & Dai, Q. (2018). Electrochemical degradation of antibiotic levofloxacin by PbO2 electrode. Kinetics, energy demands and reaction pathways. *Chemosphere*, *205*, 215–222.
Xu, C., Ge, Z., Li, C., Wan, F., & Xiao, X. (2019). Inhibition of harmful algae *Phaeocystis globosa* and *Prorocentrum donghaiense* by extracts of coastal invasive plant *Spartina alterniflora*. *Science of the Total Environment*, *696*, 133930.
Yan, J.-H., Xiao, Y., Tan, D.-Q., Shao, X.-T., Wang, Z., & Wang, D.-G. (2019). Wastewater analysis reveals spatial pattern in consumption of anti-diabetes drug metformin in China. *Chemosphere*, *222*, 688–695.

CHAPTER 4

Credibility on biosensors for monitoring contamination in aquatic environs

Akanksha Jalswar[1,*], Neha Sharma[1,*], Priyanka Basera[1], Ashish Bedi[2] and Rahul Singh Jasrotia[3]
[1]The Energy and Resources Institute, New Delhi, India
[2]School of Life and Environmental Sciences, Deakin University, Geelong, VIC, Australia
[3]Department of Biological Science, Florida State University, Tallahassee, FL, United States

4.1 Introduction

Water is the topmost valued natural resource available on the planet Earth. After air, water is the most essential element required for human survival (Ebere Enyoh, Wirnkor Verla, & Jane Egejuru, 2018). The United Nations considers that the availability of clean drinking water to humans as a human right, but people worldwide have limited access to it. It has been described in earlier studies that at least 4.3 billion (71%) of humans worldwide face some level of water scarcity during a considerable period of the year (Mekonnen & Hoekstra, 2016). It has been found that approximately 900 million out of the 7.7 billion people presently in the world do not have access to safe drinking water. A significant decrease from around 2.6 billion people in 1990 to approximately 600 billion people in 2015 was expected if the United Nations' Millennium Development Goal was achieved which aims to enhance the access to safe drinking water (Beniah & Ebere, 2019). World Water Council estimated that ~3.9 billion people might be living in water-scarce areas by 2030 (Xu et al., 2018). Also, it has been noticed that extensive urbanization, industrialization, and population explosion have increased the pressure on the prevailing surface, and groundwater resources are shrinking. Increased anthropogenic activities, climate change, water usage, and wastage will worsen the water crisis and these already exist in some regions of the world (WHO, 2005). Moreover, increased use and disposal of chemical

* Equal contribution.

Microbial Consortium and Biotransformation for Pollution Decontamination
DOI: https://doi.org/10.1016/B978-0-323-91893-0.00017-1

pollutants generating from pharmaceutical, agriculture, and household use extensively degrade the water quality and make water unfit for human consumption (Kolpin et al., 2002).

Wastewater comprises several harmful pollutants that are generated and discharged from commercial, domestic, agricultural, or commercial activities (Cornish, Mensah, & Ghesquire, 1999; Lakatos, 2018). Wastewater is composed of chemical, physical, and biological pollutants and can be characterized based on its chemical constituents, physical appearances, and the presence of microorganisms. These pollutants have adverse effects on the environment as well as on other living beings; for example, induction of genotoxicity and cytotoxicity, carcinogenic effects, and disruption of hormone and endocrine systems (Bhat et al., 2021; Korostynska, Mason, & Al-Shamma'a, 2013).

In wastewater, the concentration and composition of the pollutants is an important criterion that is taken into consideration before the selection and designing of the treatment process. Wastewater treatment and its reuse after that is an integrated part of the water resource conservation and management and can serve as an alternative to the limited natural water resources, thus mitigating the problem of freshwater availability for various applications (Pang, Xi, Li, Shi, & Hu, 2015; Peters, 2015). The treated water should be free of hazardous pollutants, and the quality of the reclaimed water should meet the standard regulations. Efficient wastewater quality management techniques are required for detecting the variable number of pollutants present in the contaminated water. Currently, several water quality indicators based on laboratory techniques are being used to determine the chemical and physical characteristics of disposed or treated water. These techniques are costly and do not give immediate results (Chong, Abdul Aziz, & Harun, 2013; Yang, Han, Lee, & Hur, 2015). Therefore there is a requirement to create new cost-effective and sensitive strategies with an easy on-site detection and quantification of numerous pollutants before, during, and after the treatment (Asadnia et al., 2016; Biswas, Karn, Balasubramanian, & Kale, 2017; Michael-Kordatou et al., 2015; Rehman et al., 2015). This chapter presents a general overview of biosensor-based approaches that are involved in monitoring the major categories of wastewater pollutants, such as heavy metals, microorganisms, and organic materials. General components of the biosensor, as well as current limitations and future prospects of the wastewater monitoring by biosensors, have also been discussed.

4.2 Major sources of water pollution

The water crisis and pollution are among the most concerning and alarming issues that we are facing today. Numerous toxic pollutants are disposed of in water bodies such as lakes, rivers, and oceans and either these get dissolved in water, remain suspended, or get deposited near the water bed. Pollution sources may arise from the point or from the nonpoint sources. Point sources include single sources such as sewage and industrial waste discharge (Wang et al., 2019). In contrast, the nonpoint source includes a combination of contaminants that enter the water bodies as runoff from gardens, agricultural lands, groundwater systems and from the atmosphere via rain (Bermúdez-Couso et al., 2013). There are several types of pollutants, that is, organic (consisting of carbon covalently bonded with other compounds); inorganic (human and animal drugs, chemicals used in industries, pesticides, heavy metals, and common household products); and radioactive (generated by power plants and uranium mining) (Fig. 4.1).

4.3 Biosensors

Cammann coined the term "biosensor," and its definition was introduced by International Union of Pure and Applied Chemistry (IUPAC) (Thâvenot, Toth, Durst, & Wilson, 1999; Thévenot, Toth, Durst, & Wilson, 2001). Leland C. Clark has invented the electrode, which can measure the amount of glucose in a person's blood, and he is also designated as the father of biosensor. His research plays a vital role in evaluating the blood glucose level in diabetic patients. A biosensor is a sensing

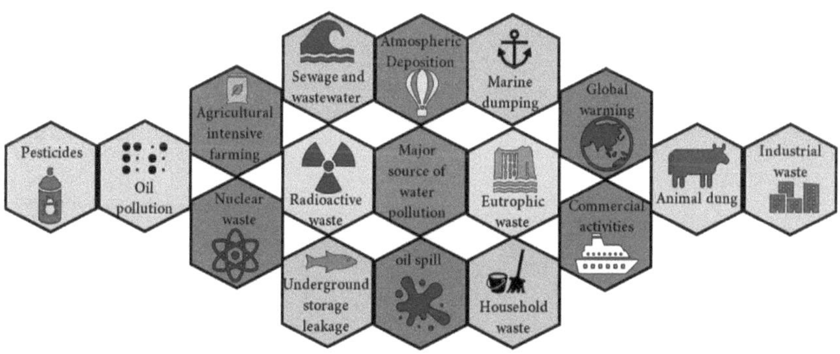

Figure 4.1 The major sources of water pollution.

device that consists of specific biological elements (enzyme, antibody, and nucleic acid) and a transducer. These biorecognition elements interact with the analyte and the biological event or response generated is interpreted and reformed into electrical signals by the transducer (Bhalla, Jolly, Formisano, & Estrela, 2016) (Fig. 4.2).

In the last few years, several types of chemical and biosensor-based monitoring techniques have been developed that serve multiple purposes. Due to numerous applications of biosensors in diverse fields, it has been established as a major sensor technology and has received significant attention in past several years. Biosensors are used in food industry to monitor the safety and quality of the food products and to detect precise glucose concentration in the saccharification process (Dar, Kamili, Dar, Chishti, & Ahmad, 2015; Scognamiglio, Arduini, Palleschi, & Rea, 2014). It is also being used in metabolic engineering for monitoring cellular metabolism within the cells. Biosensors have been widely used in medical sciences for rapid and early-stage detection of human interleukin-10 (cause of heart disease), human papilloma virus, and cancer (Lee et al., 2012; Mehrotra, 2016). The use of biosensing technologies in plant biology is to find out the missing connections required in metabolic pathways is also worth mentioning here. Some other fields in which biosensors have been applied are defense, clinical (such as measuring blood glucose level), drug discovery, agriculture, and marine applications (Mehrotra, 2016). Role of biosensors in wastewater management is substantial. Biosensors can be categorized based on the type of pollutants being monitored (organic material, heavy metals, and microorganism) in wastewater (Fig. 4.3).

4.3.1 Biosensors for the detection of heavy metals

Heavy metals are highly abundant and are natural components of the Earth's core. Numerous heavy metals are utilized in agricultural and

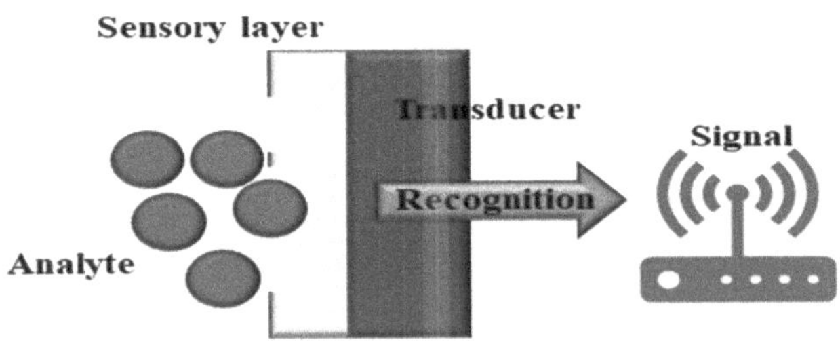

Figure 4.2 General components and working of a biosensor.

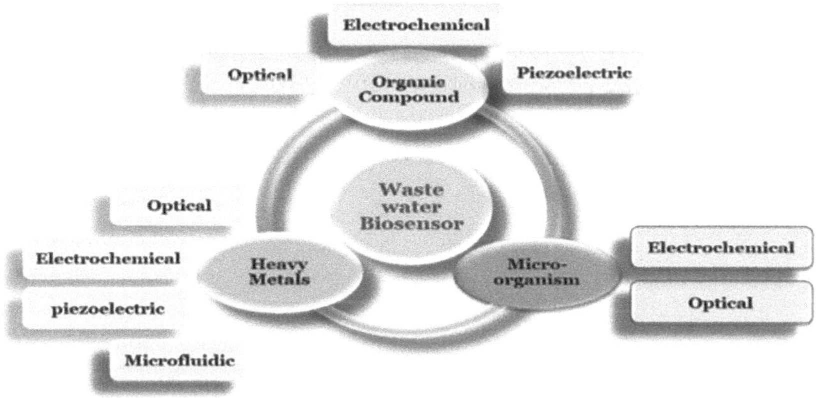

Figure 4.3 Potential applications of biosensors in wastewater monitoring.

industrial processes but due to unresponsive activities of humans, their biochemical balance has been disturbed in the atmosphere, thus affecting various geochemical cycles (Singh, Gautam, Mishra, & Gupta, 2011). Heavy metals are noxious substances and nonbiodegradable in nature, and even a small amount of them can affect the plant as well as animal systems directly or indirectly. Also, it has been observed that this kind of contamination is one of the major concerns to the human health and can lead to kidney failure, loss of appetite, anemia, loss of memory, and neurological damage (Ntihuga, 2006). Mercury is a toxic element that can easily penetrate the human food chain (Dar et al., 2016; McGhee, Loh, & Lu, 2017) and is harmful to human health, nerves, kidneys, and liver in particular (Johansson, Norman, & Burman, 2009; Vogel et al., 2014). All the metals except calcium, aluminum, potassium, sodium, and magnesium are termed as heavy metals. The atomic mass of heavy metals lies in a range of 63.5 to 200.6 and these have specific density greater than 5 g/cm^3. Some of the examples of heavy metals are mercury, lead, cadmium, arsenic, zinc, nickel, copper, chromium, and antimony (Islam, Ahmed, Raknuzzaman, Habibullah-Al-Mamun, & Islam, 2015; Odobašić, Šestan, & Begić, 2019; Zhang, Zhong, Liu, & Ouyang, 2015).

In the past, the detection of heavy metals was performed using atomic absorption spectroscopy (Vogel et al., 2014), spectrophotometry (Almela et al., 2002), atomic fluorescence spectroscopy (Bua, Annuario, Albergamo, Cicero, & Dugo, 2016), and inductively coupled plasma atomic emission spectrometry (Xia, Zhang, Lou, & Yuan, 2018). These methods have numerous drawbacks which include tedious sample preparation, long detection time, expensive equipment, high manpower cost, and less portability due to which on-site detection is difficult to perform (Nepomuscene, Daniel, & Krastanov, 2007). This leads to the development of new detection tools based on biosensing technology (Fig. 4.4).

4.3.1.1 Enzyme-based biosensors

Enzyme-based biosensors are used to monitor inorganic (e.g., heavy metals) and organic (e.g., herbicides and pesticides) substances. Numerous enzymes, including phosphatase, urease, and esterase, have been used to estimate the content of heavy metals in the wastewater on the basis of its activation and inactivation of enzyme activity (Verma & Singh, 2005). Heavy metal acted as cofactors and led to the activation of the enzyme by forming a metalloprotein complex. In a study, alkaline phosphatase apoenzyme was activated by the metal cofactor, releasing heat during the process. The enzyme was covalently immobilized and Zn (II) was detected in between 10 μM and 1.0 mM with a 3 min response time (Bontidean, 2002). The drawback of an enzyme-based biosensor is that very limited enzymes are known to be sensitive to heavy metals, hence restricting their use. Moreover, these biosensors are not cost-effective (due to enzyme extraction and purification), and the enzyme activity/stability is dependent on various factors such as pH, ionic strength, chemical inhibitors, and temperature of the samples to be analyzed (D'Orazio, 2003; Turdean, 2011; Rajpoot, 2017).

4.3.1.2 Protein-based biosensor

Various nonenzymatic proteins can naturally bind to metals or can be designed to specifically bind to certain metal ions (Afsahi et al., 2018). For example, synthetic phytochelatins or metallothioneins have been used for the identification of mercury, copper, zinc, and lead with detection limit of 100 fM (Thévenot, Toth, Durst, & Wilson, 2001).

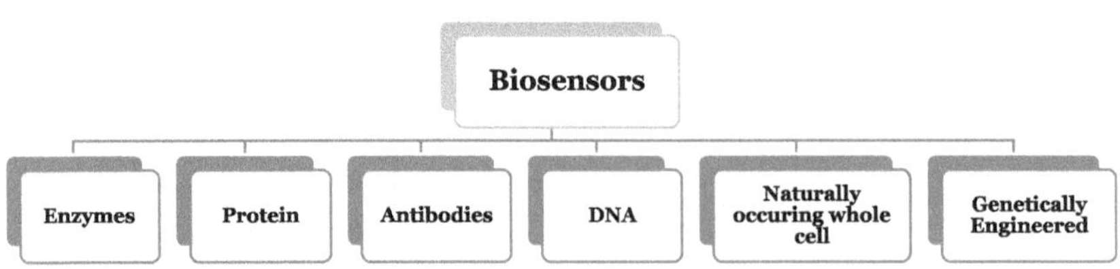

Figure 4.4 Variety of biosensors used for identification of heavy metals in the wastewater.

4.3.1.3 Antibody-based biosensor

Antibodies, also known as immunoglobulins (Ig), are produced in humans and animals by the immune system in response to a foreign moiety. The antibodies are "Y" shaped proteins that are produced by WBCs (white blood cells), also termed as B cells (B lymphocytes). Antibodies have the ability to identify and form specific complexes with antigens to make them ideal for being used as a biorecognition element in biosensors. Five classes of immunoglobulins (Ig) (IgA, IgD, IgE, IgG, and IgM) have been distinguished on the basis of their structure and function. For heavy metal detection, IgG class of antibodies is most frequently used as these show higher specificity and affinity compared to other classes of antibodies. Also, recombinant, monoclonal, or polyclonal antibodies can be utilized for the detection of various heavy metals in wastewater (Šestan, 2015).

In a study, complexes of lead–CHXDTPA, cobalt–DTPA, and cadmium–EDTA were detected with the help of monoclonal antibodies with increased sensitivities of 0.25, 10, and 6.0 nM, respectively (Blake et al., 2001).

However, this technology has several limitations, such as the regeneration of the immune surface and cross-reactivity. Moreover, some level of cross-reactivity is required to identify the different species of the same class (Rodriguez-Mozaz, Marco, Lopez De Alda, & Barceló, 2004). Moreover, the interaction between the antigen–antibody complex depends on the affinity and the orientation of the antibody, while it is immobilized to the surface (Patel, Nanda, Sahoo, & Mohapatra, 2016). Immunoassays have created an alternate strategy for detecting the metal ion as it provides substantial advantages over the other traditional detection methods, including high sensitivity, selectivity and species specificity and are theoretically applicable to any pollutant for which a suitable antibody can be generated.

4.3.1.4 Deoxyribonucleic acid-based biosensor

NABs (nucleic acid biosensors) or DNA (deoxyribonucleic acid)-based biosensors are based on the ability of the nucleic acid to form complementary base pairing with the target DNA/RNA (ribonucleic acid). It consists of short single-stranded oligonucleotides as a biorecognition element that is immobilized on the surface of the transducer (Thévenot, Toth, Durst, & Wilson, 2001). These nucleic acid elements can be denatured and reused, which makes them cost-effective. The major limitation of this technique is the time taken to prepare the sample as it requires

Polymerase chain reaction (PCR) amplification of the DNA in the sample prior to analysis (Paleček, 2002).

Scientists have reported calorimetric detection of Pb (II) that is based on a DNAzyme-directed association of gold nanoparticles. A "8−17" DNAzyme (first-generation catalytic beacon) is the metal-detecting bioelement (single-stranded oligonucleotides with catalytic activity) and displays high specificity and activity toward Pb (II). The combination of the DNAzyme, substrate, and gold nanoparticles resulted in the development of blue coloration. The sensor has the capability to sense Pb (II) in between 100 nM and 4.0 lM (Liu & Lu, 2003; Zhang, Kong, & Lu, 2011).

4.3.1.5 Naturally occurring whole-cell biosensor

This technique involves in the utilization of specific metabolic pathways occurring in microorganisms for a more efficient analysis of pollutants that cannot be estimated through simple enzymatic reactions (Riedel, Kunze, & König, 2002). The microbial electrodes have many advantages over other traditional enzyme electrodes such as the low cost of maintenance of microbial strains compared to the isolation and purification of enzymes. Further, the activity of the enzyme involved in owing to the ideal environment which is provided by the cells (D'Souza, 2001).

To date, various microbial biosensors were developed to release bioluminescence signals in the presence of heavy metals even with a low concentration (Jouanneau, Durand, & Thouand, 2012; Martín-Betancor, Rodea-Palomares, Muñoz-Martín, Leganés, & Fernández-Piñas, 2015). Bacterial sensors using recombinant *Escherichia coli* strains expressing fluorescent protein have been developed. When the bacteria come in contact with the heavy metals, the fluorescent protein expression was enhanced, which correlates the number of heavy metals (Dar, Bandh, & Kamili, 2011; Edward Raja & Selvam, 2011).

4.3.1.6 Genetic engineering-based biosensor

It has been known that proteins have some metal resistance capacity, which can be involved in the regulatory mechanism as well as genetically engineered microorganisms. These resistant capacities of proteins can be used for biomonitoring of the heavy metal-contaminated sites.

Various other portable microfluidic devices appeared to be efficient for determining the heavy metals in different fluids, majorly in the case of dark waters. To illustrate this, integration of microfluidic approach with the optical transducing system tends to develop a cost-efficient biosensor for detecting lead concentration. The mechanism of estimation depends on the suppression of immobilized urease enzyme in the presence of Pb^{2+} that was further trailed by reducing substrate hydrolysis through the colorimetric alterations of a pH-sensitive indicator (Verma, Sharma, & Kumar, 2016).

4.3.2 Biosensors for the detection of microorganisms

Microorganisms including protozoa, fungi, and bacteria are ubiquitous and play a vital role in natural decomposition, nutrient cycles, and various other biotechnological processes. However, several of them are pathogenic and harmful which tends to cause human diseases. *Legionella pneumophila* is a bacterium that attacks human macrophages and causes Legionnaires' disease (Swanson & Hammer, 2000). *Legionella* species are common waterborne pathogens, whereas *Salmonella typhimurium* and *Campylobacter jejuni* cause food poisoning in humans (Mor-Mur & Yuste, 2010). Toxins produced by bacteria in order to overcome the host defense are mainly accountable for the indications perceived in bacterial infections. It has been observed that bacterial toxins are extremely diverse, varying from small molecules (microcystin-LR) (Dar et al., 2012; Žegura, Štraser, & Filipič, 2011) to large proteins (Shiga and Shiga-like toxin) (Schmitt, Meysick, & O'Brien, 1999; Tesh & O'Brien, 1991), and even some multiprotein mixtures such as anthrax toxin which is made up of a lethal factor, an edema factor, and a protective antigen (Bhatnagar & Batra, 2001; Collier & Young, 2003). Nowadays, the monitoring of microorganisms is very important to maintain human health and safety. In the clinical environment, the finding of microorganisms is essential for the diagnosis of pathogens in patient samples, whereas in food industries, the findings of microorganisms and their toxins are necessary to maintain food safety and quality (Code of Practice on Environmental Health, 2013). In several countries, there is some permissible limit of microbial load that is allowed in the environment. For example, in all the swimming pools in Singapore, level of *E. coli* has to be maintained lower than 1 cfu/100 mL (Code of Practice on Environmental Health, 2013), whereas in cooling tower, *Legionella* bacteria is adjusted to be less than 10 cfu/mL (Code of Practice for the Control of Legionella bacteria in Cooling Towers, 2001).

The conventional methods for pathogen detection utilize expensive artificial medium and microscopic methods. These methods are not sensitive and cumbersome, involve complex sample processing, and have a low detection limit (Lee, Shannon, & Beaudette, 2006).

4.3.2.1 Optical biosensors

Optical biosensor is analytical technique which can be used for monitoring pollutants without using any extensive sample or label-free preparation (Long, Zhu, & Shi, 2013; Singh, 2017). It is based on immunoanalytical technique which required fluorescent-labeled antibodies (Ejeian et al., 2018). It has huge advantage over conventional methods such as high frequency, high specificity, real time, cost-effectiveness, and highly sensitive for monitoring of various drugs, toxins, and pathogenic bacteria in wastewater (Bae, Oh, Lee, Lee, & Choi, 2004; Lazcka, Campo, & Muñoz, 2007; Dar, Bandh, Kamili, Nazir, & Bhat, 2013; Long et al., 2013; Singh, 2017). For the detection of harmful pathogens, several optical techniques viz., surface plasmon resonance (SPR) (Guo et al., 2018) and resonant mirror (Kassal, Steinberg, & Steinberg, 2018) have been used. One of the examples is the immobilization of specific type of antibodies against *Listeria monocytogenes* and *Salmonella enteritidis* on a gold SPR which tends to provide a detection limit of 10^6 cell/mL (Koubová et al., 2001).

Noble and inert metal nanoparticles such as gold and silver have been reported as an important substrate in surface-enhanced Raman scattering (SERS) pathogen biosensors (Singh et al., 2009). Metallic nanomaterials have the tendency to detect microbial DNA. Colorimetric detection involving nanoparticles could eliminate the requirement of costly fluorescent labeling (Ivnitski, Abdel-Hamid, Atanasov, & Wilkins, 1999).

4.3.2.2 Electrochemical biosensor

Many movable devices are developed amperometrically to detect several microorganisms by their electrooxidation or electroreduction phenomena (Kokkinos & Economou, 2017). For example, an assay has been developed for the detection of *E. coli* in environmental samples using indirect amperometric quantification of the β-D-glucuronidase (GLUase) activity (Rochelet et al., 2015). Further, an amperometric biosensor for estimating the trace amounts of *Mycobacterium smegmatis* and *Bacillus cereus* has also been developed (Yemini, Levi, Yagil, & Rishpon, 2007). One of the advantages of using amperometric biosensor is that it is inexpensive and sensitive and gives rapid results (Yemini et al., 2007).

Abbaspour et al. reported the usage of electrochemical dual aptasensor to detect *Staphylococcus aureus* in real water samples. The detection procedure was performed with the help of primary aptamer that was immobilized on the magnetic beads through specific biotin–streptavidin reaction. The secondary aptamer conjugated to AgNPs was utilized to detect the existence of *S. aureus* via the sandwich immune-sensing method. The last step involved the formation of differential pulse stripping voltammetric signal by dissolved AgNPs in the acidic solution which indirectly exhibited *S. aureus* quantification in the water samples (Abbaspour, Norouz-Sarvestani, Noori, & Soltani, 2015).

4.3.3 Biosensors for the detection of organic pollutants

4.3.3.1 Organic pollutants

Pollutants are considered as the harmful particles present in the environment. These end up in the environment in an unstable form and are resistant to degradation. Environmental pollutants can be classified as organic and inorganic. The most common organic pollutants are known as persistent organic pollutants. These are robust, synthetic, and human-made organic compounds (Fig. 4.5) and tend to dissolve more in lipids than in water. Due to this property, these pollutants are highly resistant to the biodegradation process and accumulate in the environment (Schell, Knutson, & Bailey, 2012). In addition to this, they undergo a biomagnification process causing deleterious effects to the environment.

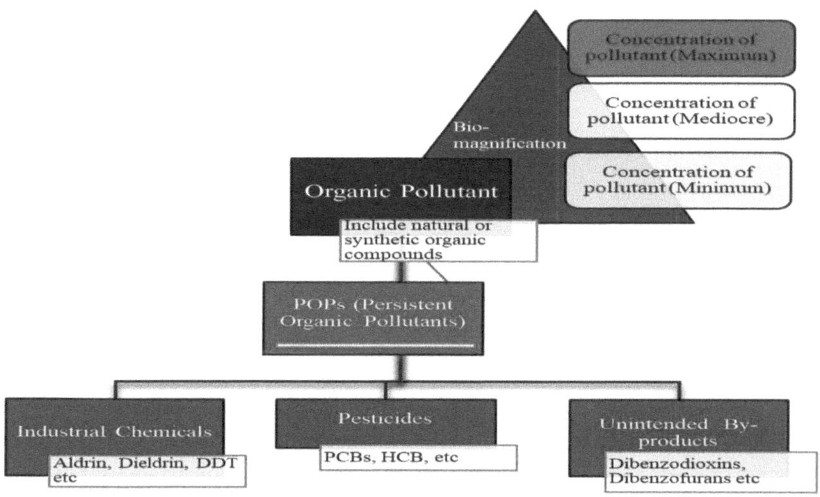

Figure 4.5 Various types of organic pollutants and their dreadful effects in nature. *DDT*, dichloro-diphenyl-trichloroethane; *HCB*, hexachlorobenzene; *PBC*, polychlorinated biphenyls.

Across the world, environmental protection from these pollutants is the primary subject of concern. In order to fulfill the human demands, a plethora of industries have been established to date. Moreover, these industries generate various hazardous agents such as toxic elements and contaminants that are released in the environment. In addition to this, essential human-made materials used globally by human society, such as pesticides, cosmetics, and pharma products are considered as the auxiliary pollutants (Claude, Houssemeddine, Andriy, & Jean-Marc, 2007). Escalating levels of pollutants in the environment on a global scale is a serious concern and therefore there is an urgent requirement to initiate the preventive measures. Modern generation sensing techniques should be designed that can detect contaminants in a precise and broad manner. The biosensors are the suitable candidate for this job. It is a rapid, sensitive, specific, and selective technique that can screen the environmental pollutants. Conventional methods such as high-performance liquid chromatography and gas chromatography coupled to mass spectrometry or capillary electrophoresis for the detection of organic pollutants were not studied these days due to their high-cost reagents and equipment, cumbersome operation, and time-taking procedure (Hassani et al., 2017; Lang, Han, Hou, Wang, & Liu, 2016). Therefore, to overcome these problems, biosensors were developed.

Based on the type of the recognition element and the transducer system, biosensors can be divided into various groups such as optical, electrochemical, piezoelectric and immunosensors, aptasensors, genosensors, and enzymatic biosensors (Turner, 2000).

4.3.3.2 Optical biosensors

Optical biosensor is an analytical device having a bioidentification sensing element in its optical transducer system. Previous studies illustrated that whole-cell biosensor technology can be applied for the detection of herbicides and organic pollutants, that is, triclosan (Atar, Eren, Yola, & Wang, 2015; Védrine, Leclerc, Durrieu, & Tran-Minh, 2003). Recently, fluorescent-labeled antibodies for sensing the organic matter are gaining attention among the scientific community. Study conducted by Olaniran, Motebejane, and Pillay (2008) illustrated the importance of whole-cell bacterial biosensors in detecting chemical oxygen demand (COD) in wastewater by evaluating the alterations in bioluminescence response.

Phenol and their derivatives are the most common organic compounds and are classified as the toxic compounds being released as a waste from various industries. Therefore enzymes such as laccase, tyrosinase, and peroxidases are utilized for the development of biosensors for their proper degradation. Diverse types of biosensors using various microorganisms were studied for sensing phenol and their derivatives from the environment and are developed either in free-state or in the immobilized form (Mehndiratta, Jain, Srivastava, & Gupta, 2013; Theron & Cloete, 2002).

4.3.3.3 Electrochemical biosensors

Amperometric is one of the most effective techniques in this class of biosensors. Nomngongo, Catherine Ngila, Msagati, Gumbi, and Iwuoha (2012) reported the use of the enzyme horseradish peroxidase to detect the organic pollutants in wastewater. Further, to estimate the biochemical oxygen demand (BOD) in the samples, a Clark-type oxygen sensor was established. Verma and Singh (2013) developed a biosensor using immobilized microorganisms on cellulose acetate membrane attached with an amperometric gold and silver oxygen probe for analyzing the BOD in industrial wastewater.

4.3.3.4 Thermal biosensors

In this class of biosensor, thermal resistors are used to assess the heat change due to the enzymatic reaction. In a study, COD was successfully measured in wastewater samples by evaluating the heat change (Verma & Singh, 2013). The biosensor used in this study demonstrated the linear range from 5 to 3000 mg/L and a low detection limit of 1.84 mg/L. Further, it was able to detect the presence of chloride ions up to 0.015 M concentration.

4.4 General limitations, challenges, and future prospects of biosensors in wastewater monitoring

Despite the substantial merits of biosensors for tracking toxins in wastewater, there are certain limitations to the currently used biosensing technologies such as toxicity to the environment, high cost of production, and low sensitivity and specificity. Moreover, most of the biosensors have not been tested in real time, and not many of them are available commercially for environmental monitoring (Bahadır & Sezgintürk, 2015).

The detection of harmful organic pollutants (herbicides, pesticides, antibiotics, hormones, antiinflammatory drugs, lipid regulators, and analgesics) generated by agricultural sewage and pharmaceutical industries is still a challenge because the available devices are sensitive and have low detection limit (Das et al., 2017; Radjenovic, Petrovic, & Barceló, 2007; Sala et al., 2017; Tschmelak, Proll, & Gauglitz, 2005; Zhou et al., 2017). Identification of appropriate and cost-effective biorecognition elements is one of the few major challenges that will be helpful in improving organic toxin detection by biosensors in wastewater. The application of microorganisms and biomolecules (enzymes, DNA, and antibody) in the biorecognition system is the one of the ways to increase the accuracy and detection limit for detecting various types of pollutants in wastewaters from different sources. However, the limited availability of appropriate bioreceptors narrows down the range of toxins that can be monitored (Yuan, Jiang, Xia, Zhang, & Zheng, 2013). Therefore there is a continuing quest for new and another type of bioreceptors that are highly sensitive. Current devices require long sample preparation and are not efficient in detecting multiple microorganisms in a single sample (Baeumner, 2003; Bazin et al., 2017). Nucleic acid-based biosensors require PCR amplification of the DNA sample to be analyzed to a readable amount to get accurate results, which is a cumbersome and time-consuming process (Patel et al., 2016).

Development of modern biosensors requires the integration of several technologies such as electrochemistry, electromagnetism, electromechanics, fluorescence, optics, nanotechnology, and synthetic biology. Hybrid devices can be developed using the combination of the aforementioned technologies (Liu, Lang, Liang, & Shi, 2017).

Technology-based protein/genetic engineering and synthetic biology offer an innovative platform to develop advanced and efficient microbial biosensors for monitoring environmental pollutants (Vigneshvar, Sudhakumari, Senthilkumaran, & Prakash, 2016). Microbial fuel cell technology-based biosensors seem a promising candidate for monitoring COD, BOD, dissolved oxygen (DO), and various toxic organic pollutants in the environment (Ivars-Barceló et al., 2018; Zhou et al., 2017).

Novel biosensors have also been developed by combining nanomaterials with electrochemical sensors (Kwon & Bard, 2012; Sun et al., 2015). The integration of nanomaterials with current devices can strongly enhance the SERS signals and can help attain the fingerprint peaks corresponding to the microorganisms. Further, the development of sensors with multistructure can give us a suitable chance to detect a variety of organisms (Theron, Eugene Cloete, & De Kwaadsteniet, 2010; Xiao, 2010). The successfully established SPR-based devices for the identification of pathogens in aqueous conditions are a great achievement in this field, but the large size of equipment, complexity, and high cost restricted its application. In the past, several potentiometric biosensors have also been used for microbial detections. The major limitations of these types of biosensors are the low detection limit and reproducibility (Lazcka et al., 2007; Zhang, Liu, Guo, & Lin, 2018).

Sensing system based on the conjugation of biosensor and drones is of great interest especially in the field of wastewater and air quality management. Some other fields in which this strategy is being used are surveillance of agricultural crops and measurement of volcanic gases. These systems provide a cost-effective, compact, and efficient system for monitoring pollutants in remote areas. Recently, this system have been deployed for monitoring air quality and water pollutants in remote places in China (Phantom 2, Shenzhen, China) (Wang, Takahashi, Du, & Anzai, 2014).

4.5 Conclusion

Rise in water pollution due to increase in urbanization, industrialization, and anthropogenic activities and effective water resource management are of critical concerns to researchers, governmental and nongovernmental organizations, and the general public. On a broader perspective, this chapter gives a general idea of biosensor-based monitoring of organic materials and inorganic (heavy metals) and biological (microorganisms) pollutants. Biosensors discussed in this chapter have the tendency to detect the lower concentration of contaminants within the water which makes them a suitable aspirant for various industrial applications. Further, the application of biosensors in various fields and their general components, limitations, and future prospects have been mentioned briefly. Recently, a lot of strategies have been developed for an effective monitoring of wastewater pollutants but still there is a need to develop a sensitive, cost-effective, and nontoxic biosensor system that can detect wide range of pollutants. Modern-era biosensors can be developed by bringing together technology from various fields such as nanotechnology, biotechnology, microelectronics, microfluidics, and telecommunication. A significant advancement has been made, yet a lot need to be explored and accomplished in this field.

References

Abbaspour, A., Norouz-Sarvestani, F., Noori, A., & Soltani, N. (2015). Aptamer-conjugated silver nanoparticles for electrochemical dual-aptamer-based sandwich detection of *Staphylococcus aureus*. *Biosensors & Bioelectronics*.

Afsahi, S., Lerner, M. B., Goldstein, J. M., Lee, J., Tang, X., Bagarozzi, D. A., ... Goldsmith, B. R. (2018). Novel graphene-based biosensor for early detection of Zika virus infection. *Biosensors & Bioelectronics*.

Almela, C., Algora, S., Benito, V., Clemente, M. J., Devesa, V., Súñer, M. A., ... Montoro, R. (2002). Heavy metal, total arsenic, and inorganic arsenic contents of algae food products. *Journal of Agricultural and Food Chemistry*.

Asadnia, M., Myers, M., Akhavan, N. D., O'Donnell, K., Umana-Membreno, G. A., Mishra, U. K., ... Parish, G. (2016). Mercury (II) selective sensors based on AlGaN/GaN transistors. *Analytica Chimica Acta*, *943*, 1–7.

Atar, N., Eren, T., Yola, M. L., & Wang, S. (2015). A sensitive molecular imprinted surface plasmon resonance nanosensor for selective determination of trace triclosan in wastewater. *Sensors and Actuators B: Chemical*.

Bae, Y. M., Oh, B. K., Lee, W., Lee, W. H., & Choi, J. W. (2004). Detection of insulin-antibody binding on a solid surface using imaging ellipsometry. *Biosensors & Bioelectronics*.

Baeumner, A. J. (2003). Biosensors for environmental pollutants and food contaminants. *Analytical and Bioanalytical Chemistry*.

Bahadır, E. B., & Sezgintürk, M. K. J. A. B. (2015). Applications of commercial biosensors in clinical, food, environmental, and biothreat/biowarfare analyses. *Analytical Biochemistry*, *478*, 107–120.

Bazin, I., Seo, H. B., Suehs, C. M., Ramuz, M., De Waard, M., & Gu, M. B. (2017). Profiling the biological effects of wastewater samples via bioluminescent bacterial biosensors combined with estrogenic assays. *Environmental Science and Pollution Research*.

Beniah, O. I., & Ebere, E. C. (2019). A review: Water pollution by heavy metal and organic pollutants: Brief review of sources, effects and progress on remediation with aquatic plants. *Analytical Methods in Environmental Chemistry Journal*.

Bermúdez-Couso, A., Fernández-Calviño, D., Álvarez-Enjo, M. A., Simal-Gándara, J., Nóvoa-Muñoz, J. C., & Arias-Estévez, M. (2013). Pollution of surface waters by metalaxyl and nitrate from non-point sources. *The Science of the Total Environment*.

Bhalla, N., Jolly, P., Formisano, N., & Estrela, P. (2016). Introduction to biosensors. *Essays in Biochemistry*.

Bhat, R. A., Singh, D. V., Qadri, H., Dar, G. H., Dervash, M. A., Bhat, S. A., ... Yousaf, B. (2021). Vulnerability of municipal solid waste: An emerging threat to aquatic ecosystems. *Chemosphere*, *287*(2022), 132223. Available from https://doi.org/10.1016/j.chemosphere.2021.132223.

Bhatnagar, R., & Batra, S. (2001). Anthrax toxin. *Critical Reviews in Microbiology*.

Biswas, P., Karn, A. K., Balasubramanian, P., & Kale, P. G. (2017). Biosensor for detection of dissolved chromium in potable water: A review. *Biosensors & Bioelectronics*.

Blake, D. A., Jones, R. M., Blake, R. C., Pavlov, A. R., Darwish, I. A., & Yu, H. (2001). Antibody-based sensors for heavy metal ions. *Biosensors and Bioelectronics*.

Bontidean, I. (2002) *Design, development and applications of highly sensitive protein based capacitive biosensors*. Lund University Dissertation Abstracts.

Bua, D. G., Annuario, G., Albergamo, A., Cicero, N., & Dugo, G. (2016). Heavy metals in aromatic spices by inductively coupled plasma-mass spectrometry. *Food Additives & Contaminants: Part B Surveillance*.

Chong, S. S., Abdul Aziz, A. R., & Harun, S. W. (2013). Fibre optic sensors for selected wastewater characteristics. *Sensors (Switzerland)*, *13*(7), 8640–8668.

Claude, D., Houssemeddine, G., Andriy, B., & Jean-Marc, C. (2007). Whole cell algal biosensors for urban waters monitoring. *Novatech*, 7, 1507–1514.

Code of Practice for the Control of Legionella bacteria in Cooling Towers, Institute of Environmental Epidemiology, Singapore 2001.

Code of Practice on Environmental Health, National Environment Agency, Singapore 2013.

Collier, R. J., & Young, J. A. T. (2003). Anthrax toxin. *Annual Review of Cell and Developmental Biology*.

Cornish, G., Mensah, E., & Ghesquire, P. (1999). *Water quality and peri-urban irrigation: an assessment of surface water quality for irrigation and its implications for human health in the peri-urban zone of Kumasi, Ghana. Project Report*. HR Wallingford Ltd.

D'Souza, S. F. (2001). Microbial biosensors. *Biosensors & Bioelectronics*.

Dar, G. H., Bandh, S. A., Kamili, A. N., Nazir, R., & Bhat, R. A. (2013). Comparative analysis of different types of bacterial colonies from the soils of Yusmarg Forest, Kashmir valley India. *Ecologia Balkanica, 5*(1), 31−35.

Dar, G. H., Kamili, A. N., Chishti, M. Z., Dar, S. A., Tantry, T. A., & Ahmad, F. (2016). Characterization of *Aeromonas sobria* isolated from fish Rohu (*Labeo rohita*) collected from polluted pond. *Journal of Bacteriology and Parasitology*, 7(3), 1−5. Available from https://doi.org/10.4172/2155-9597.1000273.

Dar, G. H., Kamili, A. N., Dar, S. A., Chishti, M. Z., & Ahmad, F. (2015). Detection and characterization of potentially pathogenic *Aeromonas sobria* isolated from fish *Hypophthalmichthys molitrix* (Cypriniformes: Cyprinidae). *Microbial Pathogenesis* (2015). Available from https://doi.org/10.1016/j.micpath.2015.10.017.

Dar, G. H., Bandh, S. A., & Kamili, A. N. (2011). Bacteriological analysis of soil from Yusmarg Health Resort of Kashmir Valley. *International Journal of Current Research, 3* (10), 53−56, 2011.

Dar, G. H., Kamili, A. N., Nazir, R., Bandh, S. A., & Bhat, R. A. (2012). A preliminary study of colony forming units of bacteria from the soils of Yusmarg forest, Kashmir valley, India. *International Journal of Current Research, 4*(12), 467−472.

Das, S., Ray, N. M., Wan, J., Khan, A., Chakraborty, T., & Ray, M. B. (2017). *Micropollutants in wastewater: Fate and removal processes. Physico-chemical wastewater treatment and resource recovery*. IntechOpen.

D'Orazio, P. J. C. C. A. (2003). *Biosensors in clinical chemistry. Clinica Chimica Acta; International Journal of Clinical Chemistry, 334*, 41−69.

Ebere Enyoh, C., Wirnkor Verla, A., & Jane Egejuru, N. (2018). pH variations and chemometric assessment of borehole water in Orji, Owerri Imo State, Nigeria. *Journal of Environmental Analytical Chemistry*.

Edward Raja, C., & Selvam, G. S. (2011). Construction of green fluorescent protein based bacterial biosensor for heavy metal remediation. *International Journal of Environmental Science and Technology*.

Ejeian, F., Etedali, P., Mansouri-Tehrani, H. A., Soozanipour, A., Low, Z. X., Asadnia, M., … Razmjou, A. (2018). Biosensors for wastewater monitoring: A review. *Biosensors & Bioelectronics*.

Guo, B., Wen, B., Cheng, W., Zhou, X., Duan, X., Zhao, M., … Ding, S. (2018). An enzyme-free and label-free surface plasmon resonance biosensor for ultrasensitive detection of fusion gene based on DNA self-assembly hydrogel with streptavidin encapsulation. *Biosensors & Bioelectronics*.

Hassani, S., Momtaz, S., Vakhshiteh, F., Maghsoudi, A. S., Ganjali, M. R., Norouzi, P., & Abdollahi, M. (2017). Biosensors and their applications in detection of organophosphorus pesticides in the environment. *Archives of Toxicology*.

Islam, M. S., Ahmed, M. K., Raknuzzaman, M., Habibullah -Al- Mamun, M., & Islam, M. K. (2015). Heavy metal pollution in surface water and sediment: A preliminary assessment of an urban river in a developing country. *Ecological Indicators*.

Ivars-Barceló, F., Zuliani, A., Fallah, M., Mashkour, M., Rahimnejad, M., & Luque, R. J. A. S. (2018). Novel applications of microbial fuel cells in sensors and biosensors. *Applied Sciences, 8*, 1184.

Ivnitski, D., Abdel-Hamid, I., Atanasov, P., & Wilkins, E. (1999). Biosensors for detection of pathogenic bacteria. *Biosensors & Bioelectronics*.

Johansson, C., Norman, M., & Burman, L. (2009). Road traffic emission factors for heavy metals. *Atmospheric Environment (Oxford, England: 1994)*.

Jouanneau, S., Durand, M. J., & Thouand, G. (2012). Online detection of metals in environmental samples: Comparing two concepts of bioluminescent bacterial biosensors. *Environmental Science & Technology*.

Kassal, P., Steinberg, M. D., & Steinberg, I. M. (2018). Wireless chemical sensors and biosensors: A review. *Sensors and Actuators: B Chemical*.

Kokkinos, C., & Economou, A. (2017). Emerging trends in biosensing using stripping voltammetric detection of metal-containing nanolabels – A review. *Analytica Chimica Acta*.

Kolpin, D. W., Furlong, E. T., Meyer, M. T., Thurman, E. M., Zaugg, S. D., Barber, L. B., & Buxton, H. T. (2002). Pharmaceuticals, hormones, and other organic wastewater contaminants in U.S. streams, 1999–2000: A national reconnaissance. *Environmental Science & Technology*.

Korostynska, O., Mason, A., & Al-Shamma'a, A. I. (2013). *Monitoring pollutants in wastewater: Traditional lab based vs modern real-time approaches. Smart Sensors, Measurement and Instrumentation*. Springer.

Koubová, V., Brynda, E., Karasová, L., Škvor, J., Homola, J., Dostálek, J., … Rošický, J. (2001). Detection of foodborne pathogens using surface plasmon resonance biosensors. *Sensors and Actuators, B: Chemical*.

Kwon, S. J., & Bard, A. J. (2012). DNA analysis by application of Pt nanoparticle electrochemical amplification with single label response. *Journal of the American Chemical Society*.

Lakatos, G. (2018) Biological wastewater treatment. *Wastewater and Water Contamination: Sources, Assessment and Remediation*.

Lang, Q., Han, L., Hou, C., Wang, F., & Liu, A. (2016). A sensitive acetylcholinesterase biosensor based on gold nanorods modified electrode for detection of organophosphate pesticide. *Talanta*.

Lazcka, O., Campo, F. J. D., & Muñoz, F. X. (2007). Pathogen detection: A perspective of traditional methods and biosensors. *Biosensors & Bioelectronics*.

Lee, D. Y., Shannon, K., & Beaudette, L. A. (2006). Detection of bacterial pathogens in municipal wastewater using an oligonucleotide microarray and real-time quantitative PCR. *Journal of Microbiological Methods*.

Lee, M., Zine, N., Baraket, A., Zabala, M., Campabadal, F., Caruso, R., … Errachid, A. (2012). A novel biosensor based on hafnium oxide: Application for early stage detection of human interleukin-10. *Sensors and Actuators: B Chemical*.

Liu, A., Lang, Q., Liang, B., & Shi, J. (2017). Sensitive detection of maltose and glucose based on dual enzyme-displayed bacteria electrochemical biosensor. *Biosensors & Bioelectronics*.

Liu, J., & Lu, Y. (2003). A colorimetric lead biosensor using DNAzyme-directed assembly of gold nanoparticles. *Journal of the American Chemical Society*.

Long, F., Zhu, A., & Shi, H. (2013). Recent advances in optical biosensors for environmental monitoring and early warning. *Sensors (Switzerland)*.

Martín-Betancor, K., Rodea-Palomares, I., Muñoz-Martín, M. A., Leganés, F., & Fernández-Piñas, F. (2015). Construction of a self-luminescent cyanobacterial bioreporter that detects a broad range of bioavailable heavy metals in aquatic environments. *Frontiers in Microbiology*.

McGhee, C. E., Loh, K. Y., & Lu, Y. (2017). DNAzyme sensors for detection of metal ions in the environment and imaging them in living cells. *Current Opinion in Biotechnology*.

Mehndiratta, P., Jain, A., Srivastava, S., & Gupta, N. (2013). Environmental pollution and nanotechnology. *Environmental Pollution (Barking, Essex: 1987)*.

Mehrotra, P. (2016). Biosensors and their applications - A review. *Journal of Oral Biology and Craniofacial Research*.

Mekonnen, M. M., & Hoekstra, A. Y. (2016). Four billion people facing severe water scarcity. *American Association for the Advancement of Science*.

Michael-Kordatou, I., Michael, C., Duan, X., He, X., Dionysiou, D. D., Mills, M. A., & Fatta-Kassinos, D. (2015). Dissolved effluent organic matter: Characteristics and potential implications in wastewater treatment and reuse applications. *Water Research.*

Mor-Mur, M., & Yuste, J. (2010). Emerging bacterial pathogens in meat and poultry: An overview. *Food and Bioprocess Technology.*

Nepomuscene, N. J., Daniel, D., & Krastanov, A. (2007). Biosensor to detect chromium in wastewater. *Biotechnology & Biotechnological Equipment.*

Nomngongo, P. N., Catherine Ngila, J., Msagati, T. A. M., Gumbi, B. P., & Iwuoha, E. I. (2012). Determination of selected persistent organic pollutants in wastewater from landfill leachates, using an amperometric biosensor. *Physics and Chemistry of the Earth.*

Ntihuga, J. N. (2006). Biosensor to detect heavy metals in waste water. In *Proceedings from the international conference on advances in engineering and technology.*

Odobašić, A., Šestan, I., & Begić, S. (2019). *Biosensors for determination of heavy metals in waters. Biosensors for environmental monitoring.* IntechOpen.

Olaniran, A. O., Motebejane, R. M., & Pillay, B. (2008). Bacterial biosensors for rapid and effective monitoring of biodegradation of organic pollutants in wastewater effluents. *Journal of Environmental Monitoring.*

WHO. (2005). *The World health report: 2005: Make every mother and child count.* World Health Organization.

Paleček, E. (2002). Past, present and future of nucleic acids electrochemistry. *Talanta.*

Pang, Y. C., Xi, J. Y., Li, G. Q., Shi, X. J., & Hu, H. Y. (2015). Prevalence of antibiotic-resistant bacteria in a lake for the storage of reclaimed water before and after usage as cooling water. *Environmental Science: Processes & Impacts.*

Patel, S., Nanda, R., Sahoo, S., & Mohapatra, E. (2016). Biosensors in health care: The milestones achieved in their development towards lab-on-chip-analysis. *Biochemistry Research International.*

Peters, E. J. (2015). Wastewater reuse in the eastern caribbean: A case study. *Proceedings of the Institution of Civil Engineers - Water Management.*

Radjenovic, J., Petrovic, M., & Barceló, D. (2007). Analysis of pharmaceuticals in wastewater and removal using a membrane bioreactor. *Analytical and Bioanalytical Chemistry.*

Rajpoot, K. J. B. J. (2017). Recent advances and applications of biosensors in novel. *Technology (Elmsford, N.Y.), 6,* 1–12.

Rehman, U., Vesvikar, M., Maere, T., Guo, L., Vanrolleghem, P. A., & Nopens, I. (2015). Effect of sensor location on controller performance in a wastewater treatment plant. *Water Science and Technology: a Journal of the International Association on Water Pollution Research.*

Riedel, K., Kunze, G., & König, A. (2002). Microbial sensors on a respiratory basis for wastewater monitoring. *Advances in Biochemical Engineering/Biotechnology.*

Rochelet, M., Solanas, S., Betelli, L., Chantemesse, B., Vienney, F., & Hartmann, A. (2015). Rapid amperometric detection of Escherichia coli in wastewater by measuring β-D glucuronidase activity with disposable carbon sensors. *Analytica Chimica Acta.*

Rodriguez-Mozaz, S., Marco, M. P., Lopez De Alda, M. J., & Barceló, D. (2004). Biosensors for environmental applications: Future development trends. *Pure and Applied Chemistry.*

Sala, S., Anton, A., McLaren, S. J., Notarnicola, B., Saouter, E., & Sonesson, U. (2017). In quest of reducing the environmental impacts of food production and consumption. *Journal of Cleaner Production.*

Schell, L. M., Knutson, K. L., & Bailey, S. (2012). Chapter 10: "Environmental effects on growth." ISBN 9780123838827. *Human growth and development* (2nd ed.). pp. 245–286.

Schmitt, C. K., Meysick, K. C., & O'Brien, A. D. (1999). Bacterial toxins: Friends or foes? *Emerging Infectious Diseases, 5,* 224–234.

Scognamiglio, V., Arduini, F., Palleschi, G., & Rea, G. (2014). Biosensing technology for sustainable food safety. *TrAC - Trends in Analytical Chemistry.*

Šestan, I. (2015). *Investigation kinetics of the reaction of the formation of metallic complexes and the biogeodistribution of heavy metals in the system sediment/ water* (Doctoral thesis).
Singh, A. K., Senapati, D., Wang, S., Griffin, J., Neely, A., Candice, P., ... Ray, P. C. (2009). Gold nanorod based selective identification of *Escherichia coli* bacteria using two-photon Rayleigh scattering spectroscopy. *ACS Nano*.
Singh, P. (2017). *Surface plasmon resonance: A boon for viral diagnostics. Reference module in life sciences*. Elsevier.
Singh, R., Gautam, N., Mishra, A., & Gupta, R. (2011). Heavy metals and living systems: An overview. *Indian Journal of Pharmacology*.
Sun, J. Z., Kingori, G. P., Si, R. W., Zhai, D. D., Liao, Z. H., Sun, D. Z., ... Yong, Y. C. (2015). Microbial fuel cell-based biosensors for environmental monitoring: A review. *Water Science and Technology: a Journal of the International Association on Water Pollution Research*.
Swanson, M. S., & Hammer, B. K. (2000). *Legionella pneumophila* pathogenesis: A fateful journey from amoebae to macrophages. *Annual Review of Microbiology*.
Tesh, V. L., & O'Brien, A. D. (1991). The pathogenic mechanisms of Shiga toxin and the Shiga-like toxins. *Molecular Microbiology*.
Thâvenot, D. R., Toth, K., Durst, R. A., & Wilson, G. S. (1999). Electrochemical biosensors: Recommended definitions and classification (Technical Report). *Pure and Applied Chemistry. Chimie Pure et Appliquee*.
Theron, J., & Cloete, T. E. (2002). Emerging waterborne infections: Contributing factors, agents, and detection tools. *Critical Reviews in Microbiology*.
Theron, J., Eugene Cloete, T., & De Kwaadsteniet, M. (2010). Current molecular and emerging nanobiotechnology approaches for the detection of microbial pathogens. *Critical Reviews in Microbiology*.
Thévenot, D. R., Toth, K., Durst, R. A., & Wilson, G. S. (2001). Electrochemical biosensors: Recommended definitions and classification. *Biosensors & Bioelectronics*.
Tschmelak, J., Proll, G., & Gauglitz, G. (2005). Optical biosensor for pharmaceuticals, antibiotics, hormones, endocrine disrupting chemicals and pesticides in water: Assay optimization process for estrone as example. *Talanta*.
Turdean, G. L. (2011). Design and development of biosensors for the detection of heavy metal toxicity. *International Journal of Electrochemistry*.
Turner, A. P. F. (2000). Biosensors - Sense and sensitivity. *Science (New York, N.Y.)*.
Védrine, C., Leclerc, J. C., Durrieu, C., & Tran-Minh, C. (2003). Optical whole-cell biosensor using Chlorella vulgaris designed for monitoring herbicides. *Biosensors & Bioelectronics*.
Verma, N., & Singh, M. (2005). Biosensors for heavy metals. *Biometals: an International Journal on the Role of Metal Ions in Biology, Biochemistry, and Medicine*.
Verma, N., & Singh, A. K. (2013). Development of biological oxygen demand biosensor for monitoring the fermentation industry effluent. *ISRN Biotechnology*.
Verma, N., Sharma, R., & Kumar, S. (2016). Advancement towards microfluidic approach to develop economical disposable optical biosensor for lead detection. *Austin Journal of Biosensors & Bioelectronics*, *2*(2), 1021.
Vigneshvar, S., Sudhakumari, C. C., Senthilkumaran, B., & Prakash, H. (2016). Recent advances in biosensor technology for potential applications - An overview. *Frontiers in Bioengineering and Biotechnology*.
Vogel, C., Adam, C., Kappen, P., Schiller, T., Lipiec, E., & McNaughton, D. (2014). Chemical state of chromium in sewage sludge ash-based phosphorus-fertilisers. *Chemosphere*.
Wang, B., Takahashi, S., Du, X., & Anzai, J. I. (2014). Electrochemical biosensors based on ferroceneboronic acid and its derivatives: A review. *Biosensors*.
Wang, P., Yao, J., Wang, G., Hao, F., Shrestha, S., Xue, B., ... Peng, Y. (2019). Exploring the application of artificial intelligence technology for identification of

water pollution characteristics and tracing the source of water quality pollutants. *The Science of the Total Environment.*

Xia, F., Zhang, X., Lou, X., & Yuan, Q. (2018). *Biosensors based on sandwich assays* Springer.

Xiao, N. (2010). *Using functionalized nanoparticles as sensors for rapid monitoring of drinking water quality. Agricultural and biosystems engineering* (p. 67) Ames, Iowa: Iowa State University.

Xu, J., Cao, Z., Zhang, Y., Yuan, Z., Lou, Z., Xu, X., & Wang, X. (2018). A review of functionalized carbon nanotubes and graphene for heavy metal adsorption from water: Preparation, application, and mechanism. *Chemosphere.*

Yang, L., Han, D. H., Lee, B. M., & Hur, J. (2015). Characterizing treated wastewaters of different industries using clustered fluorescence EEM-PARAFAC and FT-IR spectroscopy: Implications for downstream impact and source identification. *Chemosphere.*

Yemini, M., Levi, Y., Yagil, E., & Rishpon, J. (2007). Specific electrochemical phage sensing for *Bacillus cereus* and *Mycobacterium smegmatis. Bioelectrochemistry (Amsterdam, Netherlands).*

Yuan, S., Jiang, X., Xia, X., Zhang, H., & Zheng, S. (2013). Detection, occurrence and fate of 22 psychiatric pharmaceuticals in psychiatric hospital and municipal wastewater treatment plants in Beijing, China. *Chemosphere.*

Žegura, B., Štraser, A., & Filipič, M. (2011). Genotoxicity and potential carcinogenicity of cyanobacterial toxins - A review. *Mutation Research - Reviews in Mutation Research.*

Zhang, W., Liu, Q. X., Guo, Z. H., & Lin, J. S. (2018). Practical application of aptamer-based biosensors in detection of low molecular weight pollutants in water sources. *Molecules (Basel, Switzerland).*

Zhang, X., Zhong, T., Liu, L., & Ouyang, X. (2015). Impact of soil heavy metal pollution on food safety in China. *PLoS One.*

Zhang, X. B., Kong, R. M., & Lu, Y. (2011). Metal ion sensors based on DNAzymes and related DNA molecules. *Annual Review of Analytical Chemistry.*

Zhou, T., Han, H., Liu, P., Xiong, J., Tian, F., & Li, X. (2017). Microbial fuels cell-based biosensor for toxicity detection: A review. *Sensors (Switzerland).*

CHAPTER 5

Microbial systems, current trends, and future prospective: a systemic analysis

Misbah Naz[1], Muhammad Ammar Raza[2], Muhammad Rahil Afzal[3], Muhammad Jafar Jaskani[4], Muhammad Anwar[5], Muhammad Tariq[6], Muhammad Riaz[7] and Xiaorong Fan[8,9]

[1]Institute of Environment and Ecology, School of The Environment and Safety Engineering, Jiangsu University, Jiangsu Province, P.R. China
[2]Key Laboratory of Fruits and Vegetables Postharvest and Processing Technology Research of Zhejiang Province, College of Food Science and Biotechnology, Zhejiang Gongshang University, Hangzhou, P.R. China
[3]Faculty of Life Sciences, Institute of Environmental and Agricultural Sciences, University of Okara, Okara, Pakistan
[4]Faculty of Agriculture, Institute of Horticultural Sciences, University of Agriculture Faisalabad, Faisalabad, Pakistan
[5]Guangdong Key Laboratory of Plant Epigenetics, Guangdong Technology Research Center for Marine Algal Bioengineering, College of Life Sciences and Oceanography, Shenzhen, P.R. China
[6]Department of Pharmacology, Lahore Pharmacy Collage, Lahore, Pakistan
[7]School of Agriculture and Biology, Shanghai Jiao Tong University, Shanghai, P.R. China
[8]State Key Laboratory of Crop Genetics and Germplasm Enhancement, Nanjing Agricultural University, Nanjing, P.R. China
[9]Key Laboratory of Plant Nutrition and Fertilization in Lower-Middle Reaches of the Yangtze River, Ministry of Agriculture, Nanjing Agricultural University, Nanjing, P.R. China

5.1 Introduction

Nutrients are rich source of energy, support, and sustainability to micro and macro living systems of the ecosystem. The systematic damage and loss come when such traces cannot be supplied to the living systems in minute. Nutrient supplies to the entire ecosystem in a biological cycle make them functionally active (Neary, Klopatek, DeBano, & Ffolliott, 1999). Biological cycle enables traces to stick with natural systems to make them functional and sustained in the life. The traces include minerals, vitamins, and sugars which are necessary for the living of plants, animals, and human. These elements, compounds, and minerals flow from one organism to another through the biogeochemical cycle of recycling (Trichopoulou et al., 2000).

In addition, plants in nature expose to stress in the environment, leading to develop pressure and nutritional competition. Biotic and abiotic stress are mainly investigated in plant, but a lot of hidden facts are underway, including

Microbial Consortium and Biotransformation for Pollution Decontamination
DOI: https://doi.org/10.1016/B978-0-323-91893-0.00012-2

several internal and external functional units cooperate with environment to develop fuzzy situation in the plant's life (Waraich, Ahmad, & Ashraf, 2011). Current theories that the signaling cascades activated in response to biological and nonbiological stress are identical in function and travel through a same regulatory node open the door to a plethora of new options, specifically, mechanical interaction. There could be a feedback mechanism in the plant in response to these mechanical interactions and stresses (Yang et al., 2018). This information could help in recent finding that shows how plants respond to biotic and abiotic stresses. Interaction brought ecological consequences in several nutritional levels. There is several evidence that abiotic stress affects plant–microbe interaction as well as the interaction between insects (Yang et al., 2018), but its impact was less. Such interaction in the ecosystem of among plant and microbe brought climate change and nutritional challenges to survive. These are the key factors which linked plant, microbe, and nutrients in the environment for healthy living system, their better and moderate division leads to develop good agroeconomic impact on society that could came into by having extensive research in such challenges (Philippot, Raaijmakers, Lemanceau, & Van Der Putten, 2013).

However, microorganisms are used in the production food and fuel at industrial and domestic level. Yeast is used in the production of brewery via fermentation as well as of bread. *Lactic acid* bacteria are used for manufacturing yogurt, cheese, sour cream, buttermilk, and other dairy products (Bhowmik & Patil, 2018). Vinegar and other fermented foods including soy sauce, sauerkraut, dill pickles, olives, salami, cocoa, and black tea are produced at industrial level for daily use as a necessary part of our life (Lareen, Burton, & Schäfer, 2016). Yeast is also serving integral role in fermentation to transformation of carbohydrates to ethanol from corn and other vegetables. The microbes serve as the catalyst in the conversion of ferments into beer, wine, or gasoline and resourceful things. Microorganisms are also a source of waste food decomposition which is suitable for making the Earth's environment clean (Bhat et al., 2021; Garrett, Dendy, Frank, Rouse, & Travers, 2006). Microbial activities allow nutrients to be a part of natural system freshly but somehow their activity led to loss of fresh food and eatables (Dar et al., 2016; Dar, Kamili, Dar, Chishti, & Ahmad, 2015). Organic and material decomposition can be done by bacteria and fungi of various kinds (Brul & Coote, 1999). The decomposition and degradation of molecular or their decay in such a natural way is called biodegradation that makes our ecosystem healthy regardless of their harmful effects and diseases that are linked to virulent and allergic microbes (Azam & Malfatti, 2007). There are a number of products made synthetically that cannot be degraded and recycled by microbes causing air, soil, and atmospheric pollution (Dar, Bandh, & Kamili, 2011; Dar, Kamili, Nazir, Bandh, & Bhat, 2012). Microbial colonization helps to protect the environment by decomposing complex material letting nutrients and trace elements enter the ecosystem in biological and natural way (Arora et al., 2018).

5.2 Microbiology for soil health, environmental protection, and sustainable agriculture

Fungi and their effects on humans and our surrounding environment are well known, fungi represent one of five of the 16 kingdoms of living organisms, depending on the system used (Frey-Klett et al., 2011). Either way, fungi are indispensable and essential, the part of the biological world that is worthy of joint research by scientists and nonscientists. However, despite their importance to the ecosystem for human affairs, fungi are one of the most studied groups in biology (Bonfante & Anca, 2009). This is unfortunate because fungi are often very beautiful and affect everyone's life. The goal of the proposed chapter is to provide students with a familiar framework and context as well as fungi and their importance to other life forms, including humans (Stürmer, Bever, Schultz, & Bentivenga, 2021).

The most promising prospects for improving soil fertility and health are soil microbes (Dar, Bandh, Kamili, Nazir, & Bhat, 2013). Microorganisms and arbuscular mycorrhizas that enhance plant growth are employed to boost plant development and crop production under normal and stress circumstances (Lopes, Dias-Filho, & Gurgel, 2021). It can increase plant development through a range of different processes dependent on external stimuli and numerous physiological factors of plants. Plant growth regulators, the creation of various metabolites, and the conversion of atmospheric nitrogen into ammonia via direct and indirect ways are among the growth-promoting processes. Furthermore, it offers resistance to biological components (pathogens) via induced systemic resistance (ISR) and systemic acquired resistance (Vallad & Goodman, 2004). The interaction of plant microbes aids in the promotion of plant growth and disease management in a continually changing environment, as well as in achieving more sustainable agriculture without jeopardizing ecological system. Plant growth regulators keep favorable plant–microbe connections going, such as the one between plant growth-promoting rhizosphere bacteria and *arbuscular* fungi.

Soil health and production are maintained by microbial communities in the rhizosphere (Kumar & Verma, 2019). As a result, the extensive usage of plants that encourage the growth of rhizosphere bacteria aids in the achievement of long-term agricultural output under normal and stress situations. In this chapter, we would look at how microorganisms can help improve sustainable agricultural output. The application of insect microbial symbionts in the environment is a special field in environmental microbiology, which has broad prospects for future development; it may be a very important development from an economic and public perspective healthy point of view (Crotti et al., 2012). The scientific basis for the development of these fields is that they are not even universally known by all microbiologists, and they are not straightforward. A project that drafts concept papers on developments in this field can therefore be very important to actively help the formulation of regulations in this field.

The analysis of microbial populations in the environment is the main basic theme of environmental microbiology. DNA sequencing technology is constantly evolving and facts show that this strategy is a strong and necessary tool for revealing maize genetic diversity (Razin, Yogev, & Naot, 1998).

In recent years, the innovative next-generation sequencing technology, a microbe in the environment, has now been commonly used in various microbiological disciplines, notably in microbial taxonomy and ecology (Madsen, 2011). Prokaryotes, including bacteria and archaea, have entity implementation concepts for species and new technologies can be employed as a complete method for analyzing soil microbes. Next-generation metagenomics using sequencing should be helpful and broadly applicable to current microbiology and biotechnology (Xu, 2006). Considering these technologies and analyzing the information generated in these forms is critical for carrying out reliable and accurate risk/safety evaluations in this industry.

5.3 Future prospects of environmental microorganisms

The main objective of the OECD conference on environmental use of the environment microorganism is to overview the latest technology and its enlightenment to mankind (Singh, Pandey, & Singh, 2011). The biotechnology risk/safety assessment aims to provide an overview of the latest technology (Iranzo, Sainz-Pardo, Boluda, Sanchez, & Mormeneo, 2001). Use of microorganisms in the environment, with a focus on specific or anticipated developments in the field of genetically modified organisms, to support the development of the work program of the microbiology - Working Group Harmonized Biotechnology Regulatory Oversight (WG-HROB) (Fig. 5.1). (Kimman, Smit, & Klein, 2008).

There are examples of hypotheses stating that plant genotype has a massive impact on the makeup of the microbial community that is transmitted via the makeup of root exudates. There are examples of hypothesize that plant genotype has a strong influence on the composition of the microbial community, which is mediated by the composition of root exudates. Evaluate three plant genotypes with varying root exudate properties and consider three plant genotypes with different root exudate characteristics. These three plants were transplanted from the sterile system to ordinary soil at time 0 (above), where they're in immediate contact to almost the same group of microbial strains. At time 0 (above), these three plants were transplanted from the sterile system to ordinary soil, where they were close to the same set of microbial strains. At time *t* (below), these three plant genotypes attracted distinct microbial strains in their roots, resulting in varied economic expansion actions for various plants sizes at time *t* (below). These three plant genotypes have recruited unique microbial strains in their roots, giving different plant sizes and different growth-promoting effects (Jacoby, Peukert, Succurro, Koprivova, & Kopriva, 2017).

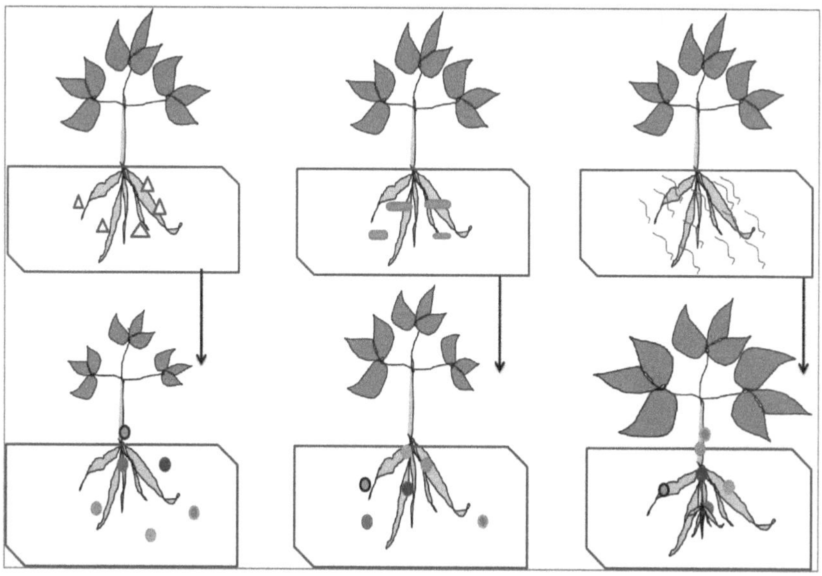

Figure 5.1 The natural variation of plants in root exudates leads to unique microbial communities.

5.4 Microbial pesticides

Microbial organisms are naturally occurring or genetically modified bacteria, fungi, algae, viruses, or protozoa. Microbial pest control agents can be efficient and used in place of chemical pesticides (Nawaz, Mabubu, & Hua, 2016). Microbial toxins are biotoxin compounds generated from microorganisms such as bacteria or fungi. The harmful effects of these microorganisms on target pests are species-specific (Usta, 2013). The activity of microbial insect pathogens is induced by penetrating the integument or gut of these insects and ensuring that their proliferation leads to the death of the host, such as insects (Rajagopal, 2009). According to research, infections create insecticidal poisons that play a crucial role in pathogenesis. The majority of toxins caused by bacterial diseases are already discovered as peptides, but their structure, toxicity, and specificity vary widely (Vey, Hoagland, & Butt, 2001).

5.5 Microorganisms' impending visions

Microbial insecticides' harmful effects usually solely target a specific insect or species. Because of this selectivity, several microbial insecticides would not directly impact beneficial insects (including natural enemies or pest parasites) in the treated area (Christou, Capell, Kohli, Gatehouse, & Gatehouse, 2006). Many microbial pesticides can be employed in conjunction with synthetic chemical pesticides if needed, because microbial products are neither inactivated nor harmed by conventional pesticide residue in most situations (Howarth, 1991). (Follow label instructions for any restrictions) Microorganisms stimulate root and plant growth by encouraging favorable soil microflora. Therefore, they help increase crop yields. Because single microbial pesticides are only toxic to specific species or groups of insects, each treatment controls only a subset of the pests found in fields and gardens, while other forms of pests will persist or may eventually cause damage in the treated area. Since typical pesticides act differently on various pests, they are susceptible to various regulations (Khater, 2012). Their position in agriculture and forestry could become more important in the future. Biological pesticides, clearly, have the power to contribute a part in the upcoming development of integrated pest management systems. We expect that in the near future, we would progressively embrace more appropriate means of dealing with biological pesticides (Usta, 2013). Chemical pesticides' short-term benefits will not lead to the destruction of biological pesticides.

5.6 Interconnections between plants and soil microorganisms

Plants are the part of a diverse ecosystem in the natural environment, which includes numerous microorganisms in the soil for a long period of time; people have recognized that certain microorganisms, such as *mycorrhizal* fungi or nitrogen-fixing symbiotic bacteria, contribute in plant growth by increasing mineral nutrition (Powlson et al., 2011). However, the entire family of plant-related microbes and their ability to replace synthetic agricultural inputs have just lately been found. In recent years, significant progress has been made in understanding the composition of the rhizosphere bacteria and their motivations (Fuhrer & Booker, 2003). There is ample proof that plants can affect the microbiome structure, with root exudates being the most likely source, and bacteria growing in the rhizosphere surrounding system. The mechanics and procedures of these exchanges are known, but the driving factor that causes the evolution of microbial populations is not (Huang et al., 2014). We are working to improve the connection between plants and root-related bacteria. Mineral nutrition, by summarizing recent data in numerous study domains, can help us better comprehend the molecular mechanisms that drive this occurrence (Schillaci, Gupta, Walker, & Roessner, 2019).

Plants are the part of a diverse ecosystem in the local habitat, including a range of microorganisms in the soil. People have long understood that some of these microorganisms (such as mycorrhizal fungi or nitrogen-fixing symbiotic bacteria) might help plants grow by enhancing mineral nutrition (Rashid et al., 2016). Nevertheless, recently, no one had found all of the microorganisms linked with plants and their ability to replace synthetic agricultural inputs. The composition and behavior of the rhizosphere microbiota have been considerably enhanced in recent years (Barrios, 2007). Plants might clearly alter the structure of the microbiome through root exudates, and bacteria have evolved a range of coping strategies to thrive in the rhizosphere environment (Berendsen, Pieterse, & Bakker, 2012).

Furthermore, the mechanisms underlying these interactions, as well as the dynamics that drive changes in the microbial community, are largely unknown (Blagodatskaya & Kuzyakov, 2008). Therefore, in this paper, we discuss the connection between plants and root-related bacteria to improve plant mineral nutrition, and we synthesize existing understanding in numerous study domains that may be aggregated to improve our understanding of what causes this phenomena and the molecular process.

5.7 Plant acquisition of nutrients: direct uptake from the soil

The relationship between microbes and plants is very important for the interaction of both microbes and plants in agriculture and crop productivity (Johansson, Paul, & Finlay, 2004). By combining topographical features in these different terrains, plants improve access and efficiency by transporting nutrients in the plant to the plant's organs (Pii et al., 2015 - soil and some nematodes, etc.). The ability to acquire other substances the nutrients that plants need for growth (Colombo, Palumbo, He, Pinton, & Cesco, 2014). Moreover, agriculturally beneficial microbes contribute directly to plant development (e.g., organic N_2 fixation, P solubilization, and phytohormone production), as well as indirectly (e.g., antimicrobial compound biosynthesis and elaboration of ISR, etc.) to crop development and fertilizer efficacy (Bargaz, Lyamlouli, Chtouki, Zeroual, & Dhiba, 2018). Several microbe-based biofertilizers have been employed to improve plant growth (Jacoby, Martyn, & Kopriva, 2018).

5.7.1 Mycorrhizal interactions with plants

Plants, in addition to symbiotic interactions with bacteria, could also form symbiotic partnerships with fungal species (Currie, 2001). According to latest projections of the prevalence of plant–mycorrhizal partnerships, around 80% of all plants develop some sort of mycorrhizal symbiosis, and numerous research works show that these interactions are millions of years old (Karandashov & Bucher, 2005; Vance, 2001). Mycorrhizae are classified into numerous types based on their structural shape, technique of colonization of plant tissue, and colonized host plants. Yet, there have been two major types that are widely considered to be the most frequent and thus the most ecologically significant (Ludwig-Müller, 2000). *Endomycorrhizae* are the fungi that form relationships with host plants by piercing the cell walls of cortical cells in plant roots. In contrary, *Ectomycorrizae* form a dense hyphae network across cortical cells but do not enter them (Gao, Delp, & Smith, 2001). *Endomycorrhizal* interactions are also most widespread between *arbuscular mycorrhizal* fungi (AMF, also known as *vesicular-arbuscular mycorrhiza*) and a range of grasses, herbs, trees, and shrubs (Abiala et al., 2013).

If phosphate is present in the soil, plants can absorb it directly through root phosphate transporters. Plants, on the other hand, are dependent on interaction with mycorrhizal fungus for phosphorus absorption under low phosphate environments (Smith, Jakobsen, Grønlund, & Smith, 2011). Substances secreted by the plants germinate mycorrhizal spores in the soil. Hyphae grow from the germinating spore and penetrate the plant root's epidermis. The hyphae branch and enter cortical cells into the root, where highly branched formations known as arbuscules grow (Grace, Cotsaftis, Tester, Smith, & Smith, 2009). *Ectomycorrhizal* fungi (EcM) create links with a wide range of tree species (birch, oak, spruce, pine, and fir), rendering these microorganisms an essential component of biodiversity in forest ecosystems, particularly temperate and boreal forests in the northern hemisphere (Cavard, Macdonald, Bergeron, & Chen, 2011). Though its molecular mechanisms behind the beginnings of EcM—plant relationships are not really understood, it is apparent that both symbiotic partners emit signaling molecules, resulting in fungal hyphae development toward compatible plant roots (Dearnaley, Perotto, & Selosse, 2016). The advantage for the fungal partners in these partnerships, as with bacterial symbioses, entails the transfer of photosynthetic carbohydrates from the plant to the microbe for metabolism and energy production (Martin, Uroz, & Barker, 2017). Furthermore, the nutritional benefit of such symbiotic partnerships for plants far exceeds the energetic cost of hosting microbes such as rhizobacteria, AMF, and EcM (Fig. 5.2) (Jansa, Finlay, Wallander, Smith, & Smith, 2011) .

Consequently, understanding of the processes influencing its richness and diversity is still quite limited. However, knowledge about the factors affecting its richness and diversity still lacks. We expect that changes in the mix of plant communities would impact root herbivores of insects since the quality of plant tissues of herbivores varies between plant species. Since the quality of plant tissues of herbivores is different among different plant species, we predict that changes in the composition of plant communities will affect the root herbivores of insects. Insect root herbivores, on the other hand, may have an impact on the AMF ecosystem by altering the consumption of photosynthetic fixed carbon. In the end, changes in the AMF community generated by insect root feeders might well be fed back into the plant community structure. On the other hand; insect root herbivores may affect the AMF community by changing the utilization of photosynthetic fixed carbon. In the final analysis, changes in the AMF community caused by insect root feeders may be fed back to the structure of the plant community. Exploration of biodiversity offers a good chance to investigate the interrelationships between plant diversity, root herbivores, and AMF over a gradient of

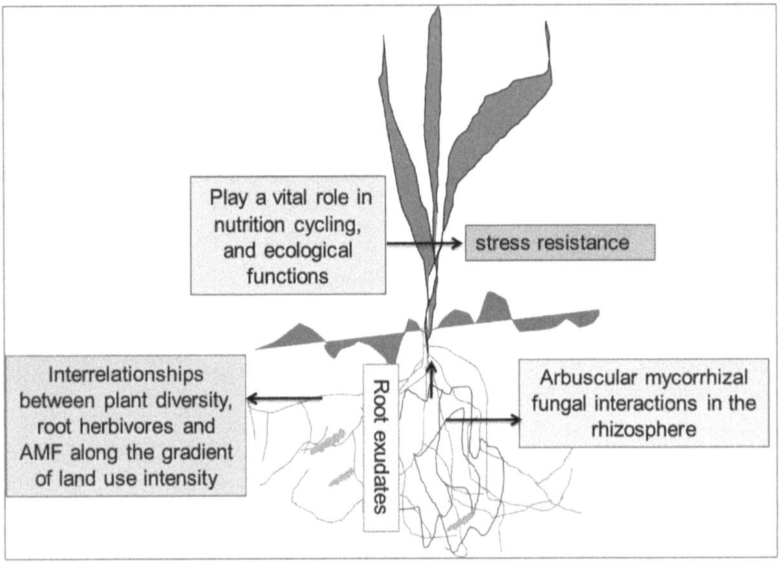

Figure 5.2 AMF have proven to be an important driver of plant diversity and productivity. *AMF*, arbuscular mycorrhizal fungi.

land use intensity. Biodiversity exploration provides an excellent opportunity to explore the interrelationships between plant diversity, root herbivores, and AMF along the gradient of land use intensity. The fundamental hypothesis is that (1) differences in plant diversity and soil nutrient levels caused by a gradient in land use intensity will affect the quantity and diversity of insect root herbivores, (2) insect root herbivores will alter AMF abundance, and (3) changes in AMF quantity and variety generated by insect roots and herbivores would be fed back into the plant community structure. The main hypothesis is (1) the difference in plant diversity and soil nutrient levels brought about by the gradient of land use intensity will affect the abundance and diversity of insect root herbivores, (2) insect root herbivore will affect the abundance of AMF, and (3) Changes in AMF abundance and diversity caused by insect roots and herbivores will feed back into the structure of plant communities, However, study is conducted on greenhouse experiments (hypotheses 1, 2 and 3) based on outdoor observations to better understand soil biology, how to interact with plants. In addition to field observations (hypothesis 1), study is conducted on greenhouse experiments (hypotheses 1, 2, and 3) to gain a deeper understanding of how these soil organisms interact with plants (Wang & Feng, 2021).

5.8 Conclusion and remark

Mineral fertilizers are used to deliver macronutrients in the majority of modern agricultural systems Nonsustainable fertilization techniques, on the other hand, have caused enormous disruptions in the Earth's biogeochemical cycle via mechanisms such as soil degradation, waterway eutrophication, and greenhouse gas emissions. This study helps us to understand how bacteria aid plant nutrition and how plants change the microbiome to maximize the nutritional benefit of this relationship. In this study, we highlight the present developments in methodologies for dissecting the interrelationships between plants and bacteria in mineral feeding, with a focus on plant and microbe metabolic capability. While analyzing this area, it is important to note that there is a substantial quantity of literature on how particular plants gain nutritional benefits through symbiosis with mycorrhizal and nodulation bacteria. These methods can improve our mechanistic understanding of plant–microorganism interactions and how to optimize these processes by lowering the usage of mineral fertilizers to drive plant feeding.

References

Abiala, M., Popoola, O., Olawuyi, O., Oyelude, J., Akanmu, A., Killani, A., ... Odebode, A. (2013). Harnessing the potentials of vesicular arbuscular mycorrhizal (VAM) fungi to plant growth - A review. *International Journal of Pure & Applied Sciences & Technology*, *14*(2).

Arora, N. K., Fatima, T., Mishra, I., Verma, M., Mishra, J., & Mishra, V. (2018). Environmental sustainability: challenges and viable solutions. *Environmental Sustainability*, *1*(4), 309–340.

Azam, F., & Malfatti, F. (2007). Microbial structuring of marine ecosystems. *Nature Reviews. Microbiology*, *5*(10), 782–791.

Bargaz, A., Lyamlouli, K., Chtouki, M., Zeroual, Y., & Dhiba, D. (2018). Soil microbial resources for improving fertilizers efficiency in an integrated plant nutrient management system. *Frontiers in Microbiology*, *9*, 1606.

Barrios, E. (2007). Soil biota, ecosystem services and land productivity. *Ecological Economics*, *64*(2), 269–285.

Berendsen, R. L., Pieterse, C. M., & Bakker, P. A. (2012). The rhizosphere microbiome and plant health. *Trends in Plant Science*, *17*(8), 478–486.

Bhat, R. A., Singh, D. V., Qadri, H., Dar, G. H., Dervash, M. A., Bhat, S. A., ... Yousaf, B. (2021). Vulnerability of municipal solid waste: An emerging threat to aquatic ecosystems. *Chemosphere*, *287*(2022), 132223. Available from https://doi.org/10.1016/j.chemosphere.2021.132223.

Bhowmik, S. N., & Patil, R. T. (2018). *Application of microbial biotechnology in food processing. Crop improvement through microbial biotechnology* (pp. 73–106). Elsevier.

Blagodatskaya, E, & Kuzyakov, Y. (2008). Mechanisms of real and apparent priming effects and their dependence on soil microbial biomass and community structure: Critical review. *Biology and Fertility of Soils*, *45*(2), 115–131.

Bonfante, P., & Anca, I.-A. (2009). Plants, mycorrhizal fungi, and bacteria: a network of interactions. *Annual Review of Microbiology, 63*, 363–383.
Brul, S., & Coote, P. (1999). Preservative agents in foods: Mode of action and microbial resistance mechanisms. *International Journal of Food Microbiology, 50*(1–2), 1–17.
Cavard, X., Macdonald, S. E., Bergeron, Y., & Chen, H. Y. (2011). Importance of mixed-woods for biodiversity conservation: Evidence for understory plants, songbirds, soil fauna, and ectomycorrhizae in northern forests. *Environmental Reviews, 19*(NA), 142–161.
Christou, P., Capell, T., Kohli, A., Gatehouse, J. A., & Gatehouse, A. M. (2006). Recent developments and future prospects in insect pest control in transgenic crops. *Trends in Plant Science, 11*(6), 302–308.
Colombo, C., Palumbo, G., He, J.-Z., Pinton, R., & Cesco, S. (2014). Review on iron availability in soil: Interaction of Fe minerals, plants, and microbes. *Journal of Soils and Sediments, 14*(3), 538–548.
Crotti, E., Balloi, A., Hamdi, C., Sansonno, L., Marzorati, M., Gonella, E., ... Alma, A. (2012). Microbial symbionts: A resource for the management of insect-related problems. *Microbial Biotechnology, 5*(3), 307–317.
Currie, C. R. (2001). A community of ants, fungi, and bacteria: A multilateral approach to studying symbiosis. *Annual Reviews in Microbiology, 55*(1), 357–380.
Dar, G. H., Bandh, S. A., & Kamili, A. N. (2011). Bacteriological analysis of soil from Yusmarg Health Resort of Kashmir Valley. *International Journal of Current Research., 3* (10), 53–56, 2011.
Dar, G. H., Bandh, S. A., Kamili, A. N., Nazir, R., & Bhat, R. A. (2013). Comparative analysis of different types of bacterial colonies from the soils of Yusmarg Forest, Kashmir VAlley India. *Ecologia Balkanica, 5*(1), 31–35.
Dar, G. H., Kamili, A. N., Chishti, M. Z., Dar, S. A., Tantry, T. A., & Ahmad, F. (2016). Characterization of *Aeromonas sobria* Isolated from fish Rohu (*Labeo rohita*) collected from polluted pond. *Journal of Bacteriology & Parasitology*, 7(3), 1–5. Available from https://doi.org/10.4172/2155-9597.1000273.
Dar, G. H., Kamili, A. N., Dar, S. A., Chishti, M. Z., & Ahmad, F. (2015). Detection and characterization of potentially pathogenic *Aeromonas sobria* isolated from fish *Hypophthalmichthys molitrix* (Cypriniformes: Cyprinidae). *Microbial Pathogenesis, 2015*. Available from https://doi.org/10.1016/j.micpath.2015.10.017.
Dar, G. H., Kamili, A. N., Nazir, R., Bandh, S. A., & Bhat, R. A. (2012). A preliminary study of colony forming units of bacteria from the soils of Yusmarg forest, Kashmir valley, India. *International Journal of Current Research, 4*(12), 467–472.
Dearnaley, J., Perotto, S., & Selosse, M. A. (2016). Structure and development of orchid mycorrhizas. *Molecular Mycorrhizal Symbiosis, 2016*, 63–86.
Frey-Klett, P., Burlinson, P., Deveau, A., Barret, M., Tarkka, M., & Sarniguet, A. (2011). Bacterial-fungal interactions: Hyphens between agricultural, clinical, environmental, and food microbiologists. *Microbiology and Molecular Biology Reviews: MMBR, 75*(4), 583.
Fuhrer, J., & Booker, F. (2003). Ecological issues related to ozone: Agricultural issues. *Environment International, 29*(2–3), 141–154.
Gao, L. L., Delp, G., & Smith, S. (2001). Colonization patterns in a mycorrhiza-defective mutant tomato vary with different arbuscular-mycorrhizal fungi. *New Phytologist, 151* (2), 477–491.
Garrett, K. A., Dendy, S. P., Frank, E. E., Rouse, M. N., & Travers, S. E. (2006). Climate change effects on plant disease: Genomes to ecosystems. *Annual Review of Phytopathology, 44*, 489–509.
Grace, E., Cotsaftis, O., Tester, M., Smith, F., & Smith, S. (2009). Arbuscular mycorrhizal inhibition of growth in barley cannot be attributed to extent of colonization, fungal phosphorus uptake or effects on expression of plant phosphate transporter genes. *New Phytologist, 181*(4), 938–949.

Howarth, F. G. (1991). Environmental impacts of classical biological control. *Annual Review of Entomology*, *36*(1), 485–509.

Huang, X.-F., Chaparro, J. M., Reardon, K. F., Zhang, R., Shen, Q., & Vivanco, J. M. (2014). Rhizosphere interactions: root exudates, microbes, and microbial communities. *Botany*, *92*(4), 267–275.

Iranzo, M., Sainz-Pardo, I., Boluda, R., Sanchez, J., & Mormeneo, S. (2001). The use of microorganisms in environmental remediation. *Annals of Microbiology*, *51*(2), 135–144.

Jacoby, R., Peukert, M., Succurro, A., Koprivova, A., & Kopriva, S. (2017). The role of soil microorganisms in plant mineral nutrition—current knowledge and future directions. *Frontiers in Plant Science*, *8*, 1617.

Jacoby, R. P., Martyn, A., & Kopriva, S. (2018). Exometabolomic profiling of bacterial strains as cultivated using *Arabidopsis* root extract as the sole carbon source. *Molecular Plant-Microbe Interactions*, *31*(8), 803–813.

Jansa, J., Finlay, R., Wallander, H., Smith, F. A., & Smith, S. E. (2011). *Role of mycorrhizal symbioses in phosphorus cycling. Phosphorus in action* (pp. 137–168). Springer.

Johansson, J. F., Paul, L. R., & Finlay, R. D. (2004). Microbial interactions in the mycorrhizosphere and their significance for sustainable agriculture. *FEMS Microbiology Ecology*, *48*(1), 1–13.

Karandashov, V., & Bucher, M. (2005). Symbiotic phosphate transport in arbuscular mycorrhizas. *Trends in Plant Science*, *10*(1), 22–29.

Khater, H. F. (2012). Prospects of botanical biopesticides in insect pest management. *Pharmacologia*, *3*(12), 641–656.

Kimman, T. G., Smit, E., & Klein, M. R. (2008). Evidence-based biosafety: A review of the principles and effectiveness of microbiological containment measures. *Clinical Microbiology Reviews*, *21*(3), 403.

Kumar, A., & Verma, J. P. (2019). *The role of microbes to improve crop productivity and soil health. Ecological wisdom inspired restoration engineering* (pp. 249–265). Springer.

Lareen, A., Burton, F., & Schäfer, P. (2016). Plant root-microbe communication in shaping root microbiomes. *Plant Molecular Biology*, *90*(6), 575–587.

Lopes, M. J. d S., Dias-Filho, M. B., & Gurgel, E. S. C. (2021). Successful plant growth-promoting microbes: inoculation methods and abiotic factors. *Frontiers in Sustainable Food Systems*, *5*, 48.

Ludwig-Müller, J. (2000). *Hormonal balance in plants during colonization by mycorrhizal fungi. Arbuscular mycorrhizas: Physiology and function* (pp. 263–285). Springer.

Madsen, E. L. (2011). Microorganisms and their roles in fundamental biogeochemical cycles. *Current Opinion in Biotechnology*, *22*(3), 456–464.

Martin, F. M., Uroz, S., & Barker, D. G. (2017). Ancestral alliances: Plant mutualistic symbioses with fungi and bacteria. *Science (New York, N.Y.)*, *356*(6340).

Nawaz, M., Mabubu, J. I., & Hua, H. (2016). Current status and advancement of biopesticides: microbial and botanical pesticides. *Journal of Entomology and Zoology Studies*, *4*(2), 241–246.

Neary, D. G., Klopatek, C. C., DeBano, L. F., & Ffolliott, P. F. (1999). Fire effects on belowground sustainability: A review and synthesis. *Forest Ecology and Management*, *122* (1–2), 51–71.

Philippot, L., Raaijmakers, J. M., Lemanceau, P., & Van Der Putten, W. H. (2013). Going back to the roots: The microbial ecology of the rhizosphere. *Nature Reviews Microbiology*, *11*(11), 789–799.

Pii, Y., Mimmo, T., Tomasi, N., Terzano, R., Cesco, S., & Crecchio, C. (2015). Microbial interactions in the rhizosphere: Beneficial influences of plant growth-promoting rhizobacteria on nutrient acquisition process. A review. *Biology and Fertility of Soils*, *51*(4), 403–415.

Powlson, D. S., Gregory, P. J., Whalley, W. R., Quinton, J. N., Hopkins, D. W., Whitmore, A. P., ... Goulding, K. W. (2011). Soil management in relation to sustainable agriculture and ecosystem services. *Food Policy*, *36*, S72–S87.

Rajagopal, R. (2009). Beneficial interactions between insects and gut bacteria. *Indian Journal of Microbiology*, *49*(2), 114–119.

Rashid, M. I., Mujawar, L. H., Shahzad, T., Almeelbi, T., Ismail, I. M., & Oves, M. (2016). Bacteria and fungi can contribute to nutrients bioavailability and aggregate formation in degraded soils. *Microbiological Research*, *183*, 26–41.

Razin, S., Yogev, D., & Naot, Y. (1998). Molecular biology and pathogenicity of mycoplasmas. *Microbiology and Molecular Biology Reviews*, *62*(4), 1094.

Schillaci, M., Gupta, S., Walker, R., & Roessner, U. (2019). *The role of plant growth-promoting bacteria in the growth of cereals under abiotic stresses. Root Biology-Growth, Physiology, and Functions* (pp. 1–21). IntechOpen.

Singh, J. S., Pandey, V. C., & Singh, D. P. (2011). Efficient soil microorganisms: a new dimension for sustainable agriculture and environmental development. *Agriculture, Ecosystems & Environment*, *140*(3–4), 339–353.

Smith, S. E., Jakobsen, I., Grønlund, M., & Smith, F. A. (2011). Roles of arbuscular mycorrhizas in plant phosphorus nutrition: Interactions between pathways of phosphorus uptake in arbuscular mycorrhizal roots have important implications for understanding and manipulating plant phosphorus acquisition. *Plant Physiology*, *156*(3), 1050–1057.

Stürmer, S. L., Bever, J. D., Schultz, P. A., & Bentivenga, S. P. (2021). Celebrating INVAM: 35 years of the largest living culture collection of arbuscular mycorrhizal fungi. *Mycorrhiza*, *31*(1), 117–126.

Trichopoulou, A., Vasilopoulou, E., Hollman, P., Chamalides, C., Foufa, E., Kaloudis, T., ... Poulima, E. (2000). Nutritional composition and flavonoid content of edible wild greens and green pies: A potential rich source of antioxidant nutrients in the Mediterranean diet. *Food Chemistry*, *70*(3), 319–323.

Usta, C. (2013). *Microorganisms in biological pest control—A review (bacterial toxin application and effect of environmental factors). Current progress in biological research* (pp. 287–317). Rijeka: IntechOpen.

Vallad, G. E., & Goodman, R. M. (2004). Systemic acquired resistance and induced systemic resistance in conventional agriculture. *Crop Science*, *44*(6), 1920–1934.

Vance, C. (2001). Symbiotic nitrogen fixation and phosphorus acquisition. Plant nutrition in a world of declining renewable resources. *Plant Physiology*.

Vey, A., Hoagland, R. E., & Butt, T. M. (2001). *Toxic metabolites of fungal biocontrol agents*, . *Fungi as biocontrol agents: progress, problems and potential* (1, pp. 311–346). .

Wang, F., & Feng, G. (2021). *Arbuscular mycorrhizal fungi interactions in the rhizosphere. Rhizosphere biology: interactions between microbes and plants* (pp. 217–235). Springer.

Waraich, E. A., Ahmad, R., & Ashraf, M. (2011). Role of mineral nutrition in alleviation of drought stress in plants. *Australian Journal of Crop Science*, *5*(6), 764–777.

Xu, J. (2006). Invited review: Microbial ecology in the age of genomics and metagenomics: Concepts, tools, and recent advances. *Molecular Ecology*, *15*(7), 1713–1731.

Yang, L., Wen, K.-S., Ruan, X., Zhao, Y.-X., Wei, F., & Wang, Q. (2018). Response of plant secondary metabolites to environmental factors. *Molecules (Basel, Switzerland)*, *23*(4), 762.

CHAPTER 6

Microbial consortia for pollution remediation—Success stories

Mir Zahoor Gul, Karuna Rupula and Sashidhar Rao Beedu
Department of Biochemistry, University College of Science, Osmania University, Hyderabad, Telangana, India

6.1 Introduction

Microorganisms are the earliest life forms that originated on the Earth some 3 billion years ago; however, researchers only began to study them in the 16th century. The microorganisms are ubiquitous and taxonomically comprise of the bacteria, fungi, viruses, protozoa, nematodes, and algae which make up the majority of biomass of the Earth ranking next to the higher organisms. According to reports, each gram of soil contains around 1 to 10 million microorganisms, while plants and animals also house several number of these organisms. The microorganisms are an essential component of the biological cycle and help to maintain a balance of all other living things on the Earth (Luo, Meng, & Gu, 2017; Smith et al., 2015). It has been well established that microorganisms perform a critical role in biogeochemical cycles [nitrogen (N), carbon (C), sulfur (S), phosphorous (P), and other metals, have been recognized to contribute to organic soil materials and provide necessary nutrients that are vital for the survival of animals and plants on the Earth (Dar, Bandh, & Kamili, 2011; Luo et al., 2017). Microorganisms with diverse physiological, ecological, biochemical, genomic, and meta-genomic bases such as methane anaerobic oxidation, photosynthesis, and biogeochemical cycles have recently been integrated into the bioremediation process (Bae et al., 2018; Varjani, 2017). Environmental biotechnology has progressed to the point where we can understand the mechanistic characterization and applications of biological systems that are relatively simple, easy, and less complicated, as well as naturally occurring diverse microbial communities that can survive in air, soil, and water habitats (Dar et al., 2012; Dar, Bandh, Kamili, Nazir, & Bhat, 2013).

Microbial Consortium and Biotransformation for Pollution Decontamination
DOI: https://doi.org/10.1016/B978-0-323-91893-0.00014-6

6.2 Bioremediation

A large percentage of potentially hazardous materials have been unleashed into the environment by multiple industries, anthropogenic activities and/or the agricultural sector in recent times. Pesticides, polychlorinated biphenyls (PCBs), polycyclic aromatic hydrocarbons (PAHs), nitrogen and phosphorus complexes, alkanes, explosives, and industrial dyestuff, as well as poisonous heavy metals, are among the most common pollutants that have harmed natural environment. Surface and ground water pollution and soil contamination by these pernicious pollutants is a severe environmental problem in a number of countries (Bilal, Adeel, Rasheed, Zhao, & Iqbal, 2019; Huang, Zhan, Bhatt, & Chen, 2019; Kästner & Miltner, 2016; Lin et al., 2020; Schwarzenbach, 2006; Kurwadkar, 2019). Moreover, complex pollution comprising multiple pollutants as well as toxic metals tends to accumulate in the environment, making it a huge concern for toxicological risk assessments of polluted field sites as well as the viability of bioremediation strategies to remediate and restore the polluted water and soil ecosystems (Mishra et al., 2020). Ecological restoration became one of the pressing priorities in environmental biotechnology, and several paradigms are driven by complex techniques that take into account multiple biological levels. To repair and restore degraded environments, physical, chemical, and biological processes involving the use of biological agents have been developed. The biological approaches have significant advantages over the chemical and physical remedial processes in *in-situ* due to their ecofriendly potentials and economical aspects. These approaches can be applied to disseminated toxins, and reduce unsafe environmental operations (Alexander, 1999; Azubuike, Chikere, & Okpokwasili, 2016). However, in *ex situ* techniques, the treatment and removal of pollutants require different and specific facility (Hamer, 1993).

Bioremediation is widely recognized as a softer and cleaner approach than conventional methods for cleaning up polluted systems in a way that safeguards both the environment and human health, whether as a natural attenuation process or as a controlled approach (Abatenh, Gizaw, Tsegaye, & Wassie, 2017; Kapahi & Sachdeva, 2019). A well-defined biodegradation technique is the eradication, reduction, or alteration of highly polluted materials under the influence of biological entities. Plants, microorganisms, enzymes, and plant–microorganism interactions are the major agents involved in bioremediation processes (Abatenh et al., 2017; Alexander, 1999; Azubuike et al., 2016). The broad diversity of microbial colonies is significant for the critical functions they play in their natural settings. Microorganisms have a higher metabolic rate and are capable of degrading diverse pollutants efficiently in both anaerobic and aerobic conditions in a variety of habitats. The utilization of these organisms in complex degrading processes brings with it several hurdles; however, it also has several advantages, both economically and environmentally. (Das & Chandran, 2011; Haritash & Kaushik, 2009; Prasad & Prasad, 2012; Varjani, 2017). Bacteria, fungi, yeast, and algae, for example, have the natural ability to break down and convert a wide range of chemicals prevalent in contaminated habitats (Balaji, Arulazhagan, & Ebenezer, 2014; Miranda et al., 2017; Wrede et al., 2014). When these microbes are exposed to adverse environmental conditions such as depletion of micronutrients, change in light, temperature, pressure, salinity, etc., they have a unique ability to acclimatize such situations by implementing various approaches such as the synthesis of biosurfactants/bioactive compounds and formation of biofilms. Furthermore, native bacteria can assist in the development of long-term bioremediation ecotechnologies. The microbial bioremediation process benefit from their small genome sizes, cellular simplicity, and rapid replication cycles. From a molecular biology standpoint, the microorganisms utilized in bioremediation processes must have a genotype that is impervious to the contaminant to be degraded (Patowary, Patowary, Kalita, & Deka, 2018; Singh, Paul, & Jain, 2006). Microorganisms acclimatize to their surroundings by altering their gene expression patterns, growth rates, enzymatic activity, and physiological adaptations. During the process of bioremediation, microorganisms that perform the breakdown of contaminants utilize these waste products as a means of energy and nourishment (Bhat et al., 2021; Das & Chandran, 2011; Johnsen, Wick, & Harms, 2005; Varjani, 2017). All through the bioremediation process, externally supplied nutrients such as nitrogen and carbon sources; electron acceptor (oxygen); substrates including toluene, methane, and phenol; or other microbes with specific catalytic potential boost the activity of these microorganisms. (Baldwin et al., 2008; Ma et al., 2007). They also have degradative enzymes that can cleave the original bonds of hazardous pollutants and transform them into innocuous inorganic forms (biotransformation) through the mineralization process. Moreover, enzymes accomplish certain tasks for which no effective chemical transformations could be designed. (Bhatt, Bhatt, Huang, Lin, & Chen, 2020; Bhatt, Huang, Zhang, Sharma, & Chen, 2020; Bhatt, Zhang, et al., 2020; Blasco, Aparicio, & Castillo, 1995; Gu et al., 2020). These characteristics make them ideal prospects for bioremediation (Chen et al., 2019). Using microbes to convert hazardous pollutants into a less pernicious or benign form is a sagacious approach. The aim of this chapter is to provide insights and highlight success stories in the use of microbial consortia to eliminate or degrade pollutants, as well as to recognize the value of different microbial consortia assessments.

6.3 Microbial consortia—a multispecialized biological system for bioremediation

Use of purified microbial species to explore certain bioprocesses is an effective strategy, as they are associated with many benefits such as degradation of pollutants, the individual microorganisms in the synchrotrophic relationship exhibit specific competences, functions and roles that would act as a key coperative links in the bioconversion cycle (Janbandhu & Fulekar, 2011; Lee, Jeong, Hur, Ko, & Jeon, 2018). Microorganism communities are ubiquitous and are implicated in processes that are pivotal to human beings, such as food digestion and environmental cleanup (Dar, Kamili, Dar, Chishti, & Ahmad, 2015; Dar et al., 2016). Microbes in these kinds of communities typically collaborate to fully exploit the available resources. The term "consortia" refers to a group of microorganisms that coexist in a specific habitat and cooperate based on their metabolic diversity, whereas "mixed culture" refers to a group of microbes that dwell but do not collaborate metabolically. Participants in various consortia are believed to not only proliferate at the highest possible pace but also regulate their numbers to fend off metabolic competitors (Taillefumier, Posfai, Meir, & Wingreen, 2017). Consortia can carry out complex functions than individual species because they are generally more resilient to environmental variations and convolutions. Microbial consortia integrate the metabolic properties of several different organisms, enhancing the performance and efficacy of intricate substrate biotransformation operations along with their ability to endure harsh conditions. Consortia have significantly wider applicability than axenic and mixed cultures because of their metabolic diversity and resistance to invasion (Jia et al., 2016). Microorganisms that constitute a consortium might indirectly improve the biodegradation process by exchanging signal molecules or metabolites that trigger cometabolic degradation (Chen & Aitken, 1999) or metabolic cross-feeding (Hu et al., 2020; Sørensen, Ronen, & Aamand, 2002). Bacterial species in a natural consortium can interchange particular signaling molecules and regulate intrapopulation events spanning from biofilm formation to antibiotic resistance (Davies, 1998; Greenberg, 2003) to virulence expression (Clarke & Sperandio, 2005;

Passador, Cook, Gambello, Rust, & Iglewski, 1993), exchanging signaling molecules of oligopeptide autoinducers in Gram-positive and acyl-HSL in Gram-negative species (Kaper & Sperandio, 2005). Physiological attributes, spatial assembly, adaptive characteristics, and the presence of natural or engineered selection factors can all affect whether participants of microbial consortia compliment or inhibit one another. (Bascompte, 2010; Fukami, 2015; Rohr, Saavedra, & Bascompte, 2014). Biofilms, microbial mats, and planktonic microalgae–bacteria ensembles are some of the prime examples of microbial consortia that exist at diverse stages of population and metabolic intricacy (Prieto-Barajas, Valencia-Cantero, & Santoyo, 2018; Ramanan, Kim, Cho, Oh, & Kim, 2016; Tan, Lee, Burmølle, Kjelleberg, & Rice, 2017). The members of microbial consortia would stand out pure cultures in their ability to execute complicated tasks independently owing to the high complex behavior among various members interacting together in a consortium (Scott & Hasty, 2016). Microbial consortia and defined cocultures have great potential to disintegrate hazardous pollutants such as PAHs, PCBs, synthetic dyes, and few others and might outperform single strains. For instance, to disintegrate 1,2,3-trichloropropane (TCP), a combination of protein and metabolic manipulation can be employed (Janssen & Stucki, 2020). Synthetic dyes and other organic pollutants have also been degraded using genetically modified bacteria (Kumar et al., 2020). Likewise, engineered *Pseudomonas* species have been found to degrade polyvinyl chloride, polypropene, polyethylene, polystyrene, and other plastic polymers, but at varying speeds (Wilkes & Aristilde, 2017). Plasmid-borne *Pseudomonas putida* strains have been found to disintegrate p-nitrophenol efficiently (Dvořák, Nikel, Damborský, & de Lorenzo, 2017)

Microbial consortia have created a completely new realm of prospects in synthetic biology in terms of system intricacy and functionality. First, such intricate consortia provide a unique microenvironment for microorganism, which may result in the stimulation of silent metabolic pathways that are not espoused under "normal" culture conditions, resulting in the finding of novel and unique chemical compounds, making microbial consortia an appealing prospect, especially in the development of new therapeutics (Netzker et al., 2018). Second, the microbial consortia use a "split of labor" strategy, which allows them to perform complicated functions in many and simpler steps, distributing the burden across the population and allowing for increased effectiveness and more intricate behavior than pure cultures (Brenner, You, & Arnold, 2008; Hays, Patrick, Ziesack, Oxman, & Silver, 2015; Zhou, Qiao, Edgar, & Stephanopoulos, 2015).

Monoculture requires a sequence of molecular engineering regulators, ribosomal binding sites, terminators, and vectors, among other things, but the modularized approach may be simple to optimize by varying the ratio of constituent species (Roell et al., 2019). Microorganisms, according to the economic concept of "the division of labor," will utilize existing resources efficiently if chemical tasks are shared between them (Taillefumier et al., 2017). The activities of multiple interacting microorganisms give the consortium sturdiness and ensure stability for its members over time, even when the environment is changing. Third, unlike monocultures, such microbial communities are made up of various functioning microorganisms that are more resistant to environmental adversities and invasion by other species as well as have greater metabolic capabilities (Burmølle et al., 2006; McCarty & Ledesma-Amaro, 2019). Due to the complexity of metabolic processes, they can even effectively withstand conditions of nutritional scarcity. (Fay, 1992). However, when nutrients are scarce, the most abundant organisms in a population might not be the most metabolically active. If a minority member is metabolically active, on which the entire consortium depends on for survival, in such scenario the minority member of the consortia may become the most dominant (LaPara, Zakharova, Nakatsu, & Konopka, 2002).

Various genetic engineering approaches have been designed and optimized to improve the desired attributes of microbial species to attain endurance in microbial pollutant disintegration (Li et al., 2020; Pieper & Reineke, 2000; Schmitz, Hartline, & Zhang, 2017). The catabolic pathways for the cleanup of persistent contaminants have been developed using genetic engineering tools. Microbial bioremediation processes are of three major types. The type 1 bioremediation process involves the use of endogenous microbial species to eradicate pollutants from the ecosystem, type 2 refers to the use of more advanced remediation strategies, while type 3 deals with production of microbial strains using cutting-edge biotechnological approaches that also include the degradation of wide range of pollutants (Dvořák et al., 2017). Since engineered strains are more effective pollution degraders than wild-type populations, microbial engineering has become a popular strategy (Jeandet et al., 2018). Engineering of whole cells, degradation pathways, and enzymatic catalysts to break down pollutants is the central objective of microbial engineering for bioremediation. In the recent past, various microbial species (e.g., *Achromobacter, Alcaligenes, Burkholderia Comamnonas, Ralstonia, Rhodococcus, Dehalococcoides, Pseudomonas*, and *Sphingomonas*) have been engineered to speed up the breakdown of hazardous pollutants (Bilal & Iqbal, 2020). These genetically modified "superbugs" offer an alternative, efficient, and environmentally beneficial method for the complete mineralization of these hazardous pollutants (Ramos et al., 2011).

6.4 Microbial consortia and degradation of pollutants

Synthetic biology advances and innovations are a stride towards the development of microbial consortia (Brenner et al., 2008). Biological strategies utilizing specialized microbes to break down a variety of harmful substances offer an environment-friendly approach of treating environmental impairment in terms of achieving sustainable and green society (Azubuike et al., 2016). The efficiency and advantages of engineered microbial consortium are always superior and exhibit robustness in the degradation of pollutants when compared to microbial monocultures. The metabolic burden of degradation of pollutants is divided among microbial species, and each organism in the consortia only degrades for a brief period (Shong, Jimenez Diaz, & Collins, 2012). Microbial consortia have been extensively used for a variety of vital processes, including the breakdown of toxic pollutants, owing to their high adaptability and diverse substrate spectrum (Brenner et al., 2008; Jia et al., 2016; Varjani, 2017; Villegas, Martínez, Rodríguez, & Amoroso, 2014). *Actinobacteria*, *Clostridium*, *Escherichia coli*, *Lactobacilli*, *Pseudomonas*, and *Salmonella* among others have been discovered to be suitable for xenobiotic bioremediation (Feng & Lim, 2005). However, single-strain efficacy for degradation and remediation of pollutants is still poor and limited, as stated earlier. (Villegas et al., 2014). Even though indigenous microbial bioremediation has been established, microbial consortia have become increasingly popular due to their inherent diverse functions, endurance, and adaptability. The graphical representation of the biodegradation potential of microbial consortia is presented in Fig. 6.1. The primary microorganisms in these consortia are mostly bacteria, fungus, yeast, and algae. They demonstrate responsiveness in the context of increasingly changing environmental conditions (light shifting patterns, pH, sea-level rise, temperature, terrestrial inputs, tropical storms, and so on). These characteristics render them ideal organisms for bioremediation process.

6.4.1 Degradation of petroleum components

Diverse microorganisms have been investigated extensively for their role in the degradation of petroleum products resulting from oil refineries (Das & Chandran, 2011; Varjani & Upasani, 2017; Varjani, 2017). There are

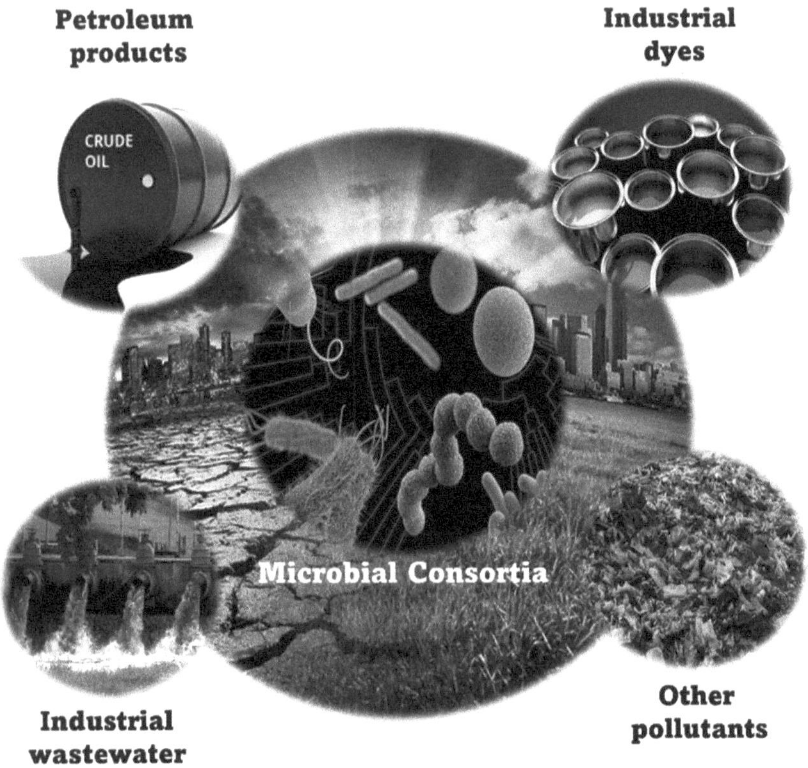

Figure 6.1 The graphical representation of the biodegradation potential of microbial consortia.

significant aboriginal microorganism species that are capable of substantial oil biodegradation over a wide range of environments, provided that the environmental factors favor these bacteria to grow and facilitate the hydrocarbon metabolic activity. Second, the competence to use hydrocarbons is spread over a large range of microbial populations. This population can be found in natural habitats and metabolizes different hydrocarbons either independently or synergistically. Multiple microorganisms that degrade petroleum hydrocarbons have now been recognized in soils and oceanic habitats, the two main habitats polluted with crude oil. The metabolic machinery of these microbes enables them to use petroleum wastes as a source of energy and carbon. The heterotrophs that are responsible for the degradation of petroleum products can utilize anaerobic or aerobic metabolic pathways to break down these hydrocarbons (i.e., they use sulfate or nitrate as an alternative electron acceptor). It has been indicated that in the soil, freshwater, and marine habitats, mixed populations as a microbial consortium with relatively broad enzymatic activities are essential to metabolize complex mixtures of hydrocarbons such as oil products (Das & Chandran, 2011; Varjani, 2017). Over the years, many researchers have used this idea to evaluate the potential of biodegradation of petroleum products by varying percentages of bacteria and yeast as a microbial consortium. In a scientific study, it was demonstrated that 87% of the diesel could be degraded by combining *Aeromonas hydrophila* Mr4 with *Yarrowia lipolytica* EH 56. In the same, 90% of the diesel could be degraded in 7 days by a consortium of *Xanthomonas maltophilia* MRP7 and *Candida maltosa* EH15 (Kaczorek, Cieślak, Bielicka-Daszkiewicz, & Olszanowski, 2013). The integration of rhamnolipids throughout the process of degradation to these microbial combinations could minimize the time required for diesel oil removal from the polluted waters (Kaczorek et al., 2013). While working on the diesel oil degraders, researchers were able to establish that cell immobilization during biofilm development can also influence the degrading efficiency of bacterial consortia (èawniczak, Kaczorek, & Olszanowski, 2011). The oil degradation efficiencies of immobilized *Bacillus* sp. E3 cells with algal materials (seaweed, *Enteromorpha*, and kelp deposits) were investigated and observed to be 65% higher post 21 days of treatment (Zhang et al., 2018). A fungus–bacteria coculture has affected the breakdown of oil hydrocarbons by mixed biosurfactants (Li & Li, 2011). The resulting oily sludge or trash is reported to be very resilient and forms intricate emulsions when surfactant additives are supplied. As a result, demulsification activity is required to separate the leftover oil from the liquid phase. In this regard, the potential for demulsification of surfactants produced by bacteria (*Bacillus* sp., *Pseudomonas aeruginosa*, and *Pseudomonas cepacia*) and yeasts (*Candida lipolytica*, *Candida guilliermondii*, and *Candida sphaerica*) has been investigated. Bacterial surfactants were able to recover approximately 65% of the brackish water emulsified with oil, whereas yeast-produced surfactants only retrieved 35%–40% (Rochae Silva et al., 2017). A scientific study evaluated four consortia of microorganisms that have a degradation potential of up to 92% of the total petroleum hydrocarbons (TPH) in 60 days and functional analysis revealed alkB, cat23, ndoB, nidA1, and xylA as the degrading genes (Omrani, Spini, Puglisi, & Saidane, 2018). Using gamma-ray irradiation, some researchers investigated the mutant strains of *P. putida* for effective deprivation of crude oil and it has been observed that degrading activity was 46.3% higher when compared to the parent strain

(Tahseen, Arslan, Iqbal, Khalid, & Afzal, 2019). *Gordonia* sp. strain 1D, a thermotolerant strain that utilizes crude oil up to 50°C, has been successfully identified from the polluted areas of oil refinery. Subsequently, the gyrB, a gene responsible for the degradation, was also isolated (Delegan et al., 2019). Microbial consortium, comprised *Bacillus subtilis* DM-04 and *P. aeruginosa* M and NM strains, were infused into the soils contaminated with petroleum wastes, and the remediation process was therefore investigated on an experimental basis. The bacterial consortium demonstrated a significant decrease in TPH levels at the end of the investigation. (Mukherjee & Bordoloi, 2011). In another scientific study, polycultures of *Serratia* sp. NB2, *Stenotrophomonas* sp. NB3, and *Arthrobacter* sp. NB1 efficiently degraded nitrobenzene than monocultures (Jin et al., 2012). Active ^{13}C-labeled phenanthrene degraders have been recently identified using the stable isotope probing approach. They discovered dominant fungi and bacteria that can degrade a variety of petroleum-based chemicals (Schwarz et al., 2018). A *P. aeruginosa* DAB recombinant strain that can degrade hydrocarbons by synthesizing rhamnolipids has been developed. (He et al., 2017). The fungal strains, *Mycelia sterilia*, *Mucor circinelloides*, *Cadophora malorum*, and *Trichoderma viride*, with the ability to break down petroleum oil sludge have been successfully isolated and investigated (Shchemelinina et al., 2017). These scientific studies indicate that *in situ* bioremediation of petroleum wastes by microbial cocultures is technically viable and that it may be advantageous in the remediation of soils polluted with crude oil. However, prudent attention must be given to the improvement of consortia as they must be capable of sustaining and biodegrading in a realistically contaminated environment.

6.4.2 Remediation of wastewater

The quantity and quality of generated waste being chucked into natural waters is a serious concern in today's world (Oh et al., 2010). To remediate primary contaminants from wastewaters, multiple approaches have been practiced. Among them, biological treatment systems, for example, rely on the use of microorganisms; microbes' unique abilities to disintegrate organic compounds, remove nutrients, and convert deleterious pollutants into harmless end products qualify them as invaluable waste removal agents. Nutrient optmization is one of the strategies in wastewater treatment and can be achieved by removal of excess amounts of elements present such as phosphorus (P), nitrogen (N), toxins, and other contaminants (Mujtaba & Lee, 2016). The existing biological processes include anaerobic digestion, nitrification, and denitrification stages; however, multiple cycles of these three stages are necessary to accomplish the accepted nutrient levels, and each stage necessitates multiple tanks and internal activated sludge recycles, leading to an overall increment in operation costs, complexity, and fuel input (Gonçalves, Pires, & Simões, 2017). Microbial consortia offer an economically viable and efficient bioremediation strategy to detoxify the contaminants from wastewaters. Microalgae and bacteria have shown enormous potential to study and explore various mechanisms in hostile habitats. Microalgae cultivation in conjunction with sewage treatment has mostly been explored for a variety of industrial and residential wastewater discharges, which include brewery, textile, and pharmaceutical wastes; slaughterhouse wastes; palm oil mill effluents; toxic metal-containing effluents; and agro-industrial effluents (Shahid et al., 2020). Since they maintain mutualism with heterotrophic bacteria, algae and bacteria consortia offer an efficient approach and play an important role in bioremediation. Microalgae–bacteria biofilms provide complimentary metabolites from algal biomass for bacteria proliferation and influence remediation. *Pseudomonas* and *Staphylococcus* cocultures were shown to be very effective at eliminating phenol than the pure cultures (Senthilvelan, Kanagaraj, Panda, & Mandal, 2014). Likewise, *Klebsiella pneumoniae* and *Ralstonia* sp. microbial consortia demonstrated improved resistance to thiocyanate, with a degradation rate of 500 mg/L/d (Chaudhari & Kodam, 2010). Integrating microalgae and bacteria together can be complimentary and synergic for improved pollutant degradation performance. On the one hand, algal photosynthesis provides O_2, which is a critical electron acceptor for pollutant-degrading heterotrophic bacteria in their degradation of organic matter. Bacteria, on the other hand, contribute CO_2 and other stimulants to assist their photoautotrophic partners in growth and development (Bordel, Guieysse, & Muñoz, 2009; Gonçalves et al., 2017; Subashchandrabose, Ramakrishnan, Megharaj, Venkateswarlu, & Naidu, 2011). This biological process's self-oxygenation, which has already been validated, can be used to successfully clean up a wide range of contaminants (Subashchandrabose et al., 2011). Microalgae can also shield bacteria from adverse environmental factors by acting as a safe harbor. (Byappanahalli et al., 2009; Mandal, Singh, & Patel, 2011; Unnithan, Unc, & Smith, 2014). Individual strains or species may have a synergistic effect when blended with other strains (like algae–bacteria), and microbial communities can execute activities which are challenging or even not possible for pure strains or cultures to perform (Escobar, Brenner, Whitmore, Kenney, & Curtis, 2008). The abilities to tolerate nutritional scarcity and adverse environmental conditions, share metabolites, and resist invasion by other species are some of the unique attributes of these cocultures. Furthermore, toxic compounds are enriched on the cell walls of microalgae, resulting in a surface zone that facilitates microbial degradation (Luo et al., 2014). The potential of cyanobacteria/microalgae—bacterial consortia as a self-sustaining system for (1) pollution detoxification and nutrient removal, (2) synthesis of high-value metabolites/by-products, and (3) CO_2 mitigation that has been thoroughly investigated (Subashchandrabose et al., 2011). There is a comprehensive list of cyanobacteria/microalgae and bacteria consortia that have been engaged in the bioremediation of multiple organic and metal pollutants. It has been reported that when microalgae *Scenedesmus dimorphus* and nitrifiers were cocultured, the elimination of N and P from wastewater was enhanced by 3.4- and 6.5-fold, respectively, compared to when nitrifiers were individually used (Choi, Han, Yoo, Cho, & Hwang, 2018). The consortia comprising *Chlorella vulgaris* (microalga), and *P. putida* (bacterium) displayed improved nutritional (N and P) and chemical oxygen demand removal efficiency than either single culture (Mujtaba, Rizwan, & Lee, 2017). The removal of 89% of nitrogen and 80% of phosphorous from starch wastewater was achieved using a coculture of *Scenedesmus* sp. with anaerobic sludge (Ren, Liu, Kong, Zhao, & Ren, 2015). The addition of activated sludge from a bacterial culture to a microalgal consortia facilitated hydrocarbon degradation in an oil-polluted habitat. Lately, an algal—bacterial coculture strategy for nitrogen-rich wastewater remediation and coproduction of lipids was established. A ratio of 1:0.75 for bacteria and microalgae was found to be optimum for the maximum lipid yield of 130 mg/L (Leong et al., 2018). Using an experimental microalga, *C. vulgaris* density culture for the remediation of marine wastewater effluent, a comparable study was conducted to evaluate the relationship between lipid synthesis and bacterial growth. When compared to the control, the lipid content was approximately two times higher (Nguyen et al., 2019). In addition to wastewater treatment and animal feed production, biomass recovered from effluent can be utilized as sustainable substrates (or feedstocks) for the manufacture of bioenergy and biomolecules. (Bhatia, Bhatia, & Yang, 2017; Salama et al., 2017; Stiles et al., 2018; Wrede et al., 2014). Despite the apparent benefits of the microalgae—bacteria/fungi cohabitation approach, land availability, adequate light intensities, and proper temperature setting remain major roadblocks to their scaling up in wastewater remediation (Osundeko, Ansolia, Gupta, Bag, & Bajhaiya, 2019). Furthermore, retrieving and processing microalgae from the culture medium continues to be a challenge due to the high energy input requirement (Osundeko et al., 2019). More functional infrastructure and effective harvesting technologies should be developed to enable microalgae treatment and minimize cell mass recovery expenses.

6.4.3 Degradation of industrial dyes

Azo dyes, are the most diversified chemical class of dyes, and are widely discharged from the textile, and dye industries, causing serious environmental pollutants due to their color, biorecalcitrance, and possible toxicity to human and animal life (Senan & Abraham, 2004). Remediation of these noxious dyes is quite challenging through available physical and chemical procedures (Khan, Bhawana, & Fulekar, 2013). Moreover, detoxification and discoloration of textile wastewaters have certain inherent drawbacks such as development of potentially colorless hazardous intermediaries that are even more challenging to break down than the parent dye and require longer time frames for the complete degradation (Solís, Solís, Pérez, Manjarrez, & Flores, 2012). Therefore, given these disadvantages, defined microbial consortia have been explored for the detoxification of diverse synthetic dyes (Achour, Khelifi, Ayed, Helal, & Bakhrouf, 2014; Ayed, Harbi, Cheref, Bakhrouf, & Achour, 2010; Khaled, 2011; Senan & Abraham, 2004). The ability of microbial consortia made up of bacterial, fungal, and algal species to biologically break down of industrial dyes and, converting them into lesser perilous versions (Chen & Yien Ting, 2015; Holkar, Jadhav, Pinjari, Mahamuni, & Pandit, 2016; Meng, Liu, Zhou, & Fu, 2014; Shah, Dave, & Rao, 2012). Since the mechanism for degradation of the industrial dyes varies, the treatment with microbial consortium consisting of suitable organisms that can provide enzymes such as manganese peroxidases, laccase, lignin, aminopyrine N-demethylase, hexane oxidase, NADH-dependent 2,6-dichloroindophenol (NADH-DCIP) reductase, and tyrosinase, among others, would be beneficial in the decomposition of these indistrial recalcitrant dyes (Solís et al., 2012). Because of the reductive breakdown of azo bonds ($-N=N-$), bacteria can degrade azo dyes, although they seldom mineralize them, and intermediate molecules are often aromatic intermediates that require further degradation (Khan, Jain, Soni, & Madamwar, 2014).

Fungal cultures, on the other hand, necessitate extended growth periods, nitrogen-restricted conditions, and huge reactor designs due to the longer time period it takes for detailed breakdown. (Anastasi et al., 2011). Thus, in comparison to axenic cultures, microbial consortia systems offer a more attractive approach for treating dyes and textile effluents. (Holkar et al., 2016; Karim, Dhar, & Hossain, 2018). Microbial consortia remediation strategies may achieve higher levels of biodegradation and mineralization due to the microbial community's apparent synergistic metabolic capabilities. (Shanmugam, Easwaran, Lakra, Deepa, & Mahadevan, 2017). Further, organic pollutants detoxifications by microbial consortia in particular ecological niches depends on the suitability in terms of temperature, O_2 levels, and all other parameters that favour the biomass production. The disintegration of dyes and textile waste by biological agents especially microbial consortia is characterized by being (1) environment-friendly, (2) cost-effective, (3) sludge-free, and (4) safe, as no menacing end products are generated. It also requires a lesser volume of water than physical, chemical, and oxidation treatment methods (Hayat, Mahmood, Pervez, Bhatti, & Baig, 2015).

Several bacterial species have been identified to be capable of decolorizing azo dyes used in the textile industries (Khehra, Saini, Sharma, Chadha, & Chimni, 2005). To optimize the dye decolorization process, *Bacillus cereus*, *Pseudomonas fluorescens*, *P. putida*, and *Stenotrophomonas acidaminiphila* were applied in various combinations. The maximum results were achieved with a microbial consortium that included all the analyzed strains and neutralized the dyes entirely within 1 day, whereas the pure culture needed more than 2 days to degrade the dye completely. Both Red HE3B and textile wastewater have been bleached off by a consortium comprising two bacterial species (Phugare, Kalyani, Patil, & Jadhav, 2011; Phugare, Kalyani, Surwase, & Jadhav, 2011). TJ-2, a new bacterial consortium, was reported to be beneficial in the mineralization of noxious aromatic amines generated during the process of decolorization (Barsing, Tiwari, Joshi, & Garg, 2011). A microbial consortium of *Enterobacter cloacae* and *Enterococcus casseliflavus* has been found to decolorize Orange II (also called acid Orange 7) completely. Interestingly, *E. cloacae* and *E. casseliflavus* can only decolorize it up to only 10% and 23%, respectively (Chan, Abdul Rash, Koay, Chang, & Tan, 2011). Similarly, scarlet red, an azo dye was completely decolorized in 3 h by a microbial consortium of *Proteus vulgaris* and *Micrococcus glutamicus*, compared to 14 h by *P. vulgaris* and 20 h by *M. glutamicus* alone (Saratale, Saratale, Kalyani, Chang, & Govindwar, 2009). Furthermore, microbial consortia of *M. glutamicus* and *P. vulgaris* decolorized the sulfonated reactive dye Green HE4BD at a significantly higher rate than independent isolates. (Saratale, Saratale, Chang, & Govindwar, 2010). A fungal–bacterial consortia comprised *Aspergillus ochraceus* NCIM-1146 and *Pseudomonas* sp. SUK1 surpassed single strains in dye decolorization (Kadam, Telke, Jagtap, & Govindwar, 2011). *P. putida* and *Shewanella oneidensis* constituted a successful synergistic consortium for the concomitant remediation of congo red in planktonic culture and improved electricity production in anode biofilms (Wang et al., 2014). Individual strains in these microbial consortia potentially target the dye molecule at different positions or even use the degradation products produced by associate strains, such as harmful aromatic amines, for further disintegration (Saratale, Saratale, Chang, & Govindwar, 2011).

6.4.4 Remediation of other organic pollutants

Microorganisms that have acclimitized to heavy metal-rich environments possess biosorption, bioprecipitation, extracellular sequestration, transport pathways, and/or chelation activities. The application of microbes in remediation strategies is based on such resistance mechanisms. Other pollutants, such as pesticides, antibiotics, and other poisons, are also decomposed by microbial consortia, indicating their unique capabilities. Microorganisms, notably bacterial and fungal species, have been investigated extensively for their ability to degrade different contaminants such as PAHs, PCBs, and DDT (Bamforth & Singleton, 2005; Varjani, 2017). *Achromobacter* sp. SSK4, *Alcaligenes* sp. SSK1B, *Microbacterium* sp. BPW, *Ralstonia* sp. BPH, and *Novosphingobium* sp. PCY are among the microbial strains that have been utilized in the degradation of multiple PAHs (Wongwongsee, Chareanpat, & Pinyakong, 2013). One of the major aspects influencing PAH attenuation in natural settings is the use of PAHs by microorganisms as a source of carbon and energy. However, no single microbe axenic culture has been reported to be able to clean up a PAH-contaminated habitat (Janbandhu & Fulekar, 2011; Mikesková, Novotný, & Svobodová, 2012; Zhong, Luan, Lin, Liu, & Tam, 2011). Under a variety of conditions, heterogeneous communities, on the other hand, can disintegrate both simple PAHs and complex composites. The degradation of PAHs in liquid media and soil was successfully investigated using fungal–bacterial consortia and bacterial consortium VUN 10,009 (Boonchan, Britz, & Stanley, 2000). Researchers have successfully isolated microbial strains from creosote-contaminated soils. *Stenotrophomonas maltophilia* VUN 10,010 and *Penicillium janthinellum* VUO 10, 201 developed a consortium that not only mineralized and thrived on benzo[a]pyrene as a primary source of carbon and energy but also degraded pyrene and benzo(a)pyrene at increased levels (Boonchan et al., 2000). Fungal–bacterial cocultures have also been proven to be effective in the degradation of PAHs. Similarly, bacterial cocultures comprising *Flavobacterium* sp., *P. putida*, and *P. aeruginosa* surpassed single bacterial strains in terms of PAH degradation (Trzesicka-Mlynarz & Ward, 1995). The study of prokaryotic consortia has advanced our understanding of microorganisms' ability to break down PAHs and the role of abiotic variables in this process. In this context, earlier studies on high-molecular-weight PAH biodegradation have been reviewed and documented (Kanaly & Harayama, 2010). Six critical abiotic parameters in the biodegradation of PAHs by the bacterial consortium C2PL05 have been optimized (Simarro, González, Bautista, Sanz, & Molina, 2011). Researchers have found that cell immobilization during the biofilm development can alter the degrading ability of bacterial consortia while investigating diesel oil degraders (èawniczak et al., 2011). The bacterial consortia were introduced into the polluted soil using a certain activated carbon material that augmented degrading bacteria to the soils for *in-situ* toxic pesticide bioremediation (Takagi, Kataoka, & Yamazaki, 2011). Chlorpyrifos, one of the most extensively used organophosphate pesticides, has been mineralized by a consortium of *Serratia* and *Trichosporon* sp. (Xu et al., 2007). The continuing discharge of unprocessed industrial and municipal pollutants into the environment has put soil and aquatic life in serious jeopardy (Liu et al., 2017; Singh, Kumar, Mittal, & Mehta, 2017). *Alcaligenes* sp., *Aquamicrobium defuvium*, *Bacillus licheniformis*, and *P. putida*, were developed to significantly boost sulfonamide decontamination in a microbial consortia (Islas-Espinoza, Reid, Wexler, & Bond, 2012).

6.5 Conclusion and future perspective

The application of microbial consortia in environmental remediation, etc. has triggered the rapid development of microbial biotechnology. Organic pollutants such as PAHs, industrial dyes, atrazine, trinitrotoluene (TNT), phenol, and pesticides have been extensively investigated for biodegradation utilizing various microbial consortia made up of diverse bacterial and fungal strains. Consortia break down organic and complex contaminants, including industrial dyes, PCBs, PAHs, and other pollutants more efficiently than pure strains due to the synergy of metabolic activity among the participating microorganisms. However, some questions remain unanswered, for example, basic understanding of metabolic pathways, microorganism local adaptability mechanisms, and the omics changes linked with them, to name a few. In this respect, omics studies will be critical in gaining a deeper insight into bioremediation processes, which will contribute to highly effective *ex situ* and *in situ* processes. The development and optimization of the consortium must be given special attention so as to improve its reproducibility and bioactivity in competitive settings. Furthermore, understanding and identifying newer microbial organisms (microcosms or scenarios) which include unstudied species of bacteria, fungi, microalgae, and other organisms must be investigated to determine their specific interactions for bioremediation goals. Enriched culture can be used in conjunction with new molecular techniques to produce inherent, multifunctional consortia for organopollutant biodegradation and to augment their normal activities. In view of above discussed succesful aaplications, the concept of microbial consortium is a promising approach in bioremediation processes and environmental clean up.

Acknowledgment

Mir Zahoor Gul gratefully acknowledges the financial assistance from the Council of Scientific and Industrial Research (CSIR-HRDG), New Delhi, India, as part of the CSIR-Research Associateship Scheme (File No. 09/132(0883)/2019-EMR-1).

References

Abatenh, E., Gizaw, B., Tsegaye, Z., & Wassie, M. (2017). The role of microorganisms in bioremediation - A review. *Open Journal of Environmental Biology*, *2*(1), 038–046. Available from https://doi.org/10.17352/ojeb.000007.

Achour, S., Khelifi, E., Ayed, L., Helal, A. N., & Bakhrouf, A. (2014). Response surface methodology for textile wastewater decolourization and biodegradation by a novel mixed bacterial consortium developed via mixture design. *Desalination and Water Treatment*, *52*(7–9). Available from https://doi.org/10.1080/19443994.2013.800254.

Alexander, M. (1999). *Biodegradation and bioremediation* (second (ed.)). San Diego, CA, USA: Academic Press. Available from http://doi.org/10.1007/978-3-662-05794-0.

Anastasi, A., Parato, B., Spina, F., Tigini, V., Prigione, V., & Varese, G. C. (2011). Decolourisation and detoxification in the fungal treatment of textile wastewaters from dyeing processes. *New Biotechnology*, *29*(1). Available from https://doi.org/10.1016/j.nbt.2011.08.006.

Ayed, L., Harbi, B., Cheref, A., Bakhrouf, A., & Achour, S. (2010). Application of the mixture design to optimise the formulation of active consortia to decolorize azo-dye methyl red. *Water Science and Technology*, *62*(12). Available from https://doi.org/10.2166/wst.2010.709.

Azubuike, C. C., Chikere, C. B., & Okpokwasili, G. C. (2016). Bioremediation techniques—classification based on site of application: principles, advantages, limitations and prospects. *World Journal of Microbiology and Biotechnology*, *32*(11), 180. Available from https://doi.org/10.1007/s11274-016-2137-x.

Bae, H.-S., Huang, L., White, J. R., Wang, J., DeLaune, R. D., & Ogram, A. (2018). Response of microbial populations regulating nutrient biogeochemical cycles to oiling of coastal saltmarshes from the Deepwater Horizon oil spill. *Environmental Pollution*, *241*. Available from https://doi.org/10.1016/j.envpol.2018.05.033.

Balaji, V., Arulazhagan, P., & Ebenezer, P. (2014). Enzymatic bioremediation of polyaromatic hydrocarbons by fungal consortia enriched from petroleum contaminated soil and oil seeds. *Journal of Environmental Biology*, *35*(3).

Baldwin, B. R., Peacock, A. D., Park, M., Ogles, D. M., Istok, J. D., McKinley, J. P., ... White, D. C. (2008). Multilevel samplers as microcosms to assess microbial response to biostimulation. *Ground Water*, *46*(2), 295—304. Available from https://doi.org/10.1111/j.1745-6584.2007.00411.x.

Bamforth, S. M., & Singleton, I. (2005). Bioremediation of polycyclic aromatic hydrocarbons: current knowledge and future directions. *Journal of Chemical Technology & Biotechnology*, *80*(7). Available from https://doi.org/10.1002/jctb.1276.

Barsing, P., Tiwari, A., Joshi, T., & Garg, S. (2011). Application of a novel bacterial consortium for mineralization of sulphonated aromatic amines. *Bioresource Technology*, *102* (2). Available from https://doi.org/10.1016/j.biortech.2010.08.098.

Bascompte, J. (2010). Structure and dynamics of ecological networks. *Science (New York, N.Y.)*, *329*(5993). Available from https://doi.org/10.1126/science.1194255.

Bhat, R. A., Singh, D. V., Qadri, H., Dar, G. H., Dervash, M. A., Bhat, S. A., ... Yousaf, B. (2021). Vulnerability of municipal solid waste: An emerging threat to aquatic ecosystems. *Chemosphere*, *287*(2022), 132223. Available from https://doi.org/10.1016/j.chemosphere.2021.132223.

Bhatia, S. K., Bhatia, R. K., & Yang, Y.-H. (2017). An overview of microdiesel—A sustainable future source of renewable energy. *Renewable and Sustainable Energy Reviews*, 79. Available from https://doi.org/10.1016/j.rser.2017.05.138.

Bhatt, P., Bhatt, K., Huang, Y., Lin, Z., & Chen, S. (2020). Esterase is a powerful tool for the biodegradation of pyrethroid insecticides. *Chemosphere*, 244. Available from https://doi.org/10.1016/j.chemosphere.2019.125507.

Bhatt, P., Huang, Y., Zhang, W., Sharma, A., & Chen, S. (2020). Enhanced cypermethrin degradation kinetics and metabolic pathway in *Bacillus thuringiensis* strain SG4. *Microorganisms*, *8*(2). Available from https://doi.org/10.3390/microorganisms8020223.

Bhatt, P., Zhang, W., Lin, Z., Pang, S., Huang, Y., & Chen, S. (2020). Biodegradation of allethrin by a novel fungus *Fusarium proliferatum* strain CF2, isolated from contaminated soils. *Microorganisms*, *8*(4). Available from https://doi.org/10.3390/microorganisms8040593.

Bilal, M., & Iqbal, H. M. N. (2020). Microbial bioremediation as a robust process to mitigate pollutants of environmental concern. *Case Studies in Chemical and Environmental Engineering*, 2. Available from https://doi.org/10.1016/j.cscee.2020.100011.

Bilal, M., Adeel, M., Rasheed, T., Zhao, Y., & Iqbal, H. M. N. (2019). Emerging contaminants of high concern and their enzyme-assisted biodegradation — A review. *Environment International*, *124*. Available from https://doi.org/10.1016/j.envint.2019.01.011.

Blasco, R., Aparicio, P. J., & Castillo, F. (1995). Photoinactivation of the detoxifying enzyme nitrophenol reductase from *Rhodobacter capsulatus*. *Archives of Microbiology*, *163* (4). Available from https://doi.org/10.1007/BF00393376.

Boonchan, S., Britz, M. L., & Stanley, G. A. (2000). Degradation and mineralization of high-molecular-weight polycyclic aromatic hydrocarbons by defined fungal-bacterial cocultures. *Applied and Environmental Microbiology*, *66*(3). Available from https://doi.org/10.1128/AEM.66.3.1007-1019.2000.

Bordel, S., Guieysse, B., & Muñoz, R. (2009). Mechanistic model for the reclamation of industrial wastewaters using algal – bacterial photobioreactors. *Environmental Science & Technology*, *43*(9). Available from https://doi.org/10.1021/es802156e.

Brenner, K., You, L., & Arnold, F. H. (2008). Engineering microbial consortia: A new frontier in synthetic biology. *Trends in Biotechnology*, *26*(9). Available from https://doi.org/10.1016/j.tibtech.2008.05.004.

Burmølle, M., Webb, J. S., Rao, D., Hansen, L. H., Sørensen, S. J., & Kjelleberg, S. (2006). Enhanced biofilm formation and increased resistance to antimicrobial agents and bacterial invasion are caused by synergistic interactions in multispecies biofilms. *Applied and Environmental Microbiology*, *72*(6). Available from https://doi.org/10.1128/AEM.03022-05.

Byappanahalli, M. N., Sawdey, R., Ishii, S., Shively, D. A., Ferguson, J. A., Whitman, R. L., & Sadowsky, M. J. (2009). Seasonal stability of *Cladophora*-associated *Salmonella* in Lake Michigan watersheds. *Water Research*, *43*(3). Available from https://doi.org/10.1016/j.watres.2008.11.012.

Chan, G. F., Abdul Rash, N. A., Koay, L. L., Chang, S. Y., & Tan, W. L. (2011). Identification and optimization of Novel NAR-1 Bacterial Consortium for the biodegradation of Orange II. *Insight Biotechnology*, *1*(1). Available from https://doi.org/10.5567/IBIOT-IK.2011.7.16.

Chaudhari, A. U., & Kodam, K. M. (2010). Biodegradation of thiocyanate using co-culture of *Klebsiella pneumoniae* and *Ralstonia* sp. *Applied Microbiology and Biotechnology*, *85*(4), 1167–1174. Available from https://doi.org/10.1007/s00253-009-2299-7.

Chen, F., Li, X., Zhu, Q., Ma, J., Hou, H., & Zhang, S. (2019). Bioremediation of petroleum-contaminated soil enhanced by aged refuse. *Chemosphere*, 222. Available from https://doi.org/10.1016/j.chemosphere.2019.01.122.

Chen, S. H., & Yien Ting, A. S. (2015). Biodecolorization and biodegradation potential of recalcitrant triphenylmethane dyes by *Coriolopsis* sp. isolated from compost. *Journal of Environmental Management*, *150*. Available from https://doi.org/10.1016/j.jenvman.2014.09.014.

Chen, S.-H., & Aitken, M. D. (1999). Salicylate Stimulates the degradation of high-molecular weight polycyclic aromatic hydrocarbons by *Pseudomonas saccharophila* P15. *Environmental Science & Technology*, *33*(3). Available from https://doi.org/10.1021/es9805730.

Choi, K.-J., Han, T. H., Yoo, G., Cho, M. H., & Hwang, S.-J. (2018). Co-culture consortium of *Scenedesmus dimorphus* and nitrifiers enhances the removal of nitrogen and phosphorus from artificial wastewater. *KSCE Journal of Civil Engineering*, *22*(9). Available from https://doi.org/10.1007/s12205-017-0730-7.

Clarke, M. B., & Sperandio, V. (2005). Events at the host-microbial interface of the gastrointestinal tract III. Cell-to-cell signaling among microbial flora, host, and pathogens: there is a whole lot of talking going on. *American Journal of Physiology-Gastrointestinal and Liver Physiology*, *288*(6). Available from https://doi.org/10.1152/ajpgi.00572.2004.

Dar, G. H., Bandh, S. A., Kamili, A. N., Nazir, R., & Bhat, R. A. (2013). Comparative Analysis of Different Types of Bacterial Colonies from the Soils of Yusmarg Forest, Kashmir valley India. *Ecologia Balkanica*, *5*(1), 31–35.

Dar, G. H., Kamili, A. N., Chishti, M. Z., Dar, S. A., Tantry, T. A., & Ahmad, F. (2016). Characterization of *Aeromonas sobria* isolated from fish Rohu (*Labeo rohita*) collected from polluted pond. *J Bacteriol Parasitol*, 7(3), 1–5. Available from https://doi.org/10.4172/2155-9597.1000273.

Dar, G. H., Kamili, A. N., Dar, S. A., Chishti, M. Z., & Ahmad, F. (2015). Detection and characterization of potentially pathogenic *Aeromonas sobria* isolated from fish *Hypophthalmichthys molitrix* (Cypriniformes: Cyprinidae). *Microbial Pathogenesis* (2015). Available from https://doi.org/10.1016/j.micpath.2015.10.017.

Dar, G. H., Bandh, S. A., & Kamili, A. N. (2011). Bacteriological analysis of soil from Yusmarg Health Resort of Kashmir Valley. *International Journal of Current Research, 3* (10), 53–56, 2011.

Dar, G. H., Kamili, A. N., Nazir, R., Bandh, S. A., & Bhat, R. A. (2012). A preliminary study of colony forming units of bacteria from the soils of Yusmarg forest, Kashmir valley, India. *International Journal of Current Research, 4*(12), 467–472.

Das, N., & Chandran, P. (2011). Microbial degradation of petroleum hydrocarbon contaminants: An overview. *Biotechnology Research International, 2011*, 1–13. Available from https://doi.org/10.4061/2011/941810.

Davies, D. G. (1998). The involvement of cell-to-cell signals in the development of a bacterial biofilm. *Science (New York, N.Y.), 280*(5361). Available from https://doi.org/10.1126/science.280.5361.295.

Delegan, Y. A., Valentovich, L. N., Shafieva, S. M., Ganbarov, K. G., Filonov, A. E., & Vainstein, M. B. (2019). Characterization and genomic analysis of highly efficient thermotolerant oil-degrading bacterium *Gordonia* sp. 1D. *Folia Microbiologica, 64*(1), 41–48. Available from https://doi.org/10.1007/s12223-018-0623-2.

Dvořák, P., Nikel, P. I., Damborský, J., & de Lorenzo, V. (2017). Bioremediation 3. 0 : Engineering pollutant-removing bacteria in the times of systemic biology. *Biotechnology Advances, 35*(7). Available from https://doi.org/10.1016/j.biotechadv.2017.08.001.

Escobar, J., Brenner, M., Whitmore, T. J., Kenney, W. F., & Curtis, J. H. (2008). Ecology of testate amoebae (thecamoebians) in subtropical Florida lakes. *Journal of Paleolimnology, 40*(2). Available from https://doi.org/10.1007/s10933-008-9195-5.

Fay, P. (1992). Oxygen relations of nitrogen fixation in cyanobacteria. *Microbiological Reviews, 56*(2), 340–373. Available from https://doi.org/10.1128/mr.56.2.340-373.1992.

Feng, J., & Lim, T.-T. (2005). Pathways and kinetics of carbon tetrachloride and chloroform reductions by nano-scale Fe and Fe/Ni particles: comparison with commercial micro-scale Fe and Zn. *Chemosphere, 59*(9). Available from https://doi.org/10.1016/j.chemosphere.2004.11.038.

Fukami, T. (2015). Historical contingency in community assembly: Integrating niches, species pools, and priority effects. *Annual Review of Ecology, Evolution, and Systematics, 46*. Available from https://doi.org/10.1146/annurev-ecolsys-110411-160340.

Gonçalves, A. L., Pires, J. C. M., & Simões, M. (2017). A review on the use of microalgal consortia for wastewater treatment. *Algal Research, 24*. Available from https://doi.org/10.1016/j.algal.2016.11.008.

Greenberg, E. P. (2003). Bacterial communication and group behavior. *Journal of Clinical Investigation, 112*(9). Available from https://doi.org/10.1172/JCI20099.

Gu, J., Chen, X., Wang, Y., Wang, L., Szlavecz, K., Ma, Y., & Ji, R. (2020). Bioaccumulation, physiological distribution, and biotransformation of tetrabromobisphenol a (TBBPA) in the geophagous earthworm *Metaphire guillelmi* – hint for detoxification strategy. *Journal of Hazardous Materials, 388*. Available from https://doi.org/10.1016/j.jhazmat.2020.122027.

Hamer, G. (1993). Bioremediation: a response to gross environmental abuse. *Trends in Biotechnology, 11*(8), 317–319. Available from https://doi.org/10.1016/0167-7799(93)90152-Y.

Haritash, A. K., & Kaushik, C. P. (2009). Biodegradation aspects of polycyclic aromatic hydrocarbons (PAHs): A review. *Journal of Hazardous Materials, 169*(1–3). Available from https://doi.org/10.1016/j.jhazmat.2009.03.137.

Hayat, H., Mahmood, Q., Pervez, A., Bhatti, Z. A., & Baig, S. A. (2015). Comparative decolorization of dyes in textile wastewater using biological and chemical treatment. *Separation and Purification Technology, 154*. Available from https://doi.org/10.1016/j.seppur.2015.09.025.

Hays, S. G., Patrick, W. G., Ziesack, M., Oxman, N., & Silver, P. A. (2015). Better together: Engineering and application of microbial symbioses. *Current Opinion in Biotechnology, 36*. Available from https://doi.org/10.1016/j.copbio.2015.08.008.

He, C., Dong, W., Li, J., Li, Y., Huang, C., & Ma, Y. (2017). Characterization of rhamnolipid biosurfactants produced by recombinant *Pseudomonas aeruginosa* strain DAB with removal of crude oil. *Biotechnology Letters, 39*(9). Available from https://doi.org/10.1007/s10529-017-2370-x.

Holkar, C. R., Jadhav, A. J., Pinjari, D. V., Mahamuni, N. M., & Pandit, A. B. (2016). A critical review on textile wastewater treatments: Possible approaches. *Journal of Environmental Management, 182*. Available from https://doi.org/10.1016/j.jenvman.2016.07.090.

Hu, B., Wang, M., Geng, S., Wen, L., Wu, M., Nie, Y., … Wu, X.-L. (2020). Metabolic exchange with non-alkane consuming *Pseudomonas stutzeri* SLG510A3-8 improves *n*-alkane biodegradation by the alkane degrader *Dietzia* sp. strain DQ12-45-1b. *Applied and Environmental Microbiology, 86*(8). Available from https://doi.org/10.1128/AEM.02931-19.

Huang, Y., Zhan, H., Bhatt, P., & Chen, S. (2019). Paraquat degradation from contaminated environments: Current achievements and perspectives. *Frontiers in Microbiology, 10*. Available from https://doi.org/10.3389/fmicb.2019.01754.

Islas-Espinoza, M., Reid, B. J., Wexler, M., & Bond, P. L. (2012). Soil bacterial consortia and previous exposure enhance the biodegradation of sulfonamides from pig manure. *Microbial Ecology, 64*(1), 140–151. Available from https://doi.org/10.1007/s00248-012-0010-5.

Janbandhu, A., & Fulekar, M. H. (2011). Biodegradation of phenanthrene using adapted microbial consortium isolated from petrochemical contaminated environment. *Journal of Hazardous Materials, 187*(1–3). Available from https://doi.org/10.1016/j.jhazmat.2011.01.034.

Janssen, D. B., & Stucki, G. (2020). Perspectives of genetically engineered microbes for groundwater bioremediation. *Environmental Science: Processes & Impacts, 22*(3). Available from https://doi.org/10.1039/C9EM00601J.

Jeandet, P., Sobarzo-Sánchez, E., Clément, C., Nabavi, S. F., Habtemariam, S., Nabavi, S. M., & Cordelier, S. (2018). Engineering stilbene metabolic pathways in microbial cells. *Biotechnology Advances, 36*(8). Available from https://doi.org/10.1016/j.biotechadv.2018.11.002.

Jia, X., Liu, C., Song, H., Ding, M., Du, J., Ma, Q., & Yuan, Y. (2016). Design, analysis and application of synthetic microbial consortia. *Synthetic and Systems Biotechnology, 1*(2). Available from https://doi.org/10.1016/j.synbio.2016.02.001.

Jin, D. F., Hu, H., Liu, D. F., Ding, H. T., Jia, X. M., & Zhao, Y. H. (2012). Optimization of a bacterial consortium for nitrobenzene degradation. *Water Science and Technology, 65*(5). Available from https://doi.org/10.2166/wst.2012.692.

Johnsen, A. R., Wick, L. Y., & Harms, H. (2005). Principles of microbial PAH-degradation in soil. *Environmental Pollution, 133*(1). Available from https://doi.org/10.1016/j.envpol.2004.04.015.

Kaczorek, E., Cieślak, K., Bielicka-Daszkiewicz, K., & Olszanowski, A. (2013). The influence of rhamnolipids on aliphatic fractions of diesel oil biodegradation by microorganism combinations. *Indian Journal of Microbiology, 53*(1). Available from https://doi.org/10.1007/s12088-012-0323-6.

Kadam, A. A., Telke, A. A., Jagtap, S. S., & Govindwar, S. P. (2011). Decolorization of adsorbed textile dyes by developed consortium of *Pseudomonas* sp. SUK1 and *Aspergillus ochraceus* NCIM-1146 under solid state fermentation. *Journal of Hazardous Materials, 189*(1–2). Available from https://doi.org/10.1016/j.jhazmat.2011.02.066.

Kanaly, R. A., & Harayama, S. (2010). Advances in the field of high-molecular-weight polycyclic aromatic hydrocarbon biodegradation by bacteria. *Microbial Biotechnology, 3*(2). Available from https://doi.org/10.1111/j.1751-7915.2009.00130.x.

Kapahi, M., & Sachdeva, S. (2019). Bioremediation options for heavy metal pollution. *Journal of Health and Pollution*, *9*(24), 191203. Available from https://doi.org/10.5696/2156-9614-9.24.191203.

Kaper, J. B., & Sperandio, V. (2005). Bacterial cell-to-cell signaling in the gastrointestinal tract. *Infection and Immunity*, *73*(6), 3197–3209. Available from https://doi.org/10.1128/IAI.73.6.3197-3209.2005.

Karim, Md. E., Dhar, K., & Hossain, Md. T. (2018). Decolorization of textile reactive dyes by bacterial monoculture and consortium screened from textile dyeing effluent. *Journal of Genetic Engineering and Biotechnology*, *16*(2). Available from https://doi.org/10.1016/j.jgeb.2018.02.005.

Kästner, M., & Miltner, A. (2016). Application of compost for effective bioremediation of organic contaminants and pollutants in soil. *Applied Microbiology and Biotechnology*, *100* (8). Available from https://doi.org/10.1007/s00253-016-7378-y.

Khaled, M. G. (2011). Statistical optimization of cultural conditions for decolorization of methylene blue by mono and mixed bacterial culture techniques. *African Journal of Microbiology Research*, *5*(15). Available from https://doi.org/10.5897/AJMR11.490.

Khan, R., Bhawana, P., & Fulekar, M. H. (2013). Microbial decolorization and degradation of synthetic dyes: A review. *Reviews in Environmental Science and Bio/Technology*, *12*(1). Available from https://doi.org/10.1007/s11157-012-9287-6.

Khan, Z., Jain, K., Soni, A., & Madamwar, D. (2014). Microaerophilic degradation of sulphonated azo dye – Reactive Red 195 by bacterial consortium AR1 through co-metabolism. *International Biodeterioration & Biodegradation*, 94. Available from https://doi.org/10.1016/j.ibiod.2014.07.002.

Khehra, M., Saini, H., Sharma, D., Chadha, B., & Chimni, S. (2005). Decolorization of various azo dyes by bacterial consortium. *Dyes and Pigments*, *67*(1). Available from https://doi.org/10.1016/j.dyepig.2004.10.008.

Kumar, A., Kumar, A., Singh, R., Singh, R., Pandey, S., Rai, A., ... Rahul, B. (2020). *Genetically engineered bacteria for the degradation of dye and other organic compounds. Abatement of environmental pollutants*. Elsevier. Available from http://doi.org/10.1016/B978-0-12-818095-2.00016-3.

Kurwadkar, S. (2019). Occurrence and distribution of organic and inorganic pollutants in groundwater. *Water Environment Research*, *91*(10). Available from https://doi.org/10.1002/wer.1166.

LaPara, T. M., Zakharova, T., Nakatsu, C. H., & Konopka, A. (2002). Functional and structural adaptations of bacterial communities growing on particulate substrates under stringent nutrient limitation. *Microbial Ecology*, *44*(4). Available from https://doi.org/10.1007/s00248-002-1046-8.

èawniczak, è., Kaczorek, E., & Olszanowski, A. (2011). The influence of cell immobilization by biofilm forming on the biodegradation capabilities of bacterial consortia. *World Journal of Microbiology and Biotechnology*, *27*(5), 1183–1188. Available from https://doi.org/10.1007/s11274-010-0566-5.

Lee, Y., Jeong, S. E., Hur, M., Ko, S., & Jeon, C. O. (2018). Construction and evaluation of a Korean Native Microbial Consortium for the bioremediation of diesel fuel-contaminated soil in Korea. *Frontiers in Microbiology*, 9. Available from https://doi.org/10.3389/fmicb.2018.02594.

Leong, W.-H., Lim, J.-W., Lam, M.-K., Uemura, Y., Ho, C.-D., & Ho, Y.-C. (2018). Co-cultivation of activated sludge and microalgae for the simultaneous enhancements of nitrogen-rich wastewater bioremediation and lipid production. *Journal of the Taiwan Institute of Chemical Engineers*, 87. Available from https://doi.org/10.1016/j.jtice.2018.03.038.

Li, M., Liu, C., Yang, J., Nian, R., Xian, M., Li, F., & Zhang, H. (2020). Common problems associated with the microbial productions of aromatic compounds and corresponding metabolic engineering strategies. *Biotechnology Advances, 41*. Available from https://doi.org/10.1016/j.biotechadv.2020.107548.

Li, Y. Y., & Li, B. (2011). Study on fungi-bacteria consortium bioremediation of petroleum contaminated mangrove sediments amended with mixed biosurfactants. *Advanced Materials Research, 183–185*, 1163–1167. Available from https://doi.org/10.4028/http://www.scientific.net/AMR.183-185.1163.

Lin, Z., Zhang, W., Pang, S., Huang, Y., Mishra, S., Bhatt, P., & Chen, S. (2020). Current approaches to and future perspectives on methomyl degradation in contaminated soil/water environments. *Molecules (Basel, Switzerland), 25*(3). Available from https://doi.org/10.3390/molecules25030738.

Liu, Y., Chang, H., Li, Z., Feng, Y., Cheng, D., & Xue, J. (2017). Biodegradation of gentamicin by bacterial consortia AMQD4 in synthetic medium and raw gentamicin sewage. *Scientific Reports*, 7(1), 11004. Available from https://doi.org/10.1038/s41598-017-11529-x.

Luo, L., Meng, H., & Gu, J.-D. (2017). Microbial extracellular enzymes in biogeochemical cycling of ecosystems. *Journal of Environmental Management*, 197. Available from https://doi.org/10.1016/j.jenvman.2017.04.023.

Luo, S., Chen, B., Lin, L., Wang, X., Tam, N. F.-Y., & Luan, T. (2014). Pyrene degradation accelerated by constructed consortium of bacterium and microalga: Effects of degradation products on the microalgal growth. *Environmental Science & Technology, 48* (23). Available from https://doi.org/10.1021/es503761j.

Ma, X., Novak, P. J., Ferguson, J., Sadowsky, M., LaPara, T. M., Semmens, M. J., & Hozalski, R. M. (2007). The impact of H_2 addition on dechlorinating microbial communities. *Bioremediation Journal, 11*(2), 45–55. Available from https://doi.org/10.1080/10889860701369490.

Mandal, S. K., Singh, R. P., & Patel, V. (2011). Isolation and characterization of exopolysaccharide secreted by a toxic dinoflagellate, *Amphidinium carterae* Hulburt 1957 and its probable role in harmful algal blooms (HABs). *Microbial Ecology, 62*(3). Available from https://doi.org/10.1007/s00248-011-9852-5.

McCarty, N. S., & Ledesma-Amaro, R. (2019). Synthetic biology tools to engineer microbial communities for biotechnology. *Trends in Biotechnology, 37*(2). Available from https://doi.org/10.1016/j.tibtech.2018.11.002.

Meng, X., Liu, G., Zhou, J., & Fu, Q. S. (2014). Effects of redox mediators on azo dye decolorization by *Shewanella* algae under saline conditions. *Bioresource Technology, 151*. Available from https://doi.org/10.1016/j.biortech.2013.09.131.

Mikesková, H., Novotný, Č., & Svobodová, K. (2012). Interspecific interactions in mixed microbial cultures in a biodegradation perspective. *Applied Microbiology and Biotechnology, 95*(4). Available from https://doi.org/10.1007/s00253-012-4234-6.

Miranda, A. F., Ramkumar, N., Andriotis, C., Höltkemeier, T., Yasmin, A., Rochfort, S., ... Mouradov, A. (2017). Applications of microalgal biofilms for wastewater treatment and bioenergy production. *Biotechnology for Biofuels, 10*(1), 120. Available from https://doi.org/10.1186/s13068-017-0798-9.

Mishra, S., Zhang, W., Lin, Z., Pang, S., Huang, Y., Bhatt, P., & Chen, S. (2020). Carbofuran toxicity and its microbial degradation in contaminated environments. *Chemosphere*, 259. Available from https://doi.org/10.1016/j.chemosphere.2020.127419.

Mujtaba, G., & Lee, K. (2016). Advanced treatment of wastewater using symbiotic co-culture of microalgae and bacteria. *Applied Chemistry for Engineering, 27*(1). Available from https://doi.org/10.14478/ace.2016.1002.

Mujtaba, G., Rizwan, M., & Lee, K. (2017). Removal of nutrients and COD from wastewater using symbiotic co-culture of bacterium *Pseudomonas putida* and immobilized microalga *Chlorella vulgaris*. *Journal of Industrial and Engineering Chemistry, 49*. Available from https://doi.org/10.1016/j.jiec.2017.01.021.

Mukherjee, A. K., & Bordoloi, N. K. (2011). Bioremediation and reclamation of soil contaminated with petroleum oil hydrocarbons by exogenously seeded bacterial consortium: A pilot-scale study. *Environmental Science and Pollution Research, 18*(3). Available from https://doi.org/10.1007/s11356-010-0391-2.

Netzker, T., Flak, M., Krespach, M. K., Stroe, M. C., Weber, J., Schroeckh, V., & Brakhage, A. A. (2018). Microbial interactions trigger the production of antibiotics. *Current Opinion in Microbiology, 45*. Available from https://doi.org/10.1016/j.mib.2018.04.002.

Nguyen, T. D. P., Nguyen, D. H., Lim, J. W., Chang, C.-K., Leong, H. Y., Tran, T. N. T., ... Show, P. L. (2019). Investigation of the relationship between bacteria growth and lipid production cultivating of microalgae *Chlorella vulgaris* in seafood wastewater. *Energies, 12*(12). Available from https://doi.org/10.3390/en12122282.

Oh, S. T., Kim, J. R., Premier, G. C., Lee, T. H., Kim, C., & Sloan, W. T. (2010). Sustainable wastewater treatment: How might microbial fuel cells contribute. *Biotechnology Advances, 28*(6), 871–881. Available from https://doi.org/10.1016/j.biotechadv.2010.07.008.

Omrani, R., Spini, G., Puglisi, E., & Saidane, D. (2018). Modulation of microbial consortia enriched from different polluted environments during petroleum biodegradation. *Biodegradation, 29*(2). Available from https://doi.org/10.1007/s10532-018-9823-3.

Osundeko, O., Ansolia, P., Gupta, S. K., Bag, P., & Bajhaiya, A. K. (2019). *Promises and challenges of growing microalgae in wastewater. Water conservation, recycling and reuse: Issues and challenges*. Singapore: Springer Singapore. Available from http://doi.org/10.1007/978-981-13-3179-4_2.

Passador, L., Cook, J., Gambello, M., Rust, L., & Iglewski, B. (1993). Expression of *Pseudomonas aeruginosa* virulence genes requires cell-to-cell communication. *Science (New York, N.Y.), 260*(5111). Available from https://doi.org/10.1126/science.8493556.

Patowary, R., Patowary, K., Kalita, M. C., & Deka, S. (2018). Application of biosurfactant for enhancement of bioremediation process of crude oil contaminated soil. *International Biodeterioration & Biodegradation*, 129. Available from https://doi.org/10.1016/j.ibiod.2018.01.004.

Phugare, S. S., Kalyani, D. C., Patil, A. V., & Jadhav, J. P. (2011). Textile dye degradation by bacterial consortium and subsequent toxicological analysis of dye and dye metabolites using cytotoxicity, genotoxicity and oxidative stress studies. *Journal of Hazardous Materials, 186*(1). Available from https://doi.org/10.1016/j.jhazmat.2010.11.049.

Phugare, S. S., Kalyani, D. C., Surwase, S. N., & Jadhav, J. P. (2011). Ecofriendly degradation, decolorization and detoxification of textile effluent by a developed bacterial consortium. *Ecotoxicology and Environmental Safety, 74*(5). Available from https://doi.org/10.1016/j.ecoenv.2011.03.003.

Pieper, D. H., & Reineke, W. (2000). Engineering bacteria for bioremediation. *Current Opinion in Biotechnology, 11*(3). Available from https://doi.org/10.1016/S0958-1669(00)00094-X.

Prasad, M. N. V., & Prasad, R. (2012). Nature's cure for cleanup of contaminated environment – A review of bioremediation strategies. *Reviews on Environmental Health, 27*(4). Available from https://doi.org/10.1515/reveh-2012-0028.

Prieto-Barajas, C. M., Valencia-Cantero, E., & Santoyo, G. (2018). Microbial mat ecosystems: Structure types, functional diversity, and biotechnological application. *In Electronic Journal of Biotechnology, 31*. Available from https://doi.org/10.1016/j.ejbt.2017.11.001.

Ramanan, R., Kim, B. H., Cho, D. H., Oh, H. M., & Kim, H. S. (2016). Algae-bacteria interactions: Evolution, ecology and emerging applications. In. *Biotechnology Advances, 34*(1). Available from https://doi.org/10.1016/j.biotechadv.2015.12.003.

Ramos, J.-L., Marqués, S., van Dillewijn, P., Espinosa-Urgel, M., Segura, A., Duque, E., ... Wittich, R.-M. (2011). Laboratory research aimed at closing the gaps in microbial bioremediation. *Trends in Biotechnology*, *29*(12). Available from https://doi.org/10.1016/j.tibtech.2011.06.007.

Ren, H.-Y., Liu, B.-F., Kong, F., Zhao, L., & Ren, N. (2015). Hydrogen and lipid production from starch wastewater by co-culture of anaerobic sludge and oleaginous microalgae with simultaneous COD, nitrogen and phosphorus removal. *Water Research*, *85*. Available from https://doi.org/10.1016/j.watres.2015.08.057.

Rochae Silva, F. C. P., Roque, B. A. C., Rochae Silva, N. M. P., Rufino, R. D., Luna, J. M., Santos, V. A., ... Sarubbo, L. A. (2017). Yeasts and bacterial biosurfactants as demulsifiers for petroleum derivative in seawater emulsions. *AMB Express*, 7(1). Available from https://doi.org/10.1186/s13568-017-0499-6.

Roell, G. W., Zha, J., Carr, R. R., Koffas, M. A., Fong, S. S., & Tang, Y. J. (2019). Engineering microbial consortia by division of labor. *Microbial Cell Factories*, *18*(1). Available from https://doi.org/10.1186/s12934-019-1083-3.

Rohr, R. P., Saavedra, S., & Bascompte, J. (2014). On the structural stability of mutualistic systems. *Science (New York, N.Y.)*, *345*(6195). Available from https://doi.org/10.1126/science.1253497.

Salama, E.-S., Kurade, M. B., Abou-Shanab, R. A. I., El-Dalatony, M. M., Yang, I.-S., Min, B., & Jeon, B.-H. (2017). Recent progress in microalgal biomass production coupled with wastewater treatment for biofuel generation. *Renewable and Sustainable Energy Reviews*, *79*. Available from https://doi.org/10.1016/j.rser.2017.05.091.

Saratale, R. G., Saratale, G. D., Chang, J. S., & Govindwar, S. P. (2010). Decolorization and biodegradation of reactive dyes and dye wastewater by a developed bacterial consortium. *Biodegradation*, *21*(6). Available from https://doi.org/10.1007/s10532-010-9360-1.

Saratale, R. G., Saratale, G. D., Chang, J. S., & Govindwar, S. P. (2011). Bacterial decolorization and degradation of azo dyes: A review. *Journal of the Taiwan Institute of Chemical Engineers*, *42*(1). Available from https://doi.org/10.1016/j.jtice.2010.06.006.

Saratale, R. G., Saratale, G. D., Kalyani, D. C., Chang, J. S., & Govindwar, S. P. (2009). Enhanced decolorization and biodegradation of textile azo dye Scarlet R by using developed microbial consortium-GR. *Bioresource Technology*, *100*(9). Available from https://doi.org/10.1016/j.biortech.2008.12.013.

Schmitz, A. C., Hartline, C. J., & Zhang, F. (2017). Engineering microbial metabolite dynamics and heterogeneity. *Biotechnology Journal*, *12*(10). Available from https://doi.org/10.1002/biot.201700422.

Schwarz, A., Adetutu, E. M., Juhasz, A. L., Aburto-Medina, A., Ball, A. S., & Shahsavari, E. (2018). Microbial degradation of phenanthrene in pristine and contaminated sandy soils. *Microbial Ecology*, *75*(4). Available from https://doi.org/10.1007/s00248-017-1094-8.

Schwarzenbach, R. P. (2006). The challenge of micropollutants in aquatic systems. *Science (New York, N.Y.)*, *313*(5790). Available from https://doi.org/10.1126/science.1127291.

Scott, S. R., & Hasty, J. (2016). Quorum sensing communication modules for microbial consortia. *ACS Synthetic Biology*, *5*(9). Available from https://doi.org/10.1021/acssynbio.5b00286.

Senan, R. C., & Abraham, T. E. (2004). Bioremediation of textile azo dyes by aerobic bacterial consortium aerobic degradation of selected azo dyes by bacterial consortium. *Biodegradation*, *15*(4). Available from https://doi.org/10.1023/B:BIOD.0000043000.18427.0a.

Senthilvelan, T., Kanagaraj, J., Panda, R. C., & Mandal, A. B. (2014). Biodegradation of phenol by mixed microbial culture: an eco-friendly approach for the pollution reduction. *Clean Technologies and Environmental Policy*, *16*(1). Available from https://doi.org/10.1007/s10098-013-0598-2.

Shah, P. D., Dave, S. R., & Rao, M. S. (2012). Enzymatic degradation of textile dye Reactive Orange 13 by newly isolated bacterial strain *Alcaligenes faecalis* PMS-1. *International Biodeterioration and Biodegradation, 69*. Available from https://doi.org/10.1016/j.ibiod.2012.01.002.

Shahid, A., Malik, S., Zhu, H., Xu, J., Nawaz, M. Z., Nawaz, S., ... Mehmood, M. A. (2020). Cultivating microalgae in wastewater for biomass production, pollutant removal, and atmospheric carbon mitigation; a review. *Science of The Total Environment, 704*. Available from https://doi.org/10.1016/j.scitotenv.2019.135303.

Shanmugam, B. K., Easwaran, S. N., Lakra, R., Deepa, P. R., & Mahadevan, S. (2017). Metabolic pathway and role of individual species in the bacterial consortium for biodegradation of azo dye: A biocalorimetric investigation. *Chemosphere, 188*. Available from https://doi.org/10.1016/j.chemosphere.2017.08.138.

Shchemelinina, T. N., Matistov, N. V., Kovaleva, V. A., Markarova, M. Yu, Gruzdev, I. V., Volodin, V. V., & Anchugova, E. M. (2017). Microfungal strains—Potential lipid producers for biodiesel. *Applied Biochemistry and Microbiology, 53*(4). Available from https://doi.org/10.1134/S0003683817040135.

Shong, J., Jimenez Diaz, M. R., & Collins, C. H. (2012). Towards synthetic microbial consortia for bioprocessing. *Current Opinion in Biotechnology, 23*(5), 798–802. Available from https://doi.org/10.1016/j.copbio.2012.02.001.

Simarro, R., González, N., Bautista, L. F., Sanz, R., & Molina, M. C. (2011). Optimisation of key abiotic factors of PAH (naphthalene, phenanthrene and anthracene) biodegradation process by a bacterial consortium. *Water, Air, & Soil Pollution, 217*(1–4), 365–374. Available from https://doi.org/10.1007/s11270-010-0593-8.

Singh, R., Paul, D., & Jain, R. K. (2006). Biofilms: Implications in bioremediation. *Trends in Microbiology, 14*(9). Available from https://doi.org/10.1016/j.tim.2006.07.001.

Singh, R., Kumar, M., Mittal, A., & Mehta, P. K. (2017). Microbial metabolites in nutrition, healthcare and agriculture. *3 Biotech*, 7(1), 15. Available from https://doi.org/10.1007/s13205-016-0586-4.

Smith, P., Cotrufo, M. F., Rumpel, C., Paustian, K., Kuikman, P. J., Elliott, J. A., ... Scholes, M. C. (2015). Biogeochemical cycles and biodiversity as key drivers of ecosystem services provided by soils. *Soil, 1*(2), 665–685. Available from https://doi.org/10.5194/soil-1-665-2015.

Solís, M., Solís, A., Pérez, H. I., Manjarrez, N., & Flores, M. (2012). Microbial decolouration of azo dyes: A review. *Process Biochemistry*, 47(12). Available from https://doi.org/10.1016/j.procbio.2012.08.014.

Sørensen, S. R., Ronen, Z., & Aamand, J. (2002). Growth in coculture stimulates metabolism of the phenylurea herbicide isoproturon by *Sphingomonas* sp. strain SRS2. *Applied and Environmental Microbiology, 68*(7). Available from https://doi.org/10.1128/AEM.68.7.3478-3485.2002.

Stiles, W. A. V., Styles, D., Chapman, S. P., Esteves, S., Bywater, A., Melville, L., ... Llewellyn, C. A. (2018). Using microalgae in the circular economy to valorise anaerobic digestate: Challenges and opportunities. *Bioresource Technology, 267*, 732–742. Available from https://doi.org/10.1016/j.biortech.2018.07.100.

Subashchandrabose, S. R., Ramakrishnan, B., Megharaj, M., Venkateswarlu, K., & Naidu, R. (2011). Consortia of cyanobacteria/microalgae and bacteria: Biotechnological potential. *Biotechnology Advances, 29*(6). Available from https://doi.org/10.1016/j.biotechadv.2011.07.009.

Tahseen, R., Arslan, M., Iqbal, S., Khalid, Z. M., & Afzal, M. (2019). Enhanced degradation of hydrocarbons by gamma ray induced mutant strain of *Pseudomonas putida. Biotechnology Letters, 41*(3). Available from https://doi.org/10.1007/s10529-019-02644-y.

Taillefumier, T., Posfai, A., Meir, Y., & Wingreen, N. S. (2017). Microbial consortia at steady supply. *ELife, 6*. Available from https://doi.org/10.7554/eLife.22644.

Takagi, K., Kataoka, R., & Yamazaki, K. (2011). Recent technology on bio-remediation of POPs and persistent pesticides. *Japan Agricultural Research Quarterly: JARQ, 45*(2), 129–136. Available from https://doi.org/10.6090/jarq.45.129.

Tan, C. H., Lee, K. W. K., Burmølle, M., Kjelleberg, S., & Rice, S. A. (2017). All together now: Experimental multispecies biofilm model systems. *Environmental Microbiology, 19*(1). Available from https://doi.org/10.1111/1462-2920.13594.

Trzesicka-Mlynarz, D., & Ward, O. P. (1995). Degradation of polycyclic aromatic hydrocarbons (PAHs) by a mixed culture and its component pure cultures, obtained from PAH-contaminated soil. *Canadian Journal of Microbiology, 41*(6). Available from https://doi.org/10.1139/m95-063.

Unnithan, V. V., Unc, A., & Smith, G. B. (2014). Mini-review: A priori considerations for bacteria–algae interactions in algal biofuel systems receiving municipal wastewaters. *Algal Research, 4*. Available from https://doi.org/10.1016/j.algal.2013.11.009.

Varjani, S. J. (2017). Microbial degradation of petroleum hydrocarbons. *Bioresource Technology, 223*. Available from https://doi.org/10.1016/j.biortech.2016.10.037.

Varjani, S. J., & Upasani, V. N. (2017). A new look on factors affecting microbial degradation of petroleum hydrocarbon pollutants. *International Biodeterioration & Biodegradation, 120*. Available from https://doi.org/10.1016/j.ibiod.2017.02.006.

Villegas, L. B., Martínez, M. A., Rodríguez, A., & Amoroso, M. J. (2014). *Microbial consortia, a viable alternative for cleanup of contaminated soils. Bioremediation in Latin America*. Cham: Springer International Publishing. Available from http://doi.org/10.1007/978-3-319-05738-5_8.

Wang, V. B., Chua, S.-L., Cai, Z., Sivakumar, K., Zhang, Q., Kjelleberg, S., ... Yang, L. (2014). A stable synergistic microbial consortium for simultaneous azo dye removal and bioelectricity generation. *Bioresource Technology, 155*, 71–76. Available from https://doi.org/10.1016/j.biortech.2013.12.078.

Wilkes, R. A., & Aristilde, L. (2017). Degradation and metabolism of synthetic plastics and associated products by *Pseudomonas* sp.: Capabilities and challenges. *Journal of Applied Microbiology, 123*(3). Available from https://doi.org/10.1111/jam.13472.

Wongwongsee, W., Chareanpat, P., & Pinyakong, O. (2013). Abilities and genes for PAH biodegradation of bacteria isolated from mangrove sediments from the central of Thailand. *Marine Pollution Bulletin, 74*(1). Available from https://doi.org/10.1016/j.marpolbul.2013.07.025.

Wrede, D., Taha, M., Miranda, A. F., Kadali, K., Stevenson, T., Ball, A. S., & Mouradov, A. (2014). Co-cultivation of fungal and microalgal cells as an efficient system for harvesting microalgal cells, lipid production and wastewater treatment. *PLoS One, 9*(11). Available from https://doi.org/10.1371/journal.pone.0113497.

Xu, G., Li, Y., Zheng, W., Peng, X., Li, W., & Yan, Y. (2007). Mineralization of chlorpyrifos by co-culture of *Serratia* and *Trichosporon* spp. *Biotechnology Letters, 29*(10). Available from https://doi.org/10.1007/s10529-007-9444-0.

Zhang, Y., Gao, W., Lin, F., Han, B., He, C., Li, Q., ... Zheng, L. (2018). Study on immobilization of marine oil-degrading bacteria by carrier of algae materials. *World Journal of Microbiology and Biotechnology, 34*(6). Available from https://doi.org/10.1007/s11274-018-2438-3.

Zhong, Y., Luan, T., Lin, L., Liu, H., & Tam, N. F. Y. (2011). Production of metabolites in the biodegradation of phenanthrene, fluoranthene and pyrene by the mixed culture of *Mycobacterium* sp. and *Sphingomonas* sp. *Bioresource Technology, 102*(3). Available from https://doi.org/10.1016/j.biortech.2010.09.113.

Zhou, K., Qiao, K., Edgar, S., & Stephanopoulos, G. (2015). Distributing a metabolic pathway among a microbial consortium enhances production of natural products. *Nature Biotechnology, 33*(4). Available from https://doi.org/10.1038/nbt.3095.

CHAPTER 7

Biological transformation as a technique in pollution decontamination

Mohamed Nouri
Faculty of Polydisciplinary, Sultan Moulay Slimane University, Beni Mellal, Morocco

7.1 Introduction

The perceived exponential industrial growth is founded on the massive production of pharmaceuticals, heavy metals, chemicals, and explosives (Jha et al., 2020). However, the wastewater discharged by their production units seriously damages the sustainability of our environment. The presence, fate, and environmental risks of pollutants in life-environment systems have been of increasing interest to the scientific community, with many outstanding questions. The statement that pollutants are a very sundry group of compounds, with many structures, poses a challenge for investigation and requires a strategy for our upcoming examinations (Jha et al., 2020).

At present, there are sundry ways to detoxify pollutants, such as ultrasonic treatment, adsorption, chemical oxidation, solvent extraction, photocatalysis, nanocomposites, electrochemical decomposition, and biological remediation (Huang et al., 2021; Nouri, 2021; Silva Conceição et al., 2021). However, the physiochemical processes may also produce more toxic secondary pollutants, which is both expensive and difficult. Using local living beings to eliminate pollutants from natural matrix has become a searching hotspot because of its high efficiency, economy, and environmental protection (Huang et al., 2021).

Compared with physicochemical treatment, bioremediation has the following benefits: no secondary pollution, low cost, no soil disturbance, in situ remediation, and uncomplicated operation, making it the most reliable pollutant remediation technology (Bhat et al., 2021; Hou et al., 2020; Xia et al., 2019). Bioremediation is based on biologically-derived materials (agricultural and forestry wastes and biochar) and biological

Microbial Consortium and Biotransformation for Pollution Decontamination
DOI: https://doi.org/10.1016/B978-0-323-91893-0.00011-0

materials (bacteria, animals, fungi, plants, and algae). The main mechanisms for contaminant bioremediation are (Jobby, Jha, Yadav, & Desai, 2018): (1) biological sorption, which includes the adsorption and amelioration of pollutants by organisms and biological developed materials, thus reducing the amount of poisons in the environment, and (2) biological transformation, which refers to the transformation of highly toxic and mobile pollutants into low toxic and nontoxic forms, making them unhazardous to the environment. Both methods possibly will diminish the bioavailability and mobility of pollutants, as well as the environmental absorption of other biota components in crops and food chains (Xia et al., 2019). Compared to biological sorption, biological transformation was the central mechanism implicated in pollutant removal (Ariste et al., 2020).

Accordingly, this chapter provides a new understanding for the biological transformation of contaminants, including the metabolic pathway, catalytic mechanism, and genetic basis of bioremediation of infected environment.

7.2 Biological transformation

Biological transformation is a green biological process. Compared with traditional physicochemical methods, biological transformation usually has the advantages of high catalytic efficiency, good selectivity, cost competitiveness, and easy operation or no downstream treatment (Ahmad, Gul, & Gul, 2020). In addition, biological transformation is passed under gentle conditions, with low energy consumption and environmental protection technology (Ripoll & Betancor, 2021; Sheldon & Woodley, 2018).

In the past decade, interest in biological processes has grown. Nevertheless, the input of biological transformation and catalysis to green chemistry is little known and has not been well translated into industry. This is a problem solved by scientists and technicians in time, which reduces the implementation speed of industrial biological transformation. This is usually due to processes with lower conversion efficiency and/or lower cost-effectiveness compared to prior art. In this regard, it is very influential to strengthen the biological method to achieve the quality standard, improve the processing efficiency, reduce the cost, and consider the environmental features in the optimal design (Leemans Martin, Peschke, Venturoni, & Mostarda, 2020; Ripoll & Betancor, 2021).

Pollutant biodegradation is a promising method and has been paid more and more research attention. Biological transformation has been verified as the most effective tool for this method (Jha et al., 2020).

Biological transformation can be carried out under aerobic, anaerobic, or electrochemically stimulated conditions (Tang et al., 2021) (Fig. 7.1). Moreover, living beings can degrade or transform pollutants by three main mechanisms: (1) growth related (metabolic), in which pollutants are used as the single energy and carbon source. (2) Nongrowth related (cometabolic), in which organisms grow using alternative sources of energy and carbon in the ecosystem and produce enzymes that catalyze the biological transformation of contaminants into various products. (3) These convert the pollutants into metabolites but do not degrade the metabolites (Pratush et al., 2020). Furthermore, biological transformation includes cleavage, reduction, oxidation, glycosylation, conjugation with glutathione, hydrolysis, and any form of fate (Ding, Wang, Yang, & Li, 2020; Jha et al., 2020; Kovačič et al., 2021; Tang et al., 2021; Wang, Li, et al., 2020) (Fig. 7.2).

The mechanisms of biological transformation and detoxification of pollutant kinds have been the subject of active research because of their importance to safety of human and environment (Hasegawa et al., 2019). The biological transformation of pollutants in living organisms has been extensively studied (Hasegawa et al., 2019).

As examples, through a series of demethylation, dehydration, decarboxylation, and deamination reactions, *Sphingobacterium mizutaii* and *Brevundimonas naejangsanensis* can transform doxycycline into a variety of biotransformation products (He et al., 2021). The main reactions concerned in the biological transformation of organic micropollutants

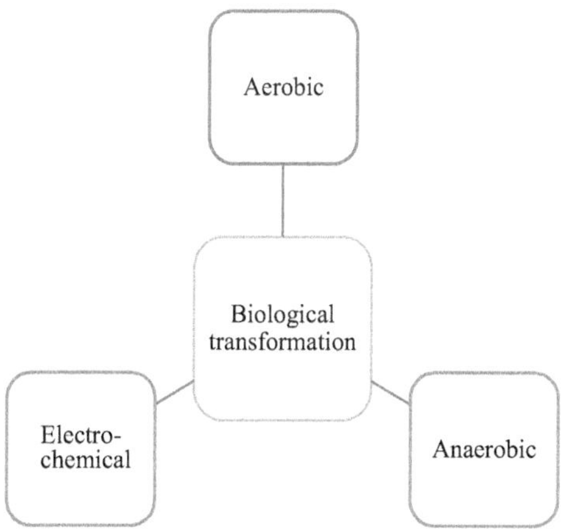

Figure 7.1 Biological transformation pathways.

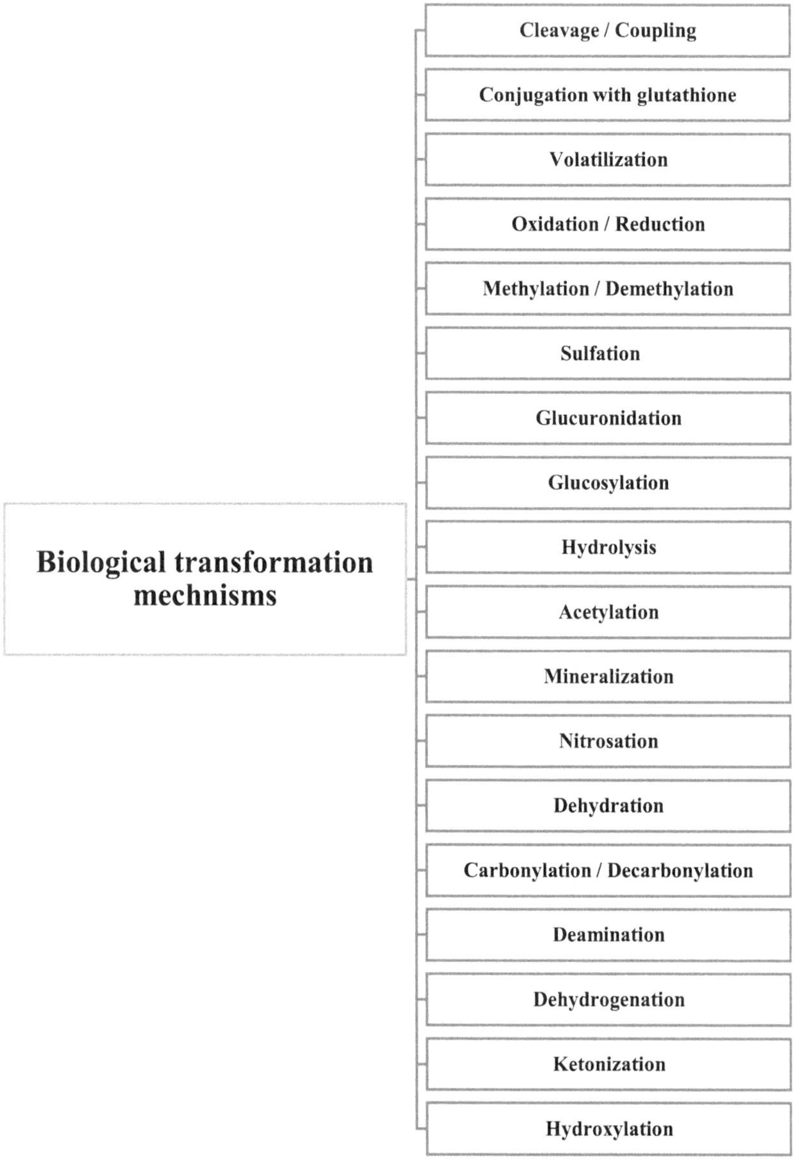

Figure 7.2 Mechanisms implicated in biological transformation.

(OMPs) under heterotrophic aerobic conditions are oxidation (dehydrogenation, hydroxylation, demethylation, and deamination); conjugation; and hydrolysis pathways (Kennes-Veiga, Vogler, Fenner, Carballa, & Lema, 2021).

Moreover, biological transformation of perfluoroalkyl acid (PFAA) precursors principally implicates cleavage of C-F liaisons and degradation of nonfluorinated divisions by defluorination, oxidation, and dealkylation to procedure petite chain PFAAs (Zhang et al., 2021).

Jobby et al. (2018) revealed that using the biomass of fungi, bacteria, algae, and plants to transform Cr (VI), it is proved that biological transformation is an efficient, environmental, and economic strategy to remediate Cr (VI) pollution.

The biological transformation of perfluorooctane sulfonamide (FOSA) in plant, soil, earthworm, and microorganisms has been verified as an important behavior in these ecosystems (Zhang et al., 2021).

7.3 Biological transformation classes

So far, a large number of pollutant-transforming living beings have been discovered and described, containing bacteria, plants, fungi, animals, and algae (Fig. 7.3) to remove harmful poisons in the ecosystem (Kumar et al., 2021).

7.3.1 Biotransformation

7.3.1.1 Biotransformation of pharmaceutical compounds

The biotransformation of three fluoroquinolone (FQ) and lincomycin (LIN) antibiotics was likely achieved by (1) hydroxylation, nitrosation, or S-oxidation; (2) demethylthioation or demethylation; and (3) division of the piperazine fragment (Zhou et al., 2021). The NH_2OH can react with LINs and FQs, while the transitional NO has been engaged in the synthesis of N-notroso elements of norfloxacin and ciprofloxacin (Zhou et al., 2021). Moreover, Zhou et al. (2021) demonstrated the biotransformation capabilities and biochemical pathways of FQs and LIN by three diverse ammonia oxidants: *Nitrospira inopinata*, *Nitrososphaera gargensis*, and *Nitrosomonas nitrosa* Nm90. The *N. gargensis* exhibited better biotransformation capabilities of LINs and FQs than *N. inopinata* and *N. nitrosa* Nm90 strains (Zhou et al., 2021).

In other study, *S. mizutaii* and *B. naejangsanensis* were capable of biotransforming doxycycline by cometabolism (He et al., 2021). Furthermore, *Vibrio diabolicus* L2−2 is qualified to transform many sulfonamides and sulfamethoxazole (Wang, Jiang, et al., 2020; Wang, Wang, et al., 2021).

7.3.1.2 Biotransformation of metals and metalloids

An iron-oxidizing bacterium, *Acidithiobacillus ferrooxidans*, was selected to promote the synthesis of tooeleite and attenuate naturally arsenic (III) in

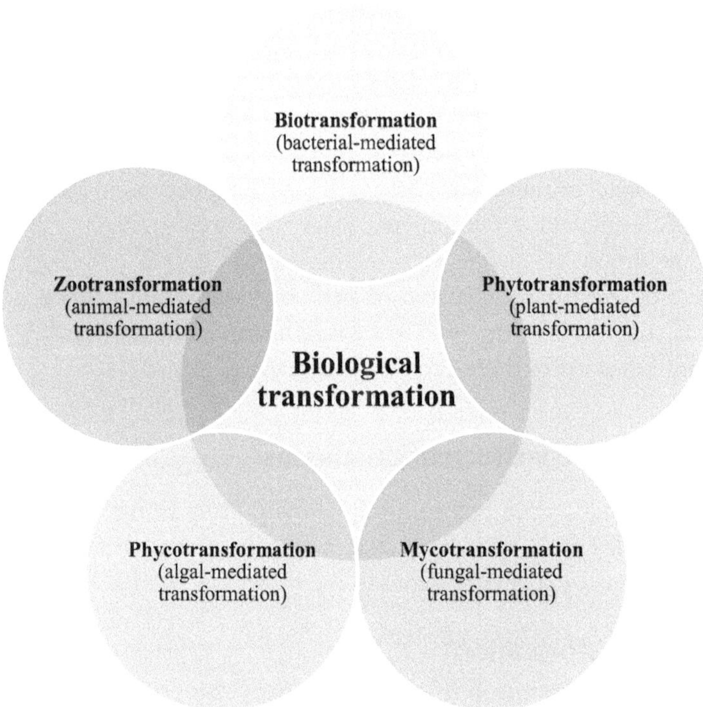

Figure 7.3 Biological transformation classes.

tooeleite (Wang, Li, et al., 2020). Under the conditions of pH 2.0, 30°C, and initial arsenic concentration of 500 mg/L, the removal rate of arsenic is 95.4% (Wang, Li, et al., 2020). The three steps of biosynthetic tooeleite crystallization include the oxidation of Fe^{2+}, the hydrolysis of Fe^{3+}, and the initial creation of Fe—As amorphous precursor, which eventually transforms into tooeleite crystal (Wang, Li, et al., 2020).

In addition, *Sinorhizobium* SAR1 can biotransform Cr (VI) to Cr (III) form with less toxicity. The bioremediation mechanism of chromium by *Sinorhizobium* SAR1 was that at first, Cr adsorbed on various components of cell wall, then Cr (VI) was transformed into Cr (III), and finally Cr (III) was released (Jobby, Jha, Gupta, Gupte, & Desai, 2019).

7.3.1.3 Biotransformation of phenol compounds

The biotransformation of phenol compounds using *Rhodococcus* sp. was an effective method to eliminate phenol from wastewater (Nor Suhaila et al., 2019; Wen, Li, Song, & Yang, 2020). *Bacillus natto* might be utilized in the biotransformation of polydatin to resveratrol in the roots of *Polygonum cuspidatum* (Fan et al., 2021).

Zhou, Zhang, Zhang, Cen, and Pan (2021) reported that microorganisms existing in drinking water systems (three strains of *Sphingomonas ursincola*, two strains of *Pseudomonas moraviensis*, one strain of *Sphingopyxis fribergensis*, one strain of *Sphingobium hydrophobicum*, and one strain of *Acidovorax soli*) can O-methylate toxic halophenols to earthy-musty haloanisoles. All bacteria might transform the five halophenols to the equivalent haloanisoles simultaneously (Zhou, Zhang, et al., 2021).

However, the bioremediation of phenol by growing cells is faced with the problems of restricted growing, short cell concentration, little reaction, and degradation rates. High concentration of phenol inhibits the development of phenol-transforming microorganisms (Nor Suhaila et al., 2019).

7.3.1.4 Biotransformation of pesticides

Bhatt, Zhou, Huang, Zhang, and Chen (2021) reported that the major allethrin degradation approach by *Sphingomonas trueperi* CW3 was division of the ester liaison, then degradation of the five-carbon rings. *S. trueperi* degraded additional commonly pyrethroids such as bifenthrin, cyphenothrin, β-cypermethrin, permethrin, tetramethrin, and chlor-empenthrin (Bhatt et al., 2021). Moreover, bioaugmentation based on strain CW3 efficiently eliminated allethrin with a considerably downgraded half-life in disinfected soil (Bhatt et al., 2021). In another study, *Pseudomonas nitroreducens* efficiently transformed a large kind of pyrethroids like a carbon source (Bhatt et al., 2021). In addition, a recent studies revealed that many bacteria were talented to transform pyrethroid (Zhan, Huang, Lin, Bhatt, & Chen, 2020) and chlorpyrifos (Huang et al., 2021).

7.3.1.5 Biotransformation of real effluents

An operative degradation of tannery effluent (TE) was achieved by a recently established consortium of bacteria (*Micrococcus lylae*, *Ochrobactrum intermedium*, and *Stenotrophomonas acidaminiphila*) in 120 h with 88.70%, 85.32%, 78.81%, 76.12%, 71.89%, 71.22%, 69.53%, and 48.59% reduction in pollution parameters such as phenol, biochemical oxygen demand (BOD), sulfate, chemical oxygen demand (COD), total dissolved solids (TDS), Cr, nitrate, and phosphate, respectively (Saxena, Purchase, Mulla, & Bharagava, 2020). Moreover, the study discovered that the toxicity of TE treated with bacteria was considerably diminished, allotting 70% of

Phaseolus aureus seeds to germinate, and thus proved the decontamination of TE (Saxena et al., 2020).

7.3.2 Phytotransformation

The phytoaccumulation and metabolic transformation of contaminants by plants are important behaviors in plant ecosystems, which have important influences on the human health and environment. In recent study, Jha et al. (2020) reported the phytotransformation of exogenous substances (phenols, drugs, pesticides, explosives, and dyes) by hairy roots of sundry plants.

7.3.2.1 Phytotransformation of fluorinated compounds

Fluorotelomeric alcohol (FTOHs) is widely considered as a precursor to perfluoroalkylate carboxylates (PFCAs), which can be absorbed by soybean from the solution and can be biodegraded in PFCAs (Zhang et al., 2016). Zhao, Zhou, et al. (2018) have revealed that N-ethyl perfluorooctane sulfonamide (N-EtFOSA) can be absorbed by crops (wheat, soybean, and pumpkin) and biologically converted into Per- and polyfluoralkyl substances (PFASs) with diverse carbon chain lengths. Moreover, in the stems and roots of soybean and pumpkin, FOSA can be converted into PFSAs with diverse carbon chain lengths, counting PFBS, PFOS, and PFHxS (Zhao, Liang, et al., 2018).

The uptake, translocation, and phytotransformation of fluorotelomer sulfonic acid (6:2 FTSA) in *Cucurbita maxima* were studied by Zhao et al. (2019). In addition, six terminal metabolites of perfluorocarboxylic acid (PFCA), counting perfluoroheptanoic acid (PFHpA), perfluoropentanoic acid (PFPeA), perfluorohexanoic acid (PFHxA), trifluoroacetic acid (TFA), perfluoropropionic acid (PFPrA), and perfluorobutanoic acid (PFBA), were recovered in pumpkin roots and shoots (Zhao et al., 2019). PFBA was the principal metabolite in shoots, while PFHpA was the highest product in roots (Zhao et al., 2019).

In new study, plant matrices (wheat grain, corn leaves and grain, Jerusalem artichoke, and ryegrass) were investigated from plants cultivated in PFAS-polluted fields (Muschket, Keltsch, Paschke, Reemtsma, & Berger, 2020). FTOH and/or perfluoroalkane sulfonamidoethanol (FASE) transformation intermediates were revealed in each experiments, with EtFOSAA and FOSA being the dominant products in corn leaves (Muschket et al., 2020). Consistent with the present results, the ending degradation metabolites PFOS and PFDA were habitually revealed

(Muschket et al., 2020). For PFCAs, based on previous results, petite chain substances usually showed the uppermost quantities (Muschket et al., 2020). However, grain of corn was a particular case, indicating the maximum amounts for PFCAs, while PFBA was not identified (Muschket et al., 2020).

The phytotransformation of 6:2 FTOH by *Glycine max* was studied using hydroponic experiments (Zhang, Wen, et al., 2020). During the 144 h of exposure, the phase I metabolites in soybeans were as follows: 6:2 FTUCA, 6:2 FTCA, 5:3 FTCA, 5:3 FTUCA, PFHxA, and PFPeA (Zhang, Wen, et al., 2020). On the completion of the exposure, 5:3 FTCA, PFPeA, and PFHxA were the three major products of the solution system. FTCA was prevalent in soybean stems and roots, while PFHxA was the greatest abundant substance in leaves (Zhang, Wen, et al., 2020). The PFBA and FTCA identified in the hydroponic milieu possible originate from the conversion of FTCA by root-related microorganisms (Zhang, Wen, et al., 2020). In addition, phase II products of FTOH were detected and supervised in tissues of soybean (Zhang, Wen, et al., 2020).

Zhou et al. (2019) were evoked that perfluorophosphinates (PFPiAs) underwent biotransformation, albeit at a small amount, to produce further persistent PFPAs in root of wheat. The finding of 1H-perfluoroalkanes (1H-PFAs) in wheat suggests that divide of the C-P liaison in PFPiAs did take place in wheat (Zhou et al., 2019).

7.3.3 Mycotransformation

Several species of fungi are advantageous for the decomposition and reutilizing of many toxic substances: these fungi comprise a complex enzyme collection, which are rapidly exuded and take action on numerous compounds (Silva Conceição et al., 2021). Fungal actions for the cleansing of ecosystem are identified as mycoremediation and rely on the huge enzymes of fungi. Even with their excellent capability to thrive in the existence of complex substances, the mechanism of mycoremediation persists feebly known (Silva Conceição et al., 2021). Moreover, the natural substance mycotransformation is an advantageous tool to find new bioactive products (Wu et al., 2020).

7.3.3.1 Mycotransformation of pesticides

The mycotransformation of the triazole herbicide, thiencarbazone-methyl (TCM), by *Aspergillus terreus*, *Penicillium chrysogenum*, *Aspergillus niger*, *Aspergillus fumigatus*, and *Aspergillus flavus* was discovered to be 98%, 95%,

81%, 74%, and 74%, respectively, in 42 days, revealing the efficacy of each strain employed in the degradation of TCM (Ahmad et al., 2020). Current research may have an impact on mycoremediation approaches that are vital for pesticide mitigation in ecological compartments (Ahmad et al., 2020). In addition, *Aspergillus* sp. (strain PYR-P2) has been employed efficaciously to clean up pyrethroid polluted liquids and soils (Kaur & Balomajumder, 2020).

The biodegradation pathways of chlorpyrifos, methyl parathion, and profenofos were studied with *Aspergillus sydowii* as biocatalyst, which provided a new way for mycoremediation in the future (Soares, Birolli, Ferreira, & Porto, 2021). Remarkably, lindane was degraded by *A. fumigatus*, as reported by (Dey et al., 2020). In addition, a recent studies revealed that numerous strains of fungi were able to transform pyrethroid (Zhan et al., 2020) and chlorpyrifos (Huang et al., 2021).

7.3.3.2 Mycotransformation of metals

The mycoremediation of metals (Ni, Cd, Cu, Pb, Cr, Zn) by *A. fumigatus* has been verified by Dey et al. (2020). In addition, under the concentration of 10–50 mg/L, the removal rate of Cr (VI) by *A. niger* was about 100% (Xu, Hao, et al., 2021). In general, 97% of the removal rate was produced by extracellular reduction and 3% by accumulation (Xu, Hao, et al., 2021). Extracellular reduction is facilitated by nonenzymatic secretion, while extracellular Cr accumulation precipitates on the surface of mycelium and is partially absorbed by cell wall (Xu, Hao, et al., 2021). Cr (VI) also entered cells and was reduced by chromate reductase in cell-free extract then accumulated in cells (Xu, Hao, et al., 2021).

Chen et al. (2019) evokes that *Phanerochaete chrysosporium* may be appropriate for the elimination of tetrabromobisphenol A from polluted waters. Elimination efficiency in the water microcosms was 63.1% within 12 h and >97% after 3 days (Chen et al., 2019).

Feng et al. (2019) indicated that *Pycnoporus sanguineus* and its extra and intracellular enzyme might efficiently degrade tetrabromobisphenol A (TBBPA). The effectiveness of TBBPA degradation by *P. sanguineus* and its enzymes diminished considerably when the Cr (VI) amount was augmented from 0 to 40 mg/L (Feng et al., 2019).

Experimental results imply that the fungus (*Trichoderma inhamatum*) is talented to transform wastewater Cr (VI) into Cr (III) under aerobic conditions (Morales-Barrera & Cristiani-Urbina, 2008). Moreover, *A. niger* has been identified as one of those potential microbes with high resistance to Cr and has shown an increased level of Cr absorption and reduction (Sriharsha, Lokesh Kumar, & Janakiraman, 2020).

7.3.3.3 Mycotransformation of pharmaceutical compounds

Tian et al. (2021) demonstrated the potential of fungal endophytes (*Penicillium oxalicum* B4) on the mycotransformation of artemisinic acid into its bioactive derivatives. Moreover, Wu et al. (2020) revealed the potential of marine fungi (*A. terreus* and *Cochliobolus lunatus*) to change the structures of natural products (cryptotanshinone) through mycotransformation.

7.3.3.4 Mycotransformation of phenol compounds

The utilization of *Trametes hirsuta* and *Pleurotus dryinus* for mycoremediation can obtain higher phenol removal rate (100% and 94%, respectively), which is helpful to remove other organic and inorganic substances and produce laccase in biorefinery wastewater (Ariste et al., 2020). Furthermore, *Aspergillus brasiliensis* effectively transformed methyl p-coumarate to two new products, 4-(3-hydroxyoxiran-2-yl)benzene-1,2-diol and 5-(oxiran-2-yl)benzene-1,2,3-triol, as well as acetophenone (Silva Conceição et al., 2021). Remarkably, the infrequent phenyloxirane fragment was produced as portion of the mycotransformation of the 4-vinylphenol transitional by oxidation of the double bond (Silva Conceição et al., 2021).

7.3.3.5 Mycotransformation of dyes

Singh and Dwivedi (2020) verified that *A. terreus* has an effective capacity for decolorization of excessive amount of direct blue-1 (DB-1) under improved conditions compared to ambient conditions via sorption and degradation. Mycodegradation of DB-1 was achieved by manganese peroxidase and laccase in an uncommon symmetric oxidative that breaks down the DB-1 dye into a less toxic product (Singh & Dwivedi, 2020). The phytotoxicity examination confirmed that the degraded substances are less toxic than the pure DB-1 dye (Singh & Dwivedi, 2020). So, *A. terreus* possibly will be an encouraging microorganism for the efficient degradation of DB-1 in carp sewage (Singh & Dwivedi, 2020).

7.3.4 Phycotransformation

Algae, as initial producers, ensure pollutant accumulation and geochemical cycling in the marine ecosystem. They have an eminent competence in absorbing pollutants from the ecosystem, which renders these microscopic species ecological and cost-effective in terms of pollutant cleaning up process (Hasegawa et al., 2019; Papry et al., 2021).

7.3.4.1 Phycotransformation of metals and metalloids

Phycotransformation of As (V) to As (III) and As methylation within phytoplankton cells happened primarily in the log phase, in contrast phycotransformation of As to organoarsenic products happened in the stationary phase (Hasegawa et al., 2019). Phytoplankton speedily liberate inorganoarsenic and methylarsenic species from their cells at the log phase, whereas organoarsenic species remain mostly within their cells (Hasegawa et al., 2019). In addition, Papry et al. (2021) exposed a diverse pathway of As phycotransformation, As accumulation, and growth of phytoplankton in several salinity stress. Two phytoplankton namely *Closterium aciculare* and *Pediastrum duplex* were talented to reduce As (V) to As (III) and methylated species, that is, dimethylarsinate. However, *Scenedesmus acutus* and *Staurastrum paradoxum* did not show a methylation reaction. Furthermore, the phytoplankton strains could be classified according to their As phycotransformation pathways as: (1) those that are effective in the phycotransformation of As (V) to As (III); (2) those that are not efficient in the phycotransformation of As, which instead preserved As (V) within their cells; and (3) those that are effective in the phycotransformation of inorganoarsenic species to methylarsenic species and complex organoarsenic species (Hasegawa et al., 2019).

The research reported by Papry et al. (2019) describes the development, phycotransformation, and speciation of As at different salinities and temperatures of six marine diatom species: *Thalassionema nitzschioides*, *Asteroplanus karianus*, *Chaetoceros didymus*, *Ditylum brightwellii*, *Skeletonema* sp., and *Nitzschia longissima*. The development rate and As phycotransformation abilities of these species in 3 weeks were considerably influenced by wide salinity and temperature (Papry et al., 2019). At optimal salinity and temperature, cell dimension and As phycotransformation were advanced for all species (Papry et al., 2019).

7.3.4.2 Phycotransformation of pharmaceutical compounds

Ding et al. (2020) confirmed that *Navicula* sp. was talented to effectively eliminate atenolol, ibuprofen, carbamazepine, and naproxen with efficiencies of $>90\%$ after 21 days of exposure. Compared to the elimination competences of each drug in the separate drug treatments, the degradation of bezafibrate, sulfamethoxazole, and naproxen was progressed in the combination treatment, while the elimination proficiencies of carbamazepine and atenolol were decreased (Ding et al., 2020). Moreover, the existence of hydrophobic pharmaceuticals (i.e., naproxen and ibuprofen) augmented the degradation of sulfamethoxazole and carbamazepine and reserved the removal of atenolol in the combination of six pharmaceuticals, whereas the addition of other compounds showed no considerable consequence on the elimination of naproxen and ibuprofen (Ding et al., 2020).

7.3.5 Zootransformation

Earthworms, rainbow trout, common carp, scallop, and fish can store and zootransform organic and inorganic contaminants through ingestion, assimilation, digestion, casts, and enzymes in the tissues (Liu et al., 2020; Saunders, Fitzsimmons, Nichols, & Gobas, 2020; Yang, Liu, Tan, Wang, & Wang, 2021; Zhao, Kang, et al., 2021; Zhao, Liu, et al., 2021).

7.3.5.1 Zootransformation of fluorinated compounds

Previous investigations verified that the precursors of PFAAs, such as N-EtFOSE (Zhao, Ma, Fang, & Zhu, 2016), N-EtFOSA (Zhao, Wang, et al., 2018), and 10:2 FTOH (Zhao & Zhu, 2017), could be accumulated by earthworms from soil and zootransformed into PFAAs. Recently, 6:2 FTSA was speedily accumulated in earthworms and zootransformed to five PFCA end metabolites, counting TFA, PFBA, PFPrA, PFHxA, and PFPeA with the exemption of PFHpA (Zhao, Liu, et al., 2021). TFA was the main stable zootransformation metabolite of 6:2 FTSA in earthworms (Zhao, Liu, et al., 2021).

Saunders et al. (2020) assessed intestinal biotransformation in *Oncorhynchus mykiss*, Erwin strain utilizing S9 fractions formulated from the upper intestinal epithelium. The results displayed that S9 fractions formulated zootransform a diversity of chemicals. Baseline metabolic actions for regular substrates of phase I and II zootransformation enzymes did not exceed twice the actions determined in hepatic S9 fractions (Saunders et al., 2020). The testosterone hydroxylation rate of intestinal S9 fraction was considerably upper than that of hepatic S9 fraction, signifying that intestinal epithelium is an imperative site for the zootransformation of fish cytochrome P450 3A substrate (Saunders et al., 2020).

In vitro and in vivo researches revealed that PFPiAs underwent zootransformation and formed PFPAs in *Cyprinus carpio* and that the zootransformation of 8:8 PFPiA was more active than that of 6:6 PFPiA (Liu et al., 2020).

It would be mentioned that perfluorooctanoic acids and perfluorohexanonate were first detected in fish as products after long exposure to PFPiAs (Liu et al., 2020).

7.3.5.2 Zootransformation of metals and metalloids

A current research revealed that exposure to As (V) and As (III) can encourage different zooaccumulation and zootransformation responses in five scallop (*Chlamys farreri*) tissues (Zhao, Kang, et al., 2021). The zootransformation of inorganic As to dimethylarsinic acid was clearly augmented and changed for diverse As species in diverse *C. farreri* tissues (Zhao, Kang, et al., 2021). Moreover, arsenobetaine was the main As species (34%–66%), and its concentration showed an upward trend, while inorganic arsenic content decreased by 84%–91% in 40 days for crucian carp (*Carassius auratus*) (Cui et al., 2020). This suggests that the zootransformation ability of arsenic increases during chronic dietary exposure to adapt to inorganic arsenic (Cui et al., 2020).

Yang et al. (2021) highlighted the position of the microbiota of gut in the biotransformation of Hg in fish (*Acanthopagrus latus*) and suggested a moderate variation of the gut microbiome as a possible solution to minimize Hg contamination in fish.

Xu, Hong, et al. (2021) presented that the zootransformation process of housefly could promote the decomposition of compost and the heavy metal inactivation and deliver a great quantity of insect protein, which had good environmental and economic profits. Moreover, the zootransformation of housefly larvae encouraged the transformation of fulvic acid (FA) to humic acid (HA) in pig manure (Xu, Hong, et al., 2021). The final increase of HA/FA was the highest in diverse kinds of pig manure, which promoted the heavy metal inactivation (Xu, Hong, et al., 2021).

7.4 Factors influencing biological transformation

Biological transformation methods are usually directly or indirectly controlled by a set of factors (Fig. 7.4). Understanding these factors is imperative to get the best results. All kinds of abiotic and biological factors can affect the biological transformation of contaminants (Fig. 7.4). Milieu physicochemical properties have effects on biological transformation. Environmental features that are vital constraints for the development and metabolic procedures of living beings differ from species to species. Most investigations have centered on environmental features counting salinity,

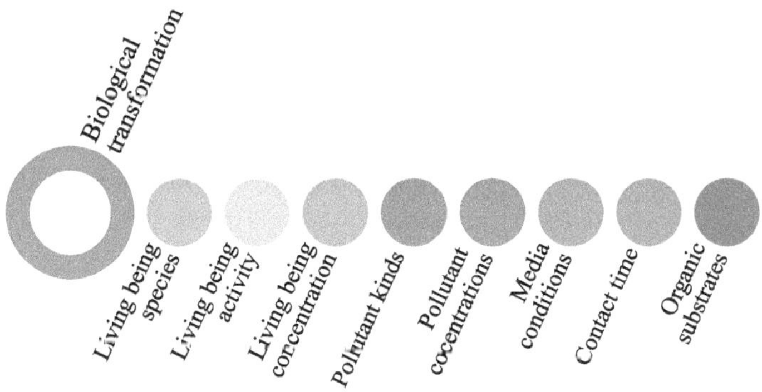

Figure 7.4 The principal factors influencing the biological transformation.

pH, temperature, light intensity, and nutrient availability. Moreover, microaerobic conditions amplified both the richness and diversity of the reactor microbiota, likely promoting the growth of oxygenase-producing microorganisms, which may have played a role in the biological transformation of parabens (do Nascimento et al., 2021).

In-depth abiotic and biological features influencing the biological transformation of contaminants can be discovered in an early review (Carneiro et al., 2020; do Nascimento et al., 2021; Hamid, Li, & Grace, 2020; Hasegawa et al., 2019; Kovačič et al., 2021; Papry et al., 2019; Papry et al., 2021; Seller, Honti, Singer, & Fenner, 2020).

7.5 Functional genes implicated in biological transformation

Transcriptome assay has considerable benefits in apprehending the molecular processes of microbiological reactions to pollutants or stress (Wang, Wang, et al., 2021).

Tang et al. (2021) summarized the functional genes implicated in organoarsenic biotransformation and the corresponding microorganisms. In addition to the genes directly responsible for organoarsenoid processing, there are various other genes that play a central role in assisting organoarsenoid processing. Based on their functions, these genes can be grouped into two groups: those that control arsenic uptake and that control arsenic detoxification (Tang et al., 2021).

In addition, *V. diabolicus* biotransformation genes could involve cytochrome c553 (*cyc-553*), arylamine N-acetyltransferases (*nat*) and acyl-CoA synthetase (*acs*) as reported by Wang, Wang, et al. (2021). At the natural amount of sulfamethoxazole (SMX) (0.1−10 μg/L), *nat* was inactivated, meaning that SMX acetylation may not take place in the ecosystem. Nevertheless, *cyc-553* was upregulated under 1 μg/L of SMX, indicating that SMX hydroxylation possibly will take place in the ecosystem (Wang, Wang, et al., 2021). Moreover, antioxidation and membrane transport of *V. diabolicus* may perhaps be stimulated under 10 μg/L SMX stress (Wang, Wang, et al., 2021). And, the *vdnatG* and *vdnatA* genes were probably to be the crucial genes in the acetylation procedure of sulfonamides (SAs), which could encode arylamine N-acetyltransferase (Wang, Jiang, et al., 2020).

7.6 Enzymes involved in biological transformation

Enzymes are essential in the biological treatment of pollutant degradation trails (Bhatt et al., 2021; Huang et al., 2021). The enzymes catalyze various reactions, which may lead to the decomposition of natural resistant polymers such as lignin and various biological mixtures, including aromatics, organic chlorides, and pesticides (Ariste et al., 2020; Huang et al., 2021; Soares et al., 2021). For example, esterases are naturally obtained in plants, animals, and microorganisms. In addition, esterases can be used for pesticide biological transformation applications (Bhatt et al., 2021). The different products of phosphatase, methyltransferase, ethyltransferase, and cytochrome P450 were detected and confirmed by synthetic standard in mycotransformation of pesticides (Soares et al., 2021).

The study performed by Kennes-Veiga, Vogler, et al. (2021) allows the selection of mono- and dioxygenases, hydrolases, dehydrogenases, and transferases as selected of the foremost enzymatic activities probably in charge for the OMPs' biotransformation under heterotrophic situations.

Cytochrome P450 (CYP450) enzymes, which are the principal family of enzyme proteins in organisms, are implicated in biological metabolism phase I and xenobiotic detoxification (Huang, Wang, Wen, Lv, & Zhang, 2019). Glutathione S-transferase (GST) is identified to be a phase II enzyme that is likewise implicated in the metabolism of several contaminants in ecosystem (Huang, Zhang, Wang, & Lv, 2013). Preceding research has revealed that GST and CYP450 are in charge for the zootransformation of PFAA precursors (Nabb et al., 2007) and phytotransformation (Zhao et al., 2019). Nevertheless, the functions of GST and CYP450 in the metabolism of FTSA to PFCA, which is decisive for apprehending the biological procedures of FTSA in earthworms, are even now undetermined (Zhao, Liu, et al., 2021). In addition, Zhao, Liu, et al. (2021) demonstrated that FTSA in earthworm homogenates and raw enzyme mixtures zootransformed at diverse levels in the subsequent order: homogenates > CYP450 > GST. GST and CYP450 have been affected in the metabolism of FTSA in earthworms.

Alcohol dehydrogenase, glutathione S-transferase, and aldehyde dehydrogenase were found to be implicated in the metabolism of FTOH in young soybean plants (Zhang, Wen, et al., 2020). For the mechanism of mycotransformation, both extracellular and intracellular enzymes were included (Ariste et al., 2020).

Furthermore, several fungi and bacteria have developed to produce lignocellulolytic enzymes talented of degrading the key lignocellulosic polymers in a synergistic system (Saldarriaga-Hernández et al., 2020).

7.7 Nanomaterial biological transformation

Biological transformation is necessary because the environment is filled with organisms, such as plants, protozoa, fungi, and bacteria/archaea. Biological transformation may include the chemical and/or physical transformation methods, which nevertheless is facilitated by biological procedures. Engineered nanomaterials (ENMs) can accumulate in biological fluids, on root surfaces or in the rhizosphere (Zhang, Guo, et al., 2020). Dissolution, degradation, redox reactions, or chelation can occur in the cell membrane, cytoplasm, vacuole, or extracellular via enzyme supported reactions or by interaction with reactive oxygen species (Zhang, Guo, et al., 2020). For example, graphene can be phytotransformed by reaction with OH, with the consequent liberation of CO_2 (Zhang, Guo, et al., 2020).

The rhizosphere is the greatest important place for the transformation of ENMs (Zhang, Guo, et al., 2020). This narrow region around the roots is controlled by complicated relations between the plant and other related organisms, counting microorganisms and invertebrates (Zhang, Guo, et al., 2020). Even though various investigations have indicated alteration of ENM in plant tissues, limited investigations have distinguished between transformation processes that happen externally (in the soil or at the root surface) and internally (in cells, sap, etc.) (Zhang, Guo, et al., 2020).

The biotransformation of manufactured carbon nanomaterials (CNs) can significantly alter their fate and toxicity in the ecosystem. So far, the mechanism of nanomaterial biotransformation is still not understood well, even though several efforts have been made (Wang, Ma, et al., 2020). Only a few bacteria have been identified as capable of biotransforming CNs and the mechanism behind this has not yet been revealed (Wang, Ma, et al., 2020).

The biotransformation of CNs [graphene oxide, reduced graphene oxide, oxidized (carboxylated), and single-walled carbon nanotubes] by *Labrys* sp., which could grow with CNs as the single carbon source, was studied (Wang, Ma, et al., 2020). Moreover, *Labrys* sp. could degrade and absorb carbon from multiwall carbon nanotubes to promote growth (Wang, Shan, et al., 2021). In addition, the fundamental CNs' biodegradation mechanism facilitated by an extracellular Fenton-like reaction was confirmed (Wang, Ma, et al., 2020). In this reaction, the synthesis of -OH was intermediated by the reduction of H_2O_2 involved in a permanent Fe(II)/Fe(III) cycle (Wang, Ma, et al., 2020). These results reveal a new mechanism of degradation by microorganisms toward high-molecular-weight substrates, which will aid to better comprehend the fate of CNs in the environment and deliver guidance for safer use of these substances (Wang, Ma, et al., 2020).

7.8 Cometabolic biological transformation

Since many pollutants are frequently existent in trace amounts, they cannot be utilized as the single sources of energy and carbon for biota metabolism and the initiation of biological transformation enzymes. Therefore cometabolism is considered crucial for their biological degradation (Trejo-Castillo, El Kassis, Cuervo-López, & Texier, 2021). In biological transformation, the substrate inserted to the system does not encourage cell development, but in the case of primary substrate, it will be transformed into other compounds. Cometabolism has been expressed as the transformation of a non-growth substance in the required existence of a growth substance or other energy source (Dalton & Stirling, 1982; Trejo-Castillo et al., 2021). The cometabolic capabilities of microbes can conduct to the adventitious transformation of pollutants that cannot be effectively employed as growth substances. In latest years, the input of cometabolism to the biological transformation of emerging organic contaminants has been extensively studied as it may correspond to a suitable occasion for biological remediation strategies (Trejo-Castillo et al., 2021). In overall, cometabolism is affected by several factors (retention time, pollutant kind, microbial strain, and nutrient natures) (Vo et al., 2020). However, the comprehending of cometabolism is still imperfect (Vo et al., 2020).

7.8.1 Cometabolic biotransformation

Trejo-Castillo et al. (2021) studied the function of cometabolism provoked by heterotrophic and ammonia oxidizing bacteria in biotransformation of benzotriazole exercising physiologically stable nitrifying sludge. The results indicated that the increasing initial ammonia nitrogen concentration (100−300 mg N/L) can obtain higher cometabolic biotransformation ability, and the benzotriazole-specific removal rates are 2.3 and 5.8 times higher. Under the condition of 300 mg NH_4^+ N/L, 40.8% of benzotriazole and 77.6% of ammonium in sludge were totally oxidized to nitrate (Trejo-Castillo et al., 2021).

The new study reported by Martínez-Quintela et al. (2021) is helpful to better understand the capability of N-damo medium to transform OMPs. Moreover, the specific removal of erythromycin, fluoxetine, roxithromycin, and sulfamethoxazole was related to the specific removal of nitrite, indicating that it was a cometabolic biotransformation (Martínez-Quintela et al., 2021).

Kennes-Veiga, Gonzalez-Gil, Carballa, and Lema (2021) showed that higher organic loading rates lead to higher rates of biotransformation of OMPs and that the removal of OMPs and organic matter occurs at the same time, demonstrating that cometabolism is the primary mechanism for OMP biotransformation under aerobic heterotrophic conditions. Moreover, organic substrate management and supervise of microbial action seem to be main factors regulating the biotransformation of OMPs (Kennes-Veiga, Gonzalez-Gil, et al., 2021).

Nguyen, Nghiem, and Oh (2018) showed that *Bradyrhizobium* sp. divide ciprofloxacin (CIP) by cometabolism in the existence of additional carbon substrate pursuing a first-order kinetic degradation reaction. In addition, CIP biotransformation by *Bradyrhizobium* sp. augmented when the quantity of the principal carbon source improved. The biodegradability of the principal carbon source likewise influenced the degradation of the CIP (Nguyen et al., 2018). Moreover, CIP biotransformation decreased with increasing initial CIP concentration. More than 70% of the biotransformation of CIP was achieved at 0.05 mg/L, while the degradation of CIP reduced to 26% at 10 mg/L (Nguyen et al., 2018).

Carneiro et al. (2020) reported that the acidogenic stage of anaerobic digestion is the secret to the biotransformation of OMPs, while carbohydrate hydrolysis has no significant involvement to the biotransformation of the studied compounds. The acidogenic condition is better than that of acetogenesis/methanogenesis for biotransformation of some OMPs

(Carneiro et al., 2020). The biotransformation of other compounds depends on the formation of acetogenesis/methanogenesis (Carneiro et al., 2020). Therefore the separation of hydrolytic/acidogenic and acetogenic/methanogenic phases in anaerobic treatment is a promising method to improve the bioconversion rate of OMPs (Carneiro et al., 2020).

The molecular structure of OMPs is an important factor affecting its anaerobic biotransformation (Carneiro et al., 2020). That is to say, the existence of electron donor group can explain the high bioconversion of some compounds, while electron absorption group may have a negative effect (Carneiro et al., 2020).

Finally, it is assumed that the biotransformation of fluorotelomer compounds is cometabolic in nature (Hamid et al., 2020).

7.8.2 Cometabolic phycotransformation

Vo et al. (2020) examined the effects of single carbon source on cometabolism of micropollutants (MPs) in *Chlorella* sp. by: (1) production of extracellular polymeric substances (EPS), peroxidase and superoxide dismutase; (2) MPs' elimination effectiveness and cometabolism rates; (3) identification of aptitude decomposition of MPs; and (4) decomposition pathways and validation to distinguish the cometabolism of *Chlorella* sp. with other microorganisms. The addition of carbon sources alone in the company of MPs augmented the concentrations of EPS and enzymes by 2 to 100 times compared to carbon sources alone (Vo et al., 2020). This proved that cometabolism of MPs had taken place. The elimination competences of bisphenol A, sulfamethoxazole, and tetracycline were 58%–99%, 32%–92%, and 16%–99%, respectively (Vo et al., 2020). By enhancing the EPS and enzyme action, the amounts of MPs stored in the microalgae cells were reduced 400-fold (Vo et al., 2020). Consequently, the appropriate microalgae carbon sources can be designated for useful purposes to address MPs in wastewater at the same time as improving biomass for multiple industries and earning revenue (Vo et al., 2020). However, microalgae cometabolism for pollutant cleanup needs considerable consideration. Numerous engineering features of microalgae cometabolism require to be adopted (Vo et al., 2020).

7.9 Conclusions and future perspectives

This chapter reviews the biological transformation pathways, the functional genes and enzymes implicated in biological transformation. Biological transformation is an effective strategy to remove environmental pollutants by exploiting bacteria, plants, fungi, algae, or animals. At present, many living organisms with pollutant transformation potential have been isolated and identified. However, there is still a big interval between the laboratory discovery and fruitful employment of biological remediation technology in the cleanup of polluted locations. The majority of laboratory investigations use contaminated synthetic solutions to evaluate the competence of pollutant remediation. Therefore it is necessary to assess the potential of biological systems to treat contaminated soil or industrial wastewater.

Moreover, some of the important research opportunities in the area of biological transformation of pollutants that could be pursued are as follows:

- It is necessary to select new species and adaptive living beings' consortia suitable for pollutant transformation to improve pollution mitigation, competitiveness, and feasibility.
- It is also important to identify other biological transformation methods and mechanisms that can alleviate the pressure of pollutants.
- The role, activity, and regulation of enzymes in biological transformation process need further study.
- In changing environmental scenarios (climate change), it is crucial to give special consideration to the performance of biological transformation. Without these studies, it is difficult to understand the life span of biological transformation treatment.
- Advances in molecular and genetics-based biological transformation technologies could be useful in elaborating recombinant products of native microbes to control the contamination difficulties.
- Advance investigation is desired to investigate the microbial degradation of pollutants using detailed molecular technology, which could deliver novel visions to the scientific community to disclose a comprehensive bioremediation process for polluted milieus.
- Future researches are required to clarify the correlation between autotrophic and heterotrophic biomass growing and the biotransformation of pollutants, to elaborate a cometabolic style that can expect the elimination of contaminants in biological treatment systems.

References

Ahmad, K. S., Gul, P., & Gul, M. M. (2020). Efficient fungal and bacterial facilitated remediation of thiencarbazone methyl in the environment. *Environmental Research, 188*, 109811. Available from https://doi.org/10.1016/j.envres.2020.109811.

Ariste, A. F., Batista-García, R. A., Vaidyanathan, V. K., Raman, N., Vaithyanathan, V. K., Folch-Mallol, J. L., ... Cabana, H. (2020). Mycoremediation of phenols and polycyclic aromatic hydrocarbons from a biorefinery wastewater and concomitant production of lignin modifying enzymes. *Journal of Cleaner Production, 253*, 119810. Available from https://doi.org/10.1016/j.jclepro.2019.119810.

Bhat, R. A., Singh, D. V., Qadri, H., Dar, G. H., Dervash, M. A., Bhat, S. A., ... Yousaf, B. (2021). Vulnerability of municipal solid waste: An emerging threat to aquatic ecosystems. *Chemosphere, 287*(2022), 132223. Available from https://doi.org/10.1016/j.chemosphere.2021.132223.

Bhatt, P., Zhou, X., Huang, Y., Zhang, W., & Chen, S. (2021). Characterization of the role of esterases in the biodegradation of organophosphate, carbamate, and pyrethroid pesticides. *Journal of Hazardous Materials*. Available from https://doi.org/10.1016/j.jhazmat.2020.125026.

Carneiro, R. B., Gonzalez-Gil, L., Londoño, Y. A., Zaiat, M., Carballa, M., & Lema, J. M. (2020). Acidogenesis is a key step in the anaerobic biotransformation of organic micropollutants. *Journal of Hazardous Materials, 389*, 121888. Available from https://doi.org/10.1016/j.jhazmat.2019.121888.

Chen, Z., Yin, H., Peng, H., Lu, G., Liu, Z., & Dang, Z. (2019). Identification of novel pathways for biotransformation of tetrabromobisphenol A by *Phanerochaete chrysosporium*, combined with mechanism analysis at proteome level. *The Science of the Total Environment, 659*, 1352–1361. Available from https://doi.org/10.1016/j.scitotenv.2018.12.446.

Cui, D., Zhang, P., Li, H., Zhang, Z., Song, Y., & Yang, Z. (2020). The dynamic effects of different inorganic arsenic species in crucian carp (*Carassius auratus*) liver during chronic dietborne exposure: Bioaccumulation, biotransformation and oxidative stress. *The Science of the Total Environment, 727*, 138737. Available from https://doi.org/10.1016/j.scitotenv.2020.138737.

Dalton, H., & Stirling, D. I. (1982). Co-metabolism. *Philosophical Transactions of the Royal Society of London. Series B, Biological Sciences, 297*, 481–496. Available from https://doi.org/10.1098/rstb.1982.0056.

Dey, P., Malik, A., Mishra, A., Singh, D. K., von Bergen, M., & Jehmlich, N. (2020). Mechanistic insight to mycoremediation potential of a metal resistant fungal strain for removal of hazardous metals from multimetal pesticide matrix. *Environmental Pollution (Barking, Essex: 1987), 262*, 114255. Available from https://doi.org/10.1016/j.envpol.2020.114255.

Ding, T., Wang, S., Yang, B., & Li, J. (2020). Biological removal of pharmaceuticals by Navicula sp. and biotransformation of bezafibrate. *Chemosphere, 240*, 124949. Available from https://doi.org/10.1016/j.chemosphere.2019.124949.

do Nascimento, J. G., da, S., de Araújo, M. H. P., dos Santos, A. B., da Silva, M. E. R., & Firmino, P. I. M. (2021). Can microaeration boost the biotransformation of parabens in high-rate anaerobic systems? *Process Safety and Environmental Protection, 145*, 255–261. Available from https://doi.org/10.1016/j.psep.2020.08.014.

Fan, X. H., Wang, L. T., An, J. Y., Zhang, S. D., Cai, Z. H., Niu, L. L., ... Fu, Yj (2021). Magnetically immobilized edible Bacillus natto for the biotransformation of polydatin to resveratrol and its bioactivity assessment. *Industrial Crops and Products, 161*, 113213. Available from https://doi.org/10.1016/j.indcrop.2020.113213.

Feng, M., Li, H., You, S., Zhang, J., Lin, H., Wang, M., & Zhou, J. (2019). Effect of hexavalent chromium on the biodegradation of tetrabromobisphenol A (TBBPA) by *Pycnoporus sanguineus*. *Chemosphere, 235*, 995–1006. Available from https://doi.org/10.1016/j.chemosphere.2019.07.025.

Hamid, H., Li, L. Y., & Grace, J. R. (2020). Effect of substrate concentrations on aerobic biotransformation of 6:2 fluorotelomer sulfonate (6:2 FTS) in landfill leachate.

Chemosphere, *261*, 128108. Available from https://doi.org/10.1016/j.chemosphere.2020.128108.

Hasegawa, H., Papry, R. I., Ikeda, E., Omori, Y., Mashio, A. S., Maki, T., & Rahman, M. A. (2019). Freshwater phytoplankton: biotransformation of inorganic arsenic to methylarsenic and organoarsenic. *Scientific Reports*, *9*, 1–13. Available from https://doi.org/10.1038/s41598-019-48477-7.

He, T., Bao, J., Leng, Y., Snow, D., Kong, S., Wang, T., & Li, X. (2021). Biotransformation of doxycycline by *Brevundimonas naejangsanensis* and *Sphingobacterium mizutaii* strains. *Journal of Hazardous Materials*, *411*, 125126. Available from https://doi.org/10.1016/j.jhazmat.2021.125126.

Hou, D., O'Connor, D., Igalavithana, A. D., Alessi, D. S., Luo, J., Tsang, D. C. W., ... Ok, Y. S. (2020). Metal contamination and bioremediation of agricultural soils for food safety and sustainability. *Nature Reviews Earth & Environment*, *1*, 366–381. Available from https://doi.org/10.1038/s43017-020-0061-y.

Huang, H., Wang, D., Wen, B., Lv, J., & Zhang, S. (2019). Roles of maize cytochrome P450 (CYP) enzymes in stereo-selective metabolism of hexabromocyclododecanes (HBCDs) as evidenced by in vitro degradation, biological response and in silico studies. *Science of the Total Environment*, *656*, 364–372. Available from https://doi.org/10.1016/j.scitotenv.2018.11.351.

Huang, H., Zhang, S., Wang, S., & Lv, J. (2013). In vitro biotransformation of PBDEs by root crude enzyme extracts: Potential role of nitrate reductase (NaR) and glutathione S-transferase (GST) in their debromination. *Chemosphere*, *90*, 1885–1892. Available from https://doi.org/10.1016/j.chemosphere.2012.10.013.

Huang, Y., Zhang, W., Pang, S., Chen, J., Bhatt, P., Mishra, S., & Chen, S. (2021). Insights into the microbial degradation and catalytic mechanisms of chlorpyrifos. *Environmental Research*. Available from https://doi.org/10.1016/j.envres.2020.110660.

Jha, P., Sen, R., Jobby, R., Sachar, S., Bhatkalkar, S., & Desai, N. (2020). Biotransformation of xenobiotics by hairy roots. *Phytochemistry*. Available from https://doi.org/10.1016/j.phytochem.2020.112421.

Jobby, R., Jha, P., Gupta, A., Gupte, A., & Desai, N. (2019). Biotransformation of chromium by root nodule bacteria *Sinorhizobium* sp. SAR1. *PLoS One*, *14*, e0219387. Available from https://doi.org/10.1371/journal.pone.0219387.

Jobby, R., Jha, P., Yadav, A. K., & Desai, N. (2018). Biosorption and biotransformation of hexavalent chromium [Cr(VI)]: A comprehensive review. *Chemosphere*. Available from https://doi.org/10.1016/j.chemosphere.2018.05.050.

Kaur, P., & Balomajumder, C. (2020). Effective mycoremediation coupled with bioaugmentation studies: An advanced study on newly isolated *Aspergillus* sp. in Type-II pyrethroid-contaminated soil. *Environmental Pollution (Barking, Essex: 1987)*, *261*, 114073. Available from https://doi.org/10.1016/j.envpol.2020.114073.

Kennes-Veiga, D. M., Gonzalez-Gil, L., Carballa, M., & Lema, J. M. (2021). The organic loading rate affects organic micropollutants' cometabolic biotransformation kinetics under heterotrophic conditions in activated sludge. *Water Research*, *189*, 116587. Available from https://doi.org/10.1016/j.watres.2020.116587.

Kennes-Veiga, D. M., Vogler, B., Fenner, K., Carballa, M., & Lema, J. M. (2021). Heterotrophic enzymatic biotransformations of organic micropollutants in activated sludge. *The Science of the Total Environment*, *780*, 146564. Available from https://doi.org/10.1016/j.scitotenv.2021.146564.

Kovačič, A., Gys, C., Gulin, M. R., Gornik, T., Kosjek, T., Heath, D., ... Heath, E. (2021). Kinetics and biotransformation products of bisphenol F and S during aerobic degradation with activated sludge. *Journal of Hazardous Materials*, *404*, 124079. Available from https://doi.org/10.1016/j.jhazmat.2020.124079.

Kumar, A., Subrahmanyam, G., Mondal, R., Cabral-Pinto, M. M. S., Shabnam, A. A., Jigyasu, D. K., … Yu, Z. G. (2021). Bio-remediation approaches for alleviation of cadmium contamination in natural resources. *Chemosphere, 268*, 128855. Available from https://doi.org/10.1016/j.chemosphere.2020.128855.

Leemans Martin, L., Peschke, T., Venturoni, F., & Mostarda, S. (2020). Pharmaceutical industry perspectives on flow chemocatalysis and biocatalysis. *Current Opinion in Green and Sustainable Chemistry*. Available from https://doi.org/10.1016/j.cogsc.2020.04.011.

Liu, M., Dong, F., Yi, S., Zhu, Y., Zhou, J., Sun, B., … Zhu, L. (2020). Probing mechanisms for the tissue-specific distribution and biotransformation of perfluoroalkyl phosphinic acids in common carp (*Cyprinus carpio*). *Environmental Science & Technology, 54*, 4932–4941. Available from https://doi.org/10.1021/acs.est.0c00359.

Martínez-Quintela, M., Arias, A., Alvarino, T., Suarez, S., Garrido, J. M., & Omil, F. (2021). Cometabolic removal of organic micropollutants by enriched nitrite-dependent anaerobic methane oxidizing cultures. *Journal of Hazardous Materials, 402*, 123450. Available from https://doi.org/10.1016/j.jhazmat.2020.123450.

Morales-Barrera, L., & Cristiani-Urbina, E. (2008). Hexavalent chromium removal by a *Trichoderma inhamatum* fungal strain isolated from tannery effluent. *Water, Air, and Soil Pollution, 187*, 327–336. Available from https://doi.org/10.1007/s11270-007-9520-z.

Muschket, M., Keltsch, N., Paschke, H., Reemtsma, T., & Berger, U. (2020). Determination of transformation products of per- and polyfluoroalkyl substances at trace levels in agricultural plants. *Journal of Chromatography. A, 1625*, 461271. Available from https://doi.org/10.1016/j.chroma.2020.461271.

Nabb, D. L., Szostek, B., Himmelstein, M. W., Mawn, M. P., Gargas, M. L., Sweeney, L. M., … Fasano, W. J. (2007). In vitro metabolism of 8-2 fluorotelomer alcohol: Interspecies comparisons and metabolic pathway refinement. *Toxicological Sciences: An Official Journal of the Society of Toxicology, 100*, 333–344. Available from https://doi.org/10.1093/toxsci/kfm230.

Nguyen, L. N., Nghiem, L. D., & Oh, S. (2018). Aerobic biotransformation of the antibiotic ciprofloxacin by *Bradyrhizobium* sp. isolated from activated sludge. *Chemosphere, 211*, 600–607. Available from https://doi.org/10.1016/j.chemosphere.2018.08.004.

Nor Suhaila, Y., Hasdianty, A., Maegala, N. M., Aqlima, A., Hazeeq Hazwan, A., Rosfarizan, M., & Ariff, A. B. (2019). Biotransformation using resting cells of Rhodococcus UKMP-5M for phenol degradation. *Biocatalysis and Agricultural Biotechnology, 21*, 101309. Available from https://doi.org/10.1016/j.bcab.2019.101309.

Nouri, M. (2021). Potentials and challenges of date pits as alternative environmental clean-up ingredients. *Biomass Conversion and Biorefinery*, 1–28. Available from https://doi.org/10.1007/s13399-020-01215-w.

Papry, R. I., Fujisawa, S., Zai, Y., Akhyar, O., Mashio, A. S., & Hasegawa, H. (2021). Freshwater phytoplankton: Salinity stress on arsenic biotransformation. *Environmental Pollution (Barking, Essex: 1987), 270*, 116090. Available from https://doi.org/10.1016/j.envpol.2020.116090.

Papry, R. I., Ishii, K., Mamun, M. A. A., Miah, S., Naito, K., Mashio, A. S., & Maki, T. (2019). Arsenic biotransformation potential of six marine diatom species: Effect of temperature and salinity. *Scientific Reports, 9*, 1–16. Available from https://doi.org/10.1038/s41598-019-46551-8.

Pratush, A., Ye, X., Yang, Q., Kan, J., Peng, T., Wang, H., … Hu, Z. (2020). Biotransformation strategies for steroid estrogen and androgen pollution. *Applied Microbiology and Biotechnology*. Available from https://doi.org/10.1007/s00253-020-10374-9.

Ripoll, M., & Betancor, L. (2021). Opportunities for the valorization of industrial glycerol via biotransformations. *Current Opinion in Green and Sustainable Chemistry*. Available from https://doi.org/10.1016/j.cogsc.2020.100430.

Saldarriaga-Hernández, S., Velasco-Ayala, C., Leal-Isla Flores, P., de Jesús Rostro-Alanis, M., Parra-Saldivar, R., Iqbal, H. M. N., & Carrillo-Nieves, D. (2020). Biotransformation of lignocellulosic biomass into industrially relevant products with the aid of fungi-derived lignocellulolytic enzymes. *International Journal of Biological Macromolecules*. Available from https://doi.org/10.1016/j.ijbiomac.2020.06.047.

Saunders, L. J., Fitzsimmons, P. N., Nichols, J. W., & Gobas, F. A. P. C. (2020). In vitro-in vivo extrapolation of hepatic and gastrointestinal biotransformation rates of hydrophobic chemicals in rainbow trout. *Aquatic Toxicology (Amsterdam, Netherlands)*, *228*, 105629. Available from https://doi.org/10.1016/j.aquatox.2020.105629.

Saxena, G., Purchase, D., Mulla, S. I., & Bharagava, R. N. (2020). Degradation and detoxification of leather tannery effluent by a newly developed bacterial consortium GS-TE1310 for environmental safety. *Journal of Water Process Engineering*, *38*, 101592. Available from https://doi.org/10.1016/j.jwpe.2020.101592.

Seller, C., Honti, M., Singer, H., & Fenner, K. (2020). Biotransformation of chemicals in water-sediment suspensions: Influencing factors and implications for persistence assessment. *Environmental Science & Technology Letters*, 7, 854–860. Available from https://doi.org/10.1021/acs.estlett.0c00725.

Sheldon, R. A., & Woodley, J. M. (2018). Role of biocatalysis in sustainable chemistry. *Chemical Reviews*. Available from https://doi.org/10.1021/acs.chemrev.7b00203.

Silva Conceição, J. C., Vieira, T. M., Miller Crotti, A. E., dos Anjos, J. P., Chagas-Paula, D. A., Vieira, A. A., & Silva, E. O. (2021). *Aspergillus brasiliensis*-mediated biotransformation of methyl p-coumarate via phenyloxiran moiety: A predictive model for environmental bioremediation. *International Biodeterioration & Biodegradation*, *158*, 105167. Available from https://doi.org/10.1016/j.ibiod.2020.105167.

Singh, G., & Dwivedi, S. K. (2020). Decolorization and degradation of Direct Blue-1 (Azo dye) by newly isolated fungus *Aspergillus terreus* GS28, from sludge of carpet industry. *Environmental Technology & Innovation*, *18*, 100751. Available from https://doi.org/10.1016/j.eti.2020.100751.

Soares, P. R. S., Birolli, W. G., Ferreira, I. M., & Porto, A. L. M. (2021). Biodegradation pathway of the organophosphate pesticides chlorpyrifos, methyl parathion and profenofos by the marine-derived fungus *Aspergillus sydowii* CBMAI 935 and its potential for methylation reactions of phenolic compounds. *Marine Pollution Bulletin*, *166*, 112185. Available from https://doi.org/10.1016/j.marpolbul.2021.112185.

Sriharsha, D. V., Lokesh Kumar, R., & Janakiraman, S. (2020). Absorption and reduction of chromium by fungi. *Bulletin of Environmental Contamination and Toxicology*, *105*, 645–649. Available from https://doi.org/10.1007/s00128-020-02979-7.

Tang, R., Wang, Y., Yuan, S., Wang, W., Yue, Z., Zhan, X., & Hu, Z. H. (2021). Organoarsenic feed additives in biological wastewater treatment processes: Removal, biotransformation, and associated impacts. *Journal of Hazardous Materials*. Available from https://doi.org/10.1016/j.jhazmat.2020.124789.

Tian, H., Li, X. P., Zhao, J., Gao, H. W., Xu, Q. M., & Wang, J. W. (2021). Biotransformation of artemisinic acid to bioactive derivatives by endophytic *Penicillium oxalicum* B4 from *Artemisia annua* L. *Phytochemistry*, *185*, 112682. Available from https://doi.org/10.1016/j.phytochem.2021.112682.

Trejo-Castillo, R., El Kassis, E. G., Cuervo-López, F., & Texier, A. C. (2021). Cometabolic biotransformation of benzotriazole in nitrifying batch cultures. *Chemosphere*, *270*, 129461. Available from https://doi.org/10.1016/j.chemosphere.2020.129461.

Vo, H. N. P., Ngo, H. H., Guo, W., Nguyen, K. H., Chang, S. W., Nguyen, D. D., ... Bui, X. T. (2020). Micropollutants cometabolism of microalgae for wastewater remediation: Effect of carbon sources to cometabolism and degradation products. *Water Research*, *183*, 115974. Available from https://doi.org/10.1016/j.watres.2020.115974.

Wang, J., Ma, Q., Zhang, Z., Li, S., Diko, C. S., Dai, C., ... Qu, Y. (2020). Bacteria mediated Fenton-like reaction drives the biotransformation of carbon nanomaterials. *The Science of the Total Environment, 746*, 141020. Available from https://doi.org/10.1016/j.scitotenv.2020.141020.

Wang, J., Shan, S., Ma, Q., Zhang, Z., Dong, H., Li, S., ... Qu, Y. (2021). Fenton-like reaction driving the degradation and uptake of multi-walled carbon nanotubes mediated by bacterium. *Chemosphere, 275*, 129888. Available from https://doi.org/10.1016/j.chemosphere.2021.129888.

Wang, Q., Jiang, Y., Wang, H., Chang, X., Lv, M., & Chen, L. (2020). Isolation and characterization of a marine bacterium *Vibrio diabolicus* strain L2-2 capable of biotransforming sulfonamides. *Environmental Research, 188*, 109718. Available from https://doi.org/10.1016/j.envres.2020.109718.

Wang, Q., Wang, H., Jiang, Y., Lv, M., Wang, X., & Chen, L. (2021). Biotransformation mechanism of *Vibrio diabolicus* to sulfamethoxazole at transcriptional level. *Journal of Hazardous Materials, 411*, 125023. Available from https://doi.org/10.1016/j.jhazmat.2020.125023.

Wang, X., Li, Q., Liao, Q., Yan, Y., Xia, J., Lin, Q., ... Liang, Y. (2020). Arsenic(III) biotransformation to tooeleite associated with the oxidation of Fe(II) via *Acidithiobacillus ferrooxidans*. *Chemosphere, 248*, 126080. Available from https://doi.org/10.1016/j.chemosphere.2020.126080.

Wen, Y., Li, C., Song, X., & Yang, Y. (2020). Biodegradation of phenol by *Rhodococcus* sp. strain SKC: Characterization and kinetics study. *Molecules (Basel, Switzerland), 25*, 3665. Available from https://doi.org/10.3390/molecules25163665.

Wu, J. S., Meng, Q. Y., Zhang, Y. H., Shi, X. H., Fu, X. M., Zhang, P., ... Wang, C. Y. (2020). Annular oxygenation and rearrangement products of cryptotanshinone by biotransformation with marine-derived fungi *Cochliobolus lunatus* and *Aspergillus terreus*. *Bioorganic Chemistry, 103*, 104192. Available from https://doi.org/10.1016/j.bioorg.2020.104192.

Xia, S., Song, Z., Jeyakumar, P., Shaheen, S. M., Rinklebe, J., Ok, Y. S., ... Wang, H. (2019). A critical review on bioremediation technologies for Cr(VI)-contaminated soils and wastewater. *Critical Reviews in Environmental Science and Technology, 49*, 1027–1078. Available from https://doi.org/10.1080/10643389.2018.1564526.

Xu, H., Hao, R. X., Xu, X. Y., Ding, Y., Lu, A. H., & Li, Y. H. (2021). Removal of hexavalent chromium by *Aspergillus niger* through reduction and accumulation. *Geomicrobiology Journal, 38*, 20–28. Available from https://doi.org/10.1080/01490451.2020.1807659.

Xu, H., Hong, C., Yao, Y., Liu, L., Wang, W., Zhu, W., ... Zhu, F. (2021). The process of biotransformation can produce insect protein and promote the effective inactivation of heavy metals. *The Science of the Total Environment, 776*, 145864. Available from https://doi.org/10.1016/j.scitotenv.2021.145864.

Yang, T. T., Liu, Y., Tan, S., Wang, W. X., & Wang, X. (2021). The role of intestinal microbiota of the marine fish (*Acanthopagrus latus*) in mercury biotransformation. *Environmental Pollution (Barking, Essex: 1987), 277*, 116768. Available from https://doi.org/10.1016/j.envpol.2021.116768.

Zhan, H., Huang, Y., Lin, Z., Bhatt, P., & Chen, S. (2020). New insights into the microbial degradation and catalytic mechanism of synthetic pyrethroids. *Environmental Research*. Available from https://doi.org/10.1016/j.envres.2020.109138.

Zhang, H., Wen, B., Hu, X., Wu, Y., Pan, Y., Huang, H., ... Zhang, S. (2016). Uptake, translocation, and metabolism of 8:2 fluorotelomer alcohol in soybean (*Glycine max* L. Merrill). *Environmental Science & Technology, 50*, 13309–13317. Available from https://doi.org/10.1021/acs.est.6b03734.

Zhang, H., Wen, B., Huang, H., Wang, S., Cai, Z., & Zhang, S. (2020). Biotransformation of 6:2 fluorotelomer alcohol by the whole soybean (*Glycine max* L. Merrill) seedlings. *Environmental Pollution (Barking, Essex: 1987)*, *257*, 113513. Available from https://doi.org/10.1016/j.envpol.2019.113513.

Zhang, P., Guo, Z., Zhang, Z., Fu, H., White, J. C., & Lynch, I. (2020). Nanomaterial transformation in the soil−plant system: implications for food safety and application in agriculture. *Small (Weinheim an der Bergstrasse, Germany)*, *16*, 2000705. Available from https://doi.org/10.1002/smll.202000705.

Zhang, W., Pang, S., Lin, Z., Mishra, S., Bhatt, P., & Chen, S. (2021). Biotransformation of perfluoroalkyl acid precursors from various environmental systems: advances and perspectives. *Environmental Pollution (Barking, Essex: 1987)*. Available from https://doi.org/10.1016/j.envpol.2020.115908.

Zhao, S., Liang, T., Zhou, T., Li, D., Wang, B., Zhan, J., & Liu, L. (2018). Biotransformation and responses of antioxidant enzymes in hydroponically cultured soybean and pumpkin exposed to perfluorooctane sulfonamide (FOSA). *Ecotoxicology and Environmental Safety*, *161*, 669−675. Available from https://doi.org/10.1016/j.ecoenv.2018.06.048.

Zhao, S., Liang, T., Zhu, L., Yang, L., Liu, T., Fu, J., … Liu, L. (2019). Fate of 6:2 fluorotelomer sulfonic acid in pumpkin (*Cucurbita maxima* L.) based on hydroponic culture: Uptake, translocation and biotransformation. *Environmental Pollution (Barking, Essex: 1987)*, *252*, 804−812. Available from https://doi.org/10.1016/j.envpol.2019.06.020.

Zhao, S., Liu, T., Zhu, L., Yang, L., Zong, Y., Zhao, H., … Zhan, J. (2021). Formation of perfluorocarboxylic acids (PFCAs) during the exposure of earthworms to 6:2 fluorotelomer sulfonic acid (6:2 FTSA). *The Science of the Total Environment*, *760*, 143356. Available from https://doi.org/10.1016/j.scitotenv.2020.143356.

Zhao, S., Ma, X., Fang, S., & Zhu, L. (2016). Behaviors of N-ethyl perfluorooctane sulfonamide ethanol (N-EtFOSE) in a soil-earthworm system: Transformation and bioaccumulation. *The Science of the Total Environment*, *554−555*, 186−191. Available from https://doi.org/10.1016/j.scitotenv.2016.02.180.

Zhao, S., Wang, B., Zhu, L., Liang, T., Chen, M., Yang, L., … Liu, L. (2018). Uptake, elimination and biotransformation of N-ethyl perfluorooctane sulfonamide (N-EtFOSA) by the earthworms (*Eisenia fetida*) after in vivo and in vitro exposure. *Environmental Pollution (Barking, Essex: 1987)*, *241*, 19−25. Available from https://doi.org/10.1016/j.envpol.2018.05.046.

Zhao, S., Zhou, T., Zhu, L., Wang, B., Li, Z., Yang, L., & Liu, L. (2018). Uptake, translocation and biotransformation of N-ethyl perfluorooctanesulfonamide (N-EtFOSA) by hydroponically grown plants. *Environmental Pollution (Barking, Essex: 1987)*, *235*, 404−410. Available from https://doi.org/10.1016/j.envpol.2017.12.053.

Zhao, S., & Zhu, L. (2017). Uptake and metabolism of 10:2 fluorotelomer alcohol in soil-earthworm (*Eisenia fetida*) and soil-wheat (*Triticum aestivum* L.) systems. *Environmental Pollution (Barking, Essex: 1987)*, *220*, 124−131. Available from https://doi.org/10.1016/j.envpol.2016.09.030.

Zhao, Y., Kang, X., Ding, H., Ning, J., Zhai, Y., & Sheng, X. (2021). Bioaccumulation and biotransformation of inorganic arsenic in zhikong scallop (*Chlamys farreri*) after waterborne exposure. *Chemosphere*, *277*, 130270. Available from https://doi.org/10.1016/j.chemosphere.2021.130270.

Zhou, J., Yang, Z., Liu, Q., Liu, Y., Liu, M., Wang, T., & Zhu, L. (2019). Insights into uptake, translocation, and transformation mechanisms of perfluorophosphinates and perfluorophosphonates in wheat (*Triticum aestivum* L.). *Environmental Science & Technology*. Available from https://doi.org/10.1021/acs.est.9b05656.

Zhou, L. J., Han, P., Zhao, M., Yu, Y., Sun, D., Hou, L., … Wu, Q. L. (2021). Biotransformation of lincomycin and fluoroquinolone antibiotics by the ammonia oxidizers AOA, AOB and comammox: A comparison of removal, pathways, and mechanisms. *Water Research*, *196*, 117003. Available from https://doi.org/10.1016/j.watres.2021.117003.

Zhou, X., Zhang, K., Zhang, T., Cen, C., & Pan, R. (2021). Biotransformation of halophenols into earthy-musty haloanisoles: Investigation of dominant bacterial contributors in drinking water distribution systems. *Journal of Hazardous Materials*, *403*, 123693. Available from https://doi.org/10.1016/j.jhazmat.2020.123693.

CHAPTER 8

Role of polyphosphate accumulating organisms in enhanced biological phosphorous removal

Fatima Akram, Amna Aqeel, Zeeshan Ahmed, Javeria Zafar and Ikram ul Haq
Institute of Industrial Biotechnology, Government College University, Lahore, Pakistan

8.1 Introduction

Phosphorus plays an integral role in the cellular metabolism since it is an indispensable energy supplier for the biosynthesis of DNA, RNA, and phospholipids. It is also a key constituent in the fabrication of numerous food products. According to an estimation in 2000, an extraction of about 19.7 MT of phosphorus was carried out from the phosphate rocks and out of total amount, 15.3 MT was consumed for the production of fertilizers. With the recent trends of population expansion, there is an upsurge in a demand of phosphorous to meet the escalating need of fertilizers in order to get a profitable agronomy. The global increase in population has led mankind to the brink of discovering innovative means to cater to the problem of food shortage, which means an increasing need for fertilizers and ultimately of phosphorous (Dar et al., 2016; Dar, Kamili, Dar, Chishti, & Ahmad, 2015; Wilfert, Kumar, Korving, Witkamp, & van Loosdrecht, 2015). However, the tallying of phosphorus as a phosphate ion is a thought-provoking environmental issue due to its engrossment in the eutrophication process of lagoons and other natural water bodies. The conceivable access of this ion to an aquatic milieu is through domiciliary sewage and industrial outflows, predominantly fertilizer and detergent diligences (Usharani & Lakshmanaperumalsamy, 2010). Approximately, 80% of the extracted inorganic phosphate has contaminated the sewage system through runoff, thus posing a significant environmental threat (Schröder, Smit, Cordell, & Rosemarin, 2011). For instance, in UK 70% of phosphorous entering into the environment comes from the

Microbial Consortium and Biotransformation for Pollution Decontamination
DOI: https://doi.org/10.1016/B978-0-323-91893-0.00003-1

sewage water and agricultural run-off (Bowes et al., 2015; Environment Agency, 2015).

Eutrophication is a worldwide stumbling block, in which the blooms of cyanobacteria and eukaryotic algae befall due to the collapse of community homeostasis from persistent effluence of oligotrophic marine water bodies with nutrients such as nitrogen (N) and phosphorus (P) at echelons which surpass growth-limiting concentrations for these photosynthetic entities (Conley, 2000; Winkler, 1984). Resultantly, these algal blooms limit the light penetration and the shrubberies croak in the littoral zones. This makes the environment extremely acidic by increasing pH and diminution in dissolved oxygen and inorganic salts (Turner & Chislock, 2010). Therefore the consumption of eutrophied water in this manner poses vulnerability on health because of some cyanobacteria that are toxicogenic in nature (Bryan, 1993). Freshwater eutrophication is gauged to charge the US $2.2 billion per annum (Dodds et al., 2009) and can prompt different issues also.

To cure this malady, municipal wastewater treatment industry is looking toward two possible approaches: chemical precipitation and biological acclimatization as these two chief technologies can remove phosphorous from wastewater (De-Bashan & Bashan, 2004). Struvite precipitation has engrossed a spectacular attention for phosphorous evacuation and recuperation (Bhat et al., 2021; Fernandes et al., 2012; Zhang & Lau, 2007). In any case, reusability of this precipitated phosphate salt is far from being obvious (Gasser et al., 2012; Huchzermeier & Tao, 2012) when it is added to the loam with alkaline condition and augmented pH as struvite in less soluble conditions. However, the aggregation of phosphorous mineral compounds, stowed as inorganic polyphosphates (poly P) by microorganisms, may become a tactic to confront forthcoming hostile natural settings (Kornberg, Rao, & Ault-Riché, 1999). Therefore an advancement in phosphate removal is brought about through enhanced biological phosphate removal (EBPR) process which is a dual-step methodology to shield coming water bodies against eutrophication (Kong, Nielsen, & Nielsen, 2005).

Enhanced biological phosphorus removal (EBPR) communities encompass the usual practice of microbial treatment established by James Banard (Wong & Beiko, 2015), which allows the removal of phosphorous and sporadically nitrogen from sewage as shown in Fig. 8.1.

The phenomenon is cost-effective and eco-friendly (Oehmen et al., 2007), typically seeded with microbial communities from an indigenous atmosphere or a seed stock. During this procedure, polyphosphate accumulating organisms (PAOs) exploit intracellular polyphosphate in

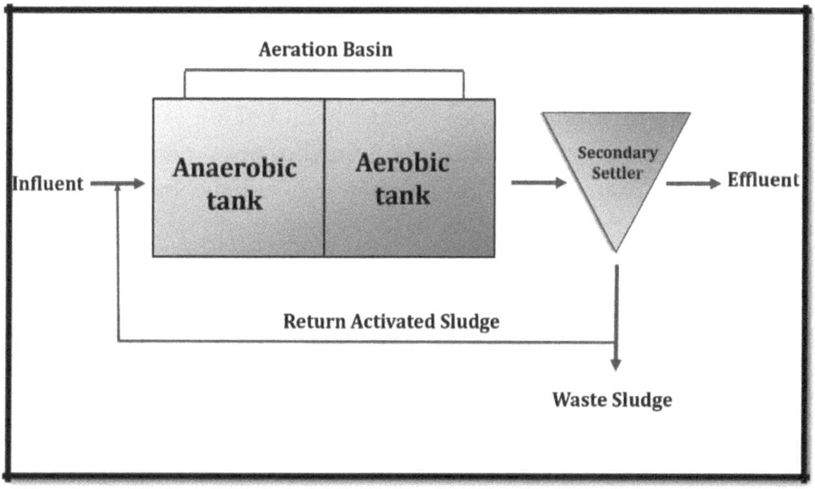

Figure 8.1 Schematic representation of enhanced biological phosphorous removal process.

anaerobic tank as an energy stock and devour carbon as a source in contiguous setting while amalgamating and amassing polyhydroxyalkanoates (PHAs) in their cells. While in the subsequent stage (aerobic), PAOs devour PHA as an energy source and eradicate phosphorus at enhanced levels and stock it in the formula of polyphosphate in their cells (Ye, Gan, & Hu, 2015). Thus P confiscation can be attained by deluxe consumption from PAOs in this phase. Various identification procedures have elucidated that PAOs in EBPR are mainly bacterial species. However, depending on the type of carbon source in the feed, PAO populace fluctuates largely (He, Gu, & McMahon, 2008). Various fungal, yeast, and microalgal (Slocombe et al., 2020; Dar, Bandh, & Kamili, 2011; Dar, Kamili, Nazir, Bandh, & Bhat, 2012; Hirota, Kuroda, Kato, & Ohtake, 2010) species have also been reported to exhibit enhanced biological phosphorous removal process. After being acquaint with an anaerobic period in a conventional biological activated sludge process, EBPR can also be attained in a modest, one-vessel, lab-scale system.

8.2 Natural occurrence of polyphosphate accumulating organisms

Saccharomyces cerevisiae is the earliest organism in which granules of polyphosphate were identified in the year 1888 (Liebermann, 1888).

After that, the ability was found in other microbes as well. These embrace some yeasts such as *Hansenula fabianii*, *Candida humicola*, and *Hansenula anomala* present in wastewaters embracing gigantic amount of phosphorous (McGrath & Quinn, 2003; Watanabe, Ozaki, Iwashita, Fujii, & Iefuji, 2008). However, the capacity of numerous yeast species of excavating inorganic phosphate from wastewater is still not completely understood. Poly P depicts about 10% to 20% of the cellular dry weight of yeast when present in their vacuoles. A very low concentration (>0.05 mM) of inorganic phosphate in the growth media is enough to support the growth of *S. cerevisiae* (Lichko, Kulakovskaya, & Kulakovskaya, 2008; Persson, Lagerstedt, & Pratt, 2003). Thus, with an upsurge in the concentration of inorganic phosphate in the complete media, the poly P content increases many fold in the cells (Vagabov, Trilisenko, & Kulaev, 2000).

Microalgae also have the competency of stowing poly P and are being widely studied for the wastewater treatment (Baker et al., 2015) due to its sustainable and eco-friendly approach (poly P). *Chlamydomonas* sp. are frequently acknowledged in sewage treatment works around the globe. In the recent years, genetic model *Chlamydomonas reinhardtii* has turned out to be a major organism to comprehend the molecular biology under the veil of P metabolism and fabrication of poly P in microalgae. However, other microalgal species have also been scrutinized for wastewater treatment (Li et al., 2019). Table 8.1 shows the list of eukaryotes and prokaryotes that are majorly involved in accumulation of polyphosphateshows

Table 8.1 Persistence of polyphosphate in various organisms.

Organisms	References
Eukaryotes	
Protozoa (Aspidisca and Trachelophyllum)	Akpor and Momba (2008)
Fungi	Kornberg et al. (1999)
Yeast (*Saccharomyces cerevisiae*)	Kornberg et al. (1999)
Plants	Kornberg et al. (1999)
Animals (rat liver, cytosol, nucleus)	Kornberg et al. (1999)
Microalgae (Chlamydomonas)	Kornberg et al. (1999)
Prokaryotes	
Bacteria (*Escherichia coli*)	Kornberg et al. (1999)
Archaea (*Sulfolobus acidocaldarius*)	Kornberg et al. (1999)

the list of eukaryotes and prokaryotes that are majorly involved in accumulation of polyphosphate.

8.3 Microbiology of EBPR and polyphosphate accumulating organisms

The microbiological analysis of organisms related to EBPR revealed the presence of clusters of microbes having coccobacillus like silhouettes as conferring to chemical staining methods such as Neisser stain (Jenkins, Richard, & Daigger, 1993). Their profusion in EBPR reactors and chemical staining profiles certified them as PAOs. The sample from EBPR was taken and diluted along with the spread plate inoculation and coccobacillus-shaped *Acinetobacter* sp. were isolated, which were concluded as PAOs (Buchan, 1983; Deinema, Habets, Scholten, Turkstra, & Webers, 1980; Fuhs & Chen, 1975). However, the pure culture phenotype tells the different tale and never matches with that of EBPR and after conduction of careful studies, it was found that *Acinetobacter* species are not PAOs (Cloete & Steyn, 1987; Hiraishi & Morishima, 1990). Apart from *Acinetobacter* species, various other pure cultures of bacteria have been found that are associated with phosphorus removal. These include *Lampropedia* (Stante, Cellamare, Malaspina, Bortone, & Tilche, 1997), *Tetrasphaera* sp. (Maszenan et al., 2000), *Micropruina glycogenica* (Shintani et al., 2000), and *Microlunatus phosphovorus* (Nakamura et al., 1995).

There are several studies that confirm their role in phosphorous removal. Fluorescence in situ hybridization (FISH) is a modus operandi used to search for real PAOs that show some level of EBPR behavior in a plant (Wagner et al., 1994). The plant was found to have 36% betaproteobacteria, 36% Actinobacteria, and 10% gammaproteobacteria. However, *Acinetobacter* sp. comprises only 3% to 6% of the total community. Later speculations were made regarding the Actinobacteria to have the matching morphology with cells containing polyphosphates, but no evidence supported the claim. A similar structure of microbial community was reported by Kämpfer et al. (1996) for an EBPR sludge and 0%−5% of the bacterial cells were Actinobacteria. Because of obstacles in obtaining the pure EBPR culture, the molecular techniques that are culture independent in nature were used for the advancement in the knowledge regarding PAO. These techniques include FISH, microautoradiography (MAR), and quinone profiles (Mara & Horan, 2003).

FISH is used for the quantification of diverse microbes in PAO- and EBPR-related studies. The shifts in microbial population and their characterization were assessed by quinone profiling. The active microbial population and their functions were studied with the help of MAR and FISH combination (MAR–FISH) (Nielsen et al., 1999). By employing FISH, twofold investigations (Crocetti et al., 2000; Hesselmann, Werlen, Hahn, Van der Meer, & Zehnder, 1999) (testified that Rhodocyclus-related organisms epitomized the mainstream of cells in the activated sludge and the organism supposed to be accountable for EBPR was Candidatus Accumulibacter phosphatis (now referenced as Accumulibacter). Through MAR–FISH studies, it was demonstrated that Accumulibacter can execute the metabolism that is archetypal for PAOs (Kong, Nielsen, & Nielsen, 2004). By employing aforementioned technique, Kong et al. (2005) and Nguyen, Le, Hansen, Nielsen, and Nielsen (2011) reported another microbial group *Tetrasphaera* that is also present in EBPR in high numbers and is capable of amassing polyphosphate but they have a diverse metabolism than Accumulibacter (Kristiansen et al., 2012). The recent proposed metabolic model of *Tetrasphaera* stated that it is also able to carry out fermentation of glucose as well as use glycogen as a storage polymer.

Moreover, bestowing to 1993 (Wagner, Amann, Lemmer, & Schleifer, 1993), betaproteobacteria have been reported as the most ample bacteria in activated sludge of various configurations when FISH was selected as the technique of analysis (Dar, Bandh, Kamili, Nazir, & Bhat, 2013; Kämpfer et al., 1996; Wagner et al., 1994). Additional PAO candidates include Actinobacteria (Christensson, Blackall, & Welander, 1998; Liu et al., 2001; Wagner et al., 1994), alphaproteobacteria (Kawaharasaki, Tanaka, Kanagawa, & Nakamura, 1999), and gammaproteobacteria (Liu et al., 2001).

8.4 Biochemistry of EBPR and phosphate accumulating organism

The increase in yield of the PAOs can be attained by optimization of operating conditions required, that is, cycles of aerobic and anoxic phases. Fig. 8.2 describes the simplified overview of the main mechanisms in aerobic and anaerobic phases of PAO metabolism. Under the anaerobic conditions, PAOs uptake VFA (volatile fatty acids) such as acetate or propionate with the secondary transport mechanism and stock them as PHA (polyhydroxyalkanoates) (Smolders, Van Loosdrecht, & Heijnen, 1995). The type of VFA gives rise to the distinct type of PHA used by PAO. Usually, VFAs are formed by fermentative bacteria in anaerobic reactor or they may already present in an effluent. However, it is

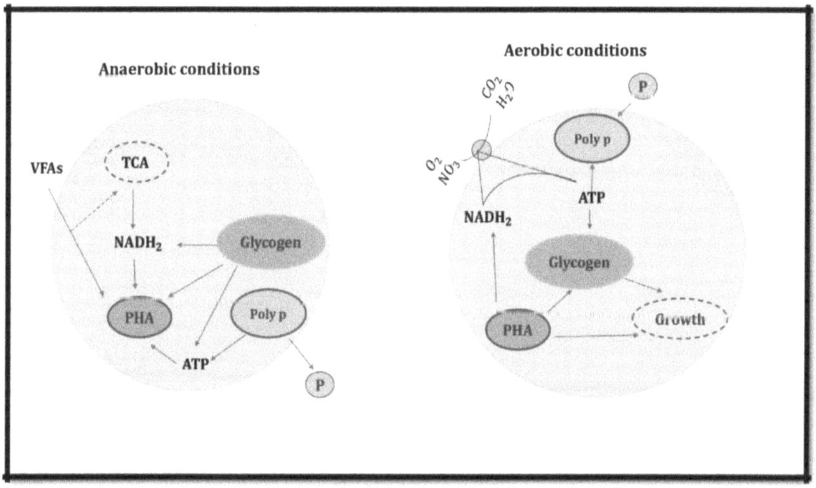

Figure 8.2 Overview of polyphosphate accumulating metabolism.

considered beneficial to provide a prefermentation tank in the process configuration to enhance VFAs' level in anaerobic zone (Oehmen et al., 2007). ATP is used as the main source of energy for this purpose ensuing from the hydrolysis of glycogen or stored polyphosphates. Polyphosphate breakdown results in the release of orthophosphate into the bulk liquid (Comeau, Hall, Hancock, & Oldham, 1986; Smolders, Van der Meij, Van Loosdrecht, & Heijnen, 1994). This synthesis of PHA under anaerobic environment requires NADH2 that is provided by tricarboxylic acid (TCA) cycle (Wentzel, Lotter, Loewenthal, & Marais, 1986) or glycolysis (Smolders et al., 1994). It was found that the genes capable for both pathways are present in *Accumulibacter* species and PAO can thus utilize both Embden–Meyerhof–Parnas pathway and TCA cycle (Zhou, Pijuan, Zeng, & Yuan, 2009). It has also been suggested that PAOs shift between the two metabolic pathways in response to the internal and external conditions (Lanham et al., 2013; Majed, Chernenko, Diem, & Gu, 2012).

In contrast to the aerobic or anoxic zone, the PHAs that were molded in the anaerobic phase are oxidized to CO_2 by using oxygen or nitrate as an electron acceptor. The energy released in this case is thus used to formulate polyphosphate and glycogen and for the exponential growth of new cells (Oehmen et al., 2007). As the amount of phosphorous accumulated by PAOs in an anaerobic phase is higher than the amount released in an aerobic zone, a net phosphorous removal is achieved (Yuan, Pratt, & Batstone, 2012).

8.5 EBPR with acetate as a carbon source

Usually, glycogen-accumulating organisms (GAOs) and PAOs are both found in EBPR. But the abundance of GAO in EBPR leads to the failure of system due to the competition with PAOs for the carbon source without contributing to phosphorous removal. The most important substrate involved is acetate which is a most common and widely studied carbon source in EBPR. Microbial community of acetate-fed EBPR systems has been studied widely (Gonzalez-Gil & Holliger, 2011; Lv et al., 2014). Accumulibacter-like bacteria are the most dominant in such systems. It is partially due to the much higher acetate uptake rate by phosphate accumulating metabolism dominating culture as compared to GAOs (Schuler & Jenkins, 2003). The EBPR model having acetate as a carbon source involves the synthesis of PHA under anaerobic conditions by using stored poly P as an energy source and the orthophosphate generated from poly P degradation is released into the bulk liquid. However, in aerobic zone the stored PHA is used as an energy source to assimilate phosphate to synthesize polyphosphate. Thus PAOs attain ascendancy under the predominant anaerobic—aerobic circumstances because they alone can cultivate aerobically in the deficiency of any exogenous source of carbon and energy, by using the PHA accumulated anaerobically (Seviour, Takashi, & Motoharu, 2003).

However, PAM may shift to GAM (glycogen-accumulating metabolism) under high acetate loadings and differing temperature and pH, which is discussed next.

8.6 EBPR metabolism with substrates other than acetate

The other concern that is encountered with such investigation is the diversity of the substrates that are dealt with in full-scale activated sludge systems. These substrates are different from acetate and other VFAs but their impact on the chemical changes are unknown albeit there are many organic compounds that support EBPR (Gerber, Mostert, Winter, & De Villiers, 1986), individually or in combined form by first changing into VFAs (Fuhs & Chen, 1975). The chemical nature of supplied VFA determines the chemical composition as well as the amount of PFA produced by anaerobic method and thus determines the phosphorous accumulated under aerobic conditions. However, no community analysis has been conducted under varying conditions (Randall & Liu, 2002).

The glucose which is used as a carbon source can cause the failure of EBPR (Cech & Hartman, 1990) and the possible microbiological reasons will be discussed later. However, the laboratory-scale studies support the biochemical models for EBPR with glucose (Sudiana, Mino, Satoh, Nakamura, & Matsuo, 1999). The example in which glucose supports stable EBPR includes the ^{13}C-NMR data (Jeon & Park, 2000) in which the bacteria capable of producing lactic acid play a vital role in anaerobic zone. This population ensures the assimilation of glucose under anaerobic condition leading to the formation of lactic acid and storage of glycogen (Wang, Peng, & Hill, 2002). In the model given by Jeon and Park (2000), it was hypothesized that PAO transforms the glycogen anaerobically into PHA, whereas the lactic acid bacteria utilized glycogen to form lactate storage polymer. This prototypical analysis permitted the conception that PAOs then uptake phosphorous anaerobically. Unfortunately, no evidence was given that tells whether the model was supported by community structure or not rather the main bacterial population whose morphological description is given in the biomass. Another model was given by Wang and Peng (2002) which was mechanistically similar to the one suggestion proposed by Mino and Comeau–Wentzel (Wang et al., 2002). However, there was still no direct biochemical or microbiological evidence for its support.

Moreover, amino acids have been known to support EBPR biomass because of their definite persistence in wastewater. The amino acids, glutamate and aspartate, could support EBPR as a sole source of carbon in a laboratory-scale process (Satoh, Mino, & Matsuo, 1990). Later, it was apprehended that anaerobic substrate can be stowed again but all the glutamate employed as substrate was not stored as the intracellular PHA. Instead, storage was supposed to implicate amalgamation of some storage polypeptide yet to be chemically characterized and recognized.

8.7 Enzymes involved in poly P metabolism

8.7.1 Poly P synthesis

There are various enzymes that have been discovered over the course of last several years for the formation as well as degradation of polyphosphates (Wood & Clark, 1988). The best among these are the polyphosphate kinases. This enzyme is responsible for catalyzing the conversion of ATP to ADP while transferring high-energy phosphate group to poly P and vice versa. The reaction is as follows:

$$ATP + PP_n \leftrightarrow ADP + PP_{n+1}$$

The enzyme polyphosphate kinase was first discovered in the yeast by Yoshida and Yamataka (1953). Later on, it was found in other organisms as well, such as *Escherichia coli* (Kornberg, Kornberg, & Simms, 1956), *Salmonella minnesota, S. cerevisiae* (Felter & Stahl, 1973), *Corynebacterium xerosis* (Muhammed, 1961), *Propionibacterium shermanii* (Robinson, Goss, & Wood, 1984), and *Arthrobacter atrocyaneus* (Levinson, Jacobs, Krulwich, & Li, 1975). The enzyme polyphosphate kinase is the major player in poly P formation in these microbes (Kulaev, 1979). They use magnesium ions (Mg2 +) for optimal activity and it is a tetramer of 69-kDa subunits (Van Wazer & Campanella, 1950). Moreover, 1,3 diphosphoglycerate that include polyphosphate phosphotransferase (Kulaev, 1979) and dolicholpyrophosphate including polyphosphate phosphotransferase (Kulaev, 1990) are the other enzymes of this nature.

8.7.2 Poly P degradation

Mycobacteria were the first organism in which the phosphorylation of glucose by poly P was discovered. The reaction proceeds is as follows:

$$Glucose + PP_n \rightarrow Glucose\, 6 - P + PP_{n-1}$$

Polyphosphate gluconate kinase, polyphosphate manno-kinase, polyphosphate depolymerase, and polyphosphate fructokinase (Kulaev & Vagabov, 1983) are polyphosphate-dependent nicotinamide adenine dinucleotide (NAD) kinases that are responsible for polyphosphate degradation. The purification of the enzyme polyphosphate glucokinase was done from *Mycobacterium phlei* as well as *Mycobacterium tuberculosis* (Szymona & Ostrowski, 1964) as reported by Szymona and Ostrowski (1964). A high concentration of salts (e.g., 0.3 M KC1) and Mg2 + ions in required concentration are used for the optimal activity of this enzyme. This enzyme consists of eight identical subunits and has a molecular mass of about 34 kDa.

The polyphosphate–AMP phosphotransferase catalyzes the degradation of poly P with the expenditure of AMP:

$$AMP + PP_n \rightarrow ADP + PP_{n-1}$$

The enzyme was observed in crude extracts of *Mycobacterium smegmatis* and purified up to 60-fold from *C. xerosis* (Dirheimer & Ebel, 1965). Later the enzyme was purified 1500-fold from *Acinetobacter johnsonii* 210A

(Bonting, Kortstee, & Zehnder, 1991). The exopolyphosphatases catalyze the hydrolysis of poly P to form orthophosphate:

$$H_2O + PP_n \rightarrow P_i + PP_{n-1}$$

Muhammed, Rodgers, and Hughes (1959) purified about 100-fold from *C. xerosis* and its properties were studied. It was active in the case of long-chain poly P but not active toward the pyrophosphate, tri-, tetra-, and hexa-metaphosphate. There were no short-chain intermediates detected during the poly P degradation which reveals that the degradation of poly P is done in a possessive manner. The inhibition of enzyme occurs with the divalent cations, that is, Mg^{2+}, and activation takes place by ethylenediaminetetraacetic acid (EDTA).

8.8 EBPR configurations

The biological removal system of phosphorous resides into three categories: mainstream, sidestream, and cycling procedures. Usually, the mainstream deals with anaerobic condition for confiscation of phosphorous in the mixed liquor and all sidestream methods revolve around treatment of return sludge under anaerobic environment (Brown & Koch, 2005; Levin & Della, 1987; Metcalf, & Eddy, Tchobanoglous, & Stensel, 2003).

8.8.1 Mainstream process

8.8.1.1 A/O or A2/O

The configuration was primarily projected by Barnard and later it was patented beneath the appellation of A/O. It is grounded on the assembly of anaerobic phase added erstwhile to the aerobic phase. It allows the selection of PAO over the other microbes without any removal of nitrogen and brings another anoxic phase amid anaerobic and aerobic phase which then concludes with the removal of nitrogen and phosphorus simultaneously. Since the confiscation of nitrogen occurred in anoxic zone, there will be a decrease in the amount of nitrate that is reimbursed back to the anaerobic phase in contrast to the A/O arrangement. Along with it, nitrate and nitrite present may cause the ordinary activity of heterotrophic organisms while the chemical oxygen demand (COD) runs short for the growth of PAO (Guerrero, Taya, Guisasola, & Baeza, 2012). There are various investigations that are conducted with different substrates on anaerobic—aerobic activated sludge process. The accretion of polyhydroxybutyrate (PHB) under anaerobic environment is also observed when acetyl-CoA is transformed into PHB using nicotinamide adenine dinucleotide hydrogen (NADH) as reducing agent (Arun, Mino, & Matsuo, 1988). A2/O systems are the most common methods used for biological nutrient removal (BNR) systems in which denitrifiers and PAOs remove phosphorous and nitrogen by using organic matter as a source of carbon (Fig. 8.3).

8.8.1.2 University of Cape Town-modified process

Just like the development of A2/O, in the University of Cape Town (UCT) configuration; returned activated sludge (RAS) is taken back to anoxic zone in spite of its return to anaerobic zone, thus averting the consumption of readily biodegradable COD (rbCOD) for denitrification phenomena. This allows the storage of PHA by PAOs. For additional enhancement of modified UCT, dual anoxic zones were added to A2/O configurations that allows for nitrate salvaging back to second anoxic zone and then mixed liquor is reimbursed from earliest anoxic zone to anaerobic precinct (Metcalf & Eddy et al., 2003; Østgaard, Christensson, & Lie, 1997). It was determined that the greater the phosphorous released in anaerobic phase, the higher the phosphorous concentration would be taken up into the resulting aerobic phase. Apart from having function in phosphorous removal, anaerobic selector of UCT (Fig. 8.4) has also been involved in forming large flocs to improve settleability (Vaiopoulou & Aivasidis, 2008). This conformation allows the complete removal of

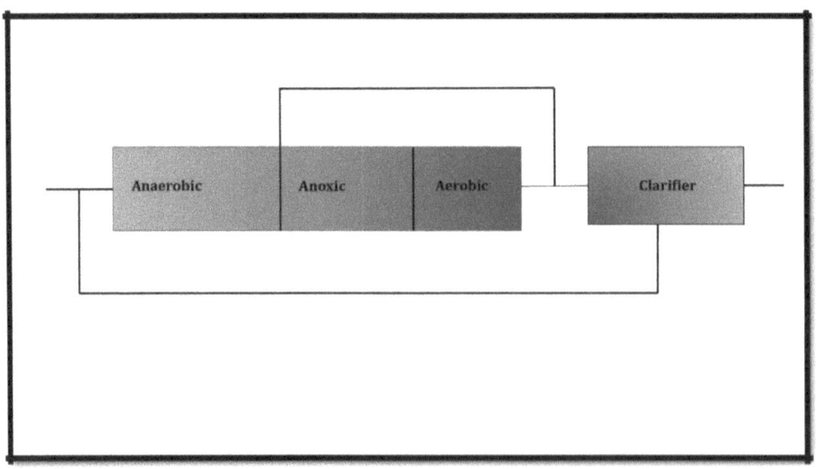

Figure 8.3 Overview of A/O or A2/O process.

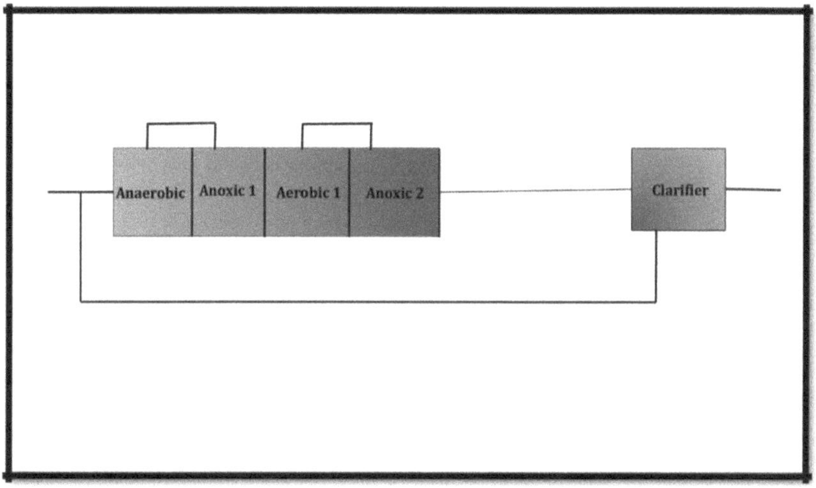

Figure 8.4 Overview of UCT process. *UCT*, University of Cape Town.

nitrate in anoxic phase and adequate period for PHA storage in anaerobic phase (Brown & Koch, 2005). Such procedures demonstrate anticipated outcomes in removal efficacies with existing adequate organic carbon source and legitimate C/N ratio.

However, due to stern limitations concerning effluent quality, various pre-existing steps have been optimized and revised. Modified UCT processes carry the advantage of bearing utterly maintained anaerobic atmosphere due to core recycle subsystem. This allows the trifling inhibitory influence of nitrate and nitrite on anaerobic phosphorus and anoxic-denitrifying phosphorous amputation with lower organic substrate requirement. This scheme provides efficient performance and thermodynamic stability because of minimized sludge recycle and primary sedimentation exclusion (Vaiopoulou & Aivasidis, 2008). However, by employing aerated-anoxic zones within UCT reactor, auxiliary contributions have been done with lesser operational budgets and marginally higher nutritional amputation (Park, Whang, Reusser, & Noguera, 2006).

The another modification of UCT, Virginia Initiative Plant (VIP) is functioned with somewhat parallel configurations carrying slight differences in mixed liquor recycle location and shorter sludge age of 5–10 days (Mogens, Van Mark, Loosdrecht, & Ekama, 2008). The VIP procedure embraces anaerobic, anoxic, and aerobic cauldrons in chain with aerobic mixed liquor recycle to the anoxic reactor, anoxic reactor mixed liquor recycle to the anaerobic reactor, and return sludge recycle to the anoxic reactor.

8.8.1.3 Johannesburg configuration

In order to curtail nitrate recycling in anaerobic zone, a single anoxic reactor is located on RAS line as a modification of UCT. The organic electron donor is either from internal storage of carbon or from external, influent carbon addition (Guerrero Camacho, 2014). Although, COD scarcity restricted denitrification capacity, the reactor is more compact with no dilution effect (Brown & Koch, 2005). However, JBH configuration has a low nitrogen removal capacity and lacks the ability of complete denitrification compared to UCT process, which is more efficient in both abovementioned processes (Barjesteh, 2016; He et al., 2008).

8.8.2 Sidestream

8.8.2.1 PhoStrip

In the early 1970s, PhoStrip was initially introduced for sidestream biological removal of phosphorous (He et al., 2008) and Levin and Shapiro were the first persons to demonstrate phosphorous uptake by this process. They demonstrated the sequential anaerobic and aerobic processes for P uptake and removal and patent their work by mounting first commercially available P removal system called as PhoStrip (Levin & Shapiro, 1965). PhoStrip works by controlling microbes available in wastewater sludge in aerated conditions to take up phosphorous and within anaerobic conditions, phosphate is on the rampage from settled sludge (Levin & Della, 1987). In this way, this configuration of chemical sidestream phosphorous removal is effective as it activates sludge from chemical sludge and minimally effects mainstream (Salehi, Cheng, Heitz, & Ginige, 2018). However, a sidestream hybrid conformation utilizes an anaerobic stripper tank for phosphate removal and thus biological and chemical processes are combined for phosphorous removal in this way.

8.8.2.2 Biological–chemical phosphorous and nitrogen removal configuration

This process is a blend of biological and chemical P removal with off-line precipitation and in-line P stripping (Van Loosdrecht, Brandse, & de Vries, 1998). It is actually a modified form of UCT process, in which a phosphorous is removed by the addition of chemical step in sludge line. Inert material is prevented from returning to activated sludge tank because sludge line is separated from chemical tank and sludge is not reimbursed back to the mainstream (Brown & Koch, 2005). A low solid retention time (SRT) preferred for attaining squat effluent concentrations of both phosphorous and nitrogen with the disadvantage of lowering biomass production by biological phosphorous removal and phosphorous which is not removed biologically can be precipitated by adding chemicals. Accessibility of COD for denitrification, by exploiting denitrifying PAOs for sanctioning P and N removal with minor COD requirements, is valuable (Wang, Peng, & Stephenson, 2009).

8.8.3 Cycling system

8.8.3.1 Biodenipho process

Nitrogen and phosphorous are simultaneously removed by alternating oxidation ditches between anoxic and aerobic conditions in a system called as Biodenipho process. To supplement nitrogen reduction procedure, the drift from anaerobic reactor is nursed to anoxic zone (Yeoman, Stephenson, Lester, & Perry, 1988). A length between anoxic and aerobic shift is kept 2-h and wastewater streams through anaerobic, anoxic, and aerobic reactors in each phase and excess sludge along with phosphorous is removed from aerobic reactor (Arvin & Kristensen, 1985). Interactions between organic substrate and nitrates are observed occasionally due to semibatch custom of this apparatus and P uptake took place in the existence of nitrate followed by a fractional P release by nitrate exhaustion (Meinhold, Pedersen, & Arnold, 1998). Moreover, in anaerobic reactor, supplementation of lactate, acetate, and propionate shows a high level of phosphate release in a full-scale Biodenipho plant (Arvin & Kristensen, 1985).

8.8.3.2 Oxidation ditch design

Initially, these designs were used for removal of only organic matter, but these systems are now used for nutrient extraction as well (Wang, Yu, & Gao, 2019a). This phenomenon is critical in mounting clarifier-less activated sludge system due to an interminable loop activated sludge reactor used in this process. Baffles are used in oxidation ditch system for clarifier role and rotating brushes and horizontal axis paddles are used for aeration and mixing purposes. However, an anaerobic reactor prior to oxidation ditch system is used for phosphorous removal (Baetens, 2000). Due to low sludge production, easy operation, and stability in effluent quality, these reactors are mainly accepted in small communities. Oxidation ditch reactors behaves differently from conventional reactors' use in nutrient removal as intense recirculation and aeration mode occurs in these systems, resulting in difficult identification of aerobic and anoxic phases (Insel, Artan, & Orhon, 2005) The absence of an anaerobic zone in oxidation ditch type reactors may result in efficient phosphorous removal, as demonstrated by using full-scale oxidation ditches for EBPR performance (Datta and Goel, 2010).

8.9 Parameters to consider in EBPR process

EBPR treatment requires proper carbon to phosphorous ratio along with suitable concentrations of other elements to effectively run EBPR process. The potential grounds of suboptimal performance or deterioration of EBPR process is an antagonism between GAOs and PAOs. For optimization of this process, it is necessary to comprehend the environmental dynamics affecting the competition between GAO and PAO along with factors affecting the metabolic rate of PAOs. In this regard, carbon source, temperature, and pH are the crucial factors which play independent roles in developing complex race between GAOs and PAOs (Lopez-Vazquez et al., 2009). Some important factors affecting EBPR process is described in the following sections.

8.9.1 Temperature

It is observed that EBPR performance can be increased by decrease in temperature (Oehmen et al., 2007) and is poles apart from biological structures which have an intensified biological kinetics with increase in temperature and thus increases EBPR process. Moreover, it was also found that at low temperature PAOs are the dominating microbes and thus efficiency of EBPR system is increased (Erdal, Erdal, & Randall, 2003; Panswad, Doungchai, & Anotai, 2003). PAOs were also the ruling microbes at 10°C irrespective of pH and carbon source, indicating temperature as an important factor for developing competitive environment between GAOs and PAOs as PAOs were the dominating microbes at low temperature (López-Vázquez, Hooijmans, Brdjanovic, Gijzen, & Van Loosdrecht, 2008).

8.9.1.1 Recent research on EBPR process in tropical conditions

It was pragmatic that tropical temperature around 30°C would result in outcompetetion of GAOs over PAOs and EBPR process does not work well under tropic conditions as tropical temperature favors the development of GAOs instead of PAOs (Lopez-Vazquez et al., 2009). Because of this assumption, only few plants were constructed in tropical areas, before this assumption was proven wrong when full-scale EBPR plants toil well at the temperature of 28°C–32°C in Malaysia and Singapore (Ong et al., 2014). Although some GAOs were present, they are so less in number that they did not cause any problem during the complete EBPR process. Furthermore, EBPR process feasibility in warm climate can be increased by using lab-scale reactors with biomass enriched from tropical EBPR plants (Cokro et al., 2017) and thus this technology can be used in tropical and warm regions effectively.

8.9.2 Carbon source and wastewater composition

Competition between PAOs and GAOs is highly affected by VFAs as GAOs only utilize sole carbon source in the form of propionate or acetate, while PAOs can also utilize a mixture of propionate and acetate (Lopez-Vazquez et al., 2009; Oehmen et al., 2007). In this way, altering the carbon sources will be helpful in obtaining a highly enriched GAOs or PAOs culture and cultures of Accumulibacter, Candidatus, and Phosphatis were obtained by outcompeting GAOs (Lu, Oehmen, Virdis, Keller, & Yuan, 2006). Although this process is beneficial in obtaining highly enriched culture at lab scale, it is not suitable for obtaining these cultures at full-scale plants and thus a controlling system which alternates the production of acetate and propionate by using prefermenter is used at large-scale level (Oehmen et al., 2007).

8.9.3 pH

EBPR system removes phosphorous effectively at high pH (7.5–8) and is suitable for the removal of phosphorous effectively at this pH range (Filipe, Daigger, & Grady, 2001; Oehmen, Teresa Vives, Lu, Yuan, & Keller, 2005). High pH is the competitive advantage of PAOs over GAOs where high pH results in shifting of population to PAO-type microbes (Oehmen et al., 2007).

8.9.4 Sludge age

In order to ensure biomass in aerobic basin to uptake phosphorous and optimum biochemical reactions, proper age of sludge is required. Short age of sludge will result in poor effluent quality with insufficient treatment, whereas long sludge age in activated sludge system will result in nitrification as an important factor for anoxic zones and denitrification of nitrates is required. Long sludge age on the other hand will result in the biomass decay by secondary phosphorous release and affects the extended aeration system, initiating problems for biochemical reactions of PAOs. In this way, 10-to 30-day sludge is considered efficient in EBPR systems (Biological Phosphorous removal - An operator's Guide, 2012).

8.9.5 Recycle of nitrates

Just like the oxygen, the recycle of nitrates to the anaerobic selector can interfere with biological removal of phosphorous. Recycle streams are denitrified ($NO2 + NO3 \rightarrow N2$) by providing anoxic zones. Moreover, recycle rates of return sludge streams or mixed liquor containing nitrates are adjusted to prevent their effect on treatment process (Biological Phosphorous removal - An operator's Guide, 2012).

8.9.6 Sludge phosphorous content

EBPR process should carry the sludge phosphorous content of $\geq 5\%$–6%. More efficient EBPR process usually carries a higher phosphorous content in the waste sludge. A waste sludge phosphorous concentration $\leq 5.0\%$ deserves a more meticulous process analysis. Since, the activated sludge phosphorus content is high in an EBPR plant, overflow of total suspended solids (TSS) must be retained low. A trivial quantity of TSS with in elevated phosphorus content could subsidize to a high total phosphorus concentration in the effluent. Sludge age in an ultimate clarifier or the reprocess of decant from an anaerobic digestion process can cause discharge of stowed phosphorous in the discarded sludge (Biological Phosphorous removal - An operator's Guide, 2012).

8.10 Criteria to monitor effective EBPR process

1. The ratio of COD to phosphorous should be greater than 45 for effective EBPR.
2. The ratio of VFAs to phosphorous should be greater than 7 for effective EBPR.
3. For an effective EBPR, pH of anaerobic and aerobic should reside in an array of 8.0 to 8.5 and 7.0 to 7.5, correspondingly.
4. Oxidation reduction potential (ORP) of the anaerobic zone should be in the range of -100 to -200 mV, while ORP in the anoxic zone should reside in a range of -50 to $+50$ mV. ORP in the aerobic zone should be in a range of $+100$ to $+300$ mV.

5. Hydraulic resistance time of the anaerobic zone should be in the range of 0.5 to 2.0 h.
6. Phosphorous content of the sludge should be >5%.
7. Nitrate heaping to the anaerobic vessel must be less than the oxygen demand of the wastewater.
8. The solids' retention time should be in the range of 15 to 25 days' contingent upon wastewater temperature (Biological Phosphorous removal - An operator's Guide, 2012).

8.11 Transfer of energy pathway genes in microbial enhanced biological phosphorous removal communities

Lateral gene transfer (LGT) is a well-established approach of evolution in bacteria that can be considered through diverse methodologies for genome sequences (Beiko, Harlow, & Ragan, 2005). LGT plays an imperative role in adaptation, for example, in heavy metal metabolism and in antibiotic resistance (Sentchilo et al., 2013). Transmissions tend to take place among adjacent lineages, but numerous instances of transfer between more distant relatives have been recounted as well. LGT is known to have befallen in sewage treatment plants, impacting antibiotic resistance genes and xenobiotic degradation. Numerous mobile genetic elements (MGEs), i.e., transposons and plasmids are responsible for xenobiotic degradation. Plasmids have been used in engineering treatment plants to augment communities for the metabolism of xenobiotics. Other contrivances of LGT also exist, such as gene transfer agents and transformation but their role in sewage treatment communities is still unknown (Wong & Beiko, 2015).

8.12 Novel and potential EBPR system

Nutrient removal optimization is an important process and novel, intricate configuration of MBR systems has been established for this purpose, that is, the development of sequencing batch moving bed membrane bioreactor system which can remove total phosphorous at 84% rate (Yang, Yang, Fu, Wang, & Lei, 2010). Although these technologies have potential of efficiency, their consistency and reliability are not yet proven for low-level removal of phosphorous and further work is needed to ensure their reliability using array of influent strengths. Furthermore, the operational, capital, and ecological costs of such technologies need to be adjusted to ensure its use even at large and small scale.

Another novel technology for P removal is the development of single system operated under anoxic—anaerobic conditions for granular sludge. This technology has the ability of simultaneous multinutrient removal and is known as AnoxAn (Díez-Montero, De Florio, González-Viar, Herrero, & Tejero, 2016). In terms of energy and mixing requirements, this system is cost-effective as upflow nature of this system is effective for this purpose and biological removal of phosphorous (Lettinga, van Velsen, Hobma, de Zeeuw, & Klapwijk, 1980). Moreover, with mechanical mixing and without compromising nitrogen removal, phosphorous exclusion rate of 89% is achieved using AnoxAn (Díez-Montero et al., 2016).

A novel adaptation of MBR technology is the membrane aerated biofilm reactors (MABRs). This technology is grounded on the electron donor or electron transfer in gaseous from across hydrophobic membrane and can be utilized commercially for nutrient removal effectively (Ivanovic & Leiknes, 2012; Nerenberg, 2016). Due to the incidence of gas exchange and addition of complex layer to treatment procedure, MABRs behave contrarily to conventional biofilms (Martin & Nerenberg, 2012). They were recently reported to remove nitrogen effectively and using sequencing batch reactor (SBR) they produce a phosphorous removal rate of 90% (Sun et al., 2015). Despite reduction in maintenance cost and energy efficiency, much work is needed for its use at the commercial level at full-scale plant for EBPR process. Further, the prerequisite for gas stowage may mean that the equipment is unsuitable for use in decentralized locations even if EBPR can be publicized to be successful in real-life settings. Finally, the whole-life costs associated with full-scale application entail additional analysis until it becomes suitable for its use at full-scale EBPR process.

8.13 Conclusion and future perspective

The universal cognizance on the hostile effects of disproportionate phosphorus utilization on plants and animals, deterrence of eutrophication, and overgrowth of toxic blue-green algae eventually ascend advanced protocols on P concentrations and more efficient P removal (Oehmen et al., 2007). Nowadays, mining is used for attaining phosphorous and it also contains potassium, sulfuric acid, and nitrogen and this source of phosphorus will be depleted in next 50—100 years due to its excessive and rapid utilization. Therefore the 21st century is marked by significant phosphorous scarcity and it will increase with the passage of time (Cordell, Rosemarin, Schroder, & Smit, 2011). Despite introducing alternate potentials, effective resource handling and prevention of overfertilization, there is a need to reach high levels of phosphorous availability by investment in scientific and economic sectors (Cordell, Drangert, & White, 2009). Wastewater treatment plants (WWTPs) prevent phosphorous from entering the environment, so it accumulates in sludge, which is then used as fertilizer for crops (Cordell et al. 2011). Most of WWTPs working today are incorporating biological and chemical procedures in combined fashion while biological removal of phosphorous will be beneficial for environment and is also cost-effective. EBPR is highly dependent on microbes as this system works by alternating aerobic, anaerobic, and anoxic phases and works biologically to remove phosphorous. Thus understanding of PAOs and GAOs and their potential requirements with optimized design, operation, and configuration offers high potential for P removal, recovery, and sustainability. Although extensive research with valuable development as well as full-scale improvements in different areas for removal and recovery of phosphorus are carried out, there are still numerous unresolved issues to reconsider and untangle.

References

Akpor, O., & Momba, M. (2008). The effects of pH and temperature on phosphate and nitrate uptake by wastewater protozoa. *African Journal of Biotechnology*, *7*.

Arun, V., Mino, T., & Matsuo, T. (1988). Biological mechanism of acetate uptake mediated by carbohydrate consumption in excess phosphorus removal systems. *Water Research*, *22*, 565–570. Available from https://doi.org/10.1016/0043-1354(88)90056-5.

Arvin, E., & Kristensen, G. H. (1985). Exchange of organics, phosphate and cations between sludge and water in biological phosphorus and nitrogen removal processes. *Water Science and Technology: A Journal of the International Association on Water Pollution Research*, *17*, 147–162.

Baetens, D. (2000). *Enhanced biological phosphorus removal: modelling and experimental design*. Belgium: Ghent University.

Baker, A., Ceasar, S. A., Palmer, A. J., Paterson, J. B., Qi, W., Muench, S. P., et al. (2015). Replace, reuse, recycle: Improving the sustainable use of phosphorus by plants. *Journal of Experimental Botany*, *66*, 3523–3540. Available from https//doi.org//10.1093/jxb/erv210.

Barjesteh, P. J. (2016). *Alternatives to lower the carbon demand of biological nutrient removal processes*. Canada, Winnipeg: University of Manitoba.

Beiko, R. G., Harlow, T. J., & Ragan, M. A. (2005). Highways of gene sharing in prokaryotes. *Proceedings of the National Academy of Sciences of the United States of America*, *102*, 14332–14337.

Bhat, R. A., Singh, D. V., Qadri, H., Dar, G. H., Dervash, M. A., Bhat, S. A., ... Yousaf, B. (2021). Vulnerability of municipal solid waste: An emerging threat to aquatic ecosystems. *Chemosphere, 287*(2022), 132223. Available from https://doi.org/10.1016/j.chemosphere.2021.132223.

Biological Phosphorous removal - An operator's Guide. 2012. https://www.miwea.org/docs/Biological%20Phosphorous%20Removal%20%20An%20Operator%20Guide.pdf.

Bonting, C. F. C., Kortstee, G. J. J., & Zehnder, A. J. B. (1991). Properties of polyphosphate:AMP phosphotransferase of Acinetobacter strain A. *Journal of bacteriology, 173*, 6484–6488.

Bowes, M. J., Jarvie, H. P., Halliday, S. J., Skeffington, R. A., Wade, A. J., Loewenthal, M., et al. (2015). Characterising phosphorus and nitrate inputs to a rural river using high-frequency concentration-flow relationships. *The Science of the Total Environment, 511*, 608–620. Available from https//doi.org//10.1016/j.scitotenv.2014.12.086.

Brown, J. A., & Koch, C. M. (2005). Biological nutrient removal (BNR) operation in wastewater treatment plants. *WEF Manual of Practice No. 29*:597. https://doi.org/10.1036/0071464158

Bryan, A. (1993). *Biological nutrient removal*. Queensland Department of Primary Industries, Brisbane, Qld.

Buchan, L. (1983). Possible biological mechanism of phosphorus removal. *Water Science and Technology, 15*, 87–103.

Cech, J. S., & Hartman, P. (1990). Glucose induced breakdown of enhanced biological phosphate removal. *Environmental Technology, 11*, 651–656.

Christensson, M., Blackall, L. L., & Welander, T. (1998). Metabolic transformations and characterisation of the sludge community in an enhanced biological phosphorus removal system. *Applied Microbiology and Biotechnology, 49*, 226–234.

Cloete, T. E., & Steyn, P. L. (1987). A combined fluorescent antibodymembrane filter technique for enumerating Acinetobacter in activated sludge. In R. Ramadori (Ed.), *Biological phosphate removal from wastewaters* (pp. 335–338). Oxford: Pergamon Press.

Cokro, A. A., Law, Y., Williams, R. B. H., Cao, Y., Nielsen, P. H., & Wuertz, S. (2017). Non-denitrifying polyphosphate accumulating organisms obviate requirement for anaerobic condition. *Water Research, 111*, 393–403.

Comeau, Y., Hall, K., Hancock, R., & Oldham, W. (1986). Biochemical model for enhanced biological phosphorus removal. *Water Research, 20*, 1511–1521. Available from https//doi.org//10.1016/0043-1354(86)90115-6.

Conley, D. J. (2000). Biogeochemical nutrient cycles and nutrient management strategies. *Hydrobiologia, 410*, 87–96.

Cordell, D., Drangert, J. O., & White, S. (2009). The story of phosphorus: Global food security and food for thought. *Global Environmental Change: Human and Policy Dimensions, 19*, 292–305. Available from https://doi.org/10.1016/j.gloenvcha.2008.10.009.

Cordell, D., Rosemarin, A., Schroder, J. J., & Smit, A. L. (2011). Towards global phosphorus security: A systems framework for phosphorus recovery and reuse options. *Chemosphere, 84*, 747–758. Available from https://doi.org/10.1016/j.chemosphere.2011.02.032.

Crocetti, G. R., Hugenholtz, P., Bond, P. L., Schuler, A., Keller, J., Jenkins, D., & Blackall, L. L. (2000). Identification of polyphosphate accumulating organisms and design of 16S rRNA directed probes for their detection and quantitation. *Applied and Environmental Microbiology, 66*(3), 1175–1182.

Dar, G. H., Bandh, S. A., & Kamili, A. N. (2011). Bacteriological analysis of soil from Yusmarg Health Resort of Kashmir Valley. *International Journal of Current Research., 3* (10), 53–56, 2011.

Dar, G. H., Bandh, S. A., Kamili, A. N., Nazir, R., & Bhat, R. A. (2013). Comparative analysis of different types of bacterial colonies from the soils of Yusmarg Forest, Kashmir Valley India. *Ecologia Balkanica, 5*(1), 31–35.

Dar, G. H., Kamili, A. N., Chishti, M. Z., Dar, S. A., Tantry, T. A., & Ahmad, F. (2016). Characterization of *Aeromonas sobria* Isolated from Fish Rohu (*Labeo rohita*) collected from polluted pond. *Journal of Bacteriology & Parasitology*, 7(3), 1–5. Available from https://doi.org/10.4172/2155-9597.1000273.

Dar, G. H., Kamili, A. N., Dar, S. A., Chishti, M. Z., & Ahmad, F. (2015). Detection and characterization of potentially pathogenic *Aeromonas sobria* isolated from fish *Hypophthalmichthys molitrix* (Cypriniformes: Cyprinidae). *Microbial Pathogenesis, 2015*. Available from https://doi.org/10.1016/j.micpath.2015.10.017.

Dar, G. H., Kamili, A. N., Nazir, R., Bandh, S. A., & Bhat, R. A. (2012). A preliminary study of colony forming units of bacteria from the soils of Yusmarg Forest, Kashmir Valley, India. *International Journal of Current Research, 4*(12), 467–472.

Datta, T., & Goel, R. (2010). Evidence and long-term feasibility of enhanced biological phosphorus removal in oxidation ditch type of aerated-anoxic activated sludge systems. *Journal of Environmental Engineering, 136*, 1237–1247. Available from https://doi.org/10.1061/(ASCE)EE.1943-7870.0000259.

De-Bashan, L. E., & Bashan, Y. (2004). Recent advances in removing phosphorus from wastewater and its future use as fertilizer (1997–2003). *Water Research, 38*, 4222–4246.

Deinema, M. H., Habets, L. H. A., Scholten, J., Turkstra, E., & Webers, H. A. (1980). The accumulation of polyphosphates in *Acinetobacter* spp. *FEMS Microbiology Letters, 9*, 275–279.

Díez-Montero, R., De Florio, L., González-Viar, M., Herrero, M., & Tejero, I. (2016). Performance evaluation of a novel anaerobic-anoxic sludge blanket reactor for biological nutrient removal treating municipal wastewater. *Bioresource Technology, 209*, 195–204. Available from https//doi.org/10.1016/j.biortech.2016.02.084.

Dirheimer, G., & Ebel, J. P. (1965). Caract6risation d'une polyphosphate-AMP phosphotransf6rase dans *Corynebacterium xerosis*. *Comptes rendus de l'Académie des Sciences Paris, 260*, s3787–s3790.

Dodds, W. K., Bouska, W. W., Eitzmann, J. L., Pilger, T. J., Pitts, K. L., Riley, A. J., & Thornbrugh, D. J. (2009). Eutrophication of U.S. freshwaters: Analysis of potential economic damages. *Environmental Science and Technology, 43*(1), 12–19.

Environment Agency (2015). *Water for life and livelihoods*: A consultation on the draft update to the river basin management plan part 2: River basin management planning overview and additional information. Bristol. https://assets.publishing.service.gov.uk/government/uploads/system/uploads/attachment_data/file/718324/Part_2_River_jbasin_management_planning_process_overview_and_additional_information.pdf.

Erdal, U., Erdal, Z., & Randall, C. (2003). The competition between PAOs (phosphorus accumulating organisms) and GAOs (glycogen accumulating organisms) in EBPR (enhanced biological phosphorus removal) systems at different temperatures and the effects on system performance. *Water Science & Technology, 47*(11), 1–8.

Felter, S., & Stahl, A. J. C. (1973). Enzymes du m6tabolisme des polyphosphates dans la levure. III. Purification et propri6t6s de la polyphosphate-ADP-phosphotransf∼rase. *Biochimie, 55*, 245–251.

Fernandes, G. W., Kunz, A., Steinmetz, R. L. R., Szogi, A., Vanotti, M., Flores, E. M. D., & Dressler, V. L. (2012). Chemical phosphorus removal: A clean strategy for piggery wastewater management in Brazil. *Environmental Technology, 33*, 1677–1683.

Filipe, C. D., Daigger, G. T., & Grady, C. L., Jr (2001). pH as a key factor in the competition between glycogen-accumulating organisms and phosphorus-accumulating organisms. *Water Environment Research*, 223–232.

Fuhs, G. W., & Chen, M. (1975). Microbiological basis of phosphate removal in the activated sludge process for the treatment of wastewater. *Microbial Ecology, 2*, 119–138.

Gasser, M. O., Chantigny, M. H., Angers, D. A., Bittman, S., Buckley, K. E., Rochette, P., & Masse, D. (2012). Plant-available and water-soluble phosphorus in soils amended with separated manure solids. *Journal of Environmental Quality*, *41*, 1290–1300.

Gerber, A., Mostert, E. S., Winter, C. T., & De Villiers, R. H. (1986). The effect of acetate and other short-chain carbon compounds on the kinetics of biological nutrient removal. *Water SA*, *12*, 7–12.

Gonzalez-Gil, G., & Holliger, C. (2011). Dynamics of microbial community structure of and enhanced biological phosphorus removal by aerobic granules cultivated on propionate or acetate. *Applied and Environmental Microbiology*, 77(22), 8041–8051. Available from https://doi.org/10.1128/AEM.05738-11.

Guerrero Camacho, F. J. (2014). *Improving EBPR stability in WWTPs aiming at simultaneous carbon and nutrient removal: from modelling studies to experimental validation*. Cerdanyola de Valle's, Barcelona: Universitat Autonoma de Barcelona, Bellaterra.

Guerrero, J., Taya, C., Guisasola, A., & Baeza, J. A. (2012). Understanding the detrimental effect of nitrate presence on EBPR systems: Effect of the plant configuration. *Journal of Chemical Technology and Biotechnology (Oxford, Oxfordshire: 1986)*, *87*, 1508–1511. Available from https://doi.org/10.1002/jctb.3812.

He, S., Gu, A. Z., & McMahon, K. D. (2008). Progress toward understanding the distribution of Accumulibacter among full-scale enhanced biological phosphorus removal systems. *Microbial Ecology*, *55*(2), 229–236.

Hesselmann, R. P. X., Werlen, C., Hahn, D., Van der Meer, J. R., & Zehnder, A. J. B. (1999). Enrichment, phylogenetic analysis and detection of a bacterium that performs enhanced biological phosphate removal in activated sludge. *Systematic and Applied Microbiology*, *22*(3), 454–465.

Hiraishi, A., & Morishima, Y. (1990). Capacity for polyphosphate accumulation of predominant bacteria in activated sludge showing enhanced phosphate removal. *Journal of Fermentation and Bioengineering*, *69*, 368–371.

Hirota, R., Kuroda, A., Kato, J., & Ohtake, H. (2010). Bacterial phosphate metabolism and its application to phosphorus recovery and industrial bioprocesses. *Journal of Bioscience and Bioengineering*, *109*(5), 423–432.

Huchzermeier, M. P., & Tao, W. D. (2012). Overcoming challenges to struvite recovery from anaerobically digested dairy manure. *Water Environment Research*, *84*, 34–41.

Insel, G., Artan, N., & Orhon, D. (2005). Effect of aeration on nutrient removal performance of oxidation ditch systems. *Environmental Engineering Science*, *22*, 802–815. Available from https://doi.org/10.1089/ees.2005.22.802.

Ivanovic, I., & Leiknes, T. O. (2012). The biofilm membrane bioreactor (BF-MBR)—A review. *Desalination and Water Treatment*, *37*, 288–295. Available from https//doi.org/10.1080/19443994.2012.661283.

Jenkins, D., Richard, M. G., & Daigger, G. T. (1993). *Manual on the causes and control of activated sludge bulking and foaming*. New York: Lewis Publishers.

Jeon, C. O., & Park, J. M. (2000). Enhanced biological phosphorous removal in a sequencing batch reactor fed with glucose as a sole carbon source. *Water Research*, *34*, 2160–2170.

Kämpfer, P., Erhart, R., Beimfohr, C., Bohringer, J., Wagner, M., & Amann, R. (1996). Characterization of bacterial communities from activated sludge culture-dependent numerical identification vs in situ identification using group- and genus-specific rRNAtargeted oligonucleotide probes. *Microbial Ecology*, *32*, 101–121.

Kawaharasaki, M., Tanaka, H., Kanagawa, T., & Nakamura, K. (1999). In situ identification of polyphosphate-accumulating bacteria in activated sludge by dual staining with rRNA-targeted oligonucleotide probes and 4,6-diamidino-2-phenylindol (DAPI) at a polyphosphate-probing concentration. *Water Research*, *33*, 257–265.

Kong, Y., Nielsen, J. L., & Nielsen, P. H. (2004). Microautoradiographic study of Rhodocyclus related polyphosphate-accumulating bacteria in full-scale enhanced biological phosphorus removal plants. *Applied and Environmental Microbiology*, *70*(9), 5383–5390.

Kong, Y., Nielsen, J. L., & Nielsen, P. H. (2005). Identity and ecophysiology of uncultured actinobacterial polyphosphate-accumulating organisms in full-scale enhanced biological phosphorus removal plants. *Applied and Environmental Microbiology*, *71*(7), 4076–4085. Available from https://doi.org/10.1128/AEM.71.7.4076-4085.2005.

Kornberg, A., Kornberg, S. R., & Simms, E. S. (1956). Metaphosphate synthesis by an enzyme from *Escherichia coli*. *Biochimica et Biophysica Acta*, *20*, 215–227.

Kornberg, A., Rao, N. N., & Ault-Riché, D. (1999). *Annual Review of Biochemistry*, *68*, 89–125.

Kristiansen, R., Nguyen, H. T. T., Saunders, A. M., Nielsen, J. L., Wimmer, R., Le, V. Q., ... Calteau, A. (2012). A metabolic model for members of the genus *Tetrasphaera* involved in enhanced biological phosphorus removal. *The ISME Journal*, 7(3), 543–554.

Kulaev, I. S. (1979). *The biochemistry of inorganic polyphosphates*. Chichester: John Wiley and Sons.

Kulaev, I. S. (1990). The physiological role of inorganic polyphosphates in microorganisms: Some evolutionary aspects. In E. A. Dawes (Ed.), *Novel biodegradable microbial polymers* (pp. 223–233). Dordrecht: Kluwer Academic Publishers.

Kulaev, I. S., & Vagabov, V. M. (1983). Polyphosphate metabolism in microorganisms. *Advances in Microbial Physiology*, *24*, 83–171.

Lanham, A. B., Oehmen, A., Saunders, A. M., Carvalho, G., Nielsen, P. H., & Reis, M. A. M. (2013). Metabolic versatility in full-scale wastewater treatment plants performing enhanced biological phosphorus removal. *Water Research*, *47*, 7032–7041.

Lettinga, G., van Velsen, A. F. M., Hobma, S. W., de Zeeuw, W., & Klapwijk, A. (1980). Use of the upflow sludge blanket (USB) reactor concept for biological wastewater treatment, especially for anaerobic treatment. *Biotechnology and Bioengineering*, *22*, 699–734. Available from https//doi.org/10.1002/bit.260220402.

Levin, G. V., & Della, S. U. (1987). *Phostrip process—A viable answer to eutrophication of lakes and coastal sea waters in Italy. Biological Phosphate Removal from Wastewaters* (pp. 249–259). Elsevier.

Levin, G. V., & Shapiro, J. (1965). Metabolic uptake of phosphorus by wastewater organisms. *Water Pollution Control Federation*, *37*, 800–821. Available from https://doi.org/10.2307/25035307.

Levinson, S. L., Jacobs, L. H., Krulwich, T. A., & Li, H. C. (1975). Purification and characterisation of a polyphosphate kinase from *Arthrobacter atrocyaneus*. *Journal of General Microbiology*, *88*, 65–74.

Li, K., Liu, Q., Fang, F., Luo, R., Lu, Q., & Zhou, W. (2019). Microalgae-based wastewater treatment for nutrients recovery: A review. *Bioresource Technology*, *291*, 121934.

Lichko, L. P., Kulakovskaya, T. V., & Kulakovskaya, E. V. (2008). Inactivation of PPX1 and PPN1 genes encoding exopolyphosphatases of *Saccharomyces cerevisiae* does not prevent utilization of polyphosphates as phosphate reserve. *Biochemistry Moscow*, *73*, 985–989. Available from https://doi.org/10.1134/S0006297908090046.

Liebermann, L. (1888). Über das Nuclein der Hefe und künstliche Darstellung eines Nucleins aus Eiweiss und Metaphosphorsäure. *Berichte der Deutschen Chemischen Gesellschaft*, *21*, 598–600.

Liu, W. T., Nielsen, A. T., Wu, J. H., Tsai, C. S., Matsuo, Y., & Molin, S. (2001). In situ identification of polyphosphate- and polyhydroxyalkanoate-accumulating traits for microbial populations in a biological phosphorus removal process. *Environmental Microbiology*, *3*, 110–122.

López-Vázquez, C. M., Hooijmans, C. M., Brdjanovic, D., Gijzen, H. J., & Van Loosdrecht, M. (2008). Factors affecting the microbial populations at full-scale enhanced biological phosphorus removal (EBPR) wastewater treatment plants in The Netherlands. *Water Research*, *42*(10), 2349–2360.
Lopez-Vazquez, C. M., Oehmen, A., Hooijmans, C. M., Brdjanovic, D., Gijzen, H. J., Yuan, Z., & Van Loosdrecht, M. C. M. (2009). Modeling the PAO–GAO competition: Effects of carbon source, pH and temperature. *Water Research*, *43*(2), 450–462.
Lu, H., Oehmen, A., Virdis, B., Keller, J., & Yuan, Z. (2006). Obtaining highly enriched cultures of Candidatus Accumulibacter phosphates through alternating carbon sources. *Water Research*, *40*(20), 3838–3848.
Lv, X., Shao, M., Li, C., Li, J., Liu, D., Gao, X., & Xia, X. (2014). Operation performance and microbial community dynamics of phosphorus removal sludge with different electron acceptors. *Journal of Industrial Microbiology and Biotechnology*, *41*(7), 1099–1108. Available from https://doi.org/10.1007/s10295-014-1444-1.
Majed, N., Chernenko, T., Diem, M., & Gu, A. Z. (2012). Identification of functionally relevant populations in enhanced biological phosphorus removal processes based on intracellular polymers profiles and insights into the metabolic diversity and heterogeneity. *Environmental Science & Technology*, *46*, 5010–5501. Available from https//doi.org/10.1021/es300044h.
Mara, D., & Horan, N. J. (2003). *Handbook of water and wastewater microbiology*. Academic Press.
Martin, K. J., & Nerenberg, R. (2012). The membrane biofilm reactor (MBfR) for water and wastewater treatment: Principles, applications and recent developments. *Bioresource Technology*, *122*, 83–94. Available from https//doi.org/10.1016/j.biortech.2012.02.110.
Maszenan, A. M., Seviour, R. J., Patel, B. K. C., Schumann, P., Burghardt, J., Tokiwa, Y., & Stratton, H. M. (2000). Three isolates of novel polyphosphate-accumulating Gram-positive cocci, obtained from activated sludge, belong to a new genus *Tetrasphaera* gen. nov., and description of two new species, *Tetrasphaera japonica* sp. nov. and *Tetrasphaera australiensis* sp. nov. *International Journal of Systematic and Evolutionary Microbiology*, *50*, 593–603.
McGrath, J. W., & Quinn, J. P. (2003). Microbial phosphate removal and polyphosphate production from wastewaters. *Advances in Applied Microbiology*, *52*, 75–100.
Meinhold, J., Pedersen, H., Arnold, E., et al. (1998). Effect of continuous addition of an organic substrate to the anoxic phase on biological phosphorus removal. *Water Science and Technology: A Journal of the International Association on Water Pollution Research*, *38*, 97–105.
Metcalf & Eddy., Tchobanoglous, G., Stensel, H. D., et al. (2003). *Wastewater engineering: treatment and resource recovery*. New York: McGraw Hill Education.
Mogens, H., Van Mark, C. M., Loosdrecht, G. A., & Ekama, D. B. (2008). *Biological wastewater treatment: principles, modelling and design*. London: IWA Pub.
Muhammed, A. (1961). Studies on biosynthesis of polymetaphosphate by an enzyme from *Corynebacterium xerosis*. *Biochimica et Biophysica Acta*, *54*, 121–132.
Muhammed, A., Rodgers, A., & Hughes, D. E. (1959). Purification and properties of a polymetaphosphatase from *Corynebacterium xerosis*. *Journal of General Microbiology*, *20*, 482–495.
Nakamura, K., Hiraishi, A., Yoshimi, Y., Kawaharasaki, M., Masuda, K., & Kamagata, Y. (1995). *Microlunatus phosphovorus* gen. nov, sp. nov, a new gram-positive polyphosphate-accumulating bacterium isolated from activated sludge. *International Journal of Systematic Bacteriology*, *45*, 17–22.
Nerenberg, R. (2016). The membrane-biofilm reactor (MBfR) as a counter-diffusional biofilm process. *Current Opinion in Biotechnology*, *38*, 131–136. Available from https://doi.org/10.1016/j.copbio.2016.01.015.

Nguyen, H. T. T., Le, V. Q., Hansen, A. A., Nielsen, J. L., & Nielsen, P. H. (2011). High diversity and abundance of putative polyphosphate-accumulating *Tetrasphaera*-related bacteria in activated sludge systems. *FEMS Microbiology Ecology*, *76*(2), 256–267.

Nielsen, A. T., Liu, W. T., Filipe, C., Grady, L., Molin, S., & Stahl, D. A. (1999). Identification of a novel group of bacteria in sludge from a deteriorated biological phosphorus removal reactor. *Applied and Environmental Microbiology*, *65*(3), 1251–1258.

Oehmen, A., Lemos, P. C., Carvalho, G., Yuan, Z., Keller, J., Blackall, L. L., et al. (2007). Advances in enhanced biological phosphorus removal: from micro to macro scale. *Water Research*, *41*, 2271–2300.

Oehmen, A., Teresa Vives, M., Lu, H., Yuan, Z., & Keller, J. (2005). The effect of pH on the competition between polyphosphate-accumulating organisms and glycogen accumulating organisms. *Water Research*, *39*(15), 3727–3737.

Ong, Y. H., Chua, A. S. M., Fukushima, T., Ngoh, G. C., Shoji, T., & Michinaka. (2014). A high-temperature EBPR process: The performance, analysis of PAOs and GAOs and the fine-scale population study of *Candidatus* "Accumulibacter phosphatis". *Water Research*, *64*, 102–112.

Østgaard, K., Christensson, M., Lie, E., et al. (1997). Anoxic biological phosphorus removal in a full-scale UCT process. *Water Research*, *31*, 2719–2726.

Panswad, T., Doungchai, A., & Anotai, J. (2003). Temperature effect on microbial community of enhanced biological phosphorus removal system. *Water Research*, *37*(2), 409–415.

Park, H. D., Whang, L. M., Reusser, S. R., & Noguera, D. R. (2006). Taking advantage of aerated-anoxic operation in a full-scale University of Cape Town Process. *Water Environment Research: A Research Publication of the Water Environment Federation*, *78*, 637–642. Available from https://doi.org/10.2175/106143006x99786.

Persson, B. L., Lagerstedt, J. O., Pratt, J. R., et al. (2003). Regulation of phosphate acquisition in *Saccharomyces cerevisiae*. *Current Genetics*, *43*(4), 225–244. Available from https//doi.org/10.1007/s00294-003-0400-9.

Randall, A. A., & Liu, Y. H. (2002). Polyhydroxyalkanoates form potentially a key aspect of aerobic phosphorus uptake in enhanced biological phosphorus removal. *Water research*, *36*(14), 3473–3478. Available from https://doi.org/10.1016/s0043-1354(02)00047-7.

Robinson, N. A., Goss, N. H., & Wood, H. G. (1984). Polyphosphate kinase from *Propionibacterium shermanii*: Formation of an enzymatically active insoluble complex with basic proteins and characterization of synthesized polyphosphate. *Biochemistry International*, *8*, 757–769.

Salehi, S., Cheng, K. Y., Heitz, A., & Ginige, M. P. (2018). Re-visiting the phostrip process to recover phosphorus from municipal wastewater. *Chemical Engineering Journal*, *343*, 390–398. Available from https://doi.org/10.1016/j.cej.2018.02.074.

Satoh, H., Mino, T., & Matsuo, T. (1990). Biochemical mechanism of anaerobic glutamate uptake in the enhanced biological phosphorus removal process. In *Proceeding of IAWPRC Biennial Conference* (531–534). Kyoto.

Schröder, J. J., Smit, A. L., Cordell, D., & Rosemarin, A. (2011). Improved phosphorus use efficiency in agriculture: A key requirement for its sustainable use. *Chemosphere*, *84*, 822–831. Available from https//doi.org/10.1016/j.chemosphere.2011.01.065.

Schuler, A. J., & Jenkins, D. (2003). Enhanced biological phosphorus removal from wastewater by biomass with different phosphorus contents, Part II: Anaerobic adenosine triphosphate utilization and acetate uptake rates. *Water Environment Research : A Research Publication of the Water Environment Federation*, *75*(6), S499–S511. Available from https://onlinelibrary.wiley.com/doi/abs/10.2175/106143003X141295.

Sentchilo, V., Mayer, A. P., Guy, L., Miyazaki, R., Green Tringe, S., Barry, K., et al. (2013). Community-wide plasmid gene mobilization and selection. *The ISME Journal*, *6*, 1173–1186.

Seviour, R., Takashi, M., & Motoharu, O. (2003). The microbiology of biological phosphorus removal in activated sludge systems. *FEMS Microbiology Reviews*, *27*, 99–127. Available from https://doi.org/10.1016/S0168-6445(03)00021-4.
Shintani, T., Liu, W. T., Hanada, S., Kamagata, Y., Miyaoka, S., Suzuki, T., & Nakamura, K. (2000). *Micropruina glycogenica* gen nov sp nov: A new Gram-positive glycogen-accumulating bacterium isolated from activated sludge. *International Journal of Systematic and Evolutionary Microbiology*, *1*, 201–207.
Slocombe, S. P., Zúñiga-Burgos, T., Chu, L., Wood, N. J., Camargo-Valero, M. A., & Baker, A. (2020). Fixing the broken phosphorus cycle: wastewater remediation by microalgal polyphosphates. *Frontiers in Plant Science*, *11*, 982. Available from https://doi.org/10.3389/fpls.2020.00982.
Smolders, G. J. F., Van der Meij, J., Van Loosdrecht, M. C. M., & Heijnen, J. J. (1994). Model of the anaerobic metabolism of the biological phosphorus removal process: stoichiometry and pH influence. *Biotechnology and Bioengineering*, *43*, 461–470. Available from https//doi.org/10.1002/bit.260430605.
Smolders, G. J. F., Van Loosdrecht, M. C. M., & Heijnen, J. J. (1995). A metabolic model for the biological phosphorus removal process. *Water Science and Technology: A Journal of the International Association on Water Pollution Research*, *31*, 79–93.
Stante, L., Cellamare, C. M., Malaspina, F., Bortone, G., & Tilche, A. (1997). Biological phosphorus removal by pure culture of *Lampropedia* spp. *Wat Res*, *31*, 1317–1324.
Sudiana, I. M., Mino, T., Satoh, H., Nakamura, K., & Matsuo, T. (1999). Metabolism of enhanced biological phosphorus removal and non-enhanced biological phosphorus removal sludge with acetate and glucose as carbon source. *Water Science and Technology: A Journal of the International Association on Water Pollution Research*, *39*, 29–35.
Sun, L., Wang, Z., Wei, X., Li, P., Zhang, H., Li, M., et al. (2015). Enhanced biological nitrogen and phosphorus removal using sequencing batch membrane-aerated biofilm reactor. *Chemical Engineering Science*, *135*, 559–565. Available from https//doi.org/10.1016/j.ces.2015.07.033.
Szymona, M., & Ostrowski, W. (1964). Inorganic polyphosphate glucokinase of *Mycobacterium phlei*. *Biochimica et Biophysica Acta*, *85*, 283–295.
Turner, A. M., & Chislock, M. F. (2010). Blinded by the stink: Nutrient enrichment impairs the perception of predation risk by freshwater snails. *Ecological Applications*, *20*, 2089–2095.
Usharani, K., & Lakshmanaperumalsamy, P. (2010). Studies on the Removal Efficiency of Phosphate from Wastewater using Pseudomonas sp YLW-7 and *Enterobacter* sp KLW-2. *Global Journal of Environmental Research*, *4*, 83–89.
Vagabov, V. M., Trilisenko, L. V., & Kulaev, I. S. (2000). Dependence of inorganic polyphosphate chain length on the orthophosphate content in the culture medium of the yeast *Saccharomyces cerevisiae*. *Biochemistry*, *65*(3), 349–354.
Vaiopoulou, E., & Aivasidis, A. (2008). A modified UCT method for biological nutrient removal: Configuration and performance. *Chemosphere*, *72*, 1062–1068. Available from https://doi.org/10.1016/j.chemosphere.2008.04.044.
Van Loosdrecht, M. C. M., Brandse, F. A., & de Vries, A. C. (1998). Upgrading of wastewater treatment processes for integrated nutrient removal-The BCFS process. *Water Science and Technology: A Journal of the International Association on Water Pollution Research*, *37*, 209–217.
Van Wazer, J. R., & Campanella, D. A. (1950). Structure and properties of the condensed phosphates IV complex ion formation in polyphosphate solutions. *Journal of the American Chemical Society*, *72*, 655–663.
Wagner, M., Amann, R., Lemmer, H., & Schleifer, K.-H. (1993). Probing activated sludge with oligonucleotides specific for proteobacteria: Inadequacy of culture-dependent methods for describing microbial community structure. *Applied and Environmental Microbiology*, *59*, 1520–1525.

Wagner, M., Erhart, R., Manz, W., Amann, R., Lemmer, H., Wedi, D., & Schleifer, K.-H. (1994). Development of an rRNA-targeted oligonucleotide probe specific for the genus Acinetobacter and its application for in situ monitoring in activated sludge. *Applied and Environmental Microbiology*, *60*, 792–800.

Wang, N., Peng, J., & Hill, G. (2002). Biochemical model of glucose induced enhanced biological phosphorus removal under anaerobic condition. *Water Research*, *36*, 49–58.

Wang, S. P., Yu, J. J., Gao, F., et al. (2019a). Particular internal recirculation frequency scope for enhancing denitrifying phosphorus removal in an oxidation ditch. *Water Science and Technology: A Journal of the International Association on Water Pollution Research*, *80*, 191–202. Available from https://doi.org/10.2166/wst.2019.265.

Wang, Y., Peng, Y., & Stephenson, T. (2009). Effect of influent nutrient ratios and hydraulic retention time (HRT) on simultaneous phosphorus and nitrogen removal in a two sludge sequencing batch reactor process. *Bioresource Technology*, *100*, 3506–3512. Available from https://doi.org/10.1016/j.biortech.2009.02.026.

Watanabe, T., Ozaki, N., Iwashita, K., Fujii, T., & Iefuji, H. (2008). Breeding of wastewater treatment yeasts that accumulate high concentrations of phosphorus. *Applied Microbiology and Biotechnology*, *80*(2), 331–338. Available from https//doi.org/10.1007/s00253-008-1529-8.

Wentzel, M. C., Lotter, L. H., Loewenthal, R. E., & Marais, G. (1986). Metabolic behaviour or *Acinetobacter* spp. in enhanced biological phosphorus removal - A biochemical model. *Water SA*, *12*, 209–224.

Wilfert, P., Kumar, P. S., Korving, L., Witkamp, G. J., & van Loosdrecht, M. C. (2015). The relevance of phosphorus and iron chemistry to the recovery of phosphorus from wastewater: A review. *Environmental Science & Technology*, *49*(16), 9400–9414.

Winkler, M. (1984). Biological control of nitrogenous pollution in wastewater. In A. Wiseman (Ed.), *Topics in enzyme and fermentation biotechnology* (pp. 31–125). Chichester: Ellis Horwood.

Wong, D. H. J., & Beiko, R. G. (2015). Transfer of energy pathway genes in microbial enhanced biological phosphorus removal communities. *BMC Genomics*, *16*, 526. Available from https://doi.org/10.1186/s12864-015-1752-5.

Wood, H. G., & Clark, J. E. (1988). Biological aspects of inorganic polyphosphates. *Annual Review of Biochemistry*, *57*, 235–260.

Yang, S., Yang, F., Fu, Z., Wang, T., & Lei, R. (2010). Simultaneous nitrogen and phosphorus removal by a novel sequencing batch moving bed membrane bioreactor for wastewater treatment. *Journal of Hazardous Materials*, *175*, 551–557. Available from https//doi.org/10.1016/j.jhazmat.2009.10.040.

Ye, Y., Gan, J., & Hu, B. (2015). Screening of phosphorus-accumulating fungi and their potential for phosphorus removal from waste streams. *Applied Biochemistry and Biotechnology*, *177* (5), 1127–1136. Available from https//doi.org/10.1007/s12010-015-1801-1.

Yeoman, S., Stephenson, T., Lester, J. N., & Perry, R. (1988). The removal of phosphorus during wastewater treatment: A review. *Environmental Pollution (Barking, Essex: 1987)*, *49*, 183–233. Available from https://doi.org/10.1016/0269-7491(88)90209-6.

Yoshida, A., & Yamataka, A. (1953). On the metaphosphate of the yeast. *Journal of Biochemistry*, *40*, 85–94.

Yuan, Z., Pratt, S., & Batstone, D. J. (2012). Phosphorus recovery from wastewater through microbial processes. *Current Opinion in Biotechnology*, *23*(6), 878–883.

Zhang, W. X., & Lau, A. (2007). Reducing ammonia emission from poultry manure composting via struvite formation. *Journal of Chemical Technology and Biotechnology*, *82*, 598–602.

Zhou, Y., Pijuan, M., Zeng, R. J., & Yuan, Z. (2009). Involvement of the TCA cycle in the anaerobic metabolism of polyphosphate accumulating organisms (PAOs). *Water Research*, *43*, 1330–1340. Available from https//doi.org/10.1016/j.watres.2008.12.008.

CHAPTER 9

Genetically engineered bacteria: a novel technique for environmental decontamination

Tarun Kumar Kumawat[1], Nalinee Kumari[2], Vishnu Sharma[1], Varsha Kumawat[3] and Manish Biyani[4]
[1]Department of Biotechnology, Biyani Girls College, University of Rajasthan, Jaipur, India
[2]Department of Zoology, DPG Degree College, Gurugram, India
[3]Naturilk Organic and Dairy Foods Pvt. Ltd., Jaipur, India
[4]Department of Bioscience and Biotechnology, Japan Advanced Institute of Science and Technology, Nomi, Japan

9.1 Introduction

Environmental pollution became increasingly severe as the number of human activities increased. Pesticides, dyes, oil, and heavy metals were all widely utilized, causing the environment to become more polluted (Wang, Zhu, Zhang, & Guan, 2014). Attempts to remove hazardous chemicals from the environment using conventional techniques have failed miserably. There has been a surge of curiosity in creating in situ techniques for remediation of environmental contaminant over the last several decades; nevertheless, because of the high financial impact of physicochemical techniques, biological approaches for remediation of these pollutants are the preferred option (Bisht et al., 2015). The bioremediation method is often considered as being cost-effective, safe, reliable, effective, and environment-friendly since it utilizes microorganisms and their metabolic activities to break down pollutants (Tyagi & Kumar, 2021). However, developing bacteria that can be utilized as biocatalysts capable of cleaning up modern-day pollutants may take a long time (de Lorenzo, 2001). There are a variety of bioremediation approaches that have been widely used to boost bacterial metabolism and degrade resistant contaminants, including bioaugmentation, chemotaxis, use of genetically engineered microorganisms (GEMs), biostimulation, and advanced omics (Pande, Pandey, Sati, Pande, & Samant, 2020).

Microbial Consortium and Biotransformation for Pollution Decontamination
DOI: https://doi.org/10.1016/B978-0-323-91893-0.00008-0

Natural species cannot break down certain toxins fast enough, therefore they must be genetically engineered (GE) to do it. These GEMs degrade contaminants considerably quicker than native microbes (Abatenh, Gizaw, Tsegaye, & Wassie, 2017). The study of microbes resulted in the creation of GEMs, which are used to fight environmental pollution (Pant et al., 2021). GEM is a microbe, which is defined as a bacterium, the DNA of which has been altered via genetic engineering based on natural or artificial processes through intermicrobial genetic exchange. This kind of creative effort combined with a scientific process is known as recombinant DNA technology (Abatenh et al., 2017). In the environment, microbes that have been GE may be able to travel through a variety of media (including air, soil, water, and groundwater) and proliferate quickly if the circumstances are right (Barnthouse & Palumbo, 1986; Dar, Bandh, & Kamili, 2011; Dar, Kamili, Nazir, Bandh, & Bhat, 2012). It has also been done to accomplish safe and effective bioremediation of polluted areas using recombinant technologies, such as suicidal-GEMs (S-GEMs) (Kumar, Dagar, Khasa, & Kuhad, 2013; Paul, Pandey, & Jain, 2005). It is based on knowledge of killer—antikiller gene(s) that are susceptible to programmed cell death after detoxification of any particular polluted region (Dar, Bandh, Kamili, Nazir, & Bhat, 2013; Pandey, Paul, & Jain, 2005). This book chapter described the major environment pollutants, construction of GEMs, and their applications in environmental remediation, including pesticide, dye, heavy metal, and oil pollution cleanup, as well as in other fields.

9.2 Environmental contaminants

Emerging contaminants that enter the environment via a variety of anthropogenic sources are dispersed across environmental matrices and to be a global concern and one of the greatest challenges facing humanity (Gavrilescu, Demnerová, Aamand, Agathos, & Fava, 2015). Inorganic, organic, and biological pollutants are the three categories of pollutants (Oliveira, Al Aukidy, & Verlicchi, 2017). Regardless of the category pollutants come into, they all receive a lot of attention because of the negative effects they have on the environment. Inorganic pollutants are usually mineral-based compounds, such as salts, minerals, and metals (Masindi & Muedi, 2018). Aside from naturally occurring materials that may be present throughout the Earth's crust, most of pollution is produced by human activities such as mining, smelting operations, metallurgical, and chemical processes (Tchounwou, Yedjou, Patlolla, & Sutton, 2012).

When organic pollutants build up in the environment, it may lead to severe issues, such as a decrease in the various aquatic ecosystems' stability. It can also have harmful consequences for environment and human health (Chekroun, Sánchez, & Baghour, 2014). Water derived from industries such as textile manufacturing, pesticide manufacturing, pharmacy, and petroleum manufacturing produces a lot of organic chemicals, including aromatic compounds, chlorinated hydrocarbons, and phenolic compounds (Anandan, Kumar Ponnusamy, & Ashokkumar, 2020). Pesticides, polychlorinated biphenyls (PCBs), and surfactants are frequently detected in the atmosphere (Farre, Kantiani, & Barcelo, 2007; Gaur, Narasimhulu, & PydiSetty, 2018). Increased levels of biological and nonbiological contaminants represent a danger to the environment's ecological systems (Rani et al., 2020). Bioaerosols are often described as airborne bacteria, fungus, viruses, pollen, and their byproducts, all of which are classified as biological pollutants (Ghanizadeh & Godini, 2018).

9.2.1 Heavy metal contamination

Natural deposits of metals exist in the Earth's crust. The geochemical cycles and biological balance of certain heavy metals have been significantly altered. Bioavailable forms of various heavy metals are released into soil and aquatic ecosystems from anthropogenic sources (Dixit et al., 2015; Gupta & Singh, 2017). The natural environment has been damaged by toxic waste products, and heavy metals. Natural catastrophes such as hurricanes and volcanic eruptions, as well as anthropogenic activities, arise from these contaminants (Bhat et al., 2021; Rajendran, Muthukrishnan, & Gunasekaran, 2003). Toxic heavy metal pollution presents danger to all living forms of ecosystem and has the potential to enter the food chain after accumulating in agricultural soils and posing danger to food safety (Ayangbenro & Babalola, 2017; Dar, Kamili, Dar, Chishti, & Ahmad, 2015; Ojuederie & Babalola, 2017). A number of heavy metals are known to cause cancer, mutation, and teratogenicity as well as endocrine disruptors, while others particularly in children induce neurological and behavioral abnormalities. Accordingly, heavy metal contamination has to be addressed (Ali, Khan, & Sajad, 2013). Heavy metal-polluted soils must be decontaminated in order to reduce risks and preserve ecological restoration and environmental health. Traditional methods for reclaiming such soils are both costly and harmful to the environment. Heavy metals and metalloids are removed from soils using chemical, physical, and biological techniques (Ashraf, Ali, Zahir, Ashraf, & Asghar, 2019; Sarwar et al., 2017).

9.2.2 Dye-based hazardous pollutants

Effluents containing dyes have the potential to pollute soils and water systems, posing a threat to the environment (Ali, 2010). Paper, cosmetic, textile, leather, and food sectors utilize dyes more extensively as industrialization expands. As a result, wastewater discharge from dye production activities would increase, posing serious risks to plants, animals, and humans (Varjani, Rakholiya, Ng, You, & Teixeira, 2020). Basic, azo, acidic, dispersion, anthraquinone-based, and metal complex dyes are all divided into structural categories (Benkhaya, M' rabet, & El Harfi, 2020; Kodoth & Badalamoole, 2020; Yang & Mcgarrahan, 2005). The use of dyes is widespread in textiles, leather tannery, paper, food processing, color photography, pharmaceuticals, cosmetics (including hair colorants), and chemical research (Couto, 2009; Dar et al., 2016; Kuhad, Sood, Tripathi, Singh, & Ward, 2004). The process of decolorization or detoxifying of hazardous dyes has emerged as a worldwide priority to avoid the catastrophic effects of their use and to save the environment (Goud, Cha, Koyyada, & Kim, 2020).

9.2.3 Radioactive compounds

Decades of nuclear activity have left a toxic legacy in the environment that must be addressed (Newsome, Morris, & Lloyd, 2014). Massive quantities of radioactive waste produced in anthropogenic sources (hospitals and nuclear power plants) were released into the environment (Jeong & Choi, 2017; Sengupta, Roy, Chowdhary, & Das, 2021). The following major anthropogenic sources may contribute to radioactive pollution: development of nuclear weapons, testing of nuclear weapons, construction of nuclear power plants, the processing and milling of uranium, storage of nuclear waste in a geological repository, and nuclear mishap (Hu, Weng, & Wang, 2010). Both the Chernobyl disaster 1986 and Fukushima Daiichi nuclear 2011 disaster wreaked havoc on the environment, generating huge amounts of radioactive waste that threatens both human and ecological health (Prakash, Gabani, Chandel, Ronen, & Singh, 2013). Plutonium-239 (^{239}Pu), Cobalt-60 (^{60}Co), Radon-222 (^{222}Rn), Radium-226 (^{226}Ra), Thorium-232 (^{232}Th), and Uranium-238 (^{238}U) are some of radionuclides that are frequently found in the environment.

Cesium-137, Strontium-90, and Thallium-201 are the most common radionuclides generated by nuclear reactors as a result of the splitting of elemental atoms, and they require a substantial amount of time to decay (Kurnaz et al., 2007). The radiotoxicity of effluents containing high-activity radionuclides may also harm the microbiological component of the system. Radioactive contamination of soils is a distinct kind of deterioration that reduces their fertility (Aleksakhin, 2009). Radionuclides may be found at thousands of waste sites throughout the globe. Cleaning up such sites with physicochemical techniques is fraught with risk and cost, therefore alternative ways are being explored (Lange, Wackett, Minton, & Daly, 1998).

9.2.4 Agricultural chemicals: herbicides, pesticides, and fertilizers

Earth is experiencing very severe consequences of anthropogenic activities such as current farming methods, industrialization, urbanization, and unhealthy struggle for dominance (Bhandari et al., 2021). Agricultural industrialization has raised the chemical burden on natural ecosystems (Nicolopoulou-Stamati, Maipas, Kotampasi, Stamatis, & Hens, 2016). Farmers are using various agrochemicals to boost crop yields (Agmas & Adugna, 2020). Chemical pesticides are frequently utilized to control pests, including insects, bacteria, fungi, weeds, nematodes, and rodents in agriculture, and residues of these pesticides can be detected in the environment (Aynalem & Muleta, 2021; Bottrell & Schoenly, 2018). Insecticides, fungicides, herbicides, rodenticides, and phytohormones are some examples of pesticides. Pesticide contamination has become a worldwide problem. Pesticides are sprayed on the agricultural crops every year, yet only 1% of them are effective, while the rest either contaminate cereal grains, vegetables, and fruits or damage the ecosystem (Sehrawat, Phour, Kumar, & Sindhu, 2021). Pesticide contamination has also been detected in milk, juices, seaweed, and other foods (Hamedi, B.G. Aghaie, & Hadjmohammadi, 2018; Kumar, Rai, Porwal, & Kumar, 2018). Chemical pesticides have been linked to various detrimental health consequences such as gastrointestinal, dermatological, carcinogenic, neurological, reproductive, pulmonary, and endocrine disorders (Salameh, 2006; Sanborn et al., 2007). In agriculture, pesticides such as organochlorines (OC), organophosphates (OP), carbamate, pyrethroids, chloronicotinyl, and others are used to protect plants against illnesses and insects (Liu, Zheng, & Li, 2013).

9.2.5 Petroleum and polycyclic aromatic hydrocarbon contaminants

The petroleum industry is the world's biggest and most diverse sector, including a broad variety of operations all over the globe (Hidalgo, Sierra-Garcia, Dellagnezze, & de Oliveira, 2020). The most serious environmental issues facing the world today is hydrocarbon pollution caused by the operations associated with the petroleum industry (Naeem & Qazi, 2020). Accidental spills of petroleum products into the environment are a particular source of worry. It has long been recognized that hydrocarbon components are members of the carcinogen and neurotoxic organic pollutant families (Das & Chandran, 2011). Environmental damage from oil spills has been occurring in recent decades and has had a long-term impact on ecosystem function and human health (Chang, Stone, Demes, & Piscitelli, 2014; French, Zhou, & Terry, 2020). Oil skimming, soil removal, soil capping, and chemical oxidation are some of the methods now used to remove crude oil from the vicinities (Doshi, Sillanpaa, & Kalliola, 2018). Polycyclic aromatic hydrocarbons (PAHs) are universally present and are harmful to all forms of life owing to their poisonous, carcinogenic, and mutagenic properties (Ji, Ripp, Layton, Sayler, & DeBruyn, 2013; Sakshi, Singh, & Haritash, 2020; Xu et al., 2021). These are dangerous contaminants generated by incomplete ignition of organic molecules (Gupta, Singh, & Mishra, 2021; Sakshi & Haritash, 2020; Samanta, Singh, & Jain, 2002). The majority of PAHs bond to soil and sediment particles, making them less accessible for biological absorption because of their hydrophobic nature (Peng et al., 2008). PAHs are degraded via adsorption, photolysis, chemical degradation, volatilization, and microbiological. Various microorganisms such as algae, bacteria, and fungus are biotransforming PAHs into H_2O, CO_2, or CH_4 under anaerobic environments (Premnath et al., 2021). The toxicity of PAHs increases with the increase in their molecular weight. These refractory hydrocarbons settle in the coastal environment due to their hydrophobic character (Tirkey, Ram, & Mishra, 2021).

9.2.6 Polychlorinated biphenyls

Pollution mitigation has been acknowledged as posing greater economic and environmental problems throughout the world. Numerous studies over the past few decades have highlighted uncontrolled anthropogenic activities as the main cause of pollution in the environment (Adetunji & Anani, 2021). PCBs are typical organic contaminants and a synthetic category of chlorinated aromatic hydrocarbons that are extensively used worldwide (Xiang, Xing, Liu, Qin, & Huang, 2020; Zenteno-Rojas et al., 2020). PCBs can persist in environment for a prolonged period because of their low stability and reactivity under severe environmental conditions (Elangovan, Pandian, S., & Joshi, 2019). PCBs are a family of 209 congeners with chemical characteristics that make them suitable for use in electrical equipment industry and operations (Furukawa & Fujihara, 2008). Electronic trash is referred to as e-waste, and it contributes to environmental contamination when it is recycled. PCBs are a major e-waste contaminant and are used in transformers and capacitors to exchange heat and hydraulic fluids (Khalid, Hashmi, Jamil, Qadir, & Ali, 2021). However, owing to their recalcitrance, accumulation potential, widespread contamination, and probable cancer-causing effects, PCBs are presenting long-term and detrimental risks to the environment and human health (Bilal, Iqbal, & Barceló, 2019). Foods grown in PCB-rich environments may cause developmental and reproductive toxicity, neurotoxicity, cutaneous toxicity, endocrinological and hepatic impacts, carcinogenesis, and stimulation of various drug-metabolizing enzymes (Munawar, Akram, Javed, & Shahid, 2021).

9.3 Genetically engineered bacteria and their construction

Microorganisms express their genetic system differently against various harmful chemicals to adapt to their environment. Adding appropriate genes may enhance adaption (Brooks, Turkarslan, Beer, Yin Lo, & Baliga, 2011; Cavicchioli et al., 2019; Poursat, van Spanning, de Voogt, & Parsons, 2019). Recent years have seen the development of a range of genetic tools and techniques for optimizing the expression of genes encoding particular enzymes and metabolic pathways, thus aiding the efficient breakdown of target compounds (Pieper & Reineke, 2000). In the late 1980s genetic engineering was used to develop catabolic pathways for refractory compounds and several research groups have continued to work on this problem (Singh, Abhilash, Singh, Singh, & Singh, 2011). A genetically modified microbe that can destroy OPs and be immediately detected and tracked in the environment was reported by Li, Wang, Chen, Li, and Wu (2014). The GEM was constructed utilizing an enhanced green fluorescent protein fusion method. Yang et al. (2021) created genetically engineered bacteria (GEB) that adsorb and self-flocculate nitrate from water by integrating the genes of MO2.1 from *Moringa oleifera* and FLO5 from *Saccharomyces cerevisiae*. Li, Li, Kang, and Wu (2020) developed an engineered bacterium that can break down pesticide residues in the environment and generate green fluorescence using a dual plasmid expression technique.

9.4 Genetically engineered bacteria for a sustainable environment

Scientists are able to construct GEMs for applications in a number of areas (environment, agriculture, and other industries), thanks to the advances in biotechnology over the past 25 years (Forabosco, Lohmus, Rydhmer, & Sundstrom, 2013; Jafari, Danesh, Goltapeh, & Varma, 2013). Microbial systems for the management of contaminated soils and waterways are employed in bioremediation initiatives. In instances when native populations are insufficient or inappropriate for use in contaminated remediation, it may be essential to employ GEMs for in situ bioremediation (Garbisu & Alkorta, 1999).

Recent advances in microbial research and synthetic biology have inspired researchers to create a wide spectrum of engineered bacteria expressing useful environmental decontamination applications. The applications of GEMs including synthetic dyestuffs, heavy metals, petroleum hydrocarbons, PCBs, and agricultural chemicals were summarized in this section (Fig. 9.1).

9.4.1 Remediation of toxic heavy metals

The growing concern about the dangers of heavy metal pollution to human health has sparked a renewed interest in creating or developing technologies that may remove or neutralize the detrimental impacts of heavy metal pollution in soil, and wastewater (Valls & de Lorenzo, 2002). An alternate approach is to use microbial biomass to remove heavy metal. Microorganisms play fundamental role in the bioremediation of toxic heavy metals (Ezezika & Singer, 2010; Verma & Kuila, 2019). Environmentally safe and economically viable strategies for the removal of metal and metalloid contaminants are based on methods using GEMs which are produced by recombinant DNA and RNA technologies (Sharma, Sirohi, Tong, Kim, & Pandey, 2021). In situ bioremediation of heavy metal-contaminated regions using engineered bacteria requires knowledge of microbiology, ecology, biochemical processes, and field engineering designs (Singh et al., 2011). Some bacterial species that aid in heavy metal bioremediation include *Flavobacterium*, *Bacillus*, *Arthrobacter*, *Corynebacterium*, *Rhodococcus*, *Mycobacterium*, and *Nocardia*.

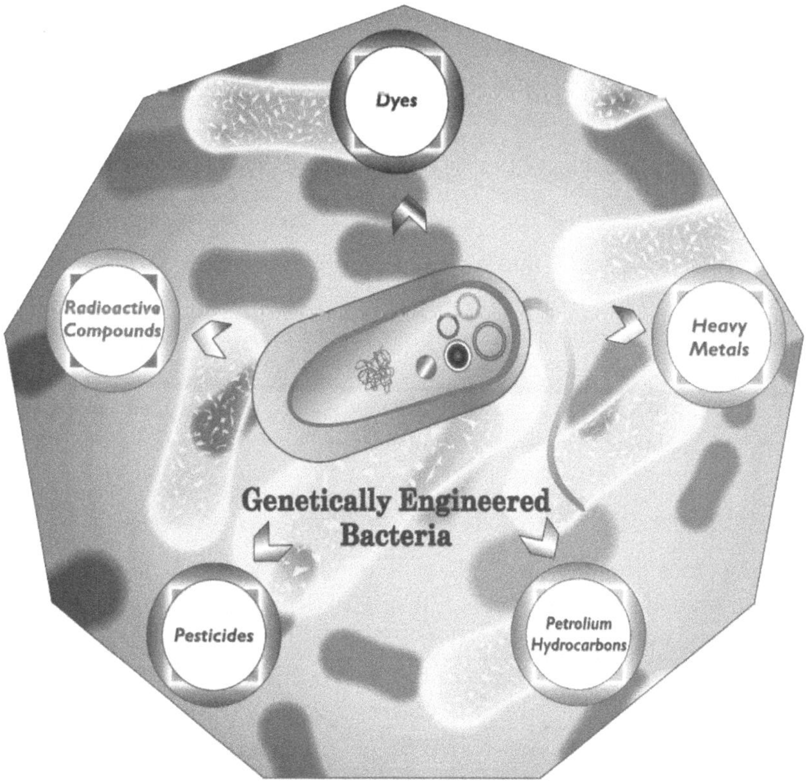

Figure 9.1 Applications of genetically engineered bacteria.

One of the more quickly expanding topics of environmental restoration is bioremediation using microorganisms to eliminate mercury from polluted sites (Kiyono & Pan-Hou, 2006). Kiyono & Pan-Hou (2006) engineered (genetically) *Escherichia coli* to produce mercury transport system and organomercurial lyase enzyme concurrently as well as overexpress polyphosphate, a powerful chelator of important divalent metals, to construct bacteria capable of selectively accumulating mercury. It was reported that Ruiz, Alvarez, Gonzalez-Ruiz, and Torres (2011) created a bacterial system (transgenic) that effectually expressed the metallothionein (*mt-1*) and polyphosphate kinase genes for mercury accumulation. Researchers have devised a technique for bioremediation of mercury that relies on the production of the bacteria's *mer* genes (Jackson & Summers, 1982). Zhao et al. (2005) investigated the prospective for mercury bioaccumulation of the GE *E. coli* JM109, which expressed the merT−merP protein and metallothionein. In addition, lead (Pb) has no known biological activity vital to human life because it is pervasive, and persistent in all realms of soil, air, and water. Pb causes severe health and environmental issues that need remediation (Mitra, Chatterjee, Kataki, Rastogi, & Gupta, 2021). Jia, Li, Xu, and Wu (2020) demonstrated selective adsorption of lead ions (Pb^{2+}) and bioremediation of lead in soil using *Cupriavidus metallidurans* strain CH34 proteins PbrR, PbrR691, and PbrD. Nowadays, remediation of a wide number of heavy metals using GEB is gaining attention due to its environmentally benign character and lowering risk to human health compared to other physicochemical techniques. It was discovered that genetically modified *E. coli* SE5000 strain that expressed both metallothionein and nickel transport system was capable of accumulating Ni^{2+} from an aqueous solution (Deng, Li, Lu, He, & Jiang, 2005).

9.4.2 Bioremediation of dye by engineered bacteria

In comparison to physicochemical processes, biological processes have attracted researchers' attention because of their economic efficiency, and environmental friendliness (Stolz, 2001). When it comes to attaining total dye degradation, GEMs play a significant role (Varjani et al., 2020). Liu, Zhou, Qu, and Ma (2007) reported the decolorization of sulfonated azo dyes by GE strain (*E. coli* YB) and wild-type photosynthetic bacteria (*Rhodobacter sphaeroides* AS1.1737 and *Rhodopseudomonas palustris* AS1.2352). Chang and Lin (2001) reported that engineered *E. coli* CY1 decolorized azo dye (C.I. Reactive Red 22), while the host strain (*E. coli* DH5α) exhibited no color-removing activity. The activity of *E. coli* JM109 (pGEX-AZR) to break down the azo dye Direct Blue 71 (DB71) was also examined (Jin, Yang, Zhang, Wang, & Liu, 2009). GEMs have garnered interest for biodegradation research and have been extensively used in bioaugmentation systems. Bioaugmentation is the technique of introducing indigenous GEMs into a bioreactor or polluted site to accelerate biological processes and enhance results (Limbergen, Top, & Verstraete, 1998).

9.4.3 Bioremediation of radionuclides

There is currently no effective method available that can safely store radioactive wastes. Numerous research paths have contributed to recent advancements in understanding the function of microorganisms in the bioremediation of hazardous radionuclides (Stephen & Macnaughtont, 1999). As a result, bioremediation of radionuclides is an inescapable need that has been attempted with little success utilizing biotransformation, bioaccumulation, biosorption, biostimulation, and bioaugmentation (Kumar, Singh, & Singh, 2007). Increased awareness of environmental contamination over the last several decades has resulted in the exploitation of bacterial metabolic capacity via the development of various GEMs for bioremediation applications (Paul et al., 2005). Efforts have been made to create GEB that detoxify radionuclides present in radioactive waste sites (Prakash et al., 2013) (Fig. 9.2). Thanks to bioremediation technology once again!

Deinococcus radiodurans, an organism that belongs to the bacterial family "Deinococcaceae" is extremely resistant to ionizing radiation, is now being developed for the treatment of both metal and organic pollutants utilizing genetic methods (Lloyd & Lovley, 2001; Makarova et al., 2001). This bacterium also shows potential in the bioremediation of mixed radioactive wastes (Sghaier et al., 2012). *D. radiodurans* is nonpathogenic, genetically modifiable, and known for its gamma radiation tolerance (Brim, Venkateswaran, Kostandarithes, Fredrickson, & Daly, 2003; Brim et al., 2006). Also, modified *Deinococcus geothermalis* strains have been created for high-temperature bioremediation of radioactive mixed waste

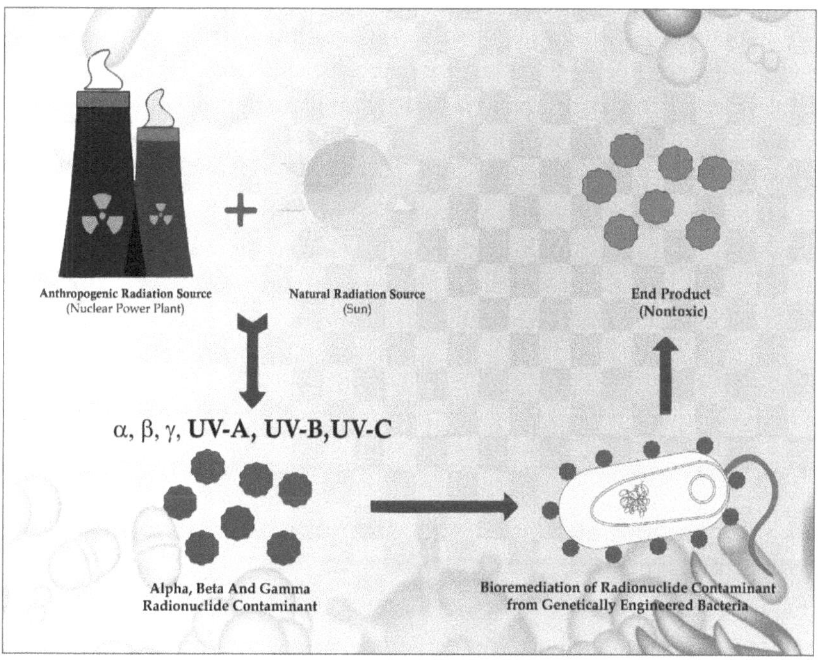

Figure 9.2 Bioremediation of radioactive compounds from genetically engineered microbes.

(Brim et al., 2003). Engineered strains of *D. radiodurans* (Appukuttan, Rao, & Apte, 2006), *E. coli* (Basnakova, Stephens, Thaller, Rossolini, & Macaskie, 1998), and *Pseudomonas veronii* (Powers et al., 2002) have also been shown to precipitate uranyl phosphates (Martinez et al., 2007).

9.4.4 Bioremediation of agricultural chemicals: herbicides, pesticides, and fertilizers

OP and OC, which are significant components of pesticides, build in agricultural soils and eventually enter aquatic bodies via agricultural runoff (Panelli et al., 2017). Recombinant DNA technology has created several genetically modified microorganisms that can degrade pesticides (Ingham, Doyle, & Hendricks, 1995; Lan et al., 2006). Genetically modified bacteria are effective in the remediation of methyl parathion (OP) and γ-hexachlorocyclohexane (OC) (Gong et al., 2016). Furthermore, using GE *Pseudomonas putida* KT2440, researchers have investigated the possibility of bioremediation of OP and pyrethroids (Zuo et al., 2015). Li et al. (2020) used a dual plasmid expression method and created a GEB that can break down pesticide residues in the environment and emit green fluorescence. Yuanfan et al. (2010) used a mini-Tn-transposon method to insert *mpd* (methyl parathion hydrolase) gene into the chromosome of *Sphingobium* sp. JQL4−5 strain to expand its substrate degradation range. JQL4−5-mpd was developed to break down methyl parathion and fenpropathrin concurrently. Samin et al. (2014) reported that transposon delivery system was utilized to transfer the dehalogenase gene (dhaA31) into the genome of *P. putida* strain MC4 for aerobic bioremediation of a refractory chlorinated hydrocarbon. Liu, Hong, Xu, Jun, and Li (2006) revealed that genetically modified *Sphingomonas* sp. CDS-1 was capable of mineralizing OP and carbamates. *Acinetobacter calcoaceticus* BD413 was genetically altered to produce the xenobiotic-metabolizing enzyme CYP105D1 from *Streptomyces griseus* ATCC 13273, allowing the microbes to flourish on agricultural pesticides (Lamb, Kelly, Masaphy, Jones, & Kelly, 2000).

9.4.5 Petroleum and polycyclic aromatic hydrocarbons contaminants

There are many types of soil remediation technologies that are frequently employed. These include mechanical methods, evaporation, burying, washing, and dispersion. These methods, on the other hand, are costly and have the potential to result in the partial breakdown of pollutants

(Das & Chandran, 2011). Bacteria may be seen as pathogens to target with new medicines or as biocatalysts for large-scale industrial, agricultural, or environmental activities (bioremediation of hazardous material in situ or ex situ). With advent of genetic engineering in the late 1970s, scientists could create recombinant bacteria that would biodegrade toxic chemicals (de Lorenzo, 2009). In 1970 the first GEM "Superbug" was created to break down oil via the transfer of plasmids that were capable of using a variety of hazardous organic compounds such as octane, hexane, xylene, toluene, camphor, and naphthalene (Kulshreshtha, 2013). Petroleum degrading bacteria are abundant in nature and may utilize petroleum hydrocarbons as carbon and energy sources. Because of their capacity to digest petroleum oil, these bacteria are frequently employed in bioremediation.

Recently, microbial remediation technology has advanced at a breakneck pace, yielding significant improvements (Xu et al., 2018). It is possible to increase and enhance the bioremediation activity of *Pseudomonas* sp. by using molecular methods to modify the genetic makeup. Catechol 2,3-dioxygenase (C23O) is produced by *Pseudomonas* sp., and this enzyme is capable of using and decomposing PAHs (Mardani et al., 2017). The pilot research conducted by French et al. (2020) has shown that transferring catabolic genes involved in petroleum degradation from *E. coli* DH5α to indigenous bacteria may be a feasible solution. In 1996 soil lysimeter research began to evaluate the in situ performance of *Pseudomonas fluorescens* HK44 (HK44) which includes the luminescence genes (luxCDABE) from *Vibrio fischeri* that have been inserted into naphthalene degradation pathway gene (nahG), and it luminesces during the breakdown of 2- or 3-ring PAHs (Ripp et al., 2000a; Trogl et al., 2007). Table 9.1 provides some instances of how genetic engineering technology may be used to enhance the bioremediation of pollutants utilizing bacteria in the context of this study.

Bioremediation of PAH's polluted sites is usually extremely sluggish since effective bioremediation is dependent on a variety of biotic and abiotic variables (Ghosal, Ghosh, Dutta, & Ahn, 2016). Recent advances in molecular biology have made gene conversion, duplication, and transposon or plasmid distribution possible. These technologies may be critical in enhancing the biodegrading capacity of naturally existing microbes (Ang, Zhao, & Obbard, 2005; Megharaj, Ramakrishnan, Venkateswarlu, Sethunathan, & Naidu, 2011). GEMs have emerged as a desired need for improved bioremediation effectiveness and environmental sustainability of

Table 9.1 Biodegradation of contaminants from genetically engineered bacteria.

Name of bacteria	Name of pollutant	References
Pseudomonas putida	4-Ethylbenzoate	Ramos, Wasserfallen, Rose, and Timmis (1987)
Pseudomonas sp. FRI	Chloro- and methyl aromatics	Rojo, Pieper, Engesser, Knackmuss, and Timmis (1987)
Pseudomonas sp. strain CB15	3-Chlorobiphenyl	Adams, Huang, Higson, Brenner, and Focht (1992)
Pseudomonas sp. LB400	PCB	Erickson and Mondello (1993)
Pseudomonas pseudoalcaligenes KF707-D2	TCE, toluene, benzene	Suyama et al. (1996)
P. putida KT2442	Toluene/benzoate	Panke, Sanchez-Romero, and de Lorenzo (1998)
Comamonas testosteroni VP44	4-Chlorobenzoate	Hrywna, Tsoi, Maltseva, Quensen, and Tiedje (1999)
Pseudomonas fluorescens HK44	Naphthalene	Ripp et al. (2000a), Ripp, Nivens, Werner, and Sayler (2000b)
P. fluorescens 4F39	Ni	Lopez, Lazaro, Morales, and Marques (2002)
Escherichia coli	As	Kostal, Yang, Wu, Mulchandani, and Chen (2004)
E. coli JM109	Hg	Zhao et al. (2005)
E. coli JM109	Acid red GR C.I. Direct Blue 71	Jin, Zhou, Zhang, and Wang (2008), Jin et al. (2009)
Methylococcus capsulatus	Cr^{6+}	Al Hasin et al. (2010)
P. putida KT2440	Organophosphates and pyrethroids	Zuo et al. (2015)

contaminated sites with PAHs. GEMs are an ecologically friendly and cost-effective alternative to certain traditional methods (Wu, Li, Yi, & Ge, 2021). GEM known as *Pseudomonas* sp. B13 SN45RE can degrade chloro- and methyl aromatics that would otherwise be deadly to microbial communities by using a newly discovered ortho-cleavage pathway (Erb, Eichner, Wagner-Dobler, & Timmis, 1997; Pipke, Wagner-Döbler, Timmis, & Dwyer, 1992). It has been shown that genetically modified *P. putida* strains harboring the XYL, NAH, CAM, and OCT plasmids breakdown octane, camphor, naphthalene, salicylate, and among other compounds (Sayler & Ripp, 2000). A genetically modified *E. coli* strain encoding catechol 2,3-dioxygenase (C230) was developed, and its capacity to biodegrade phenanthrene and pyrene in spiked soil was studied using high-performance liquid chromatography (Ahankoub et al., 2020).

9.4.6 Bioremediation of polychlorinated biphenyls

The removal/mitigation of PCBs may be greatly aided by microbial cultures produced via genetic manipulation (Zaynab et al., 2021). Recombinant DNA technologies, such as the development of new expression vectors, the alteration of gene regulatory networks, and the reporting of genetically modified organisms (GMOs), may be used to create innovative systems for bioremediation (Sanghvi, Thanki, Pandey, & Singh, 2020). Genetic insights are being gained via the use of omics technologies (transcriptomics, genomics, metabolomics, and, proteomics) (Malik, Arora, Chaturvedi, & Paul, 2021). The use of GEMs derived from native strains would significantly impact the overall effectiveness of in situ bioremediation initiatives, along with their active participation in pollution reduction via reuse and recycling (Adetunji & Anani, 2021). A GE and nitrogen-fixing soil bacterium *Sinorhizobium meliloti* was shown to have increased capacity to dechlorinate the 2′,3,4-trichlorobiphenyl with alfalfa (*Medicago sativa*) plants (Chen, Adam, Toure, & Dutta, 2005). Through the use of shuffling techniques, strains carrying portion of the bph gene were developed from *Burkholderia cepacia* strain LB400 and *Rhodococcus globerulus* P6 and resulting in higher PCB breakdown efficiency (Barriault, Plante, & Sylvestre, 2002). Aerobic cometabolization of lower chlorinated biphenyls is possible. Toxic PCB congeners may also allow some aerobes to develop as their sole carbon/energy source. A transition of 4-chlorobiphenyl to chlorobenzoates results in the growth of *B. cepacia* (Field & Sierra-Alvarez, 2008). The bph locus of *Burkholderia xenovorans* LB400 was inserted into the genome of *Cupriavidus necator* JMP134-X3 to mineralize PCB congeners. These bph genes were inserted into JMP134-X3 strain using mini-Tn5 transposons and biparental mating (Saavedra, Acevedo, González, & Seeger, 2010).

9.5 Factors affecting bioremediation from genetically engineered bacteria

The usage of GEB requires strains capable of successfully resolving problems via the use of bioremediation methods for the removal of ecologically harmful refractory substances (Janssen & Stucki, 2020). Biosafety is sometimes interpreted as a reason for slow development in the use of GMOs for environmental remediation (Gong et al., 2017). Handling of GEMs in a responsible manner requires that the scientists have accurate knowledge over the use of the GEM to prevent negative consequences and improve public acceptance of this technology (Ford, Sayler, & Burlage, 1999). Environmental variables (e.g., temperature, moisture of soil, and organic substance content) may have a significant effect on the efficiency with which catabolic plasmids are transferred, the expression of horizontally acquired genes, and eventually pollutant degrading activity (Garbisu, Garaiyurrebaso, Epelde, Grohmann, & Alkorta, 2017). Numerous variables, particularly cell adherence to soil organic matter, may significantly restrict bacteria' dispersion throughout the soil matrix. In the case of bioaugmentation, the most significant abiotic and biotic variables affect the efficiency (Tahri, Bahafid, Sayel, & El Ghachtouli, 2013; Wolski, Murialdo, & Gonzalez, 2007).

9.6 Limitations and challenges of in-field release of genetically engineered bacteria

GEMs have shown bioremediation potential in soil, groundwater, and activated sludge with enhanced degradative capabilities. Though, most research on genetically engineered microbial bioremediation relies on laboratory data (Sayler & Ripp, 2000). There may be unforeseen complications emerging from the technology (Sarma, 1989). Although it is enticing to use GEMs in bioremediation, this approach faces major environmental laws, policies, and regulatory obstacles since the release of GMOs for effective field bioremediation tests into the environment is prohibited in the United States and Europe (Davison, 2005; Urgun-Demirtas, Stark, & Pagilla, 2006). In addition, environmental and ecological risk assessment is required for the introduction of GEMs into the environment in a safe manner (Gorlach, 1994). First, GEB approved for bioremediation field testing in the United States is *P. fluorescens* HK44. (Ripp et al., 2000a). Many GEMs have been created, but their field uses are restricted due to the risk of additional environmental contamination. To reduce possible hazards, microorganisms discharged into the environment must be monitored and controlled (Li & Wu, 2008). There is a great deal of public concern about the deliberate in-field release of GEMs. GEMs have managed to the development of biocontainment schemes in which bacteria are released and then destroyed using a controlled suicide process mechanism (Garbisu & Alkorta, 1999).

9.7 Survivability and sustenance of genetically engineered bacteria

The rate of multiplication of GEMs, the efficiency with which resources are used, the existence of nutritional materials, the physicochemical properties of the environment, the microorganisms' density, and the accessibility of competitive microbes are all important factors in their survival (Kumar et al., 2020). The use of GEMs is often limited owing to the danger of genes spreading from engineered bacteria to another species. As a result, a mechanism to track down and manage the survival of GEMs released into the fields is required (Li et al., 2020). Genetically modified *P. putida* PpY101 strains were tested in aquatic microcosms for survival with a recombinant DNA plasmid pSR134 encoding mercury resistance (Iwasaki, Uchiyama, & Yagi, 1993). Fujita (1994) examined the survivability of GEMs carrying recombinant plasmid pBH500 which encodes the catechol 2,3-oxygenase gene. The green fluorescence protein gene was used to monitor the survival of *E. coli* bacteria. This method allowed for the counting of GEMs by plating and microscopy at the same time (Leff & Leff, 1996). The stability of modified plasmids used in the process of bioremediation is determined by vector type, plasmid number, replication rate, the concentration of oxygen, the nutrient profile, and the climatic conditions.

9.8 Conclusion

As a consequence of anthropogenic activities, the natural environment has been severely contaminated. PCBs, PAHs, and pesticides are all resistant to degradation and provide a long-term toxicological danger to both animals and people. Conventional treatment processes for removing harmful chemicals from the environment have failed miserably that can only be overcome via the use of biological technologies. The employment of helpful microorganisms in this scenario, according to environmental perspectives, has a high chance of being a successful strategy for bioremediation. The applications of GEB offer a new frontier in research with far-reaching consequences for public health. These bacteria are more potent and degradative than native strains, as well as having the capacity to rapidly adapt to a broad variety of contaminants. A long and worthwhile journey awaits the use of genetic engineering to produce such microbes for the benefit of the environment and, eventually, the public's health.

Acknowledgments

We are thankful to the Director of Biyani Girls College, Jaipur, India, for support and encouragement.

Abbreviations

CBAs	chlorobenzoates
GE	genetically engineered
GEB	genetically engineered bacteria
GEMs	genetically engineered microbes
mt-1	metallothionein
OC	organochlorines
OP	organophosphates
PAHs	polycyclic aromatic hydrocarbons
PCBs	polychlorinated biphenyls
S-GEMs	suicidal-genetically engineered microbes

References

Abatenh, E., Gizaw, B., Tsegaye, Z., & Wassie, M. (2017). The role of microorganisms in bioremediation – A review. *Open Journal of Environmental Biology*, *2*(1), 038–046.

Adams, R. H., Huang, C. M., Higson, F. K., Brenner, V., & Focht, D. D. (1992). Construction of a 3-chlorobiphenyl-utilizing recombinant from an intergeneric mating. *Applied and Environmental Microbiology*, *58*(2), 647–654.

Adetunji, C. O., & Anani, O. A. (2021). Recent advances in the application of genetically engineered microorganisms for microbial rejuvenation of contaminated environment. In C. O. Adetunji, D. G. Panpatte, & Y. K. Jhala (Eds.), *Microbial rejuvenation of polluted environment* (pp. 303–324). Singapore: Springer.

Agmas, B., & Adugna, M. (2020). Attitudes and practices of farmers with regard to pesticide use in NorthWest Ethiopia. *Cogent Environmental Science*, *6*(1), 1791462.

Ahankoub, M., Mardani, G., Ghasemi-Dehkordi, P., Mehri-Ghahfarrokhi, A., Doosti, A., Jami, M. S., Allahbakhshian-Farsani, M., et al. (2020). Biodecomposition of phenanthrene and pyrene by a genetically engineered *Escherichia coli*. *Recent Patents on Biotechnology*, *14*(2), 121–133.

Al Hasin, A., Gurman, S. J., Murphy, L. M., Perry, A., Smith, T. J., & Gardiner, P. H. E. (2010). Remediation of Chromium(VI) by a methane-oxidizing bacterium. *Environmental Science & Technology*, *44*(1), 400–405.

Aleksakhin, R. M. (2009). Radioactive contamination as a type of soil degradation. *Eurasian Soil Science*, *42*(12), 1386–1396.

Ali, H. (2010). Biodegradation of synthetic dyes—A review. *Water, Air, & Soil Pollution*, *213*(1–4), 251–273.

Ali, H., Khan, E., & Sajad, M. A. (2013). Phytoremediation of heavy metals—Concepts and applications. *Chemosphere, 91*(7), 869–881.

Anandan, S., Kumar Ponnusamy, V., & Ashokkumar, M. (2020). A review on hybrid techniques for the degradation of organic pollutants in aqueous environment. *Ultrasonics Sonochemistry, 67*, 105130.

Ang, E. L., Zhao, H., & Obbard, J. P. (2005). Recent advances in the bioremediation of persistent organic pollutants via biomolecular engineering. *Enzyme and Microbial Technology, 37*(5), 487–496.

Appukuttan, D., Rao, A. S., & Apte, S. K. (2006). Engineering of *Deinococcus radiodurans* R1 for bioprecipitation of uranium from dilute nuclear waste. *Applied and Environmental Microbiology, 72*(12), 7873–7878.

Ashraf, S., Ali, Q., Zahir, Z. A., Ashraf, S., & Asghar, H. N. (2019). Phytoremediation: Environmentally sustainable way for reclamation of heavy metal polluted soils. *Ecotoxicology and Environmental Safety, 174*, 714–727.

Ayangbenro, A., & Babalola, O. (2017). A new strategy for heavy metal polluted environments: A review of microbial biosorbents. *International Journal of Environmental Research and Public Health, 14*(1), 94.

Aynalem, B., & Muleta, D. (2021). Microbial biosensors as pesticide detector: An overview. *Journal of Sensors, 2021*, 5538857.

Barnthouse, L. W., & Palumbo, A. V. (1986). Assessing the transport and fate of bioengineered microorganisms in the environment. In J. Fiksel, & V. T. Covello (Eds.), *Biotechnology risk assessment* (pp. 109–128). Elsevier.

Barriault, D., Plante, M. M., & Sylvestre, M. (2002). Family shuffling of a targeted bphA region to engineer biphenyl dioxygenase. *Journal of Bacteriology, 184*(14), 3794–3800.

Basnakova, G., Stephens, E. R., Thaller, M. C., Rossolini, G. M., & Macaskie, L. E. (1998). The use of *Escherichia coli* bearing a phoN gene for the removal of uranium and nickel from aqueous flows. *Applied Microbiology and Biotechnology, 50*(2), 266–272.

Benkhaya, S., M' rabet, S., & El Harfi, A. (2020). A review on classifications, recent synthesis and applications of textile dyes. *Inorganic Chemistry Communications, 115*, 107891.

Bhandari, S., Poudel, D. K., Marahatha, R., Dawadi, S., Khadayat, K., Phuyal, S., Shrestha, S., et al. (2021). Microbial enzymes used in bioremediation. *Journal of Chemistry, 2021*, 1–17.

Bhat, R. A., Singh, D. V., Qadri, H., Dar, G. H., Dervash, M. A., Bhat, S. A., … Yousaf, B. (2021). Vulnerability of municipal solid waste: An emerging threat to aquatic ecosystems. *Chemosphere, 287*(2022), 132223. Available from https://doi.org/10.1016/j.chemosphere.2021.132223.

Bilal, M., Iqbal, H. M. N., & Barceló, D. (2019). Persistence of pesticides-based contaminants in the environment and their effective degradation using laccase-assisted biocatalytic systems. *Science of The Total Environment, 695*, 133896.

Bisht, S., Pandey, P., Bhargava, B., Sharma, S., Kumar, V., & Sharma, K. D. (2015). Bioremediation of polyaromatic hydrocarbons (PAHs) using rhizosphere technology. *Brazilian Journal of Microbiology, 46*(1), 7–21.

Bottrell, D. G., & Schoenly, K. G. (2018). Integrated pest management for resource-limited farmers: Challenges for achieving ecological, social and economic sustainability. *The Journal of Agricultural Science, 156*(3), 408–426.

Brim, H., Osborne, J. P., Kostandarithes, H. M., Fredrickson, J. K., Wackett, L. P., & Daly, M. J. (2006). *Deinococcus radiodurans* engineered for complete toluene degradation facilitates Cr(VI) reduction. *Microbiology (Reading, England), 152*(8), 2469–2477.

Brim, H., Venkateswaran, A., Kostandarithes, H. M., Fredrickson, J. K., & Daly, M. J. (2003). Engineering *Deinococcus geothermalis* for bioremediation of high-temperature radioactive waste environments. *Applied and Environmental Microbiology, 69*(8), 4575–4582.

Brooks, A. N., Turkarslan, S., Beer, K. D., Yin Lo, F., & Baliga, N. S. (2011). Adaptation of cells to new environments. *Wiley Interdisciplinary Reviews: Systems Biology and Medicine*, *3*(5), 544–561.

Cavicchioli, R., Ripple, W. J., Timmis, K. N., Azam, F., Bakken, L. R., Baylis, M., Behrenfeld, M. J., et al. (2019). Scientists' warning to humanity: Microorganisms and climate change. *Nature Reviews. Microbiology*, *17*(9), 569–586.

Chang, J. S., & Lin, C. Y. (2001). Decolorization kinetics of a recombinant *Escherichia coli* strain harboring azo-dye-decolorizing determinants from *Rhodococcus* sp. *Biotechnology Letters*, *23*, 631–636.

Chang, S. E., Stone, J., Demes, K., & Piscitelli, M. (2014). Consequences of oil spills: A review and framework for informing planning. *Ecology and Society*, *19*(2), 26.

Chekroun, B. K., Sánchez, E., & Baghour, M. (2014). The role of algae in bioremediation of organic pollutants. *International Research Journal of Public and Environmental Health*, *1*, 19–32.

Chen, Y., Adam, A., Toure, O., & Dutta, S. K. (2005). Molecular evidence of genetic modification of *Sinorhizobium meliloti*: enhanced PCB bioremediation. *Journal of Industrial Microbiology & Biotechnology*, *32*(11–12), 561–566.

Couto, S. R. (2009). Dye removal by immobilised fungi. *Biotechnology Advances*, *27*(3), 227–235.

Dar, G. H., Bandh, S. A., & Kamili, A. N. (2011). Bacteriological analysis of soil from Yusmarg health resort of Kashmir Valley. *International Journal of Current Research*, *3*(10), 53–56.

Dar, G. H., Bandh, S. A., Kamili, A. N., Nazir, R., & Bhat, R. A. (2013). Comparative analysis of different types of bacterial colonies from the soils of Yusmarg Forest, Kashmir Valley India. *Ecologia Balkanica*, *5*(1), 31–35.

Dar, G. H., Kamili, A. N., Chishti, M. Z., Dar, S. A., Tantry, T. A., & Ahmad, F. (2016). Characterization of *Aeromonas sobria* isolated from fish Rohu (*Labeo rohita*) collected from polluted pond. *Journal of Bacteriology & Parasitology*, 7(3), 1–5. Available from https://doi.org/10.4172/2155-9597.1000273.

Dar, G. H., Kamili, A. N., Dar, S. A., Chishti, M. Z., & Ahmad, F. (2015). Detection and characterization of potentially pathogenic *Aeromonas sobria* isolated from fish *Hypophthalmichthys molitrix* (Cypriniformes: Cyprinidae). *Microbial Pathogenesis*. Available from https://doi.org/10.1016/j.micpath.2015.100.017.

Dar, G. H., Kamili, A. N., Nazir, R., Bandh, S. A., & Bhat, R. A. (2012). A preliminary study of colony forming units of bacteria from the soils of Yusmarg Forest, Kashmir Valley, India. *International Journal of Current Research*, *4*(12), 467–472.

Das, N., & Chandran, P. (2011). Microbial degradation of petroleum hydrocarbon contaminants: An overview. *Biotechnology Research International*, *2011*, 1–13.

Davison, J. (2005). Risk mitigation of genetically modified bacteria and plants designed for bioremediation. *Journal of Industrial Microbiology & Biotechnology*, *32*(11–12), 639–650.

de Lorenzo, V. (2001). Cleaning up behind us. *EMBO Reports*, *2*(5), 357–359.

de Lorenzo, V. (2009). Recombinant bacteria for environmental release: What went wrong and what we have learnt from it. *Clinical Microbiology and Infection*, *15*, 63–65.

Deng, X., Li, Q. B., Lu, Y. H., He, N., & Jiang, J. (2005). Genetic engineering of *E. coli* SE5000 and its potential for Ni^{2+} bioremediation. *Process Biochemistry*, *40*(1), 425–430.

Dixit, R., Wasiullah., Malaviya, D., Pandiyan, K., Singh, U., Sahu, A., Shukla, R., et al. (2015). Bioremediation of heavy metals from soil and aquatic environment: An overview of principles and criteria of fundamental processes. *Sustainability*, 7(2), 2189–2212.

Doshi, B., Sillanpaa, M., & Kalliola, S. (2018). A review of bio-based materials for oil spill treatment. *Water Research*, *135*, 262–277.

Elangovan, S., Pandian, S. B. S., S., J. G., & Joshi, S. J. (2019). Polychlorinated Biphenyls (PCBs): Environmental fate, challenges and bioremediation. In P. K. Arora (Ed.), *Microbial metabolism of xenobiotic compounds* (pp. 165–188). Singapore: Springer.
Erb, R. W., Eichner, C. A., Wagner-Dobler, I., & Timmis, K. N. (1997) Bioprotection of microbial communities from toxic phenol mixtures by a genetically designed pseudomonad. *Nature Biotechnology, 15*(4), 378–382.
Erickson, B. D., & Mondello, F. J. (1993). Enhanced biodegradation of polychlorinated biphenyls after site-directed mutagenesis of a biphenyl dioxygenase gene. *Applied and Environmental Microbiology, 59*(11), 3858–3862.
Ezezika, O. C., & Singer, P. A. (2010). Genetically engineered oil-eating microbes for bioremediation: Prospects and regulatory challenges. *Technology in Society, 32*(4), 331–335.
Farre, M., Kantiani, L., & Barcelo, D. (2007). Advances in immunochemical technologies for analysis of organic pollutants in the environment. *TrAC Trends in Analytical Chemistry, 26*(11), 1100–1112.
Field, J. A., & Sierra-Alvarez, R. (2008). Microbial transformation and degradation of polychlorinated biphenyls. *Environmental Pollution, 155*(1), 1–12.
Forabosco, F., Lohmus, M., Rydhmer, L., & Sundstrom, L. F. (2013). Genetically modified farm animals and fish in agriculture: A review. *Livestock Science, 153*(1–3), 1–9.
Ford, C. Z., Sayler, G. S., & Burlage, R. S. (1999). Containment of a genetically engineered microorganism during a field bioremediation application. *Applied Microbiology and Biotechnology, 51*(3), 397–400.
French, K. E., Zhou, Z., & Terry, N. (2020). Horizontal 'gene drives' harness indigenous bacteria for bioremediation. *Scientific Reports, 10*(1), 15091.
Fujita, M. (1994). Operation parameters affecting the survival of genetically engineered microorganisms in activated sludge processes. *Water Research, 28*(7), 1667–1672.
Furukawa, K., & Fujihara, H. (2008). Microbial degradation of polychlorinated biphenyls: Biochemical and molecular features. *Journal of Bioscience and Bioengineering, 105*(5), 433–449.
Garbisu, C., & Alkorta, I. (1999). Utilization of genetically engineered microorganisms (GEMs) for bioremediation. *Journal of Chemical Technology & Biotechnology, 74*(7), 599–606.
Garbisu, C., Garaiyurrebaso, O., Epelde, L., Grohmann, E., & Alkorta, I. (2017). Plasmid-mediated bioaugmentation for the bioremediation of contaminated soils. *Frontiers in Microbiology*, 8.
Gaur, N., Narasimhulu, K., & PydiSetty, Y. (2018). Recent advances in the bioremediation of persistent organic pollutants and its effect on environment. *Journal of Cleaner Production, 198*, 1602–1631.
Gavrilescu, M., Demnerová, K., Aamand, J., Agathos, S., & Fava, F. (2015). Emerging pollutants in the environment: Present and future challenges in biomonitoring, ecological risks and bioremediation. *New Biotechnology, 32*(1), 147–156.
Ghanizadeh, F., & Godini, H. (2018). A review of the chemical and biological pollutants in indoor air in hospitals and assessing their effects on the health of patients, staff and visitors. *Reviews on Environmental Health, 33*(3), 231–245.
Ghosal, D., Ghosh, S., Dutta, T. K., & Ahn, Y. (2016). Current state of knowledge in microbial degradation of Polycyclic Aromatic Hydrocarbons (PAHs): A review. *Frontiers in Microbiology*, 7.
Gong, T., Liu, R., Zuo, Z., Che, Y., Yu, H., Song, C., & Yang, C. (2016). Metabolic engineering of *Pseudomonas putida* KT2440 for complete mineralization of methyl parathion and γ-hexachlorocyclohexane. *ACS Synthetic Biology, 5*(5), 434–442.

Gong, T., Xu, X., Che, Y., Liu, R., Gao, W., Zhao, F., Yu, H., et al. (2017). Combinatorial metabolic engineering of *Pseudomonas putida* KT2440 for efficient mineralization of 1,2,3-trichloropropane. *Scientific Reports*, 7(1), 7064.

Gorlach, K. (1994). Problems in the introduction of genetically engineered microorganisms into the environment. *Acta Microbiologica Polonica*, *43*(2), 121–131.

Goud, B. S., Cha, H. L., Koyyada, G., & Kim, J. H. (2020). Augmented biodegradation of textile azo dye effluents by plant endophytes: A sustainable, eco-friendly alternative. *Current Microbiology*, 77(11), 3240–3255.

Gupta, A., Singh, A., & Mishra, V. K. (2021). Sources, fate, and treatment of polycyclic aromatic hydrocarbons from the polluted environment. In A. Kumar, V. K. Singh, P. Singh, & V. K. Mishra (Eds.), *Microbe mediated remediation of environmental contaminants* (pp. 369–379). Woodhead Publishing.

Gupta, S., & Singh, D. (2017). Role of genetically modified microorganisms in heavy metal bioremediation. In R. Kumar, A. K. Sharma, & S. S. Ahluwalia (Eds.), *Advances in environmental biotechnology* (pp. 197–214). Singapore: Springer.

Hamedi, R., B.G. Aghaie, A., & Hadjmohammadi, M. R. (2018). Magnetic core micelles as a nanosorbent for the efficient removal and recovery of three organophosphorus pesticides from fruit juice and environmental water samples. *Journal of Separation Science*, *41*(9), 2037–2045.

Hidalgo, K. J., Sierra-Garcia, I. N., Dellagnezze, B. M., & de Oliveira, V. M. (2020). Metagenomic insights into the mechanisms for biodegradation of Polycyclic Aromatic Hydrocarbons in the oil supply chain. *Frontiers in Microbiology*, 11.

Hrywna, Y., Tsoi, T. V., Maltseva, O. V., Quensen, J. F., & Tiedje, J. M. (1999). Construction and characterization of two recombinant bacteria that grow on ortho- and para-substituted chlorobiphenyls. *Applied and Environmental Microbiology*, *65*(5), 2163–2169.

Hu, Q. H., Weng, J. Q., & Wang, J. S. (2010). Sources of anthropogenic radionuclides in the environment: A review. *Journal of Environmental Radioactivity*, *101*(6), 426–437.

Ingham, E. R., Doyle, J. D., & Hendricks, C. W. (1995). Assessing interactions between the soil foodweb and a strain of *Pseudomonas putida* genetically engineered to degrade 2,4-D. *Applied Soil Ecology*, *2*(4), 263–274.

Iwasaki, K., Uchiyama, H., & Yagi, O. (1993). Survival and impact of genetically engineered *Pseudomonas putida* harboring mercury resistance gene in aquatic microcosms. *Bioscience, Biotechnology, and Biochemistry*, *57*(8), 1264–1269.

Jackson, W. J., & Summers, A. O. (1982). Biochemical characterization of $HgCl_2$-inducible polypeptides encoded by the *mer* operon of plasmid R100. *Journal of Bacteriology*, *151*(2), 962–970.

Jafari, M., Danesh, Y. R., Goltapeh, E. M., & Varma, A. (2013). Bioremediation and genetically modified organisms. In E. M. Goltapeh, Y. R. Danesh, & A. Varma (Eds.), *Fungi as bioremediators* (pp. 433–451). Berlin, Heidelberg: Springer.

Janssen, D. B., & Stucki, G. (2020). Perspectives of genetically engineered microbes for groundwater bioremediation. *Environmental Science: Processes & Impacts*, *22*(3), 487–499.

Jeong, S. W., & Choi, Y. J. (2017). Research perspective of an extremophilic bacterium, *Deinococcus radiodurans* on bioremediation of radioactive wastes. *Applied Chemistry for Engineering*, *28*(2), 133–140.

Ji, X., Ripp, S. A., Layton, A. C., Sayler, G. S., & DeBruyn, J. M. (2013). Assessing long term effects of bioremediation: Soil bacterial communities 14 years after polycyclic aromatic hydrocarbon contamination and introduction of a genetically engineered microorganism. *Journal of Bioremediation & Biodegradation*, *04*(08), 209.

Jia, X., Li, Y., Xu, T., & Wu, K. (2020). Display of lead-binding proteins on *Escherichia coli* surface for lead bioremediation. *Biotechnology and Bioengineering, 117*(12), 3820–3834.

Jin, R., Yang, H., Zhang, A., Wang, J., & Liu, G. (2009). Bioaugmentation on decolorization of C.I. Direct Blue 71 by using genetically engineered strain *Escherichia coli* JM109 (pGEX-AZR). *Journal of Hazardous Materials, 163*(2–3), 1123–1128.

Jin, R., Zhou, J., Zhang, A., & Wang, J. (2008). Bioaugmentation of the decolorization rate of acid red GR by genetically engineered microorganism *Escherichia coli JM109* (pGEX-AZR). *World Journal of Microbiology and Biotechnology, 24*(1), 23–29.

Khalid, F., Hashmi, M. Z., Jamil, N., Qadir, A., & Ali, M. I. (2021). Microbial and enzymatic degradation of PCBs from e-waste-contaminated sites: A review. *Environmental Science and Pollution Research, 28*(9), 10474–10487.

Kiyono, M., & Pan-Hou, H. (2006). Genetic engineering of bacteria for environmental remediation of mercury. *Journal of Health Science, 52*(3), 199–204.

Kodoth, A. K., & Badalamoole, V. (2020). Silver nanoparticle-embedded pectin-based hydrogel for adsorptive removal of dyes and metal ions. *Polymer Bulletin,* 77(2), 541–564.

Kostal, J., Yang, R., Wu, C. H., Mulchandani, A., & Chen, W. (2004). Enhanced arsenic accumulation in engineered bacterial cells expressing ArsR. *Applied and Environmental Microbiology, 70*(8), 4582–4587.

Kuhad, R. C., Sood, N., Tripathi, K. K., Singh, A., & Ward, O. P. (2004). Developments in microbial methods for the treatment of dye effluents. *Advances in Applied Microbiology, 56*, 185–213.

Kulshreshtha, S. (2013). Genetically engineered microorganisms: A problem solving approach for bioremediation. *Journal of Bioremediation & Biodegradation, 04*, e133.

Kumar, A., Kumar, A., Singh, R., Singh, R., Pandey, S., Rai, A., Singh, V. K., et al. (2020). Genetically engineered bacteria for the degradation of dye and other organic compounds. In P. Singh, A. Kumar, & A. Borthakur (Eds.), *Abatement of environmental pollutants* (pp. 331–350). Elsevier.

Kumar, D., Rai, D., Porwal, P., & Kumar, S. (2018). Compositional quality of milk and its contaminants on physical and chemical concern: A review. *International Journal of Current Microbiology and Applied Sciences,* 7(05), 1125–1132.

Kumar, R., Singh, S., & Singh, O. V. (2007). Bioremediation of radionuclides: Emerging technologies. *OMICS: A Journal of Integrative Biology, 11*(3), 295–304.

Kumar, S., Dagar, V. K., Khasa, Y. P., & Kuhad, R. C. (2013). Genetically modified microorganisms (GMOs) for bioremediation. In R. C. Kuhad, & A. Singh (Eds.), *Biotechnology for environmental management and resource recovery* (pp. 191–218). India: Springer.

Kurnaz, A., Kucukomeroglu, B., Keser, R., Okumusoglu, N. T., Korkmaz, F., Karahan, G., & Cevik, U. (2007). Determination of radioactivity levels and hazards of soil and sediment samples in Firtina Valley (Rize, Turkey). *Applied Radiation and Isotopes: Including Data, Instrumentation and Methods for Use in Agriculture, Industry and Medicine, 65*(11), 1281–1289.

Lamb, D. C., Kelly, D. E., Masaphy, S., Jones, G. L., & Kelly, S. L. (2000). Engineering of heterologous cytochrome P450 in *Acinetobacter* sp.: Application for pollutant degradation. *Biochemical and Biophysical Research Communications, 276*(2), 797–802.

Lan, W. S., Gu, J. D., Zhang, J. L., Shen, B. C., Jiang, H., Mulchandani, A., Chen, W., et al. (2006). Coexpression of two detoxifying pesticide-degrading enzymes in a genetically engineered bacterium. *International Biodeterioration & Biodegradation, 58*(2), 70–76.

Lange, C. C., Wackett, L. P., Minton, K. W., & Daly, M. J. (1998). Engineering a recombinant *Deinococcus radiodurans* for organopollutant degradation in radioactive mixed waste environments. *Nature Biotechnology*, *16*(10), 929–933.

Leff, L. G., & Leff, A. A. (1996). Use of green fluorescent protein to monitor survival of genetically engineered bacteria in aquatic environments. *Applied and Environmental Microbiology*, *62*(9), 3486–3488.

Li, Q., Li, J., Kang, K. L., & Wu, Y. J. (2020). A safety type of genetically engineered bacterium that degrades chemical pesticides. *AMB Express*, *10*(1), 33.

Li, Q., Wang, P., Chen, R., Li, W., & Wu, Y. J. (2014). Construction of genetically engineered bacteria that degrades organophosphorus pesticide residues and can be easily detected by the fluorescence. *Environmental Technology*, *35*(5), 556–561.

Li, Q., & Wu, Y. (2008). Progress of environmental monitoring and biological containment system on genetically engineered microorganisms. *Shengwu Gongcheng Xuebao/ Chinese Journal of Biotechnology*, *24*(3), 355–362.

Limbergen, H. Van, Top, E. M., & Verstraete, W. (1998). Bioaugmentation in activated sludge: Current features and future perspectives. *Applied Microbiology and Biotechnology*, *50*(1), 16–23.

Liu, G., Zhou, J., Qu, Y., & Ma, X. (2007). Decolorization of sulfonated azo dyes with two photosynthetic bacterial strains and a genetically engineered *Escherichia coli* strain. *World Journal of Microbiology and Biotechnology*, *23*(7), 931–937.

Liu, S., Zheng, Z., & Li, X. (2013). Advances in pesticide biosensors: Current status, challenges, and future perspectives. *Analytical and Bioanalytical Chemistry*, *405*(1), 63–90.

Liu, Z., Hong, Q., Xu, J.-H., Jun, W., & Li, S.-P. (2006). Construction of a genetically engineered microorganism for degrading organophosphate and carbamate pesticides. *International Biodeterioration & Biodegradation*, *58*(2), 65–69.

Lloyd, J. R., & Lovley, D. R. (2001). Microbial detoxification of metals and radionuclides. *Current Opinion in Biotechnology*, *12*(3), 248–253.

Lopez, A., Lazaro, N., Morales, S., & Marques, A. M. (2002). Nickel biosorption by free and immobilized cells of *Pseudomonas fluorescens* 4F39: A comparative study. *Water, Air, and Soil Pollution*, *135*, 157–172.

Makarova, K. S., Aravind, L., Wolf, Y. I., Tatusov, R. L., Minton, K. W., Koonin, E. V., & Daly, M. J. (2001). Genome of the extremely radiation-resistant bacterium *Deinococcus radiodurans* viewed from the perspective of comparative genomics. *Microbiology and Molecular Biology Reviews*, *65*(1), 44–79.

Malik, G., Arora, R., Chaturvedi, R., & Paul, M. S. (2021). Implementation of genetic engineering and novel omics approaches to enhance bioremediation: A focused review. *Bulletin of Environmental Contamination and Toxicology*, 3. Available from https://doi.org/10.1007/s00128-021-03218-.

Mardani, G., Mahvi, A. H., Hashemzadeh-Chaleshtori, M., Naseri, S., Dehghani, M. H., & Ghasemi-Dehkordi, P. (2017). Application of genetically engineered dioxygenase producing *Pseudomonas putida* on decomposition of oil from spiked soil. *Jundishapur Journal of Natural Pharmaceutical Products*, *12*(3 Supp), e64313.

Martinez, R. J., Beazley, M. J., Taillefert, M., Arakaki, A. K., Skolnick, J., & Sobecky, P. A. (2007). Aerobic uranium (VI) bioprecipitation by metal-resistant bacteria isolated from radionuclide- and metal-contaminated subsurface soils. *Environmental Microbiology*, *9*(12), 3122–3133.

Masindi, V., & Muedi, K. L. (2018). Environmental contamination by heavy metals. In H. M. Saleh, & R. F. Aglan (Eds.), *Heavy metals*. IntechOpen.

Megharaj, M., Ramakrishnan, B., Venkateswarlu, K., Sethunathan, N., & Naidu, R. (2011). Bioremediation approaches for organic pollutants: A critical perspective. *Environment International*, *37*(8), 1362–1375.

Mitra, A., Chatterjee, S., Kataki, S., Rastogi, R. P., & Gupta, D. K. (2021). Bacterial tolerance strategies against lead toxicity and their relevance in bioremediation application. *Environmental Science and Pollution Research*, *28*(12), 14271–14284.

Munawar, A., Akram, M. S., Javed, M. T., & Shahid, M. (2021). Polychlorinated biphenyls (PCBs): Characteristics, toxicity, phytoremediation, and use of transgenic plants for PCBs degradation. In M. Hasanuzzaman, & M. N. V. Prasad (Eds.), *Handbook of bioremediation* (pp. 677–687). Academic Press.

Naeem, U., & Qazi, M. A. (2020). Leading edges in bioremediation technologies for removal of petroleum hydrocarbons. *Environmental Science and Pollution Research*, *27* (22), 27370–27382.

Newsome, L., Morris, K., & Lloyd, J. R. (2014). The biogeochemistry and bioremediation of uranium and other priority radionuclides. *Chemical Geology*, *363*, 164–184.

Nicolopoulou-Stamati, P., Maipas, S., Kotampasi, C., Stamatis, P., & Hens, L. (2016). Chemical pesticides and human health: The urgent need for a new concept in agriculture. *Frontiers in Public Health*, 4.

Ojuederie, O., & Babalola, O. (2017). Microbial and plant-assisted bioremediation of heavy metal polluted environments: A review. *International Journal of Environmental Research and Public Health*, *14*(12), 1504.

Oliveira, T. S., Al Aukidy, M., & Verlicchi, P. (2017). Occurrence of common pollutants and pharmaceuticals in hospital effluents. In P. Verlicchi (Ed.), *Hospital wastewaters* (pp. 17–32). Cham: Springer.

Pande, V., Pandey, S. C., Sati, D., Pande, V., & Samant, M. (2020). Bioremediation: An emerging effective approach towards environment restoration. *Environmental Sustainability*, *3*(1), 91–103.

Pandey, G., Paul, D., & Jain, R. K. (2005). Conceptualizing "suicidal genetically engineered microorganisms" for bioremediation applications. *Biochemical and Biophysical Research Communications*, *327*(3), 637–639.

Panelli, S., Capelli, E., Comandatore, F., Landinez-Torres, A., Granata, M. U., Tosi, S., & Picco, A. M. (2017). A metagenomic-based, cross-seasonal picture of fungal consortia associated with Italian soils subjected to different agricultural managements. *Fungal Ecology*, *30*, 1–9.

Panke, S., Sanchez-Romero, J. M., & de Lorenzo, V. (1998). Engineering of quasi-natural *Pseudomonas putida* strains for toluene metabolism through an ortho -cleavage degradation pathway. *Applied and Environmental Microbiology*, *64*(2), 748–751.

Pant, G., Garlapati, D., Agrawal, U., Prasuna, R. G., Mathimani, T., & Pugazhendhi, A. (2021). Biological approaches practised using genetically engineered microbes for a sustainable environment: A review. *Journal of Hazardous Materials*, *405*, 124631.

Paul, D., Pandey, G., & Jain, R. K. (2005). Suicidal genetically engineered microorganisms for bioremediation: Need and perspectives. *Bioessays: News and Reviews in Molecular, Cellular and Developmental Biology*, *27*(5), 563–573.

Peng, R. H., Xiong, A. S., Xue, Y., Fu, X. Y., Gao, F., Zhao, W., Tian, Y. S., et al. (2008). Microbial biodegradation of polyaromatic hydrocarbons. *FEMS Microbiology Reviews*, *32*(6), 927–955.

Pieper, D. H., & Reineke, W. (2000). Engineering bacteria for bioremediation. *Current Opinion in Biotechnology*, *11*(3), 262–270.

Pipke, R., Wagner-Döbler, I., Timmis, K. N., & Dwyer, D. F. (1992). Survival and function of a genetically engineered Pseudomonad in aquatic sediment microcosms. *Applied and Environmental Microbiology*, *58*(4), 1259–1265.

Poursat, B. A. J., van Spanning, R. J. M., de Voogt, P., & Parsons, J. R. (2019). Implications of microbial adaptation for the assessment of environmental persistence of chemicals. *Critical Reviews in Environmental Science and Technology*, *49*(23), 2220–2255.

Powers, L. G., Mills, H. J., Palumbo, A. V., Zhang, C., Delaney, K., & Sobecky, P. A. (2002). Introduction of a plasmid-encoded phoA gene for constitutive overproduction of alkaline phosphatase in three subsurface *Pseudomonas* isolates. *FEMS Microbiology Ecology*, *41*(2), 115–123.

Prakash, D., Gabani, P., Chandel, A. K., Ronen, Z., & Singh, O. V. (2013). Bioremediation: A genuine technology to remediate radionuclides from the environment. *Microbial Biotechnology*, *6*(4), 349–360.

Premnath, N., Mohanrasu, K., Guru Raj Rao, R., Dinesh, G. H., Prakash, G. S., Ananthi, V., Ponnuchamy, K., et al. (2021). A crucial review on polycyclic aromatic Hydrocarbons – Environmental occurrence and strategies for microbial degradation. *Chemosphere*, *280*, 130608.

Rajendran, P., Muthukrishnan, J., & Gunasekaran, P. (2003). Microbes in heavy metal remediation. *Indian Journal of Experimental Biology*, *41*(9), 935–944.

Ramos, J., Wasserfallen, A., Rose, K., & Timmis, K. (1987). Redesigning metabolic routes: Manipulation of TOL plasmid pathway for catabolism of alkylbenzoates. *Science (New York, N.Y.)*, *235*(4788), 593–596.

Rani, P., Kumar, V., Singh, P. P., Matharu, A. S., Zhang, W., Kim, K.-H., Singh, J., et al. (2020). Highly stable AgNPs prepared via a novel green approach for catalytic and photocatalytic removal of biological and non-biological pollutants. *Environment International*, *143*, 105924.

Ripp, S., Nivens, D. E., Ahn, Y., Werner, C., Jarrell, J., Easter, J. P., Cox, C. D., et al. (2000a). Controlled field release of a bioluminescent genetically engineered microorganism for bioremediation process monitoring and control. *Environmental Science & Technology*, *34*(5), 846–853.

Ripp, S., Nivens, D. E., Werner, C., & Sayler, G. S. (2000b). Bioluminescent most-probable-number monitoring of a genetically engineered bacterium during a long-term contained field release. *Applied Microbiology and Biotechnology*, *53*(6), 736–741.

Rojo, F., Pieper, D., Engesser, K., Knackmuss, H., & Timmis, K. (1987). Assemblage of ortho cleavage route for simultaneous degradation of chloro- and methylaromatics. *Science (New York, N.Y.)*, *238*(4832), 1395–1398.

Ruiz, O. N., Alvarez, D., Gonzalez-Ruiz, G., & Torres, C. (2011). Characterization of mercury bioremediation by transgenic bacteria expressing metallothionein and polyphosphate kinase. *BMC Biotechnology*, *11*(1), 82.

Saavedra, J. M., Acevedo, F., González, M., & Seeger, M. (2010). Mineralization of PCBs by the genetically modified strain *Cupriavidus necator* JMS34 and its application for bioremediation of PCBs in soil. *Applied Microbiology and Biotechnology*, *87*(4), 1543–1554.

Sakshi., & Haritash, A. K. (2020). A comprehensive review of metabolic and genomic aspects of PAH-degradation. *Archives of Microbiology*, *202*(8), 2033–2058.

Sakshi., Singh, S. K., & Haritash, A. K. (2020). Evolutionary relationship of polycyclic aromatic hydrocarbons degrading bacteria with strains isolated from petroleum contaminated soil based on 16S rRNA diversity. *Polycyclic Aromatic Compounds*, 1–14.

Salameh, P. (2006). Respiratory diseases and pesticide exposure: A case-control study in Lebanon. *Journal of Epidemiology & Community Health*, *60*(3), 256–261.

Samanta, S. K., Singh, O. V., & Jain, R. K. (2002). Polycyclic aromatic hydrocarbons: Environmental pollution and bioremediation. *Trends in Biotechnology*, *20*(6), 243–248.

Samin, G., Pavlova, M., Arif, M. I., Postema, C. P., Damborsky, J., & Janssen, D. B. (2014). A *Pseudomonas putida* strain genetically engineered for 1,2,3-trichloropropane bioremediation. *Applied and Environmental Microbiology*, *80*(17), 5467–5476.

Sanborn, M., Kerr, K. J., Sanin, L. H., Cole, D. C., Bassil, K. L., & Vakil, C. (2007). Non-cancer health effects of pesticides: Systematic review and implications for family doctors. *Canadian Family Physician Medecin de Famille Canadien*, *53*(10), 1712–1720.

Sanghvi, G., Thanki, A., Pandey, S., & Singh, N. K. (2020). Engineered bacteria for bioremediation. In V. C. Pandey, & V. Singh (Eds.), *Bioremediation of pollutants* (pp. 359–374). Elsevier.
Sarma, V. (1989). Environmental issues in the planned release of genetically-engineered organisms. *Australian Journal of Biotechnology*, *3*(1), 13–16.
Sarwar, N., Imran, M., Shaheen, M. R., Ishaque, W., Kamran, M. A., Matloob, A., Rehim, A., et al. (2017). Phytoremediation strategies for soils contaminated with heavy metals: Modifications and future perspectives. *Chemosphere*, *171*, 710–721.
Sayler, G. S., & Ripp, S. (2000). Field applications of genetically engineered microorganisms for bioremediation processes. *Current Opinion in Biotechnology*, *11*(3), 286–289.
Sehrawat, A., Phour, M., Kumar, R., & Sindhu, S. S. (2021). Bioremediation of pesticides: An eco-friendly approach for environment sustainability. In D. G. Panpatte, & Y. K. Jhala (Eds.), *Microbial rejuvenation of polluted environment* (pp. 23–84). Singapore: Springer.
Sengupta, S., Roy, U., Chowdhary, S., & Das, P. (2021). New bioremediation technologies to remove heavy metals and radionuclides. In M. P. Shah (Ed.), *Removal of emerging contaminants through microbial processes* (pp. 23–45). Singapore: Springer.
Sghaier, H., Bouchami, O., Desler, C., Lazim, H., Saidi, M., Rasmussen, L. J., & Ben Hassen, A. (2012). Analysis of the antimicrobial susceptibility of the ionizing radiation-resistant bacterium *Deinococcus radiodurans*: Implications for bioremediation of radioactive waste. *Annals of Microbiology*, *62*(2), 493–500.
Sharma, P., Sirohi, R., Tong, Y. W., Kim, S. H., & Pandey, A. (2021). Metal and metal (loids) removal efficiency using genetically engineered microbes: Applications and challenges. *Journal of Hazardous Materials*, *416*, 125855.
Singh, J. S., Abhilash, P. C., Singh, H. B., Singh, R. P., & Singh, D. P. (2011). Genetically engineered bacteria: An emerging tool for environmental remediation and future research perspectives. *Gene*, *480*(1–2), 1–9.
Stephen, J. R., & Macnaughtont, S. J. (1999). Developments in terrestrial bacterial remediation of metals. *Current Opinion in Biotechnology*, *10*(3), 230–233.
Stolz, A. (2001). Basic and applied aspects in the microbial degradation of azo dyes. *Applied Microbiology and Biotechnology*, *56*(1–2), 69–80.
Suyama, A., Iwakiri, R., Kimura, N., Nishi, A., Nakamura, K., & Furukawa, K. (1996). Engineering hybrid pseudomonads capable of utilizing a wide range of aromatic hydrocarbons and of efficient degradation of trichloroethylene. *Journal of Bacteriology*, *178*(14), 4039–4046.
Tahri, N., Bahafid, W., Sayel, H., & El Ghachtouli, N. (2013). Biodegradation: Involved microorganisms and genetically engineered microorganisms. In R. Chamy, & F. Rosenkranz (Eds.), *Biodegradation – Life of science*. IntechOpen.
Tchounwou, P. B., Yedjou, C. G., Patlolla, A. K., & Sutton, D. J. (2012). Heavy metal toxicity and the environment. *Experientia supplementum (2012)*, *101*, 133–164.
Tirkey, S. R., Ram, S., & Mishra, S. (2021). Bacteria-mediated remediation: Restoring polycyclic aromatic hydrocarbon (PAHs) contaminated marine ecosystems. In M. P. Shah (Ed.), *Advanced Oxidation Processes for Effluent Treatment Plants* (pp. 119–136). Elsevier.
Trogl, J., Kuncova, G., Kubicova, L., Parik, P., Halova, J., Demnerova, K., Ripp, S., et al. (2007). Response of the bioluminescent bioreporter *Pseudomonas fluorescens* HK44 to analogs of naphthalene and salicylic acid. *Folia Microbiologica*, *52*(1), 3–14.
Tyagi, B., & Kumar, N. (2021). Bioremediation: Principles and applications in environmental management. In G. Saxena, V. Kumar, & M. P. Shah (Eds.), *Bioremediation for environmental sustainability* (pp. 3–28). Elsevier.
Urgun-Demirtas, M., Stark, B., & Pagilla, K. (2006). Use of genetically engineered microorganisms (GEMs) for the bioremediation of contaminants. *Critical Reviews in Biotechnology*, *26*(3), 145–164.

Valls, M., & Delorenzo, V. (2002). Exploiting the genetic and biochemical capacities of bacteria for the remediation of heavy metal pollution. *FEMS Microbiology Reviews, 26* (4), 327–338.

Varjani, S., Rakholiya, P., Ng, H. Y., You, S., & Teixeira, J. A. (2020). Microbial degradation of dyes: An overview. *Bioresource Technology, 314*, 123728.

Verma, S., & Kuila, A. (2019). Bioremediation of heavy metals by microbial process. *Environmental Technology & Innovation, 14*, 100369.

Wang, J. H., Zhu, D., Zhang, S. S., & Guan, J. F. (2014). Construction and application of genetic engineering bacteria in contaminated environment bioremediation. *Advanced Materials Research, 955–959*, 1935–1938.

Wolski, E., Murialdo, S., & Gonzalez, J. (2007). Effect of pH and inoculum size on pentachlorophenol degradation by *Pseudomonas* sp. *Water SA, 32*(1), 93–97.

Wu, C., Li, F., Yi, S., & Ge, F. (2021). Genetically engineered microbial remediation of soils co-contaminated by heavy metals and polycyclic aromatic hydrocarbons: Advances and ecological risk assessment. *Journal of Environmental Management, 296*, 113185.

Xiang, Y., Xing, Z., Liu, J., Qin, W., & Huang, X. (2020). Recent advances in the biodegradation of polychlorinated biphenyls. *World Journal of Microbiology and Biotechnology, 36*(10), 145.

Xu, M., Wu, M., Zhang, Y., Zhang, H., Liu, W., Chen, G., Xiong, G., et al. (2021). Biodegradation of polycyclic aromatic hydrocarbons (PAHs) by bacterial mixture. *International Journal of Environmental Science and Technology*.

Xu, X., Liu, W., Tian, S., Wang, W., Qi, Q., Jiang, P., Gao, X., et al. (2018). Petroleum hydrocarbon-degrading bacteria for the remediation of oil pollution under aerobic conditions: A perspective analysis. *Frontiers in Microbiology*, 9.

Yang, C., & Mcgarrahan, J. (2005). Electrochemical coagulation for textile effluent decolorization. *Journal of Hazardous Materials, 127*(1–3), 40–47.

Yang, Q. Z., Zhou, B., Liu, J. W., Shen, W. R., Jia, X. D., He, X. J., & Zhao, H. Z. (2021). Nitrate removal from water via self-flocculation of genetically engineered bacteria. *Chemical Engineering Science, 242*, 116750.

Yuanfan, H., Jin, Z., Qing, H., Qian, W., Jiandong, J., & Shunpeng, L. (2010). Characterization of a fenpropathrin-degrading strain and construction of a genetically engineered microorganism for simultaneous degradation of methyl parathion and fenpropathrin. *Journal of Environmental Management, 91*(11), 2295–2300.

Zaynab, M., Fatima, M., Sharif, Y., Sughra, K., Sajid, M., Khan, K. A., Sneharani, A. H., et al. (2021). Health and environmental effects of silent killers Organochlorine pesticides and polychlorinated biphenyl. *Journal of King Saud University - Science (New York, N.Y.), 33*(6), 101511.

Zenteno-Rojas, A., Martinez-Romero, E., Castaneda-Valbuena, D., Rincon-Molina, C. I., Ruiz-Valdiviezo, V. M., Meza-Gordillo, R., Villalobos-Maldonado, J. J., et al. (2020). Structure and diversity of native bacterial communities in soils contaminated with polychlorinated biphenyls. *AMB Express, 10*(1), 124.

Zhao, X. W., Zhou, M. H., Li, Q. B., Lu, Y. H., He, N., Sun, D. H., & Deng, X. (2005). Simultaneous mercury bioaccumulation and cell propagation by genetically engineered *Escherichia coli*. *Process Biochemistry, 40*(5), 1611–1616.

Zuo, Z., Gong, T., Che, Y., Liu, R., Xu, P., Jiang, H., Qiao, C., et al. (2015). Engineering *Pseudomonas putida* KT2440 for simultaneous degradation of organophosphates and pyrethroids and its application in bioremediation of soil. *Biodegradation, 26* (3), 223–233.

CHAPTER 10

An eco-friendly approach for the degradation of azo dyes and their effluents by *Pleurotus florida*

Manmeet Kaur, Shivani Sharma and H.S. Sodhi
Department of Microbiology, Punjab Agricultural University, Ludhiana, Punjab, India

10.1 Introduction

The microenvironment challenges brought by the profound industry have won most extreme mindfulness for quite a few years since textile industry is one of the quintessential sire of dirtied effluents, which overwhelmingly stir up through various ongoing processes and are mostly interrelated with the contaminated water which is provoked by the release of crude or ineffectively handled effluents (Bhat et al., 2021; Robinson, McMullan, Marchant, & Nigam, 2001). Wastewater from the yarn industry can be figured as the most fouling and their treatment is generally backbreaking. The quickening release and unseemly lead of liquid and solid modern squanders have made a colossal worry among industrialists regarding their monetary treatment and safe removal. The squeezed off effluents from these processes has hostile impacts in terms of biological oxygen demand (BOD), chemical oxygen demand (COD), salinity, color, suspended solids, and the recalcitrance of synthetic compounds, such as azo dyes. The disposed waste contains a wide percentage of inappropriate nonbiodegradable organic matter which showed that the ratio of COD/BOD ranges from 0.5 to 0.2, respectively (Kuberan, Anburaj, Sundaravadivelan, & Kumar, 2011). For instance, the coloration of 1 kg cotton with respect to reactive dyes required a large concentration of sodium chloride and dyestuff, and the wastewater generated contains a admixture of salt components, unfixed reactive dyes, and dyeing auxiliaries under the appropriate range of 2000 ppm (Babu, Parande, Raghu, & Kumar, 2007). With the spending year, almost 280,000 ton of material colors are released in

Microbial Consortium and Biotransformation for Pollution Decontamination
DOI: https://doi.org/10.1016/B978-0-323-91893-0.00006-7

mechanical effluents over the world. The azo colors delivered in the water have unfavorable effect on the Earth, for example, their inhibitory impact on amphibian photosynthesis and capacity to exhaust the breakdown of oxygen. The anaerobic breakdown of colors produces aromatic amines, which are extremely harmful, cancer-causing, and mutagenic, and prompts the decreasing of verdure. Azo colors are the most generally utilized colors and speak to over 60% of the complete colors. The white decay organism, basidiomycete *Phanerochaete chrysosporium*, can decolorize industrial dyes and polymeric colors. Other ligninolytic fungi, for example, *Pleurotus florida*, *Pleurotus ostreatus*, *Trametes hispida*, *Trametes versicolor*, *Irpex lacteus*, *Heterobasidion annosum*, and a lot more have likewise demonstrated the limit with regards to color decolorization. The white decay fungus *P. florida* can deliver peroxidases [manganese peroxidase (MnP)], laccase (phenoloxidases), and lignin peroxidase (LiP), which furthermore discharge low subatomic weight go between, which increments the scope of intensifies that they can diminish a wide scope of toxic substances looking like chemical synthetics, including azo colors without the development of aromatic amines (Chagas & Durrant, 2001). This audit sums up the ongoing achievements in the organism-based advances which are produced for the expulsion of azo colors utilizing yarn industry effluents. In addition, the fungus decoloration innovation is talked about with factors that influence color expulsion are examined with respect to the investigation of harmfulness of the metabolic consequences of azo tint bleach and the frameworks used to evaluate this toxicity.

10.2 White-rot fungi

These fungi have filamentous growths that possess the wood of dead and passing on trees. They are "higher fungi," most of which are from the *Basidiomycota* [e.g., *P. chrysosporium* or *Trametes* (*Coriolus*) *versicolor*], while some are in the *Ascomycota* (e.g., *Xylaria polymorpha* or *Xylaria hypoxylon*). The trademark highlight of white decays is their capacity to debase lignin, a property that they share with synchronous decay growths and some "litter breaking down fungus." White decays are supposed in view of the white appearance of the spoiled wood, which is believed to be caused somewhat by the nonappearance of lignin and halfway oxidized aromatic lignin subordinates. It is commonly viewed as that white decays can't degrade cellulose without earlier expulsion of lignin, while synchronous decays eliminate the two polymers simultaneously. The three-dimensional polyaromatic polymer lignin has an auxiliary function in woody plants and furthermore has a defensive impact against contagious microorganisms. It is the primary hindrance against enzymatic debasement of cellulose and it has been demonstrated more than once that degradation of lignocellulosic materials is eased back somewhere around the presence of lignin, evacuation of which prompts quickened cellulolysis. The extracellular chemical arrangement of the white decay fungi licenses these organisms to endure upgraded centralization of contagion pollutants (Knapp, Vantoch-Wood, & Zhang, 2001).

10.2.1 Oyster mushroom or *Pleurotus florida*

Pleurotus includes 39 species in all, but in literature only 5 species, namely, *Pleurotus cystidiosus*, *P. florida*, *P. flabellatus*, *P. sajor-caju*, *P. ostreatus*, are commonly referred to as cultivable types. All the species are edible with requisite culinary properties. This mushroom (edible fungus), which is popularly known as "Dhingri" in Northern India, is already being grown on a commercial scale in Japan, Italy, Switzerland, France, South Korea, and China. The *P. florida* is soft and whitish in color. The pileus is usually shell and tongue shaped. The color intensity may alter due to substrate as well environmental conditions such as light and temperature. The size of the fruit body is 5 to 30 cm in circumference. Gills are whitish or gray. Stipe is generally eccentric or lateral. The classification of *P. florida* is given in Table 10.1.

P. florida produce different forms of extracellular enzymes, including lignin peroxidase, laccase, and manganese peroxidase, which are involved in debasement of lignin in their regular lignocellulosic substrates and legitimately associated with the degradation of different xenobiotic mixes and colors. The lignin-modifying enzymes lead to the disintegration of covalent bonds and chiefly releases bioactive molecules such as phenolic compounds (Carlile, Watikinson, & Gooday, 2001). The organism *P. florida*

Table 10.1 Classification of *Pleurotus florida*.

Kingdom	Fungi
Phylum	Basidiomycota
Class	Agaricomycetes
Order	Agaricales
Family	Pleurotaceae
Genus	*Pleurotus*

can adequately degrade extremely low centralizations of toxins to nonrecognizable levels. The utmost substrates for the development of organisms in nature are lignocellulosic substrates. Thus reasonable waste improved with lignocellulosics, such as straw, sawdust, nut shell, and corn cobs, can be added as nutrient adjuncts to the spoiled destinations to achieve upgraded debasement of pollutants by *P. florida*. The fungi additionally produce oxygen revolutionaries, for example, -OH which is equipped for crumbling biomolecules, for example, proteins and DNA, and results in the passing of other serious organisms under supplement inadequate conditions. Toward the end, the fungus can arrive at pollutants in the dirt through hyphal expansion; from numerous points of view, the same number of different living beings can't build up (Christian, Shrivastava, Shukla, Modi, & Vyas, 2005).

10.3 Textile dyes

Color has become a crucial prepossessing envoy in the textile world. The colorants comprise various colors and shades which depend on two significant compounds: azo and anthraquinone. Between the abovementioned dyes, azo colors are broadly utilized in fabric industries and innumerable of some hints enter effluvium treatment premises (Sudhakar, Palaniappan, & Gowrisankar, 2002). These tinctures flourish prominent remonstrance to crud decadence in bilge treatment frameworks and maintain their shading, auxiliary respectability atop divulgence to daylight, soil, and microorganisms. Roughly, 100,000 industry-based accessible colors (7×105 metric ton) are created every year and the manufactured textures such as rayon, nylon, and polyester are involved in the creation of advanced colors that can tie with the these materials firmly (Dar, Bandh, & Kamili, 2011; Zollinger, 2003). Fiber-receptive azo colors can covalently tie to the substrates which oblige them unrivaled in high-wet speed. Schematic chart of fiber responsive azo colors is shown in Fig. 10.1.

Various colors were utilized in textile industry, including azo, reactive, disperse, direct, sulfur, acid, metal complex, vat, mortant, and basic (Vijaya, Padmavathy, & Sandhya, 2003). These colors are normally fragrant and heterocyclic mixes with some poisonous and cancer-causing properties (Vyas & Molitoris, 1995). Azo colors contain $N=N$ twofold bonds and are appended with aromatic heterocyclic or enolizable aliphatic gatherings besides likewise connected with benzene and naphthalene rings (Zollinger, 2003). On the basis of the compound structure, the colors are

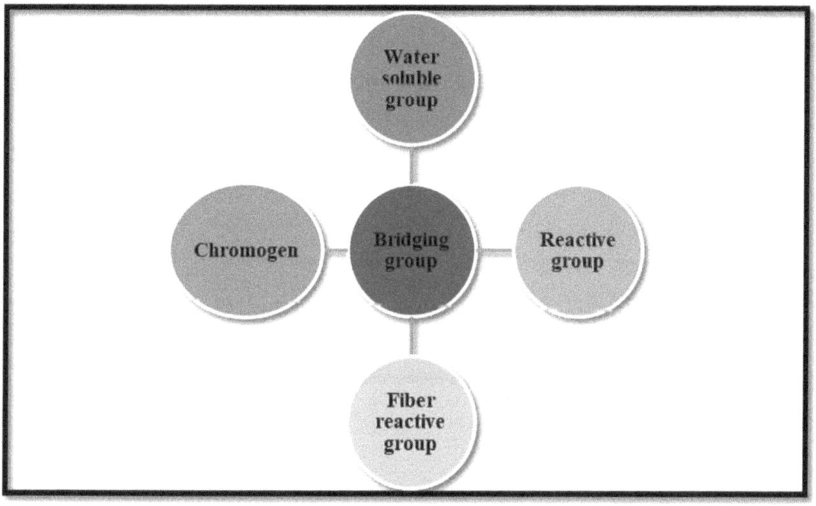

Figure 10.1 Flow diagram of fiber-reactive azo dye.

Table 10.2 Division of synthetic dyes related to their chemical structure and termed as color index (Wesenberg et al., 2003).

Chemical class	Code
Nitro	10,300
Disazo	20,000
Polyazo	35,000
Azoic	37,000
Xanthene	45,000
Acridine	46,000
Methine	48,000
Azine	50,000
Aminoketone	56,000
Hydroxyketone	57,000
Anthraquinone	58,000
Natural	75,000
Inorganic	77,000

grouped methodically and given as color index (Wesenberg, Kyriakides, & Agathos, 2003) (Table 10.2).

Aromatic and heterocyclic side chains with enolizable are essential for bestowing the color shading along with various shades and splendor. Color atoms probably regularly depicted as chromogen, their nucleophiles often known as auxochromes, and the aromatic bunches are alluded a chromophore. Blending of synthetic colors happens initially through diazotization of an essential primary amine by coupling with nucleophiles. The most ordinarily utilized coupling segments are amino and hydroxyl gatherings. Receptive colors are the shaded mixes with a couple of practical gatherings and are equipped for shaping covalent bonds with the dynamic destinations in filaments. Around 80%–95% of the receptive colors depend on azochromogen (Edwards, 2000). The C- and P-related particles pioneered in the color atom will be predicated with hydroxyl radical. This ordering additionally allocates features on the biodegradability of the colors. Manufactured colors as indicated by their most prevalent synthetic structures are given in Table 10.3.

Table 10.3 Arrangement of dyes based on their chemical structures.

S. no.	Dyes	Structures
1	Reactive Orange 96 (N = N)	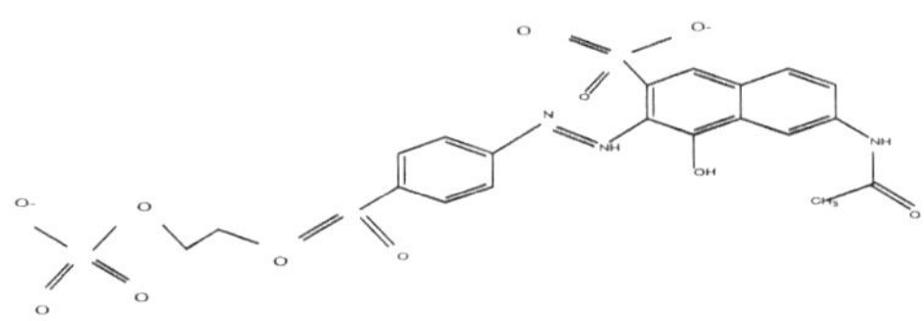
2	Remazol Brilliant Blue	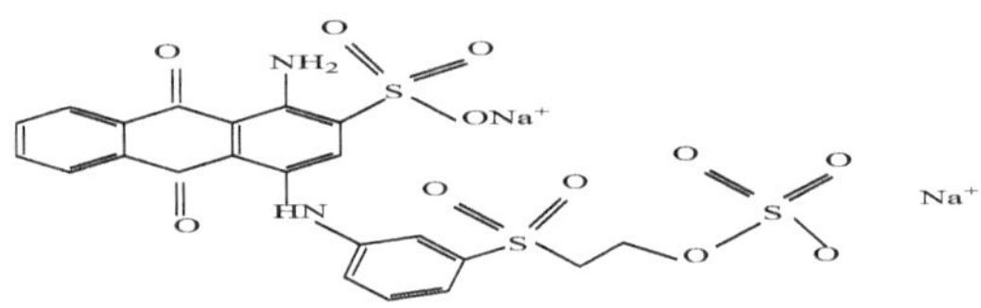
3	Bromothymol Blue	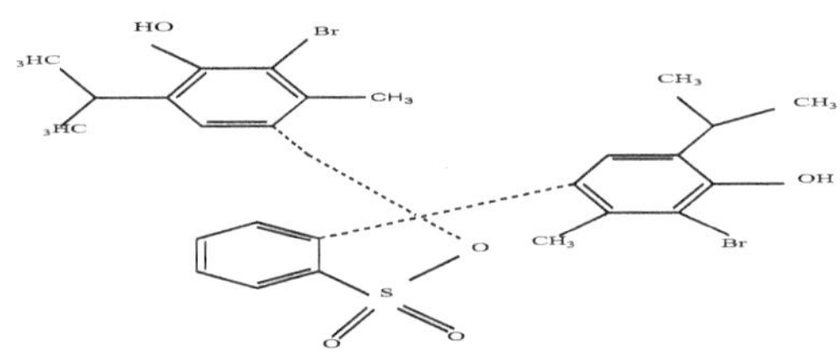
4	Congo Red	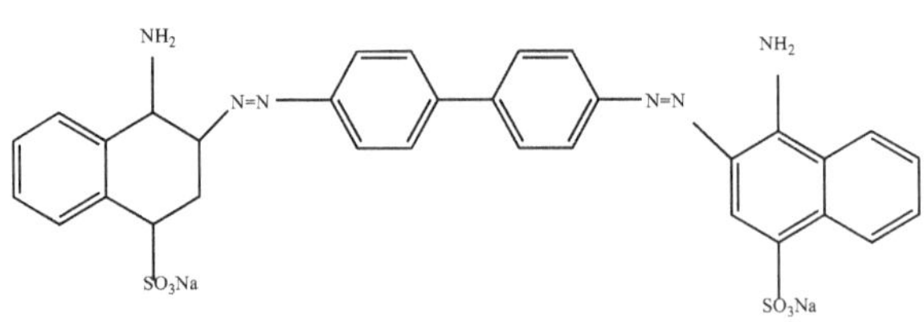

(*Continued*)

Table 10.3 (Continued)

S. no.	Dyes	Structures
5	Amaranth dye	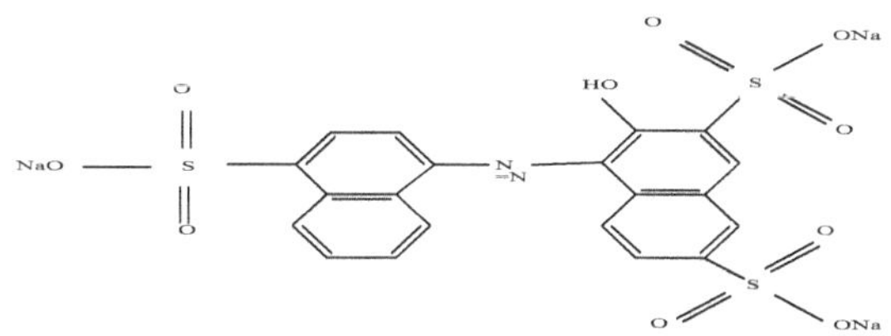

10.3.1 Description of dyes

Differentiation of dyes takes place on the application of textile fibers for coloring purposes or on the properties. Table 10.4 gives this classification of dyes according to its usage. The dyes can be subgrouped as:

1. anionic dyes: consists of direct, reactive, and acid dyes
2. nonanionic dyes: dispersed dyes
3. cationic dyes: preferably known as basic dyes.

10.4 Scenario of textile dyes utilized in India

Indian coloring industry generates huge amounts of pigments (approximate 60,000 metric ton); from all colorants, only 6.6% of that are utilized worldwide (Teli, 2008). The different colors and dyestuffs utilized in India are azo dyes, scatter colors, imbue colors, naphthols, tank colors, receptive colors, shade emulsion, sulfur colors, and other different colors. The sort of colors and synthetics utilized in the textile industry are found to contrast contingent upon the textures made (Table 10.5). The amount of bilge created from concrete ventures (cotton and fabricated materials) was assessed as 840 m^3/day. The material handling and mixing industry releases different proportionate of dyes per day separately, while woolen industry utilizes maximum range of the dyes around 2700 m^3/day (Sarayu & Sandhya, 2012).

10.5 Explication of dyeing process in textile industries

Colors have solid recoloring features and are often even at a centralization of 1 ppm. The amount of wastewater generated from yarn industry is reaching the sky and further released into sewage, running or stale water

Table 10.4 Categorization of textile dyes with respective to its characteristics.

S. no.	Dye class	Definition/properties	Major substrates	Metals contained in dyes	Method of application	Advantages	Disadvantages	References
1	Acid dyes	Chemically bound to sodium salt of a sulfuric, carboxylic acid	Nylon, wool, paper, leather	Copper, lead, zinc, chromium, cobalt	From neutral to acidic dye baths	Highly solubility, adherence with fiber cationic sites through ionic bonding	Descend wash fastness and highly expensive	Waring and Hallas (2013)
2	Azo dyes	60%–70% of the dye groups fall into group, i.e., the chromophore (N = N, C = C, C = O) and the auxochrome (-OH, $-NH_2$)	Cotton, rayon, cellulose acetate polyester paper	Copper, lead, zinc, chromium	Fiber drenched with coupling aggregate and reckoned with a solution of sturdy dia-azonium salts	Comfort fastens and give bright hellacious colors but not so potent in dyeing colors	Cost-effective	Nguyen and Juang (2013)
3	Basic dyes	Color base is not water-soluble but become soluble when treated with alcohol	Acrylic, modified nylon, polyester inks	Lead, mercury, zinc	Applied from acidic dye baths	Good brightness, high tinctorial strength	High acid content, poor shade stability	Nguyen and Juang (2013)
4	Direct dyes	Tinctures in which color atoms are pulled in by physical powers at the subatomic level to the material colors	Cotton, rayon, paper	Copper, lead, zinc	Applied from nonpartisan or somewhat soluble particle containing electrolyte	Water-soluble	Need electrolyte for exhaustion	Ventura-Camargo and Marin-Morales (2013)

5	Disperse dyes	Dye polyesters and acetate fibers	Polyester polyamide, cellulose acetate	None	Fine, fluid scatterings applied by higher weight, temperature or low temperature transporter techniques	Small dyes molecules and fast penetration	Nonsolvent and not easy to warm	Ventura-Camargo and Marin-Morales (2013)
6	Optical brightness	Excellent whitening	Soap, detergents, fibers, oil	None	Solution dispersion	Temperature resistance, color–light	Poor weather resistance	Waring and Hallas (2013)
7	Reactive dyes	A pigment is the chromophore containing substituent that is actuated and permitted to legitimately respond to the outside of the substrates	Cotton, wool, rayon, silk	Copper, chromium, lead	Responsive site on color responds with utilitarian gathering on the fiber to tie color covalently affected by pH and temperature	Good fastens properties and bonding during dyeing	Alkaline condition is required	Ventura-Camargo and Marin-Morales (2013)
8	Sulfide dyes	A colorant enriched with sulfur and is highly resolvable in 0.25%–0.50% sulfide solution	Cotton and rayon	Copper, lead, zinc, chromium	Aromatic substrate valued with sodium sulfide and reoxidized to insoluble sulfur	Water-insoluble, good fastens	Poor color value and poor wash	Nguyen and Juang (2013)
9	Vat dyes	The dyes are based on indigo dyes produced synthetically	Cotton, rayon, fleece	None	Water-insoluble colors solubilized by diminishing with sodium hydro sulfite	Color fastness	Poor rubbing fastness	Nguyen and Juang (2013)

(Continued)

Table 10.4 (Continued)

S. no.	Dye class	Definition/properties	Major substrates	Metals contained in dyes	Method of application	Advantages	Disadvantages	References
10	Solubilized vat dyes	Superior fastness property	Cotton, wool, cellulose, and protein fiber silk	None	No alkali is involved when impregnated fiber is treated with an oxidized operator, normally sulfuric corrosive and sodium nitrite for cotton dichromate fleece. This class is material to cellulose and protein strands	Cross-staining fastness, highly resistance to sunlight fading	Highly cost-effective	Waring and Hallas (2013)

Table 10.5 Various sorts of colors and synthetic substances utilized by the Indian material ventures (Ghaly, Ananthashankar, Alhattab, & Ramakrishnan, 2014).

S. no.	Colorants	Fabric used
1	Acid and reactive	Silk
2	Cellulosics and pigments	Any kind of material
3	Acid dyes (azo dye, anthraquinone dyes)	Protein fibers
4	Dispersed dyes (Disperse Yellow 218) and basic dyes (Basic Red 1)	Synthetic fibers
5	Reactive dyes (Remazol), direct dyes (Congo Red), naphthol dyes (Fast Blue B), and indigo dyes (Indigo White, Indigo Carmine)	Cellulosic fibers
6	Cationic	Acrylics
7	Reactive, direct, vat, sulfur, indigo	Cotton

bodies, and bordering waste terrains without appropriate treatment. Because of high synthesis, assortment, and shading power, liquid-filled garbage from the industry can't be dealt with agreeably. The physico-chemical attributes of untreated material and color enterprises' gushing demonstrate that in coming years the side effects have impact on soil and its natural properties. In any case, because of rigid natural guidelines and mindfulness, squander water of the color ventures is treated at singular processing plant premises or common effluent treatment plants or at singular processing plant premises before reusing. The accompanying advances are engaged with the tinting cycle (Fig. 10.2).

1. Scouring: The crude material is revealed to the scouring activity for killing the debasements by heating up the crude material in a vessel at higher temperature (80°C) for 3 h.
2. Bleaching: Hydrogen peroxide and NaOH of 3% quality are blended altogether in the vessel preceding completing the scouring activity.
3. Neutralization and washing: First, the substance are washed with the distilled water. Second, for maintaining the neutralization by the acid, washing activity is repeated. Lastly, the coloring activity is completed by placing the utilizing substance into the coloring tank.
4. Tinting: The coloring arrangement is finalized by dissolving the colors in which alongside the salts, for example, sodium chloride, soda ash debris is required for weariness and fixing the colors, separately. The coloring activity is retained for 3 hat 70°C to accomplish the ideal pigment.

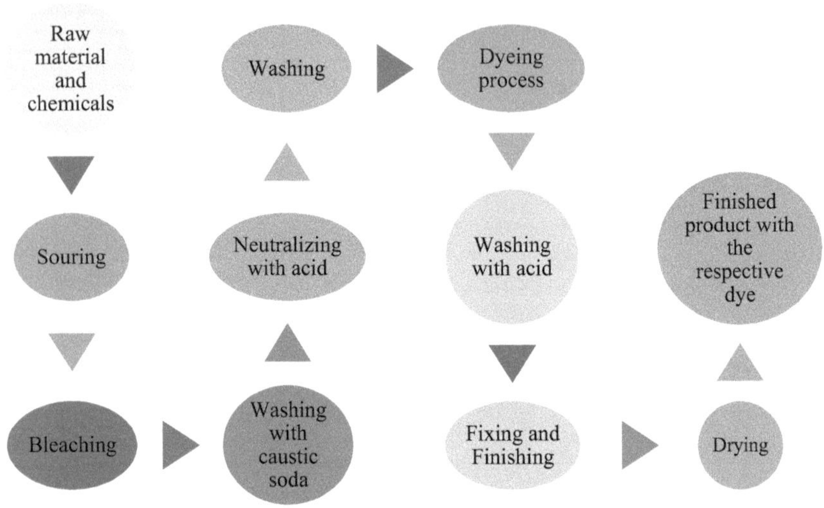

Figure 10.2 Dyeing process in textile industries.

10.6 Hallmarks of wastes effected by the textile industry

The poison highlights of yarn splurges contrast broadly among different natural substances. The constituent's significant boundaries of material wastewater, for example, temperature, solids, smell, shading, and the level of treatment depend upon its gushing and influent attributes (Upadhye & Joshi, 2012). Colors not just contain poisonous mixes, for example, hefty metals which can amass in condition and living beings; however, they likewise meddle with daylight infiltration into water, impede photosynthesis, and influence gas dissolvability in shallow (Daassi, Mechichi, Nasri, & Rodriguez-Couto, 2013). Notwithstanding, colors as well as different effluents, for example, solids in material bilge from stringy substrate and process synthetic compounds may likewise conceivably restrain the development of vegetation in water courses. Some of engineered colors, for example, azo, phthalocyanine, and anthraquinone, are generally intended to be impervious to light, natural movement, ozone, and other degradative condition so as to expand the nature of the material. These opposition qualities are identified with the substance structure of color as such responsive cleavage of its azo gathering (Yao, Jia, Zheng, & Wang, 2013) or mutagenic development of anilines (Martins, Ferreira, Santos, Queiroz, & Lima, 2001). Wet preparation in material industry is imperative to

Table 10.6 Pollutants produced in the wastewater from yarn dry and wet processing (Ghaly et al., 2014).

Process	Gas emission	Pollutants in wastewater	Type of waste
Slashing	Volatile organic compound	Starch, waxes, polyvinyl alcohol, carboxymethyl cellulose	Fiber, yarn waste, packaging waste, high in BOD and COD
Desizing	Carbon monoxide, sulfides, nitrogen oxides	Ammonia waxes, lubricant wastes, resins, fats, pectins	Oily waste, undesirable appearance, toxic products, high total solids
Bleaching	Chlorine, chlorine dioxide	Hydrogen peroxide, high pH sodium	High alkalinity
Dyeing	Carriers H_2S aniline vapors	Color metals, salts, surfactants, sulfide, formaldehyde	Strongly colored, high in BOD and COD
Finishing	Formaldehydes transits, slicking oils, low-molecular-weight polymers	Chlorinated compounds, acetate, solvents, stearate	Toxic products
Printing	Hydrocarbons, ammonia	Urea, solvents, color metals, starch	Highly colored oily appearance
Waving	Slight	Compact	Yarn and fabric scrap
Tufting	Brief	Little or none	Wrapping waste

BOD, biological oxygen demand; *COD*, chemical oxygen demand.

improve its taste, comfort, and utilitarian peculiarities to the fibers (Saxena, Raja, & Arputharaj, 2017). Lamentably in the cycle, some of the hues are fixed to the materials, and the unfixed colors (primarily receptive colors) are released in huge sum into wastewater along the cycle. A nitty-gritty outline of natural contaminations in wastewater created by material industry is shown in Table 10.6.

10.7 Impact of textile dyes on environment

The release of wastewater is the fundamental driver of the harmful effect of the textile industry. The industry represents 66% of the all dyestuff advertised and devours enormous volumes of water and synthetic substances for wet preparation of textiles (Riu, Schonsee, & Barcelo, 1988). The substances utilized in the textile industry are fluctuated in synthetic structure, extending from inorganic mixes to polymers and natural items, and besides that they rely upon the idea of the crude material and product. Effluents from material industry are described by their high noticeable shading (3000—4500 units), alkaline pH scope of 9—11, and COD (800—1600 mg/L). They additionally have enormous measure of natural synthetics, low biodegradability, and absolute solids in the scope of approximately 6500 mg/L (Manu & Chaudhari, 2002). Robinson et al. (2001) assessed that material colors are released into streams as gushing and effluents released from such ventures comprise high salts, sulfate, hardness, and cancer-causing color ingredients. The significant contamination by textile wastewater originates from coloring and wastewater is described by COD, shading, highly dissolved solids, and heat. The presence of colors in fluid biological system lessens daylight infiltration into more profound layers reducing photosynthetic action and decreases the water quality, bringing down the gas dissolvability which causes delicate poisonous impacts on oceanic widely varied vegetation. Therefore the arrival of unsafe colors in the environment can be a toxic hazard and can influence people through the natural pecking order (Van der Zee & Villaverede, 2005). Among various material colors utilized, azo colors and nitrated polycyclic sweet-smelling carbonic compounds are the two gatherings of synthetic substances that are bountiful in the Earth. They cause thorough defilement in stream and groundwater in the encompassing zone of coloring industries (Riu et al., 1988). The effect of azo colors in food industry has caused worry over various years through their corrupted items on human well-being (Dar et al., 2016; Dar, Kamili, Dar, Chishti, & Ahmad, 2015). In addition, azo colors have been connected to human bladder malignancy, splenic sarcomas, hepatocarcinomas and atomic abnormalities in exploratory creatures, and chromosomal variations in mammalian cells. Benzidine-based azo colors are broadly utilized in color producing, material coloring, shading, and paper printing and perceived as a cancer-causing agent (Haley, 1975). The diminished sugar, protein, and green pigment chlorophyll substance in the plants show the poisonous idea of the color industry effluents and also had lessen the pace of seed germination and development of yield plants (Nirmalarani & Janardhanan, 1988).

10.8 Dye decolorization methods

Emanating release from yarn and fiber stuff enterprises into wastewater treatment frameworks is causing critical well-being worries to natural administrative offices. Different methods that can be used to eliminate pigment from dye-containing wastewater are portrayed in Fig. 10.3 and Table 10.7.

10.8.1 Physical method

Adsorption alludes to an approach in which a substance is assembled at a strong surface from its fluid of vaporous trade. There are two kinds of adsorption dependent on the sort of fascination between the strong surface and the adsorbed particles. This attraction powers are because of synthetic holding, the cycle is called chemical adsorption (chemisorption). Adsorption procedures have been picked up recently because of their proficiency in the evacuation of contaminations excessively stable for conventional strategies (Bansal & Goyal, 2005). Chitosan dabs and cetyltrimethylammonium bromide (CTAB)-altered titanium dioxide nanoparticles were additionally utilized for decolorization of blend of three commercial colors during adsorption treatment (Zolgharnein et al., 2014). Various parameters, for example, measure of adsorbent, color, evacuation

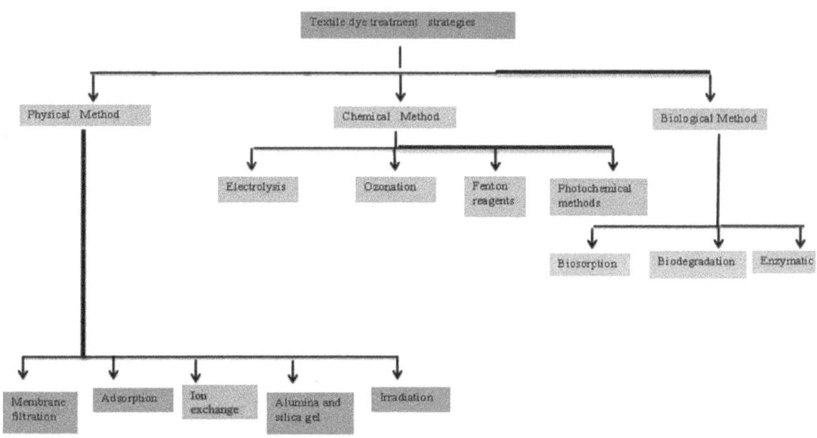

Figure 10.3 Techniques involved in decolorization of dyes.

Table 10.7 Techniques used for decolorization of dyes.

Methods	Techniques	Dyes	Decolorization (%)	References
Physical method	Adsorption	Alizarin Red and Alizarin Yellow	99 and 78.5	Zolgharnein, Asanjrani, Bagtash, and Azimi (2014)
		Direct Fast Scarlet, Reactive Violet K-3R, Eosin Y	95	Kang et al. (2009)
		Eriochrome Black T, Phenol Red, Methyl Blue	100	Iqbal and Ashiq (2007)
Chemical method	Electrolysis	Remazol Black B, Remazol Red RB, Drimaren Yellow K-2R	100	Chatzisymeon, Xekoukoulotakis, Coz, Kalogerakis, and Mantzavinos (2006)
		Reactive yellow 84, Reactive Red 120, Reactive orange 16	100	Rajkumar and Kim (2006)
		Bromophenol Blue, Crystal Violet, Methyl Orange	100	Sanroman, Pazos, Ricart, and Cameselle (2004)
	Ozonation	Super Black G, Terasil Red	90	Wijannarong, Aroonsrimorakot, Thavipoke, and Sangjan (2013)
		Golden Yellow (GY), Remazol Yellow (RY)	95–99	Sarayu, Swaminathan, and Sandhya (2007)
	Photo-Fenton method	Cationic dyes	100	Chergui, Oturan, Khalaf, and Oturan (2012)
		Remazol Red RR, Remazol Blue RR	100	Punzi, Mattiasson, and Jonstrup (2012)
		Congo Red, Methylene Blue, and Yellow Drimaren	97	Lahkimi, Oturan, Oturan, and Chaouch (2007)
	Photochemical method	Methyl Orange, Methylene Blue	90–98	Wetchakun, Chaiwichain, Inceesungvorn, and Pingmuang (2012)
		Crystal Violet and Aniline Blue	100	Shanthi and Padmavathi (2012)
		Six Azo dyes	100	Palacio et al. (2012)

Biological method	Biosorption	Brilliant Blue R, Rubine GFL	94.73	Kabra, Khandare, Waghmode, and Govindwar (2012)
		Basic Fuchsin, Methyl Violet	100	Bhole, Ganguly, Madhuram, Deshpande, and Joshi (2004)
	Fungal degradation	Azo–anthraquinone dye mixture	100	Taha, Adetutu, Shahsavari, Smith, and Ball (2014)
		Brilliant Blue, Methyl Red, Remazol Red, Scarlet RR, Rubine GFR, Golden Yellow HER	88	Waghmode, Kurade, and Govindwar (2011)
	Enzymatic degradation	Disperse Black 9, Disperse Red 19	82.1	Jamal, Singh, Khatoon, and Mehrotra (2013)
		Reactive Red 198, Reactive Orange	80	Chhabra, Mishra, and Sreekrishnan (2008)
		Reactive Blue 4, Reactive Red 11, Reactive Orange 4, Reactive Yellow 84	87–98	Khan and Husain (2007)

productivity, color expulsion effectiveness, residual color dissemination, and adsorption energy were examined. Layer filtration strategy has been generally utilized for drinking and wastewater treatment. The cycle of filtration comprises nanofiltration, ultrafiltration, microfiltration, and converse assimilation. Particle trade methods are viable in decolorizing cationic and anionic colors and have not been utilized in broadly for color wastewater therapy because of the supposition that particle exchangers can't oblige a wide scope of dyes. The light is an another factor needed to eliminate color from wastewater effluents in a double cylinder percolating reactor with huge volumes of disintegrated oxygen for natural substances which are further to be stalled adequately by radiation (Huang, Hsueh, Huang, Su, & Chen, 2007).

10.8.2 Chemical method

The electrochemical method is exceptionally productive to eliminate color from dye wastewater. Electrochemical expulsion of colors from wastewater is a moderately new cycle displaying effective color evacuation and degradation of refractory toxins. In this cycle, electrochemical oxidation of the material emanating and engineered color arrangement was finished by utilizing titanium−tantalum−platinum−iridium anode and quantitative decolorization was acquired in 20 min. Rajkumar and Kim (2006) considered that a blend of nine receptive colors from azo, triazine, and anthraquinone classes of colors was diminished in chlorine interceded electrochemical oxidation. The overall etiolation was achieved with the evacuation of COD at 120 min of the treatment. Ozonation treatment was analyzed for the expulsion of receptive colors from material coloring mechanical effluents in a cluster reactor of 35°C. Impacts of pH and reaction time on the decolorization effectiveness were assessed. The shading force of the wastewater was decreased with the time. The decolorization productivity expanded from 33% to 57% which indicated that the 90% of the color was taken out as the time advances following 6 h (Wijannarong et al., 2013). Debasement of blend of three responsive colors was concentrated by electro-Fenton measure. In the process, cathode coated with carbon paper altered with carbon nanotubes was utilized and the impact of starting grouping of oxidants, i.e., Fe^{+3} and H_2O_2 on the color degradation was examined (Chergui et al., 2012). Electro-Fenton strategy is additionally utilized for decolorization of colors in a nonstop air pocket reactor. The different boundaries such as iron measurements and pH assume a significant part in expanding the proficiency. At ideal conditions, a few colors were decolorized utilizing this method. For color blend, about 43% of decolorization was acquired following 21 h with total organic carbon decrease of 46% (Rosales, Pazos, Longo, & Sanroman, 2009). Photochemical or photocatalytic strategy debases color particles to carbon dioxide and water by UV treatment within the sight of hydrogen peroxide. Decadence is brought about by the creation of high groupings of hydroxyl revolutionaries. Copper sulfate coupled with clinoptilolite zeolite was utilized as a catalyst during photodegradation of different colors under daylight (Ejhieh & Moazzeni, 2013). The nanocomposites, namely, bismuth vanadate and cerium dioxide were also used as light photocatalysts in the desolation of cocktail of Methyl Orange and Methylene Blue colors (Wetchakun et al., 2012).

10.8.3 Biological method

Biological method is commonly viewed as the best, eco-accommodating, and less vital concentrated to eliminate the greater part of poisons from wastewater. The total disintegration of colors by fungal microorganism is done in two phases (Imran et al., 2015). In the main stage, minimal enzymatic severance of diazene bond brings about creation of lackluster metabolites. Furthermore, it is trailed by the disintegration stage, in which the oxidized condition is required. Specifically cycle of color decolorization is intervened by growths; the process is normally accomplished through the mechanisms such as biosorption, bioaccumulation, and biodegradation as summed up in Table 10.8 (Shree Nath, 2014).

Adsorption intensity of *Saccharomyces cerevisiae* was read for its biosorption limit with respect to a blend of colors, in which most extreme sorption was accomplished by the Freundlich isotherm model at ideal acidic pH (Ghaedi, Hajati, Barazesh, Karimi, & Ghezelbash, 2013). For debasement of cocktail of colors, *Portulaca grandiflora* was grown in a passage and penetrated through polyvinyl chloride pipe in which the color blend and profluent was allowed to attempt up to 91% decolorization (Khandare, Kabra, Kurade, & Govindwar, 2011). The wastewater containing two synthetic colors was decolorized both individually and with a cocktail of colors by utilizing immobilized *P. chrysosporium* (Pakshirajan & Singh, 2010). Sumathi and Manju (2000) confined fungus, *Aspergillus foetidus* which successfully decolorized up to 80% diamerene material color in the presence of higher concentration of chromium and NaCl. Chhabra et al. (2008)

Table 10.8 Fungal mechanisms involved in dye decolorization.

	Biosorption	Biodegradation	Bioaccumulation
Definition	A process which involves the binding of biosorbents to the surface of living or dead fungal biomass	In biodegradation, the breakdown of recalcitrant dyes from actively growing cells to simple molecules through enzymatic actions takes place	In this process, the defilements accelerate innermost cytoplasm of the diligently growing cells
Cellular growth	Not occur	Occur	Occur
Toxicity of pollutants	No harmfulness of pollutants toxic effect	Toxicity occurs when the coupling of pollutants and released byproducts takes place	Hazardness of pollutants toxic effects
Metabolic pathway	Not inhibited by metabolism	Restrained through metabolism	Deliberated by metabolism
The rate of process	Fast	Slow	Slow
Type of mechanism	Adsorption	Adsorption followed by degradation	Adsorption
Nutrient requirement	Not required	Required	Required
Stage of process	Single-step processing	Double-stage process	Two-stage step

demonstrated that *Cyathus bulleri*-producing laccase enzyme has the ability to diminish the acidic colors and mixture of respective pigments.

10.9 Oxidative and hydrolytic enzymes of *Pleurotus florida* used in decolorization of azo dyes

P. florida discomfit the lignocellulosic substrate consisting of cellulose, lignin, and hemicelluoses with the help of oxidative and hydrolytic enzymes such as lignin peroxidase (LiP), laccase, manganese peroxidase (MnP), lignin peroxidase (LiP). These enzymes and *P. florida* have the ability to degrade the wide range of xenobiotics.

10.9.1 Laccase (E.C 1.10. 3.2)

Laccases represent the heaviest subgroup of blue multicopper oxidases (MCO) which are broadly dispersed among eukaryotes (fungus, plants) and prokaryotes (microbes). Laccase is a monomeric protein comprising a β-barrel molded design which is like specific blue copper proteins (azurin or plastocyanin). Laccases are knowingly capable of two significant functions:

1. lignin polymerization (plants)
2. lignin depolymerization by *P. florida*

Laccases utilize their unmistakable redox capacity of copper particles, in which the oxidation of different aromatic substrates results in the breakdown of subatomic oxygen to water. Laccases can be termed as dimeric or tetrameric glycoproteins and mostly rely on copper for their reactant activity. Moreover, in view of the sorts of copper particle, laccases are delegated contrast in their trademark electronic paramagnetic reverberation signals as:

1. Type I (blue Cu place): It organizes with four amino acids as ligands: 2 his,1 cys, and 1 meth. Type-I copper containing laccases are commonly a dark blue shading, which can be distinguished by its absorbance at 600 nm frequency.
2. Type II (ordinary Cu community): Type-II copper facilitates with 2 his and water as ligands.
3. Type III (coupled binuclear Cu place): Type-III copper arranges with 3 his and a hydroxyl connect which confers antiferromagnetic coupling between the sort III copper particles.

Laccases are isolated into high- and low-redox expected compounds on the properties of copper particles. Microscopic organisms and plants emit negative redox potential laccases, while white decay fungi discharge positive redox potential laccases. Laccase catalysis can be differentiated and divided into three primary stages: Following the inside electron move between the sort I, type-II and type-III Cu groups, the type-I particle is diminished by reducing the substrate. Last, the reduction of the O_2 takes place at the sort II and III Cu groups and releasing water molecule (Fig. 10.4). In vitro the lignin corruption by laccase partially oxidizes the phenolic hydroxyl groups to phenoxy compounds which rearrange to

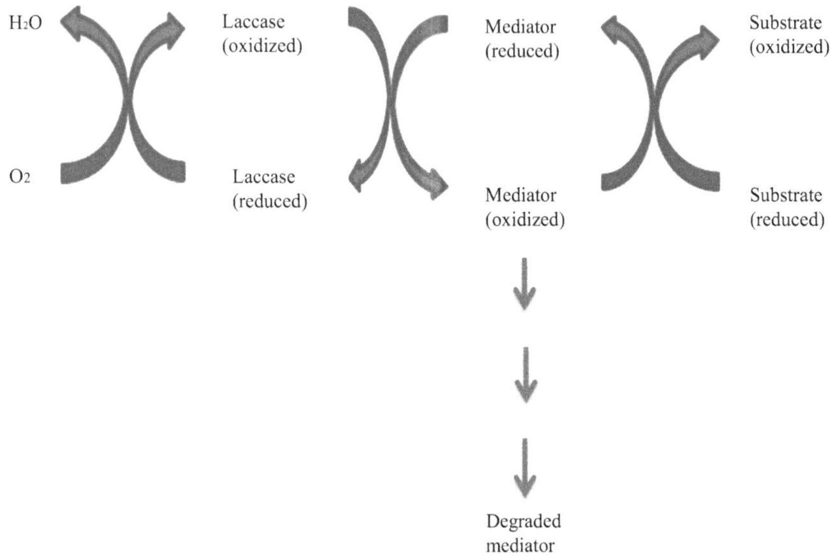

Figure 10.4 Laccase catalytic cycle.

sever the alkyl side chains. Laccase can also degrade the β-1 and β-O-4 dimer linkages between Cα and Cβ and further leads to Cα oxidation and aryl–alkyl cleavage (Wesenberg et al., 2003).

Legerska, Chmelova, and Ondrejovic (2018) assessed the bleaching of two different azo colors (Orange 2 and Orange 6) by contagious laccase from *T. versicolor* and demonstrated the decolorization of azonaphthol Orange 2, with 73%, was degraded more quickly than the azobenzene Acid Orange 6, with 46%. Kanagaraj, Senthilvelan, and Panda (2014) uncovered that the cell-free concentrates of laccase compound demonstrated the debasement for CI Acid Dark 210 is 96% and for the CI Acid dark 234 is 91%. Dhaarani, Priya, Rajan, and Kartic (2012) got the laccase compound from *P. ostreatus* and demonstrated high productivity against Malachite Green, Fundamental Red, and Corrosive Magenta with decolorization limits of 97%, 94%, and 94% separately.

10.9.2 Manganese peroxidase (E.C. 1.11.1.13)

The chemical component of MnP is like that of heme-containing peroxidases, in which the manganese acts as a reducing substrate. MnP oxidizes Mn (II) to Mn (III), which catalyzes the oxidation of monomeric phenols, lignin-related phenolic mixes, and colors. MnP has a manganese restricting site which consists of acidic amino acids and a heme propionate gathering. In this process, one electron moves to Compound I from bound Mn^{+2}. Further, Mn^{+3} in the presence of chelators such as oxalate is transferred from the dynamic site resulting in the incommensurate Mn^{+2} and insoluble Mn^{+4}. This response is required for the exchange of increasing highly active intensity of MnP to Mn^{3+}, which readily dissolves into the lignified cell. A significant component of MnP is to oxidize the low porous lignocellulose network, and chelators (oxalate) increment the electron thickness on Mn^{+3} which acts as an oxidant. Mn^{+3} chelates are not playing a significant role in lignolysis as they can only assault uncommon phenolic structures which are the released compounds of lignin. The resulting pattern of MnP starts when the hydrogen peroxide acts on the ferric protein resulting in the formation of an iron–peroxide complex (Wesenberg et al., 2003) (Fig. 10.5).

Zhang, Zhang, Zhang, and Geng (2018) purified the manganese peroxidase from *Cerrena unicolor* BBP6 which could adequately decolorize numerous kinds of colors, including Congo Red (54%), Bromophenol Blue (62%), Methyl Orange (78%), Gem Violet (81%), and Remazol Brilliant Blue R (80%). Li, Zhang, Tang, Zhang, and Mao (2015) demonstrated that the decolorization of Indigo Carmine by 90% takes place by the production of manganese peroxidase (MnP) from *P. chrysosporium*.

10.9.3 Lignin peroxidase

LiP was first found in the extracellular vehicle of *P. chrysosporium* under nitrogen-restricted conditions and considered as the most contemplated

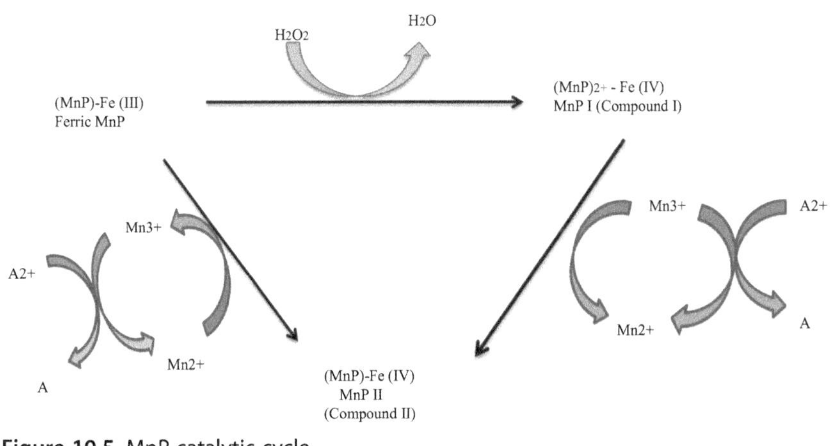

Figure 10.5 MnP catalytic cycle.

lignin depolymerizing compounds. The overall reaction mechanism of LiP is:

$$1,2\text{-bis}(3,4\text{-dimethoxyphenyl})\text{propane-}1,3\text{-diol} + H_2O_2 \rightarrow 3,4\text{-dimethoxybenzaldehyde} + 1\text{-}(3,4\text{-dimethoxyphenyl})\text{ethane- }1,2\text{-diol} + H_2O$$

H_2O_2 oxidizes the LiP (Compound I) in which iron is available as Fe (IV) leaving a free revolutionary on the tetrapyrrole ring. Compound I oxidizes the substrate to further Compound II (Fig. 10.6). The similar substrate particle in the reaction leads to the movement of catalyst to its resting state. The more ideal substrate "veratryl alcohol" for Compound II has the ability to transform the resting protein which leads to the completion of reactant cycle. LiP can oxidize 1, 2, 4, 5-tetra methoxybenzene, anilines, and phenols, which results in the breakdown of transitory cation intermediate.

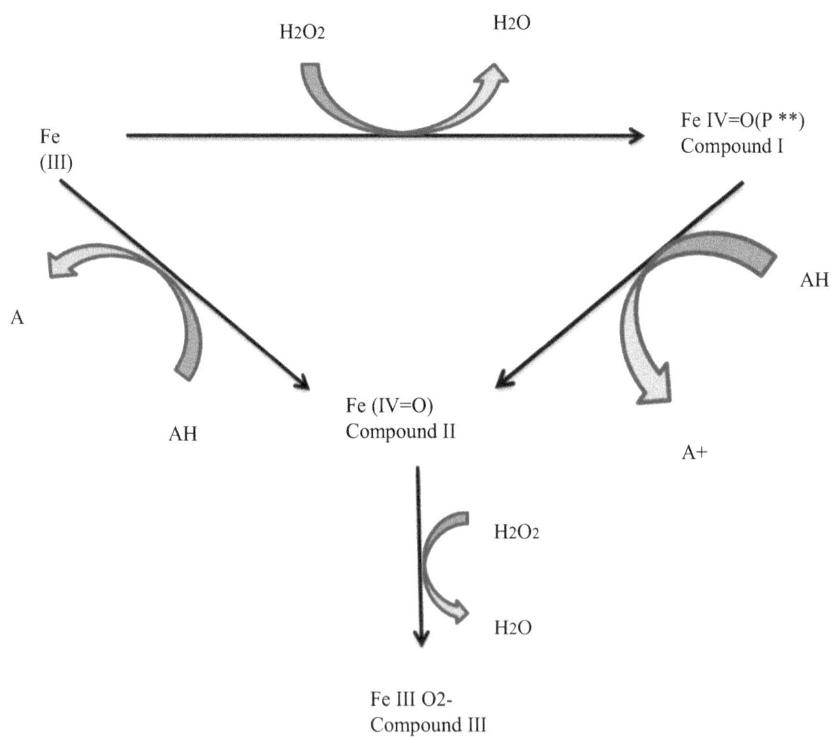

Figure 10.6 LiP catalytic cycle.

Bilal and Iqbal (2019) examined that the calcium alginate-immobilized LiP is applied for the decolorization of Remazol Brilliant Blue R (RBBR) up to 80% after the fifth cycle in a pressed bed reactor framework.

10.10 Factors influencing the dye decolorization

10.10.1 Influence of pH and temperature

White spoils ordinarily show ideal development at acidic pH esteems. Moreover, development in sugar-containing media by and large causes fermentation of the medium, the degree of which relies upon the carbon source and the sort and measure of buffering present. It might be exceptionally antacid, unbiased, or acidic relying upon the idea of the salts and colors. Azo color (RB5) wastewater is generally nonpartisan to soluble. Forgacs, Cserhati, and Oros (2004) stated that the *Phanerochaete sordida* under acidic conditions had yielded 100% decoloration, whereas *P. chrysosporium* reached up to 93% under basic conditions. Selvakumar, Manivasagan, and Chinnappan (2013) examined that wastewater treated with *Ganoderma lucidum* leads to decoloration of 81% and a COD decrease of 90% under enhanced conditions. Consequently, weakening and diminished pH of the first profluent were vital for most extreme decoloration and COD decrease. A high temperature is important to expand the proficiency of obsession of the colors. Most of the microbial species debase colors at the quicker rate at a temperature scope of above 35°C.

10.10.2 Impact of nitrogen source

The nitrogen utilized in the media effected the formulation of laccase. Peralta-Zamora et al. (2003) confined four white decay organisms and among them *Lentinus edodes* had shown the best decolorization capacity. The shading expulsion by contagious hyphae is predominantly through degradation by extracellular and intracellular catalysts. D'Souza-Ticlo, Verma, Mathew, and Raghukumar (2006) examined the impact of potassium nitrate, glycine, corn steep alcohol, and glutamic acid under fixed conditions. The results were predominated when glutamic acid was utilized as nitrogen source. The nitrogen-restricted medium expanded the creation of lignin peroxide and manganese peroxide resulting in enlarging the pace of decolorization (Singhal & Rathore, 2001). Hence, oxidation of the lignin catalyzes the breakdown of aromatic colors either through precipitation or by alterations taking place in the complex ring structure, and thus no vitality source is required to the fungus (Husain, 2010).

10.10.3 Influence of carbon source

The carbon source animates the development and gives oxidants to the fungus to "fuel" the decolorization cycle. For instance, glucose can be oxidized by fungus oxidases and release gluconic acid and H_2O_2. Blanquez, Sarra, and Vicent (2008) indicated that *T. versicolor* constantly etiolate 50% spent coloring shower from a material processing plant under nonsanitized conditions. Hence, *T. versicolor* effectively treated the effluvium in nonstop mode for a period of 15 days. Phenol-debasing white decay parasite *T. versicolor* confined from 14C paper industry, in which lignin mineralization demonstrated that 56% degradation of protocatechoic acid utilizing fructose was observed during the incubation period of 5 days (Udayasoorian & Prabu, 2005).

10.10.4 Concentration of dye

Anastasi et al. (2010) revealed that under nonsterile conditions and with no supplement expansion, the decoloration of effluents by the white decay fungus *Bjerkandera adusta* takes place and the growth stays successful during four patterns of decoloration. Osorio Echavarria, Vidal Benavides, Quintero, and Diaz (2011) detailed the decoloration of wastewater by the white decay fungus *Bjerkandera* sp. and resulted that the gushing was decolorized by 65% under sterile conditions and the poisonousness effectivity was decreased by 58%. Ma et al. (2014) uncovered that the *Ganoderma* sp. had extreme potential and resilience to decolorize and detoxify the sulfonated azo color dye under in vitro conditions. According to Senthilkumar, Perumalsamy, and Prabhu (2014), *P. chrysosporium* showed greatest decolorization of 98% while utilizing the various concentrations of emanating and found that gushing focus hinders the decolorization rate, whereas the inducers applied to the treatment also increase the pace of decolorization.

10.10.5 Consequence of redox mediators

The protein interceded debasement of colors (in vivo or in vitro) is adjustable and influential. The most appropriate redox arbiters utilized are violuric acid, hydroxybenzotriazole, veratryl liquor, and 2-methoxyphenothiazone which can be compromised by two components. The oxidation of a substrate proceeds one electron within the sight of a redox arbiter by a laccase or peroxidase. Moreover, the hydrogen particle released from the substrate is converted into a radical and thus making the substrate cooxidize (Fabbrini, Galli, & Gentili, 2002).

10.10.6 Repercussion of azo dye structure

The atomic structure of pigment relatively affects the degree of decolorization. Spadaro, Gold, and Renganathan (1992) observed that fragrant rings with substituents mainly considered as hydroxyl, acetamido, and amino capacities were mineralized more heavily than the unsubstituted rings in color decolorization by *P. chrysosporium*. Azo colors are electron-insufficient atoms which can go through degradation through the amide bond reduction. The primary capacity of the peroxidase/redox in this manner comprises oxidatively delivering the azo color more powerless to promote nucleophilic assault and nitrogen is disposed of in subatomic structure. Solid electron pulling group, for example, sulfo bunches at explicit situations in an azo color are all the more effectively biodegradable when contrasted with the ones with a carboxyl gathering.

10.11 Toxicity of decolorization products and evaluation methods

Toxicological investigation is utilized for the release of treated modern wastewater before released into water bodies. They can be utilized to gauge the ability of sea-going biotopes and as an "early notice" framework for the checking and screening of surface water (Asgher, Kausar, Bhatti, Shah, & Ali, 2008). Harmfulness test is quick and delicate strategy. It is utilized to screen impacts of color effluents. Poisonousness examines are gradually being consolidated into ecological checking of remediation locales, with additional substance portrayal utilizing by and large gas chromatographic–mass spectrometric techniques (Dominici, Cerbone, Villarini, Fatigoni, & Moretti, 2010). Solis, Solis, Perez, Manjarrez, and Flores (2012) stated that the metabolites created from color degradation are highly poisonous than the dominant color. It was concluded that the results of the oxidation of Indigo Blue through electroincineration, the utilization of laccase enzyme, and the coagulation with aluminum sulfate are more poisonous than the parent color. Consequently, it turns out to be significant for bioremediation innovation to evaluate the harmfulness

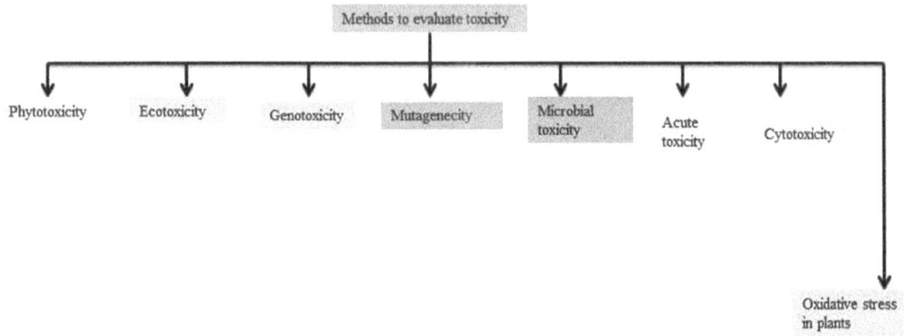

Figure 10.7 Techniques applied to appraise the toxicity of synthetic dyes and their decadence products present in processed confluents.

of the toxins and metabolites shaped after color debasement so as to contemplate the practicality of the strategy. Poisonousness has been assessed utilizing different systems as for phytotoxicity, mutagenicity, cytotoxicity, ecotoxicity, microbial poisonousness, genotoxicity, and toxicity on spineless creatures (Fig. 10.7) (Jadhav, Phugare, Patil, & Jadhav, 2011).

Biswas and Khuda-Bukhsh (2005) stated that mitotic alterations due to genotoxic and cytotoxic had huge impacts on the chromosome distortions and micronucleus test in bone marrow cells and spermatozoids of rodents is caused by azo color p-dimethylaminobenzene (p-DAB). Al-Sabti (2000) examined mutagenic impacts for a material azo color, the "chlorotriazine-responsive azo red 120," by the enlistment of micronuclei in erythrocytes of fishes. Tsuboy, Angeli, Mantovani, Umbuzeiro, and Ribeiro (2007) demonstrated that genotoxicity and mutagenicity of the Scattered Blue 291 dye in mammalian cells resulted in fracture of the DNA and alterations in arrangement of micronuclei. Matsuoka et al. (2001) indicated that antecedent azo colors are cytotoxic for the hamster cells prompting the development of micronucleated cells, multilobulated cores, and profoundly dense and binucleated cells. Most likely, these mixes influence the DNA, yet in addition auxiliary and administrative proteins associated with the cell division measure. Sudan azo colors prompt genotoxic impacts and the monoazo Sudan 1 color is viewed as a cancer-causing agent for well-evolved creatures and of mutagenic potential for people, while Sudan 2 is considered genotoxic for hepatic cells of rodents. Late toxicological examinations with the azo color Red HE3B (Reactive Red 120) and bacterial medicines indicated that the color had the option to instigate oxidative pressure and a high recurrence of chromosome distortions, when contrasted with the impacts brought about by its metabolites showing that the microbe-based medicines were paragon for the detoxification of Red HE3B (Dar, Bandh, Kamili, Nazir, & Bhat, 2013; Dar, Kamili, Nazir, Bandh, & Bhat, 2012). Bafana, Jain, Agrawal, and Chakrabarti (2009) demonstrated that the toxicity of the azo color Direct Red 28 is identified by metabolites, namely, benzidine and 4-aminobiphenyl, further actuating a high recurrence of DNA damage and apoptosis in human leukemia cells.

10.12 Conclusion

Wastewater released by the yarn developing industry is a perplexing blend of a few substances that go with a tremendous assortment of colors with differing compound structures. The utilization of *P. florida* in crude material wastewater can be compelling for accomplishing a huge decrease in the color substance of wastewater regardless of the colossal measures of salts, metals, and different pollutants. The white decay organism *P. florida* can lapidify xenobiotic compounds to carbon dioxide and water through oxidative and hydrolytic enzymes. Azo colors establish the biggest and most adaptable class of engineered colors utilized in the textile industry. The presence of colors grants an exceptional shading to effluents which prompts the ecological and esthetic problems. Treatment of color-containing effluents is a difficult assignment. Microbial and enzymatic decolorization and debasement of azo colors can possibly help in this issue because of their ecological amicable and economical nature. The fungus *P. florida* enzymatic process is extremely encouraging for the decolorization of manufactured azo colors. The oxidative and hydrolytic enzymes are contemplated as infinitesimal accoutrements for bioremediation of synthetic dyes. Further, toxicity contemplates must be done to guarantee the well-being of the decolorized wastewater.

References

Ahmad, A., Harris, W., & Ooi, B. (2012). Removal of dye from wastewater of textile industry using membrane technology. *Jurnal Teknologi*, *36*, 31–44.

Al-Sabti, K. (2000). Chlorotriazine reactive azo red 120 textile dye induces micronuclei in fish. *Ecotoxicology and Environmental Safety*, *47*, 149–155.

Anastasi, A., Spina, F., Prigione, V., Tigini, V., Giansanti, P., & Varese, G. C. (2010). Scale-up of a bioprocess for textile wastewater treatment using *Bjerkandera adusta*. *Bioresource Technology*, *101*, 3067–3075.

Asgher, M., Kausar, S., Bhatti, H. N., Shah, S. A. H., & Ali, M. (2008). Optimization of medium for decolorization of solar golden yellow R direct textile dye by *Schizophyllum commune* IBL-06. *International Biodeterioration & Biodegradation*, *61*, 189–193.

Babu, B. R., Parande, A. K., Raghu, S., & Kumar, T. P. (2007). Cotton textile processing: waste generation and effluent treatment. *Journal of Cotton Science*, *11*, 141–153.
Bafana, A., Jain, M., Agrawal, G., & Chakrabarti, T. (2009). Bacterial reduction in genotoxicity of Direct Red 28 dye. *Chemosphere*, *74*, 1404–1406.
Bansal, R. C., & Goyal, M. (2005). *Activated carbon adsorption* (pp. 351–353). London: Taylor and Francis Group.
Bhat, R. A., Singh, D. V., Qadri, H., Dar, G. H., Dervash, M. A., Bhat, S. A., ... Yousaf, B. (2021). Vulnerability of municipal solid waste: An emerging threat to aquatic ecosystems. *Chemosphere*, *287*(2022), 132223. Available from https://doi.org/10.1016/j.chemosphere.2021.132223.
Bhole, B. D., Ganguly, B., Madhuram, A., Deshpande, D., & Joshi, J. (2004). Biosorption of methyl violet, basic fuchsin and their mixture using dead fungal biomass. *Current Science*, *86*, 1641–1644.
Bilal, M., & Iqbal, N. M. H. (2019). Lignin peroxidase immobilization on Ca-alginate beads and its dye degradation performance in a packed bed reactor system. *Biocatalysis and Agricultural Biotechnology*, *20*, 101–205.
Biswas, S. J., & Khuda-Bukhsh, A. R. (2005). Cytotoxic and genotoxic effects of the azo-dye pdimethylaminoazobenzene in mice: A time-course study. *Mutation Research*, *587*, 1–8.
Blanquez, P., Sarra, M., & Vicent, T. (2008). Development of a continuous process to adapt the textile wastewater treatment by fungi to industrial conditions. *Process Biochemistry*, *43*, 1–7.
Carlile, M. J., Watikinson, S. C., & Gooday, G. (2001). *The fungi*. London: Academic Press.
Chagas, P. E., & Durrant, R. L. (2001). Decolorization of azo dyes by *Phanerochaete chrysosporium* and *Pleurotus sajor-caju*. *Enzyme and Microbial Technology*, *29*, 473–477.
Chatzisymeon, E., Xekoukoulotakis, N. P., Coz, A., Kalogerakis, N., & Mantzavinos, D. (2006). Electrochemical treatment of textile dyes and dyehouse effluents. *Journal of Hazardous Materials*, *137*, 998–1007.
Chergui, S. B., Oturan, N., Khalaf, H., & Oturan, M. A. (2012). A photo-fentontreatment of a mixture of three cationic dyes. *Procedia Engineering*, *33*, 181–187.
Chhabra, M., Mishra, S., & Sreekrishnan, T. R. (2008). Mediator assisted decolorization and detoxification of textile dyes/ dye mixture by *Cyathus bulleri* laccase. *Applied Biochemistry and Biotechnology*, *151*, 587–598.
Christian, V., Shrivastava, R., Shukla, D., Modi, H. A., & Vyas, B. R. M. (2005). Degradation of xenobiotic compounds by lignin-degrading white-rot fungi: Enzymology and mechanisms involved. *Indian Journal of Experimental Biology*, *43*, 301–312.
D'Souza-Ticlo, D., Verma, A. K., Mathew, M., & Raghukumar, C. (2006). Effect of nutrient nitrogen on laccase production, its isozyme pattern and effluent decolorization by the fungus NIOCC# 2a, isolated from mangrove wood. *Indian Journal of Marine Sciences*, *34*, 364–372.
Daassi, D., Mechichi, T., Nasri, M., & Rodriguez-Couto, S. (2013). Decolorization of the metal textile dye Lanaset Grey G by immobilized white-rot fungi. *Journal of Environmental Management*, *129*, 324–332.
Dar, G. H., Bandh, S. A., Kamili, A. N., Nazir, R., & Bhat, R. A. (2013). Comparative analysis of different types of bacterial colonies from the soils of Yusmarg Forest, Kashmir Valley India. *Ecologia Balkanica*, *5*(1), 31–35.
Dar, G. H., Bandh, S. A., & Kamili, A. N. (2011). Bacteriological analysis of soil from Yusmarg Health Resort of Kashmir Valley. *International Journal of Current Research.*, *3* (10), 53–56, 2011.
Dar, G. H., Kamili, A. N., Chishti, M. Z., Dar, S. A., Tantry, T. A., & Ahmad, F. (2016). Characterization of *Aeromonas sobria* isolated from Fish Rohu (*Labeo rohita*) collected

from polluted pond. *Journal of Bacteriology & Parasitology*, 7(3), 1–5. Available from https://doi.org/10.4172/2155-9597.1000273.
Dar, G. H., Kamili, A. N., Dar, S. A., Chishti, M. Z., & Ahmad, F. (2015). Detection and characterization of potentially pathogenic *Aeromonas sobria* isolated from fish *Hypophthalmichthys molitrix* (Cypriniformes: Cyprinidae). *Microbial Pathogenesis, 2015*. Available from https://doi.org/10.1016/j.micpath.2015.10.017.
Dar, G. H., Kamili, A. N., Nazir, R., Bandh, S. A., & Bhat, R. A. (2012). A preliminary study of colony forming units of bacteria from the soils of Yusmarg Forest, Kashmir Valley, India. *International Journal of Current Research*, *4*(12), 467–472.
Dhaarani, S., Priya, K. A., Rajan, V. T., & Kartic, N. D. (2012). Degradation of various dyes using laccase enzyme. *Journal of Environmental Science and Engineering*, *54*, 489–494.
Dominici, L., Cerbone, B., Villarini, M., Fatigoni, C., & Moretti, M. (2010). In vitro testing for genotoxicity of indigo naturalis assessed by micronucleus test. *Natural Product Communications*, *5*, 1039–1042.
Edwards J. C. (2000). *Investigation of color removal by chemical oxidation for three reactive textile dyes and spent textile dye wastewater* (Doctoral dissertation). Virginia Tech.
Ejhieh, A. N., & Moazzeni, N. (2013). Sunlight photodecolorization of a mixture of Methyl Orange and Bromocresol Green by CuS incorporated in a clinoptilolite zeolite as a heterogeneous catalyst. *Journal of Industrial and Engineering Chemistry*, *19*, 1433–1442.
Fabbrini, M., Galli, C., & Gentili, P. (2002). Comparing the catalytic efficiency of some mediators of laccase. *Journal of Molecular Catalysis B Enzymatic*, *16*, 231–240.
Forgacs, E., Cserhati, T., & Oros, G. (2004). Removal of synthetic dyes from wastewaters: A review. *Environment International*, *30*, 953–971.
Ghaedi, M., Hajati, S., Barazesh, B., Karimi, F., & Ghezelbash, G. (2013). *Saccharomyces cerevisiae* for the biosorption of basic dyes from binary component systems and the high order derivative spect rophotometric method for simultaneous analysis of Brilliant green and Methylene blue. *Journal of Industrial and Engineering Chemistry*, *19*, 227–233.
Ghaly, A., Ananthashankar, R., Alhattab, M., & Ramakrishnan, V. (2014). Production, characterization and treatment of textile effluents: A critical review. *Journal of Chemical Engineering & Process Technology*, *5*, 1–19.
Haley, T. J. (1975). Benzidine revisited; A review of the literature and its congeners. *Clinical Toxicology*, *8*, 13.
Huang, Y. H., Hsueh, C. L., Huang, C. P., Su, L. C., & Chen, C. Y. (2007). Adsorption thermodynamic and kinetic studies of Pb(II) removal from water onto a versatile Al2O3-supported iron oxide. *Separation Purification Technology*, *55*, 23–29.
Husain, Q. (2010). Peroxidase mediated decolorization and remediation of wastewater containing industrial dyes: A review. *Reviews in Environmental Science and Biotechnology*, *9*(2), 117–140.
Imran, M., Crowley, D. E., Khalid, A., Hussain, S., Mumtaz, M. W., & Arshad, M. (2015). Microbial biotechnology for decolorization of textile wastewaters. *Reviews in Environmental Science and Biotechnology*, *14*, 73–92.
Iqbal, M. J., & Ashiq, M. N. (2007). Adsorption of dyes from aqueous solutions on activated charcoal. *Journal of Hazardous Materials*, *139*, 57–66.
Jadhav, S. B., Phugare, S. S., Patil, P. S., & Jadhav, J. P. (2011). Biochemical degradation pathway of textile dye Remazol Red and subsequent toxicological evaluation by cytotoxicity, genotoxicity and oxidative stress studies. *International Biodeterioration & Biodegradation*, *65*, 733–743.
Jamal, F., Singh, S., Khatoon, S., & Mehrotra, S. (2013). Application of immobilized pointed gourd (*Trichosanthes dioica*) peroxidase-concanavalin A complex on calcium alginate pectin gel in decolorization of synthetic dyes using batch processes and continuous two reactor system. *Journal of Bioprocessing & Biotechniques*, *3*, 1–5.

Kabra, A. N., Khandare, R. V., Waghmode, T. R., & Govindwar, S. P. (2012). Phytoremediation of textile effluent and mixture of structurally different dyes by *Glandularia Pulchella* (Sweet) Tronc. *Chemosphere*, *87*, 265–272.

Kanagaraj, J., Senthilvelan, T., & Panda, C. R. (2014). Degradation of azo dyes by laccase: Biological method to reduce pollution load in dye wastewater. *Clean Technologies and Environmental Policy*, *17*, 1443–1456.

Kang, Q., Zhou, W., Li, Q., Gao, B., Fan, J., & Shen, D. (2009). Adsorption of anionic dyes on poly (epicholorohydrindimethylamine) modified bentonite in single and mixed dye solutions. *Applied Clay Science*, *45*, 280–287.

Khan, A. A., & Husain, Q. (2007). Decolorization and removal of textile and non-textile dyes from polluted wastewater and dyeing effluent by using potato (*Solanum tuberosum*) soluble and immobilized polyphenol oxidase. *Bioresource Technology*, *98*, 1012–1019.

Khandare, R. V., Kabra, A. N., Kurade, M. B., & Govindwar, S. P. (2011). Phytoremediation potential of *Portulaca grandiflora* Hook. (Moss-Rose) in degrading a sulfonated diazo reactive dye Navy Blue HE2R (Reactive Blue 172). *Bioresource Technology*, *102*, 6774–6777.

Knapp, J. S., Vantoch-Wood, E. J., & Zhang, F. (2001). *Use of wood-rotting fungi for the decolorization of dyes and industrial effluents. Fungi in bioremediation*. Cambridge University Press.

Kuberan, T., Anburaj, J., Sundaravadivelan, C., & Kumar, P. (2011). Biodegradation of azo dye by Listeria sp. *International Journal of Environmental Science*, *1*, 1760–1770.

Lahkimi, A., Oturan, M. A., Oturan, N., & Chaouch, M. (2007). Removal of textile dyes from water by the electro-Fenton process. *Environmental Chemistry Letters*, *5*, 35–39.

Legerska, B., Chmelova, D., & Ondrejovic, M. (2018). Decolourization and detoxification of monoazo dyes by laccase from the white-rot fungus *Trametes versicolor*. *Journal of Biotechnology*, *285*, 84–90.

Li, H., Zhang, R., Tang, L., Zhang, J., & Mao, Z. (2015). Manganese peroxidase production from cassava residue by Phanerochaete chrysporium in solid state fermentation and its decolorization of indigo carmine. *Chinese Journal of Chemical Engineering*, *23*, 227–233.

Ma, L., Zhuo, R., Liu, H., Yu, D., Jiang, M., Zhang, X., & Yang, Y. (2014). Efficient decolorization and detoxification of the sulfonated azo dye Reactive Orange 16 and simulated textile wastewater containing Reactive Orange 16 by the white-rot fungus *Ganoderma* sp. En3 isolated from the forest of Tzu-chin Mountain in China. *Biochemical Engineering Journal*, *82*, 1–9.

Manu, B., & Chaudhari, S. (2002). Anaeobic decolourization of simulated textile wastewater containing azo dyes. *Bioresource Technology*, *82*, 225–231.

Martins, M., Ferreira, I., Santos, I., Queiroz, M., & Lima, N. (2001). Biodegradation of bioaccessible textile azo dyes by *Phanerochaete chrysosporium*. *Journal of Biotechnology*, *89*, 91–98.

Matsuoka, A., Tada, A., Terao, Y., Nukaya, H., Onfelt, A., & Wakabayashi, K. (2001). Chromosomal effects of newly identified water pollutants PBTA-1 and PBTA-2 and their possible mother compounds (azo dyes) and intermediates (non-CIPBTAs) in two Chinese hamster cell lines. *Mutation Research*, *493*, 75–85.

Nguyen, T. A., & Juang, R. S. (2013). Treatment of waters and wastewaters containing sulfur dyes: A review. *Chemical Engineering Journal*, *219*, 109–117.

Nirmalarani, J., & Janardhanan, K. (1988). Effect of south India Viscose factory effluent on seed germination seedling growth and chloroplast pigments content in five varieties of Maize (*Zea mays* I). *Madras Agricultural Journal*, *75*, 41.

Osorio Echavarria, J., Vidal Benavides, A. I., & Quintero Diaz, J. C. (2011). Decolorization of textile wastewater using the white rot fungi anamorph R1 of *Bjerkandera* sp. *Revista Facultad de Ingenieria Universidad de Antioquia*, *57*, 85–93.

Pakshirajan, K., & Singh, S. (2010). Decolorization of synthetic wastewater containing azo dyes in a batch operated rotating biological contactor reactor with the immobilized fungus *Phanerochaete chrysosporium*. *Industrial & Engineering Chemistry Research, 49*, 7484–7487.

Palacio, S. M., Quinones, F. R. E., Modenes, A. N., Manenti, D. R., Oliveira, C. C., & Garcia, J. C. (2012). Optimised photocatalytic degradation of a mixture of azo dyes using a TiO2/H2O2/UV process. *Water Science and Technology, 65*, 1392–1398.

Peralta-Zamora, P., Pereira, C. M., Tiburtius, E. R., Moraes, S. G., Rosa, M. A., Minussi, R. C., & Duran, N. (2003). Decolorization of reactive dyes by immobilized laccase. *Applied Catalysis B: Environmental, 42*, 131–144.

Punzi, M., Mattiasson, B., & Jonstrup, M. (2012). Treatment of synthetic text ile wastewater by homogeneous and heterogeneous photo-Fenton oxidation. *Journal of Photochemistry and Photobiology: Chemistry, 248*, 30–35.

Rajkumar, D., & Kim, J. G. (2006). Oxidation of various reactive dyes with in situ electro-generated active chlorine for textile dyeing industry wastewater treatment. *Journal of Hazardous Materials, 136*, 203–212.

Riu, J., Schonsee, I., & Barcelo, D. (1988). Determination of sulfonated azo dyes in ground water and industrial effluent by automated solid-phase extraction followed by capillary electrophoresis/mass spectrometry. *Journal of Mass Spectrometry, 33*, 653.

Robinson, T., McMullan, G., Marchant, R., & Nigam, P. (2001). Remediation of dyes in textile effluent: a critical review on current treatment technologies with a proposed alternative. *Bioresource Technology*, 77, 247–255.

Rosales, E., Pazos, M., Longo, M. A., & Sanroman, M. A. (2009). Electro-Fenton decoloration of dyes in a continuous reactor: a promising technology in colored wastewater treatment. *Journal of Chemical & Engineering Data, 155*, 62–67.

Sanroman, M. A., Pazos, M., Ricart, M. T., & Cameselle, C. (2004). Electrochemical decolorisation of structurally different dyes. *Chemosphere, 57*, 233–239.

Sarayu, K., & Sandhya, S. (2012). Current technologies for biological treatment of textile wastewater—A review. *Applied Biochemistry and Biotechnology, 167*, 645–661.

Sarayu, K., Swaminathan, K., & Sandhya, S. (2007). Assessment of degradation of eight commercial reactive azo dyes individually and in mixture in aqueous solution by ozonation. *Dyes and Pigments, 75*, 362–368.

Saxena, S., Raja, A. S. M., & Arputharaj, A. (2017). Challenges in sustainable wet processing of textiles. In S. S. Muthu (Ed.), *Textiles and clothing sustainability: Sustainable textile chemical processes* (pp. 43–79). Singapore: Springer.

Selvakumar, S., Manivasagan, R., & Chinnappan, K. (2013). Biodegradation and decolourization of textile dye wastewater using *Ganoderma lucidum*. *Biotechnology Advances, 3*, 71–79.

Senthilkumar, S., Perumalsamy, M., & Prabhu, H. J. (2014). Decolourization potential of white-rot fungus *Phanerochaete chrysosporium* on synthetic dye bath effluent containing Amido black 10B. *Journal of Saudi Chemical Society, 18*, 845–853.

Shanthi, S., & Padmavathi, V. (2012). Zinc oxide catalyzed photo degradation of crystal violet and aniline blue dyes from aqueous solutions of their binary mixture, using sun light and microwaves – A comparative study of nano particles of ZnO with AR ZnO. *International Journal of Chemistry and Applications, 4*, 351–358.

Shree Nath, S. (2014). *Microbial degradation of synthetic dyes in wastewaters*. Cham: Springer International Publishing.

Singhal, V., & Rathore, V. S. (2001). Effects of Zn^{2+} and Cu^{2+} on growth, lignin degradation and ligninolytic enzymes in *Phanerochaete chrysosporium*. *World Journal of Microbiology and Biotechnology, 17*, 235–240.

Solis, M., Solis, A., Perez, H. I., Manjarrez, N., & Flores, M. (2012). Microbial decoloration of azo dyes: A review. *Process Biochemistry, 47*, 1723–1748.

Spadaro, J. T., Gold, M. H., & Renganathan, V. (1992). Degradation of azo dyes by the lignin degrading fungus *Phanerochaete chrysosporium*. *Applied and Environmental Microbiology, 58*, 2397–2401.

Sudhakar, P., Palaniappan, R., & Gowrisankar, R. (2002). Degradation of azo dye (Black-E) by an indigenous bacterium *Pseudomonas* sp. BSP-4. *Asian Journal of Microbiology Biotechnology & Environmental Sciences*, *4*, 203−208.

Sumathi, S., & Manju, B. S. (2000). Uptake of reactive textile dyes by *Aspergillus foetidus*. *Enzyme and Microbial Technology*, *27*, 347−355.

Taha, M., Adetutu, E. M., Shahsavari, E., Smith, A. T., & Ball, A. S. (2014). Azo and anthraquinone dye mixture decolourization at elevated temperature and concentration by a newly isolated thermophilic fungus, *Thermomucorindicae seudaticae*. *Journal of Environmental Chemical Engineering*, *2*, 415−423.

Teli, M. (2008). Textile coloration industry in India. *Coloration Technology*, *124*, 1−13.

Tsuboy, M. S., Angeli, J. P. F., Mantovani, M. S., Umbuzeiro, G. A., & Ribeiro, L. R. (2007). Genotoxic, mutagenic and cytotoxic effects of the commercial dye CI Disperse Blue 291 in the human hepatic cell line HepG2. *Toxicology in vitro*, *21*, 1650−1655.

Udayasoorian, C., & Prabu, P. (2005). Biodegradation of phenols by ligninolytic fungus *Trametes versicolor*. *The Journal of Biological Sciences*, *5*, 558−561.

Upadhye, V., & Joshi, S. S. (2012). Advances in wastewater treatment − A review. *International Journal of Chemical Sciences*, *3*, 264−268.

Van der Zee, F. P., & Villaverde, S. (2005). Combined anaerobic aerobic treatment of azo dyes-A short review of bioreactor studies. *Water Research*, *39*, 1425.

Ventura-Camargo, B. C., & Marin-Morales, M. A. (2013). Azo dyes: Characterization and toxicity − A review. *Textiles and Light Industrial Science and Technology (TLIST)*, *2*, 85−103.

Vijaya, P., Padmavathy, P., & Sandhya, S. (2003). Decolourization and biodegradation of reactive azo dyes by mixed culture. *Indian Journal of Biotechnology*, *2*, 259−263.

Vyas, B., & Molitoris, H.-P. (1995). Involvement of an extracellular H_2O_2-dependent ligninolytic activity of the white rot fungus *Pleurotus ostreatus* in the decolorization of Remazol brilliant blue R. *Applied and Environmental Microbiology*, *61*, 3919−3927.

Waghmode, T. R., Kurade, M. B., & Govindwar, S. P. (2011). Time dependent degradation of mixture of structurally different azo and non azo dyes by using *Galactomyces geotrichum* MTCC 1360. *International Biodeterioration & Biodegradation*, *65*, 479−486.

Waring, D. R., & Hallas, G. (2013). *The chemistry and application of dyes* (pp. 17−47). New York/London: Plenum Press/Springer US.

Wesenberg, D., Kyriakides, I., & Agathos, S. N. (2003). White-rot fungi and their enzymes for the treatment of industrial dye effluents. *Biotechnology Advances*, *22*, 161−187.

Wetchakun, N., Chaiwichain, S., Inceesungvorn, B., & Pingmuang, K. (2012). $BiVO_4$/CeO_2 nanocomposites with high visible-light-induced photocatalytic activity. *ACS Applied Materials & Interfaces*, *4*, 3718−3723.

Wijannarong, S., Aroonsrimorakot, S., Thavipoke, P., & Sangjan, S. (2013). Removal of reactive dyes from textile dyeing industrial effluent by ozonation process. *APCBEE Procedia*, *5*, 279−282.

Yao, J., Jia, R., Zheng, L., & Wang, B. (2013). Rapid decolorization of azo dyes by crude manganese peroxidase from *Schizophyllum* sp. F17 in solid-state fermentation. *Biotechnology and Bioprocess Engineering*, *18*, 868−877.

Zhang, H., Zhang, J., Zhang, X., & Geng, A. (2018). Purification and characterization of a novel manganese peroxidase from white rot fungus *Cerrena unicolor* BBP6 and its application in dye decolorization and denim bleaching. *Process Biochemistry*, *66*, 222−229.

Zolgharnein, J., Asanjrani, N., Bagtash, M., & Azimi, G. (2014). Multi-response optimization using Taguchi design and principle component analysis for removing binary mixture of alizarin red and alizarin yellow from aqueous solution by nano a-alumina. *Spectrochimica Acta Part A: Molecular and Biomolecular Spectroscopy*, *126*, 291−300.

Zollinger, H. (2003). *Color chemistry: syntheses, properties, and applications of organic dyes and pigments*. John Wiley & Sons.

CHAPTER 11

Endophytic Microbes: Bioremediation of soil contaminants

Bhat Mohd Skinder[1], Masarat Nabi[1], Bashir Ahmad Sheer Gojree[1], Gowhar Hamid Dar[2] and Bashir Ahmad Ganai[3]
[1]Department of Environmental Science/Centre of Research for Development (CORD), University of Kashmir, Srinagar, Jammu & Kashmir, India
[2]Department of Environmental Science, Sri Pratap College, Cluster University Srinagar, Srinagar, Jammu & Kashmir, India
[3]Centre of Research for Development (CORD), University of Kashmir, Srinagar, Jammu & Kashmir, India

11.1 Introduction

The contaminants generated by different anthropogenic activities such as inappropriate utilization of synthetic insecticides and growing industrialization have severe consequences on human well-being; the natural ecosystem of soil, water, and plants; and the natural microbiota in multiple approaches (Dar, Bandh, & Kamili, 2011; Dar, Kamili, Nazir, Bandh, & Bhat, 2012). Transitioning climate patterns and an increase in mean temperature have emerged as a serious worldwide problem in recent years; and, surprisingly, the primary forces liable for these shifts are related to the aggregation and deposition of different natural contaminants into the atmosphere. Different forms of natural contaminants have been encountered by humans in their everyday life in recent years as a result of unregulated use of natural resources (for instance, water and soil); anthropogenic practices such as accelerated population growth, discharge of domestic sludge, and effluents from public transportation; amplified greenhouse gases such as carbon dioxide (CO_2), nitrogen oxides (NO_X), and sulfur dioxide (SO_2) emissions; and use of enormous quantity of pesticides, insecticides, fertilizers, and other chemicals in agriculture (Dar, Bandh, Kamili, Nazir, & Bhat, 2013; Gupta et al., 2020). Furthermore, emerging mining sites, industrial effluents, and natural disasters are all prevalent causes that tend to accumulate an array of xenobiotic substances in the area. The existence or accretion of these contaminants has an impact on ecosystem, such as in the soil, where it affects both microbiota

Microbial Consortium and Biotransformation for Pollution Decontamination
DOI: https://doi.org/10.1016/B978-0-323-91893-0.00002-X

and crop yields owing to important alterations in soil structure. Likewise, the existence of these pollutants in water has been shown to have an effect on its purity and cleanliness (Cheng, 2003). Managing the increased rates of contaminants once they reach the environment is a complicated task, and therefore it appears to be a key challenge for scientist and researcher communities globally (Ferronato & Torretta, 2019). Contaminants may be natural or anthropogenic in nature or either different types of pollutants have adverse effects on the environment; however, natural pollutants can have elevated toxic effects, persistence, and a lower degree of biodegradation than synthetic pollutants. Organic substances such as organochlorine pesticides, phenolic compounds, polycyclic aromatic hydrocarbons (PAHs), petroleum products, and herbicides have an elevated perseverance in nature, except in extremely minimal concentrations, and therefore can impede soil dynamics and efficiency and also indigenous microbial communities. Moreover, owing to their cancer causing and teratogenic origin, these organic contaminants have exposed a wide range of health-associated problems in both animals and humans. To replace the conventional methods for removing environmental pollutants, it is important to investigate environmentally sustainable and long-term strategies for the control of environmental contaminants (Juwarkar, Singh, & Mudhoo, 2010). In this background, bioremediation appears as an appropriate substitute to physiochemical remediation method for managing environmental contaminants.

11.2 Endophytic microbes

Plants connect with a wide range of microbe communities, and these microbes communicate with plants in a complicated manner. Endophytic microbes are a form of microbe that lives in plant tissue without showing any visible signs or inflicting harm to the host and is thought to expend some part of its life there (Compant et al., 2016). The availability of high quantity of chemical exudates secreted by roots such as carbs, lipids, and amino acids in the rhizosphere of the plant makes it a hotspot for diversified microbiota. (Bais, Weir, Perry, Gilroy, & Vivanco, 2006). These exudates reach a huge number of microbiota; a few of them can inhabit the root surface and invade plant tissue, acting as endophytes (Gupta et al., 2020). Endophytic microbes and their hosts may develop a strong mutual association that benefits both the microbes and the host. A range of stressors within the host plant provide nutrition and healthy environment for the resident endophytes, which in exchange produce a variety of growth-promoting agents and biologically active constituents that aid in plant defense, growth stimulation, and stress resistance (Backer et al., 2018). Endophytes, like other plant growth-promoting (PGP) microbes, can play a direct or indirect role in plant growth promotion, phytopathogen biocontrol, and stress reduction (Rana, Kour, & Kaur, 2020). Indole-3-acetic acid (IAA), gibberellins, enzyme ACC deaminase activity, nitrogen fixation, and phosphate solubilization are the examples of direct mechanisms that are specifically associated in the production of growth hormones. In sustainable agriculture, microbes with PGP characteristics could be used as biofertilizers (Bhardwaj, Ansari, & Sahoo, 2014). Antibiotic development, hydrolytic enzymes, hydrogen cyanide, and siderophores are the examples of some mechanisms that indirectly aid in plant growth and disease control (Olanrewaju, Glick, & Babalola, 2017). Further, they increase plant productivity and have been able to control both biotic and abiotic stresses in addition to growth stimulation and disease prevention (Kumar, Singh, Tripathi, Singh, & Singh, 2018). Because of their ability to adapt in the intrinsic environment of the host plants, endophytic microorganisms are thought to be highly stress resistant than rhizospheric or phyllospheric microbial variants. Pollutant-degrading receptors found in endophytic microorganisms aid in the removal of or detoxifying harmful environmental contaminants (Meena et al., 2017). The mutual relationship between plants and microbes during the breakdown of organic contaminants has been demonstrated. The beneficial interaction has been identified as improving the ability of consortia to turn complex organic contaminants into simpler ones, thereby providing an appropriate forum for genetic modification (Haldar & Sengupta, 2016). Since microorganisms have a particular multiplication rate and growth criteria, choosing microbial consortia for bioremediation is a critical process that is heavily influenced by the polluted sites' specific growth rate and PGP characteristics (Endeshaw, Birhanu, Zerihun, & Misganaw, 2017).

11.3 Plant growth-promoting bacteria

"Plant beneficial bacteria" [i.e., plant growth-promoting bacteria (PGPB)] is a term used to describe bacteria that help plants to grow and develop. Through their inhibitory action against plant pathogens, PGPB can function as biological control agents and indirectly help in growth and development of plants. By increasing available mineral nutrients or regulating phytohormone rates, PGPB, on the other hand, may directly boost the growth and development of plants. They can produce plant hormones, including auxins, cytokinins, and gibberellins, which can help plants, developing at different levels; produce siderophores, which are iron chelators that improve plant Fe nutrition or suppress the activity of phytopathogens; synthesize 1-aminocyclopropane-1-carboxylate (ACC) deaminase, which can lower ethylene concentration in plants (Glick, 2005); and dissolve minerals such as phosphorous, making them easily available for plant development (Olanrewaju et al., 2017). Either some or all of these mechanisms could be employed by PGPB to aid growth and development of plants, and a single PGPB could use various mechanisms depending on the environment. When plants grow in inadequate or environmental stresses, such as salinity and heavy metals, the positive results of PGPB on the growth of plants are usually more prominent (Shilev, 2020).

11.4 Mechanisms involving endophyte-mediated phytoremediation enhancement

Endophytic microbes acclimate to particular physiological environments within plants and the rhizosphere by forming tiny tissues or site specificity. Endophytes, plants, as well as soilborne microbial communities have complex associations with symbiotic, mutualistic, commensalistic, and trophobiotic dialogs emerging. Endophytes promote phytoremediation in a variety of ways, both direct and indirect (Fig. 11.1). They may contain responsive biological molecules and actively involved in pollutant deterioration, cometabolization, extraction, and hyperaccumulation mechanisms directly. Furthermore, endophytes might help to promote mixed pollutants' phytoremediation by increasing plant nutrient availability, regulating plant growth, mitigating toxic strains, and communicating with microbiota.

11.4.1 Direct ways of phytoremediation via endophytes

11.4.1.1 Biodegradation

Functional proteins or enzymes produced by endophytes may directly improve phytoremediation. Hydrolases, glutathione-*S*-transferases, cytochrome P450 monooxygenases, dehalogenases, catalases, laccases, nitroreductases, peroxidases, and polyphenol peroxidases are among the enzymes involved in the degradation, stabilization, deposition, and detoxification of mixed pollutants (Schwitzguebel, 2017). While incorporating polar groups

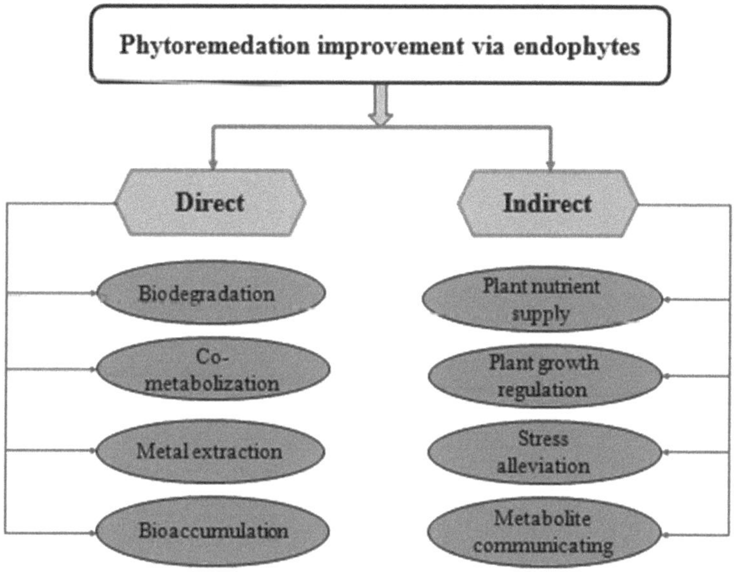

Figure 11.1 Phytoremedation assisted via endophytes.

such as sulfides, malonic acid, carbohydrates, amino acids, or glutathione, these enzymes can oxygenate, dehydrogenate, and covalently alter mixed organic contaminants, weakening solid chemical bonds, improving polarity and water solubility, and turning the pollutants degradable or assimilable to endophytes and plants. The treatment of contaminants with a basic structural system has been the focus of recent endophyte-assisted phytoremediation investigations. To identify efficient approaches to enhance soil mixed pollutants' phytoremediation, more work is needed to investigate the metabolic mechanisms and evaluate deteriorating enzymes. Biodegradation of some organic contaminants via endophytes is presented in Table 11.1.

11.4.1.2 Cometabolization

Plants and other soil microbes may cometabolize complicated contaminants with endophytes. Each counterpart can have specific metabolic characteristics and use their respective degradation mechanisms to disintegrate refractory large molecular organic contaminants into smaller ones. They can also bind one or more of the mixed pollutants independently, resulting in greater phytoremediation efficiency than if plant or soil microbes were used individually. As a result, mixing various endophytes

Table 11.1 Biodegradation of some organic contaminants via endophytes.

Organic contaminant	Endophytic species	Host plant species	References
Toluene	*Burkholderia cepacia*	*Zea mays* and *Triticum aestivum*	Wang et al. (2010)
Diesel	*Pseudomonas* sp. strain ITRI53 and *Rhodococcus* sp. strain ITRH43	*Lolium multiflorum* L.	Andria et al. (2009)
Fenpropathrin	*Klebsiella terrigena* E42; *Pseudomonas* sp. E46	*Spirodela polyrhiza*	Xu et al. (2015)
Chlorobenzoic acids	*Pseudomonas aeruginosa* R75; *Pseudomonas savastanoi* CB35	*Elymus dauricus*	Siciliano et al. (1998), Gupta et al. (2020)
2,4-Dichlorophenoxyacetic acid	*Pseudomonas putida* VM1450	*Pisum sativum*	Germaine et al. (2006)
Trichloroethylene	*Pseudomonas putida* W619-TCE	*Populus* sp.	Weyens et al. (2010)
Alkanes	*Pseudomonas* sp.	*Lolium multiflorum* L.	
Polychlorinated biphenyls	*Pseudomonas putida* Flav1−1 and *Pseudomonas putida* PML2	*Arabidopsis thaliana*	Narasimhan et al. (2003)
Pyrene	*Staphylococcus* sp. BJ106	*Alopecurus aequalis*	Sun et al. (2014)
Hydrocarbons	*Bacillus* sp. and *Pseudomonas* sp.	*Azadirachta indica*	Singh and Padmavathy (2015)
Diesel	*Enterobacter ludwigii*	*Lolium multiflorum* L., *Lotus corniculatus* and *Medicago sativa*	Yousaf et al. (2011)

with their host plants increases the likelihood of mixed pollutant cometabolism. For instance, soil polluted by crude oil was treated using phytoremediation, which involved inoculating two different grass species with two oil-degrading endophytic bacterial strains. Both strains had the alkane hydroxylase gene (alkB) and were capable of degrading both long-chain alkanes and the aromatic hydrocarbons. Plant roots were also essential in absorbing contaminants and degrading them with native cytochrome P450. Ultimately, each member of the symbiotic system partnered to boost phytoremediation of crude oil-polluted soil. Likewise, *Armoracia rusticana* (horseradish) was inoculated with endophytes, and various mechanisms were triggered. The acridine and glutathione conjugation mechanisms collaborated to enhance carbamazepine elimination. Such findings are actually the only ones attempting to explain the cometabolism of plants and endophytes (Sauvetre et al., 2018). The degree to which the members of this system perform a significant part, who take the control and also how environmental factors influence this cometabolism are all unclear.

11.4.1.3 Metal extraction

Most endophytes, as well as their host plants, can generate extracellular organic matter during carbon/nitrogen metabolism. Tiny molecular organic acids, siderophores, and biosurfactants are the examples of extracellular chelating compounds, which are usually known as acidic long-chain carbohydrates. For instance, the development of acidic organic matter can assist soil demineralization and nickel solubilization (Ma, Rajkumar, & Luo, 2011). Endophytic fungi's hypha can carry metallic elements to plant roots over long distances owing to their filamentous structures. The availability of chelating compounds is likely to improve, not only pollutant deposition, but also the inactivation, and significant efflux of metal pollutants through the formation of insoluble phosphate composites (He, Megharaj, Wu, Subashchandrabose, & Dai, 2020).

11.4.1.4 Bioaccumulation

Endophytes, in association with root hairs, have the ability to absorb soil contaminants. Porins, transporter proteins, heavy metal ATPases, binding proteins, and metallothioneins are some membrane channel proteins. Plants are shielded from phytotoxicity due to excessive contaminant deposition by these microbial protein receptors, which struggle with plants for pollutant uptake (He et al., 2020).

11.4.2 Indirect ways to promote phytoremediation via endophytes

11.4.2.1 Plant nutrient supply

To deal with contaminants and reduce phytotoxicity, plants need more nutrition and energy. Owing to the degradation of the microecology and nutritional deficiencies, it is often challenging for plants to fulfill their requirements for their growth in polluted soils in realistic field implementations As a consequence, plant growth is inhibited, and soil phytoremediation is reduced. Endophytic microbes can help plants get more nutrients by developing metabolic pathways with diverse roles. For instance, SBER3 is a *Bacillus* sp. strain isolated from *Populus deltoides* that has been shown to mineralize PAHs into various molecular masses (Bisht, Pandey, & Kaur, 2014). Also, endophytes have been shown to help plants absorb organic acid and minerals from the soil.

11.4.2.2 Plant growth regulation

Endophytes contain phytohormones such as ethylene, acetoin, 2,3-butanediol, cytokinins, gibberellins, and indole acetic acid (IAA), which have been shown to delay the aging of plant cells and maintain endophytic symbioses in plants. Also, they can direct their host plants to produce phytohormones in response to pollutant stress. Plant physiology and metabolism changes may lead to mixed pollutants' phytoextraction or biodegradation in an indirect way. Many endophytes, for instance, have been shown to produce IAA, which regulates multiple cellular developing systems and enhances plant growth even when exposed to salt. IAA-producing endophytes can also promote plant pollutant absorption by increasing root length, root hair number, root exudates, and nitrogen fixation. Likewise, they may also influence the synthesis of ethylene, acetoin, cytokinins, gibberellins, and 2,3-butanediol, all of which are essential for plant growth. As a result, the contaminated plants can extract mixed pollutants more effectively from the soil.

11.4.2.3 Stress alleviation

Some of the advantageous characteristics of endophytes have been linked to antioxidant development and the activation of the plant antioxidant mechanism through endophytes. Plant antioxidant enzymes such as peroxidase, superoxide dismutase, ascorbate peroxidase, and glutathione peroxidise, and lipid peroxidation are activated by endophytes, which help plants withstand oxidants triggered by pollutants (He et al., 2020). Most endophytes that produce 1-aminocyclopropane-1-carboxylic acid (ACC) deaminase help to relieve mixed pollutant-related stress in an indirect way. It has been documented that inoculating plants with endophytes activates systematic resistance to abiotic stresses and stimulates the development of nonenzymatic antioxidants, including proline, glutathione, ascorbate, and tocopherol (Khan, Hamayun, & Kang, 2012). Plants deliver improved growth capacity, chlorophyll biosynthesis, and photosynthetic performance, most of which facilitate phytoremediation of pollutants.

11.4.2.4 Biocommunication and metabolite sharing

An ecological network connects endophytes, plants, and highly diverse soil microorganisms that could be influenced by soil pollutants and vice versa. Endophytes can play supporting roles in coordinating with plants or other native microbes, helping the community to effectively manage with various contaminants, even if they are involved indirectly in the contaminant deterioration pathway. Studies suggest that endophytes communicate with host plants and their respective microbial communities through quorum-sensing metabolites that amplify microbiota, synergistic or antagonistic interactions, and gene expressions. Even though the role and molecular configuration of most endophytic metabolites are unknown, their pathogenic sensitivity or biological control abilities should not be overlooked while using them for phytoremediation of pollutants. For instance, endophytes and their metabolites can be assessed via advanced analytical techniques and cutting-edge technology. Stable isotope probing coupled with nanoscale secondary ion mass spectrometry techniques (NanoSIMS) paired with advanced microscopy may be very important for understanding in situ plant—endophyte—contaminant associations, particularly when the symbiont lives in a complicated environment contaminated with a variety of chemical-based substances.

11.5 Functions of endophytes in pollutant bioremediation

The microbes living in the soil and roots of plants play a significant role in the accumulation of metals from polluted environments. These microbes help to break down intricate natural and synthetic substances into simple ones, activate metal ions, and improve plant bioavailability. Bacteria, such as rhizobacteria, encourage plant growth and biomass generation as well as increase plants' ability to absorb hazardous contaminants and mitigate metal toxicity. Endophytes thrive in the plant's tissues, enhancing the plant's capacity to deal with various biotic and abiotic pressures. In general, endophytic bacteria protect plants from biotic stress, relieve abiotic stress, and sustain them by supplying nutrients, including nitrogen, phosphorus, and other vital elements. Plants that have been pre-inoculated with endophytes are less susceptible to bacterial, fungal, and viral pathogens, as well as insect and nematode destruction. Endophytes can be obligate or facultative endophytes, depending on their interaction with the host plant. In stressful conditions, endophytic microbes can assist the plant in reducing stress through a combination of pathways. Siderophore formation, antimicrobial metabolites, phosphate-solubilizing agents, nitrogen fixation capacities, and phytohormones are the examples of direct pathways. Bioremediation and biocontrol are two indirect processes. This mechanism works against various pathogenic bacteria by prohibiting the infected bacteria from causing noticeable symptoms of the disease in the host plant. They have been shown to promote the xenobiotic deterioration and their auxiliary metabolites by expressing catabolic genes. Besides, endophytic microbes may help to alleviate metal stress by enhancing photosynthesis and antioxidative enzymatic activities and altering heavy metal ion efflux and accumulation. For instance, maize inoculated with *Gaeumannomyces cylindrosporus* produced considerably higher yields under lead stress. Inoculation with *Pseudomonas aeruginosa* improves crop cadmium tolerance (Cd) and its deposition and efflux in inoculated plants. The increased levels of hazardous substances can induce toxic effects in macrophytes, reducing their ability to remediate pollutants. PGP features and deterioration genes are found in endophytic microbes, which help the plant to cope with a variety of environmental strains. The symbiotic relationship between plants and epiphytic microbes aids in the remediation of blended pollutants by deterioration and increases metal efflux. The key benefit of employing endophytic microbes to enhance remediation of xenobiotics is that it is simpler for genetically modified microbes compared to plants to achieve optimal deterioration of contaminants.

11.6 Role of endophytes in plant growth promotion

Endophytes have been documented to provide a variety of benefits to their host plant, including preventing herbivores by producing alkaloids toxic to grasses, allowing dicots to grow in hot springs, and protecting dicots from pests. Endophytes are said to have a lot in common with pathogens that live within the host plant. Recent findings suggest that pathogen–endophyte interactions take place in various aspects in different plant hosts and that disrupted plant physiology can suppress pathogen development, alter nutrient distribution in favor of endophytes, or activate the defense system of plants (Busby, Ridout, & Newcombe, 2016). The presence of endophytic fungi on the plant provides enhanced protection against nematodes that invade plants. This is a stressful condition and the antagonistic pathway used by endophytes is still unclear. Therefore endophytic microbes play a critical role in the pathosystem's productivity, as well as the diversity, survival, and conservation of plants (Fadiji & Babalola, 2020). They have a wide range of effects on plants, and although their exact roles are unknown, the existence of endophytic microbes in the tissues of host plants is highly beneficial. Furthermore, they can also promote plant growth directly or indirectly through various pathways and their metabolites contribute different health to host plants by initiating plant survival in abiotic and biotic conditions, as well as plant growth promotion.

11.6.1 Biofertilization

Biofertilizers are an environment-friendly, cost-effective, and sustainable source of nutrients for plants that contribute to minimize our reliance on synthetic fertilizers while also improving nutrient availability and thus boosting plant growth. Endophytes have been used to stimulate crop growth and yield in sustainable agriculture. Endophytes have long been known to colonize host tissue and develop a strong symbiotic relationship with their plant hosts. Endophytes produce a variety of phytohormones, including cytokinins, auxins, and gibberellic acids. They have become more important in plant growth promotion owing to their potential to improve soil nitrogen levels. Endophytic bacteria such as *Rhizobium* spp. and nonnodulating strains such as *Brevibacillus choshinensis*, *Microbacterium trichothecenolyticum*, *Micromonospora* spp., and *Endobacter medicaginis* have been found in plant root nodules. Endophyte-produced phytohormones can increase plant growth and reduce reliance on synthetic fertilizers.

11.6.2 Potential source of bioactive constituents

Endophytes may produce bioactive constituents that help plants to defend themselves against various pathogens, and most of these have been used in drug development. According to a recent analysis, endophytes are the source of several organic products, including terpenoids, flavonoids, alkaloids, and steroids. Antibiotics, anticancer, antidiabetic, immune-suppressants, antiviral, and biocontrol agents are only a few of the properties ascribed to endophytes' bioactive compounds. Maytansinoids, such as geldanamycin and rifamycin, belong to the ansamycin family of polyketide macrolactams and are formed by three closely related plant families namely, *Rhamnaceae*, *Celastraceae*, and *Euphorbiaceae*, as well as bacteria isolates, including *Actinosynnema pretiosum* and mosses. It's been suggested that rhizospheric microorganisms play a role in the biosynthesis of maytansinoids in plants. In addition, *Taxus chinensis* developed an endophyte called *Metarhizium anisopliae*, which was discovered to be the source of taxol. Furthermore, the leaves of the medicinal plant *Justicia gendarussa* contain an endophyte called *Colletotrichum gloeosporioides*, which is known to produce taxol.

11.6.3 Biocontrol activities

Endophytes are often recognized as having certain biocontrol properties, making them a viable alternative to synthetic chemicals. They are useful not just for managing conifers, as well as for insect herbivory. Insect pathogens such as borer insects, which primarily target sorghum and coffee seedlings, have been found to be managed by a fungal endophyte known as *Beauveria bassiana*. *Botrytis cinerea* is a fungus that induces rot in tomato, shortening their life span and lowering their postharvest quality. In vitro experiments on *B. cinerea* revealed that endophytic bacteria known as *Bacillus subtilis*, which was found residing in the *Speranskia tuberculata* tissues, had a strong inhibitory action. Besides, *Enterobacter cloacae*, which contains the PtA gene, was reported to be an effective bioinsecticide against the white-backed planthopper, *Sogatella furcifera*. However, the use of recombinant endophytic microbes as biological control agents is critical because they manufacture antipest protein-like substances using a new method for controlling plant pests and can colonize crop plants effectively. Thus endophytes' biocontrol ability improves crop disease tolerance and decreases reliance on pesticides.

11.6.4 Nutrient cycling

Endophytes play a significant role in the decomposition of host plant debris. In the biodegradation process, endophytic microbes colonize the plant first and then cause saprophytic organisms to act on it via an antagonistic reaction, resulting in increment in litter degradation. Almost all endophytic microbes have the ability to decompose organic matter, including hemicellulose, lignin, and cellulose, which are all useful in disintegrating various types of organic matter.

11.6.5 Biodegradation and bioremediation

The majority of endophytes can decompose composite organic substances. Bioremediation is the method of using microbes to remove contaminants from the environment (Zhao & Poh, 2008; Bhat et al., 2021; Juwarkar et al., 2010). In the degradation of waste materials, it is a biological process that heavily relies on microbes. This is possible attributed to the abundance of microbes found in nature. The efficacy of certain endophytic fungi to degrade the plastic polymer polyester polyurethane (PUR) has been investigated. Several microbes were found to be capable of degrading PUR in both liquid and solid media, but the *Pestalotiopsis* genus revealed the best results. When subjected to anaerobic and aerobic environments, two isolates of *Pestalotiopsis* microspora efficiently employed PUR as their sole carbon source. Once molecular analysis was performed, an enzyme called serine hydrolase was expected to be responsible for this characteristic; this enzyme may help the plant's strain resistance potentials. Most heavy metals can be extracted with the help of endophytic bacteria. Several investigations have been performed on how endophytic bacteria can eliminate heavy metals, suggesting that endophytes can help plants improve their strain resistance potential. These endophyte-mediated biodegradation and bioremediation may help plants survive in harsh environments.

11.7 Conclusion and future perspective

Bioremediation is a promising pollutant abatement technique for coping with the environment's ever escalating pollutant load. Endophytes have gained a great deal of interest in bioremediation field over the past few years owing to their genetic variability, PGP behaviors, and strong tolerance to both the abiotic and biotic pressures. Further, the mutual association of a plant with endophytes may unlock novel approaches for their growth and development of plants responsible for bioenergy production, allowing for remediation of a harmful and hazardous environmental pollutant. Plants have been recorded to have a varied endophytic microbiota that may be exceptionally capable of improving plant species' contaminant removal performance. Despite the fact that several studies have recorded failure in natural field conditions, possibly because of considerable alteration in soil properties as well as microbiota from one location to another, further exploration in this field would be beneficial in addressing the limitations. Several obstacles of phytoremediation, on the other hand, may be addressed via creating genetically modified plants and endophytic microbes with the potential to overexpress pollutant-degrading enzymes.

References

Andria, V., Reichenauer, T. G., & Sessitsch, A. (2009). Expression of alkane monooxygenase (alkB) genes by plant-associated bacteria in the rhizosphere and endosphere of Italian ryegrass (Lolium multiflorum L.) grown in diesel contaminated soil. *Environmental Pollution*, *157*(12), 3347–3350. Available from https://doi.org/10.1016/j.envpol.2009.08.023.

Backer, R., Rokem, J. S., Ilangumaran, G., Lamont, J., Praslickova, D., Ricci, E., ... Smith, D. L. (2018). Plant growth-promoting rhizobacteria: Context, mechanisms of action, and roadmap to commercialization of biostimulants for sustainable agriculture. *Frontiers in Plant Science*, *9*, 1473. Available from https://doi.org/10.3389/fpls.2018.01473.

Bais, H. P., Weir, T. L., Perry, L. G., Gilroy, S., & Vivanco, J. M. (2006). The role of root exudates in rhizosphere interactions with plants and other organisms. *Annual Review of Plant Biology*, *57*, 233–266.

Bhardwaj, D., Ansari, M. W., Sahoo, R. K., et al. (2014). Biofertilizers function as key player in sustainable agriculture by improving soil fertility, plant tolerance and crop productivity. *Microbial Cell Factories*, *13*, 66. Available from https://doi.org/10.1186/1475-2859-13-66.

Bhat, R. A., Singh, D. V., Qadri, H., Dar, G. H., Dervash, M. A., Bhat, S. A., ... Yousaf, B. (2021). Vulnerability of municipal solid waste: An emerging threat to aquatic ecosystems. *Chemosphere*, *287*(2022), 132223. Available from https://doi.org/10.1016/j.chemosphere.2021.132223.

Bisht, S., Pandey, P., Kaur, G., et al. (2014). Utilization of endophytic strain Bacillus sp SBER3 for biodegradation of critical reviews in biotechnology 13 polyaromatic hydrocarbons (PAH) in soil model system. *European Journal of Soil Biology*, *60*, 67–76.

Busby, P. E., Ridout, M., & Newcombe, G. (2016). Fungal endophytes: Modifiers of plant disease. *Plant Molecular Biology*, *90*(6), 645–655.

Cheng, S. (2003). Heavy metals in plants and phytoremediation. *Environmental Science and Pollution Research*, *10*, 335–340.

Compant, S., Saikkonen, K., Mitter, B., Campisano, A., & Mercado-Blanco, J. (2016). Editorial special issue: soil, plants and endophytes. *Plant and Soil*, *405*(1), 1–11.

Dar, G. H., Bandh, S. A., & Kamili, A. N. (2011). Bacteriological analysis of soil from yusmarg health resort of Kashmir Valley. *International Journal of Current Research*, *3*(10), 53–56, 2011.

Dar, G. H., Bandh, S. A., Kamili, A. N., Nazir, R., & Bhat, R. A. (2013). Comparative analysis of different types of bacterial colonies from the soils of Yusmarg Forest, Kashmir valley India. *Ecologia Balkanica*, *5*(1), 31–35.

Dar, G. H., Kamili, A. N., Nazir, R., Bandh, S. A., & Bhat, R. A. (2012). A preliminary study of colony forming units of bacteria from the soils of Yusmarg Forest, Kashmir Valley, India. *International Journal of Current Research*, *4*(12), 467–472.

Endeshaw, A., Birhanu, G., Zerihun, T., & Misganaw, W. (2017). Application of microorganisms in bioremediation-review. *Journal of Applied & Environmental Microbiology, 1* (1), 2–9.

Fadiji, E., & Babalola, O. O. (2020). Exploring the potentialities of beneficial endophytes for improved plant growth. *Saudi Journal of Biological Sciences, 27*(12). Available from https://doi.org/10.1016/j.sjbs.2020.08.002.

Ferronato, N., & Torretta, V. (2019). Waste mismanagement in developing countries: A review of global issues. *International Journal of Environmental Research and Public Health, 16*(6), 1060. Available from https://doi.org/10.3390/ijerph16061060.

Germaine, K. J., Liu, X., Cabellos, G. G., Hogan, J. P., Ryan, D., & Dowling, D. N. (2006). Bacterial endophyte-enhanced phytoremediation of the organochlorine herbicide 2, 4-dichlorophenoxyacetic acid. *FEMS microbiology ecology, 57*(2), 302–310. Available from https://doi.org/10.1111/j.1574-6941.2006.00121.x.

Glick, B. R. (2005). Modulation of plant ethylene levels by the bacterial enzyme ACC deaminase. *FEMS Microbiology Letters, 251*(1), 1–7.

Gupta, A., Singh, S. K., Singh, V. K., Singh, M. K., Modi, A., Zhimo, V. Y., & Kumar, A. (2020). *Endophytic microbe approaches in bioremediation of organic pollutants. Microbial Endophytes* (pp. 157–174). Woodhead Publishing.

Haldar, S., & Sengupta, S. (2016). *Microbial ecology at rhizosphere: Bioengineering and future prospective. Plant-microbe interaction: An approach to sustainable agriculture* (pp. 63–96). Singapore: Springer.

He, W., Megharaj, M., Wu, C. Y., Subashchandrabose, S. R., & Dai, C. C. (2020). Endophyte-assisted phytoremediation: Mechanisms and current application strategies for soil mixed pollutants. *Critical Reviews in Biotechnology, 40*(1), 31–45.

Juwarkar, A. A., Singh, S. K., & Mudhoo, A. (2010). A comprehensive overview of elements in bioremediation. *Reviews in Environmental Science and Bio/Technology, 9*, 215–288. Available from https://doi.org/10.1007/s11157-010-9215-6.

Khan, A. L., Hamayun, M., Kang, S. M., et al. (2012). Endophytic fungal association via gibberellins and indole acetic acid can improve plant growth under abiotic stress: An example of *Paecilomyces formosus* LHL10. *BMC Microbiology, 12*, 3.

Kumar, A., Singh, V. K., Tripathi, V., Singh, P. P., & Singh, A. K. (2018). Plant growth-promoting rhizobacteria (PGPR): Perspective in agriculture under biotic and abiotic stress. *Crop Improvement Through Microbial Biotechnology*, 333–342. Available from https://doi.org/10.1016/B978-0-444-63987-5.00016-5.

Ma, Y., Rajkumar, M., Luo, Y. M., et al. (2011). Inoculation of endophytic bacteria on host and non-host plants – Effects on plant growth and Ni uptake. *Journal of Hazardous Materials, 195*, 230–237.

Meena, K. K., Sorty, A. M., Bitla, U. M., Choudhary, K., Gupta, P., Pareek, A., … Minhas, P. S. (2017). Abiotic stress responses and microbe-mediated mitigation in plants: The omics strategies. *Frontiers in Plant Science, 8*, 172. Available from https://doi.org/10.3389/fpls.2017.00172.

Narasimhan, K., Basheer, C., Bajic, V. B., & Swarup, S. (2003). Enhancement of plant-microbe interactions using a rhizosphere metabolomics-driven approach and its application in the removal of polychlorinated biphenyls. *Plant Physiology, 132*(1), 146–153. Available from https://doi.org/10.1104/pp.102.016295.

Olanrewaju, O. S., Glick, B. R., & Babalola, O. O. (2017). Mechanisms of action of plant growth promoting bacteria. *World Journal of Microbiology & Biotechnology, 33*(11), 197. Available from https://doi.org/10.1007/s11274-017-2364-9.

Rana, K. L., Kour, D., Kaur, T., et al. (2020). Endophytic microbes: Biodiversity, plant growth-promoting mechanisms and potential applications for agricultural sustainability. *Antonie Van Leeuwenhoek, 113*, 1075–1107. Available from https://doi.org/10.1007/s10482-020-01429-y.

Sauvêtre, A., May, R., Harpaintner, R., Poschenrieder, C., & Schröder, P. (2018). Metabolism of carbamazepine in plant roots and endophytic rhizobacteria isolated from Phragmites australis. *Journal of hazardous materials*, *342*, 85–95.

Schwitzguebel, J. P. (2017). Phytoremediation of soils contaminated by organic compounds: Hype, hope and facts. *Journal of Soils and Sediments*, *17*, 1492–1502.

Shilev, S. (2020). Plant-growth-promoting bacteria mitigating soil salinity stress in plants. *Applied Sciences*, *10*(20), 7326.

Siciliano, S. D., Goldie, H., & Germida, J. J. (1998). Enzymatic activity in root exudates of Dahurian wild rye (Elymus dauricus) that degrades 2-chlorobenzoic acid. Journal of agricultural and food chemistry, 46(1), 5-7. Available from https://doi.org/10.1021/jf9708195.

Singh, M. J., & Padmavathy, S. (2015). *Hydrocarbon Biodegradation by endophytic bacteria from neem leaves*. Available from https://doi.org/10.5958/2319-1198.2015.00004.4.

Sun, K., Liu, J., Jin, L., & Gao, Y. (2014). Utilizing pyrene-degrading endophytic bacteria to reduce the risk of plant pyrene contamination. *Plant and Soil*, *374*(1), 251–262.

Wang, Y., Li, H., Zhao, W., He, X., Chen, J., Geng, X., & Xiao, M. (2010). Induction of toluene degradation and growth promotion in corn and wheat by horizontal gene transfer within endophytic bacteria. *Soil Biology and Biochemistry*, *42*(7), 1051–1057. Available from https://doi.org/10.1016/j.soilbio.2010.03.002.

Weyens, N., Truyens, S., Dupae, J., Newman, L., Taghavi, S., van der Lelie, D., & Vangronsveld, J. (2010). Potential of the TCE-degrading endophyte Pseudomonas putida W619-TCE to improve plant growth and reduce TCE phytotoxicity and evapotranspiration in poplar cuttings. *Environmental Pollution*, *158*(9), 2915–2919. Available from https://doi.org/10.1016/j.envpol.2010.06.004.

Xu, X. J., Sun, J. Q., Nie, Y., & Wu, X. L. (2015). Spirodela polyrhiza stimulates the growth of its endophytes but differentially increases their fenpropathrin-degradation capabilities. *Chemosphere*, *125*, 33–40. Available from https://doi.org/10.1016/j.chemosphere.2014.12.084.

Yousaf, S., Afzal, M., Reichenauer, T. G., Brady, C. L., & Sessitsch, A. (2011). Hydrocarbon degradation, plant colonization and gene expression of alkane degradation genes by endophytic Enterobacter ludwigii strains. *Environmental pollution*, *159*(10), 2675–2683. Available from https://doi.org/10.1016/j.envpol.2011.05.031.

Zhao, B., & Poh, C. L. (2008). Insights into environmental bioremediation by microorganisms through functional genomics and proteomics. *Proteomics*, *8*(4), 874–881.

CHAPTER 12

Fungi, eukaryotic microorganisms involved in bioremediation of contaminated environments

Luis Fernando García-Ortega[1,*], Claudia Geraldine León-Ramírez[1,*], Yesenia Ithaí Ángeles-López[2,*], Juan Antonio Cervantes-Montelongo[3], John Martin Velez-Haro[3], Rosario Razo-Belman[1], Lucila Ortiz-Castellanos[1], Candy Carranza-Álvarez[4] and Domingo Martínez-Soto[5,6]

[1]Departamento de Ingeniería Genética, Unidad Irapuato, Centro de Investigación y de Estudios Avanzados del Instituto Politécnico Nacional, Irapuato Guanajuato, México
[2]Department of Entomology and Plant Pathology, North Carolina State University, Raleigh North Carolina, United States
[3]Departamento de Ingeniería Bioquímica, Tecnológico Nacional de México, Instituto Tecnológico de Celaya, Celaya Guanajuato, México
[4]Facultad de Estudios Profesionales Zona Huasteca, Universidad Autónoma de San Luis Potosí, Ciudad Valles San Luis Potosí, México
[5]Department of Biochemistry and Molecular Biology, University of Massachusetts Amherst, Amherst Massachusetts, United States
[6]Ingeniería en Innovación Agrícola Sustentable, Instituto Tecnológico Superior de Los Reyes, Tecnológico Nacional de México, Los Reyes Michocán, México

12.1 Introduction

Environmental contamination is one of the most important problems that affect the world. The development of industrial, agricultural, and clinical processes often leads to contamination, owing to lack of planning and analysis of the environmental impacts that these cause (Domínguez-Gual, 2015). Currently, it is widely recognized that contaminated land is a potential risk to human health. Therefore, in recent years, international efforts have increased to recover these places and reenable them for their use (Vidali, 2001). Environmental biotechnology plays an important role in counteracting the multiple anthropogenic effects, using a wide range of technologies such as biopesticides, renewable energy, bioremediation, and biofertilizers (Cota-Ruíz, Núñez-Gastelúm, Delgado-Ríos, & Martínez-Martínez, 2019). Bioremediation emerges as a branch of environmental biotechnology that

* These authors contributed equally to this work.

Microbial Consortium and Biotransformation for Pollution Decontamination
DOI: https://doi.org/10.1016/B978-0-323-91893-0.00016-X

searches to solve pollution problems through biological systems that use organisms such as bacteria, fungi, plants, or a combination of them. These organisms are capable of degrading or transforming compounds that cause an imbalance in the environment, for example, sediments, mud, petroleum, etc. (Cavero-Zegarra & Pari-Quispe, 2018; Cota-Ruíz et al., 2019; Dar, Bandh, & Kamili, 2011; Dar et al., 2012; Dar, Bandh, Kamili, Nazir, & Bhat, 2013). In relation to other technologies, bioremediation usually has lower costs and causes less intrusion into the contaminated site, thus consequently less significant environmental damage in the process of destructing the pollutants (Garzón, Rodríguez-Miranda, & Hernández-Gómez, 2017). Therefore understanding the mechanisms used by fungi to degrade organic compounds is important owing to multiple applications in bioremediation such as decontamination and detoxification of contaminated environments, as well as human health-related fields (Ostrem-Loss & Jae-Hyuk, 2018).

12.2 Environmental contamination

Environmental contamination is defined as the presence of any agent, physical, chemical, biological, or combination of several of them in places, forms, and concentrations that are harmful to the health, safety, or welfare of any living organism.

The environmental pollution process is developed in three fundamental steps: emission, processes, and effects. First, the contaminant is emitted at a speed or level of emission; then, contaminants undergo the processes of dispersion, transport, and transformation; and finally, they are deposited on the receiving surfaces, resulting in environmental effects. The sources of contamination can be natural or anthropogenic. On the one hand, natural sources are usually related to the composition of soil and water, and they represent a low proportion of the public health problems due to contamination. On the other hand, anthropogenic sources of contamination are more alarming due to the variety and nature of the contaminants they emit, and the public health problems they cause (Albert, 2004).

12.3 Types of environmental contamination

Environmental contamination can be classified in several ways; the main ones are mentioned in this section. The first classification depends on the process that causes the contamination (Fig. 12.1). When contamination is the result of a process in which humans do not participate and they

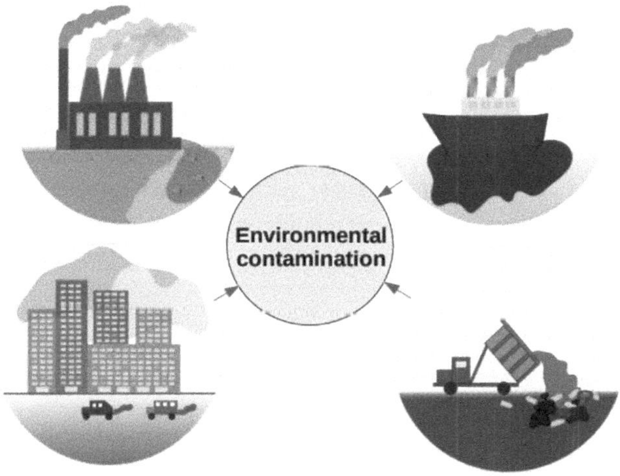

Figure 12.1 Different types of environmental contamination of air, soil, and water.

cannot control it, it is of natural origin. However, if the contaminant (regardless of its nature) is found in an environment to which it does not belong or in concentrations higher than natural due to human activity, the contamination is of anthropogenic origin (Albert, 2004). Furthermore, contaminants can be classified as biological, physical, or chemical. Biologic contaminants can only be of natural origin, while physical and chemical contaminants can be both natural and anthropogenic. Naturally derived chemicals are integrated into local biogeochemical cycles, dispersed, or degraded and do not cause long-term problems. However, the chemicals of anthropogenic origin are considered harmful in any concentration and medium present, because they should not naturally be present in any substrate (Albert, 2004). Finally, pollution can be classified according to the place where the contaminant is deposited, so that regardless of its origin, it is considered that there are three types of contamination: air, water, and land pollution.

12.3.1 Air contamination

Air contamination involves any natural or artificial chemical element capable of remaining or being dragged through the air, which may affect the normal development of plants and animals as well as involves a risk, damage, or nuisance to human health (Encinas-Malagón, 2011). The atmospheric contaminant has three ways of reaching the receiving surfaces and depends on the physical state of the contaminant and the mechanism of transportation to the surface. These can be wet, dry, and occult deposition. Wetting involves the incorporation of contaminants into clouds or raindrops and their subsequent precipitation, whether in the form of rain or snow. In dry deposition, contaminants are transported to the receiving surface, either by turbulent transport (wind) or by sedimentation (gravity), and once there, they hit the receiver without first being dissolved. Lastly, the occult deposition consists of the incorporation of contaminants into the mist droplets and their subsequent interception by the receiver (Encinas Malagón, 2011).

12.3.2 Water contamination

Water contamination is the incorporation of foreign matter into the water, such as microorganisms, chemicals, industrial waste, and other types or wastewater. These materials deteriorate the quality of the water and make it useless for the intended uses.

Natural sources are very dispersed and do not cause high concentrations, except in some very specific places. Anthropogenic contamination is concentrated in specific areas (industries, cities, etc.) (Encinas Malagón, 2011). There are four main sources of anthropogenic contamination: industrial, urban dumping, shipping, and agriculture. All these sources of contamination alter some of the physical, chemical, and biological properties of water.

12.3.3 Soil contamination

It is the incorporation into the soil of foreign matter, such as chemicals and industrial waste. Soil contamination produces a physical, chemical, and biological imbalance that negatively affects plants, animals, and humans, causing loss of soil productivity.

In soil, a distinction must be made between endogenous and exogenous contamination. Endogenous contamination is caused by a natural soil constituent, it occurs when an extreme action causes an imbalance in the soil and causes a natural constituent to change its chemical form, buildup, and become toxic. Exogenous occurs when contaminants come from outside (Encinas Malagón, 2011).

The effects of the contaminants produced in the soil depend on the characteristics of the contaminants and the soil. The main factors that influence the response of soil to contaminants are soil variability, contaminant bioavailability, contaminant mobility, contaminant persistence, and critical load (Encinas Malagón, 2011). Also other factors such as temperature, pH, humidity, light, heat, and the nature of the organisms present in the environment and their interactions with the substance can be considered (Albert, 2004).

It is now widely recognized that contaminated land is a potential threat to human health, and its continual discovery over recent years has led to international efforts to recover many of these sites. Either as a response to the risk of adverse health or environmental effects caused by contamination or to enable the site to be redeveloped for use.

Given the problems that arise when a contaminant accumulates in a given environment, it has been discovered that microorganisms play a decisive role in the degradation of contaminants. They can carry out processes of transformation of contaminants molecules in a natural way under favorable conditions. The microorganisms take the contaminant as the main source of energy for their metabolism and produce the transformation or degradation of the contaminant.

Microorganisms can be isolated from almost any environmental conditions, due to their adaptability, that is, some of them can grow at subzero temperatures, as well as extreme heat, desert conditions, in water, with an excess of oxygen, and in anaerobic conditions, with the presence of hazardous compounds or any waste steam. Due to this adaptability, microbes can be used to degrade or remediate contaminated and hazardous environments (Bhat et al., 2021; Vidali, 2001).

Thus the term "bioremediation" arises, being a promising technology that uses biological systems, strictly microorganisms or their products, to reduce, eliminate, or transform toxic compounds present in soils, sediments, water, and/or air, into "safe" substances or less contaminated (Cota-Ruíz et al., 2019).

12.4 Bioremediation

The term "bioremediation," coined in the early 1980s, comes from the concept of remediation that refers to the application of physical–chemical strategies to avoid damage and contamination of the soil. The American Academy of Microbiology defines bioremediation as "the use of living organisms to reduce or eliminate environmental risks resulting from the accumulation of toxic chemicals and other hazardous wastes" (Gibson & Sayler, 1992). These microorganisms use their enzymatic potential to degrade contaminant compounds to intermediate products, in an aerobic or anaerobic environment (González-Rojas, 2011). Different techniques are employed depending on certain characteristics, for example, the degree of saturation and aeration of the area. Most bioremediation systems are applied in aerobic conditions; however, when systems are applied in anaerobic conditions, these can allow microbial organisms to degrade otherwise recalcitrant molecules (Shukla, Singh, & Sharma, 2010).

Bioremediation techniques can be classified according to the treatment and phase used in situ and ex situ (Cavero-Zegarra & Pari-Quispe, 2018).

12.4.1 In situ bioremediation

These techniques are generally the most desirable options due to lower cost and fewer disturbances since they provide on-site treatment while avoiding excavation and transport of contaminants.

In situ bioremediation is divided into five methods: natural attenuation (intrinsic process, acts without human intervention), bioventing (supplying air and nutrients to an area contaminated by pipes), biostimulation (increases the activity of endogenous bacteria from contaminated soils by providing them with nutrients, oxygen, etc.), bioaugmentation (added microorganisms to the contaminated environment), and biosparging (combines the effect of ventilation with the use of indigenous microorganisms to degrade organic compounds) (Cavero-Zegarra& Pari-Quispe, 2018; Ocampo-Hernández, 2021).

12.4.1.1 Natural attenuation

Natural attenuation refers to the intrinsic process of bioremediation, depending on physical, chemical, and biological factors, which, under favorable conditions, act without human intervention, reducing the mass, volume, toxicity, or concentration of contaminants (Cota-Ruíz et al., 2019; González-Rojas, 2011).

12.4.1.2 Bioventing

Bioventing is a treatment that consists in supplying, by injection, air and nutrients to the contaminated area to stimulate the aerobic metabolism of microorganisms. Due to the aeration of the soil, the degradation of hydrocarbons is promoted for two reasons: first, the volatilization facilitates the migration of the volatile phase of contaminants and, second, the biodegradation increases soil oxygenation stimulating the microbial activity (Cota-Ruíz et al., 2019; González-Rojas, 2011; Vidali, 2001).

12.4.1.3 Biostimulation

Biostimulation consists in supplementing the environment in which microorganisms are found with nutrients or appropriate conditions, for example, the adjustments of nitrates, phosphates, electron acceptors (oxygen), or pH (González-Rojas, 2011). However, this method has some limitations because sometimes the soils do not have the appropriate microorganisms to degrade the contaminants, or the concentrations of them are not the required (Cota-Ruíz et al., 2019). In this case the growth and activity of microorganisms must be stimulated (Vidali, 2001).

12.4.1.4 Bioaugmentation

Bioaugmentation is the process by which microorganisms are added to the contaminated environment to optimize bioremediation (González-Rojas, 2011). This method can be carried out through the enrichment or isolation of microorganisms from the target site and their subsequent culture and reinoculation and also can be carried out through the exclusion of microorganisms and the introduction of genetically modified microorganisms (Cota-Ruíz et al., 2019).

12.4.1.5 Biosparging

This involves the injection of air and nutrients pressurized in the saturated zone to improve the activity of the microorganisms. Air vaporizes contaminants dissolved in the water or trapped by capillarity in the pores. The vaporized pollutants are carried by the air bubbles to the unsaturated zone, where they are biodegraded and/or recovered with a vapor extraction system (Cavero-Zegarra & Pari-Quispe, 2018; Vidali, 2001). Biosparging increases the mixing in the saturated zone and thereby increases the contact between soil and groundwater. This technique is used for cleaning organic compounds in soil and groundwater.

12.4.2 Ex situ bioremediation

Ex situ bioremediation is when the contaminant is removed and broken down at another site under controlled laboratory conditions. However, this is an expensive process (Cavero-Zegarra & Pari-Quispe, 2018). This method can be divided into five submethods, such as composting (degradation of organic waste through anaerobic microbial metabolism), landfarming (contaminated soil is excavated and spread over a prepared bed and periodically tilled until pollutants are degraded), biopiles (mix contaminated soils with treated soils and then aerate them), bioreactors (treats solid materials or contaminated water), and bioslurry (controlled soil treatment in a bioreactor).

12.4.2.1 Composting

Composting is a primitive form of bioremediation where waste is accumulated in containers allowing its degradation by microorganisms (Cavero-Zegarra & Pari-Quispe, 2018).

This is a process through which organic wastes are degraded by microorganisms, usually at elevated temperatures, in the range of 55°C–65°C. Microorganisms produce heat during the degradation of organic material in the waste, increasing the temperature. The optimal operation of this process required the optimization of the aeration, temperature, humidity, and pH. The product obtained, the "compost," is a good slow-release fertilizer and an exceptional organic soil regenerator (González-Rojas, 2011; Shukla et al., 2010).

12.4.2.2 Landfarming

Landfarming comprises removing the contaminated soil and pouring it onto a previously prepared bed at the bottom of which is a geomembrane (used as an insulator to prevent seepage to the subsoil). Periodically, nutrients are plowed, watered, and added. Once the soil has been emptied into the bed, microbial consortia are stimulated to begin with the degradation of contaminants. This technique is cheap and effective for a wide range of hydrocarbons, typical in refineries or large workspaces in production areas (Cota-Ruíz et al., 2019; González-Rojas, 2011).

12.4.2.3 Biopiles

Biopiles are a hybrid of landfarming and composting. Essentially, engineered cells are constructed as aerated compost piles. They provide a favorable environment for indigenous aerobic and anaerobic microorganisms (Vidali, 2001).

It is a full-scale technology in which excavated soils are mixed with soil amendments, placed on a treatment area, and bioremediated using forced aeration. The contaminants are reduced to carbon dioxide and water. The basic biopile system includes a treatment bed, an aeration system, an irrigation/nutrient system, and a leachate collection system. Moisture, heat, nutrients, oxygen, and pH are controlled to enhance biodegradation (Shukla et al., 2010).

12.4.2.4 Bioreactors

Bioreactors are used to treat solid materials (soil, sediment, sludge) or contaminated water. Contaminated matrixes are transferred to containers with inoculums, nutrients, and controlled parameters for bioremediation. Suspended solid bioreactors are among the most used to bioremediate contaminated soils due to their ability to control operating conditions and to increase microbial activity. Suspended solids are containers equipped to create three-phased conditions (solid, liquid, and gas) and to increase the bioremediation rates of contaminants by increasing the bioavailability of contaminants (Cota-Ruíz et al., 2019).

12.4.2.5 Bioslurry

Slurry reactors or aqueous reactors are used for ex situ treatment of contaminated soil and water pumped up from a contaminated plume. Bioremediation in reactors involves the processing of contaminated solid material (soil, sediment, sludge) or water through a containment system. Within the bioreactor, the presence of nutrients, microorganisms, oxygen, and pH can be controlled to promote biodegradation. Once the process is completed, the resulting sludge is dried, and the treated soils can be returned to their original site (González-Rojas, 2011; Vidali, 2001).

Bioremediation can be integrated with other chain technologies, favoring the treatment of mixed and complex waste and can safely reduce or decompose contaminants resulting from the recovery process.

12.5 Fungi and its significant role in bioremediation

Currently available methods for treating effluent include physicochemical methods such as electroflotation, sedimentation, adsorption, and/or filtration, which produce compounds difficult to treat. Most of these methods used to treat recalcitrant organic contaminants are inefficient, costly, and not eco-friendly. Biosorption and bioaccumulation are two mechanisms in the bioremediation processes used with several fungi and microorganisms. Fungi have an exceptional capacity in the bioremediation of metal; their cell wall has a rapid biosorption capacity due to the presence of various functional groups. They show great adaptability in their metabolic pathways and some enhanced mechanisms to bioaccumulate toxins (Deshmukh, Khardenavis, & Purohit, 2016; Zapana-Huarache, Romero-Sánchez, Dueñas Gonza, Torres Huaco, & Lazarte Rivera, 2020).

Fungi play key roles in contaminant transformation in wastewater treatment plants and, besides having heavy metal tolerance, they have a natural biocontrol function, for example, the stimulation plant defense systems against contaminants by trace elements, pesticides, among others (Javaid, Ashiq, & Tahir, 2016; Vilela, 2021; Zhang, Fan, & Meng, 2020). In addition, they possess a versatile metabolomic mechanism that can secrete a wide range of extracellular enzymes. Several fungi species have shown their ability to absorb and degrade a variety of environmental contaminants, such as insecticides, dyes, some hazardous chemicals, and heavy metals (Zhang et al., 2020), carrying out the transformation of contaminants, representing an effective tool for a sustainable bioremediation of pesticides in water and soil. Several studies revealed an untapped potential of these microorganisms, observing biotechnological developments regarding the application of fungi to improve environmental quality and also human health in recent years (Spina et al., 2018).

Microorganisms, such as fungi, have the ability to turn both chemically and physically substances, conducting to structural changes or to complete degradation of the molecule, introducing minor structural changes in the molecule, and becoming less toxic molecules that can be released into the soil for further degradation by other organisms (Hai et al., 2012). Russo, Ceci, Maggi et al. (2019) observed the potential of fungi in their fundamental role in the soil, and where they were isolated from in situ pilot reactors installed to a contaminated site in Czech Republic with a high concentration of hexachlorocyclohexane (HCH), highlighting two main species, *Penicillium simplicissimum* and *Trichoderma harzianum*. Their research suggests that the isolated fungal species may provide opportunities for new eco-friendly and integrated and cost-effective solutions for environmental management and remediation.

Rhodes (2014) analyzed white-rot fungi, for cleaning contaminated land noting that they are so effective in degrading a wide range of organic molecules, due to their release of extracellular lignin-modifying enzymes. They also noted that the degradation can be augmented by adding carbon sources such as straw, sawdust, and corn cob at contaminated sites. Russo, Ceci, Pinzari et al. (2019) used saprotrophic fungal strains to decontaminate agricultural soils of dichlorodiphenyltrichloroethane (DDT), an insecticide used for decades in soils for agricultural soils in Poland. Two of these fungi, *Trichoderma hamatum* and *Rhizopus arrhizus*, were tested for tolerance in DDT presence, using two indices, the fungal growth rate and biomass production, showing a clear tolerance to DDT for both. Despite the presence of DDT, both strains used most of the substrates provided in experiment, resulting in a high metabolic versatility by these organisms, plus an ecological functionality in the use of carbon sources. Similarly, Ceci et al. (2019) investigated the roles of saprotrophic fungi in biodegradation and transformation of inorganic and organic pollutants in contaminated sites.

The pollution of heavy metals (e.g., Pb[II], Cu[II], and Cd[II]) is very prevalent in surface water, sediments, and soil. The contamination of Pb is given in the air, and it is mainly caused by industrial activities such as mining, electricity generation, sewage application, fertilization, and petrochemical. Interestingly, the method of biosorption by live or dead culture of yeast and fungi (as *Trichoderma asperellum*) is widely used. And their biomass can be used as an effective biosorbent to effectively remove and detoxify the contaminated soil (Zhu, Song, & Dong, 2018).

Other use of fungal bioremediation is the treatment of diuron-contaminated waters, a pesticide (a phenylurea herbicide) used in modern global agricultural, and the bioremediation of selenium-contaminated industrial and municipal wastewaters (Hu et al., 2020; Sabuda et al., 2020). Fungi possess morphological structures such as mycelia that facilitate deep penetration into soil, besides a better surface area for absorption of pesticides. In addition, extracellular enzymes from fungi, such as laccases, peroxidases, hydrolases, and esterases, exhibit broad substrate specificity by catalyzing oxidation of a variety of phenolic compounds, playing a significant role in the degradation process. The mycoremediation of chlorinated pesticides is an example where fungal ezymes are used for detoxificating and degradating of phenolics and halogenated compounds (Bokade, Purohit, & Bajaj, 2021).

Fungal communities are a group of microorganisms that are actively involved in dibenz (a,h) and anthracene (DBahA) removal, a pollutant in the field removed by bioremediation and known for its potential carcinogenic, tumorigenic, and mutagenic effects. There are few works on the biotic elimination of this compound, despite the higher persistence of DBahA in natural environments. It has been observed that the different factors influencing the fungal degradation of DBahA in sediments and soils could be for the presence of electron acceptors, molecular structure, added nutrients, degrading population, bioavailability, pH, and temperature (Vasudevan, Gayathri, & Krishnan, 2018).

Other bioremediation potential by fungi is their utilization to degrade hydrocarbon as crude oil isolated from selected marine substrates, such as mangrove sediments, sea sediments, and tarballs. It has been observed that a group of fungal isolates, such as *Aspergillus*, *Fusarium*, *Penicillium*, and *Acremonium*, found to degrade mainly the long-chain n-alkanes as opposed to short-chain n-alkanes from the crude oil (Barnes, Khodse, Lotlikar, Meena, & Damare, 2018). Even the use of two indigenous fungal isolates for in situ bioremediation of crude oil in the presence of silver nanoparticles showed promising crude oil-degrading abilities (Al-Zaban, Mahmoud, AlHarbi, & Bahatheq, 2020). Another example is the use of basidiomycete fungus *Ganoderma lucidum* to remove toxic metals from an abandoned battery slag dumpsite soil (Ipeaiyeda, Adenipekun, & Oluwole, 2020).

Moreover, the use of fungi with secretion of tyrosinase-like enzyme or polyphenol oxidases (PPOs). These enzymes have potential as bioremediation agents is the degradation of polyphenols, substituted phenols, catechols, and methoxyphenols, which are mostly associated with the postharvest in fruits and vegetables. *Thermothelomyces thermophila* is a cosmopolitan and environmental fungus, found in soil and heat environments. Their genome has genes that encode extracellular enzymes that could oxidize phenolic compounds (Nikolaivits et al., 2018).

The fungi cultures that have been isolated from agricultural fields have a biodegradation efficiency of some herbicide, with acetochlor as active ingredient, suggesting their use and application in agricultural fields. *Tolypocladium geodes* and *Cordyceps cicadae* fungi might be used in acetochlor bioremediation (Erguven, 2018).

Now, recent advances focused on fungal bioremediation of distinct contaminant aromatic amines and their derivatives (de Lima et al., 2018; Silar et al., 2011). The potential application of fungi in biotransformation and bioremediation of wastewaters with tannins (polyphenolic compounds) has also been studied. These are one of the major components of wastewaters and may cause serious environmental pollution (Prigione et al., 2018). It has been observed the bioremediation based in fungal endophytes of grass (*Agrostis stolonifera*) stimulates the growth of *Festuca arundinacea* in lead-contaminated soil (Soldi, Casey, Murphy, & Hodkinson, 2020), and these fungi can grant mercury resistance to plants, such as *Aeschynomene fluminensis* and *Zea mays* (Pietro-Souza et al., 2020), or even find a potential application in the synthetic textile dye decolorization (Agrawal, Upadhyay, Shrivastava, Sharma, & Garlapati, 2021; Nyanhongo et al., 2002).

12.6 Types of fungi involved in bioremediation

Fungi are ubiquitous and have been found to inhabit numerous diverse habitats. For example, they are pathogens of important crops, live inside plant tissues (endophytes), act as decomposers, are important in nutrient cycling within the ecosystem, are associated with roots as arbuscular mycorrhizal fungi (AMF), and its use has focused on agroforestry systems (play vital roles in forestry health), medicine, environmental research, and biotechnology (da Silva, Silva, Thomazini, Schaefer, & Rosa, 2019; Hyde, Jeewon, & Chen, 2020). In addition to the already mentioned and of their robust morphology, together with their numerous metabolic capacities, fungi play crucial roles in bioremediation, due to its the ecological and biochemical capabilities to reduce organic and inorganic chemicals, decreasing the risk associated with contaminants, either by chemical alteration or by manipulating chemical bioavailability in a contaminated environment (Mawad et al., 2020; Tomer et al., 2021).

Although about 100,000 accepted species of fungi and fungus-like taxa are known, which are between 2.6% and 4.5% of the estimated species in the world (Hyde, et al., 2020). Overall, it has been observed that the most abundant phyla in contaminated samples are Ascomycota, Basidiomycota, and Mucoromycota (see Fig. 12.2), with average relative similar (see Ani, Adekunle, Kadiri, & Njoku, 2021; Ceci et al., 2018; Kameshwar & Qin, 2019; Srivastava et al., 2011; Zhang et al., 2020; among others).

Over the past years, a higher number of fungi isolates capable of degrading pesticides and antibiotics or used in detoxification processes of contaminants have been analyzed. Among the highlighted fungi, especially ligninolytic fungi is the most promising group of organisms for performing the treatment in situ without improperly disturbing native ecosystems, being able to transform recalcitrant compounds through a set of extracellular oxidative enzymes (Harms, Wick, & Schlosser, 2017; Spina et al., 2018).

Quintella, Mata, and Lima (2019) documented a review of bioremediation technologies, based on published articles and patent documents, including different types of contaminated matrices, contaminants, and bioremediation agents. The fungi most commonly found are *Chrysosporium* spp., *Phanerochaete* spp., *Aspergillus* spp., *Acremonium* spp. and *Penicillium* spp. The white-rot fungi showed bioremediation potential of 2,3,7,8-tetrachlorodibenzo-p-dioxin, among other contaminants (Dao et al., 2019; Pointing, 2001).

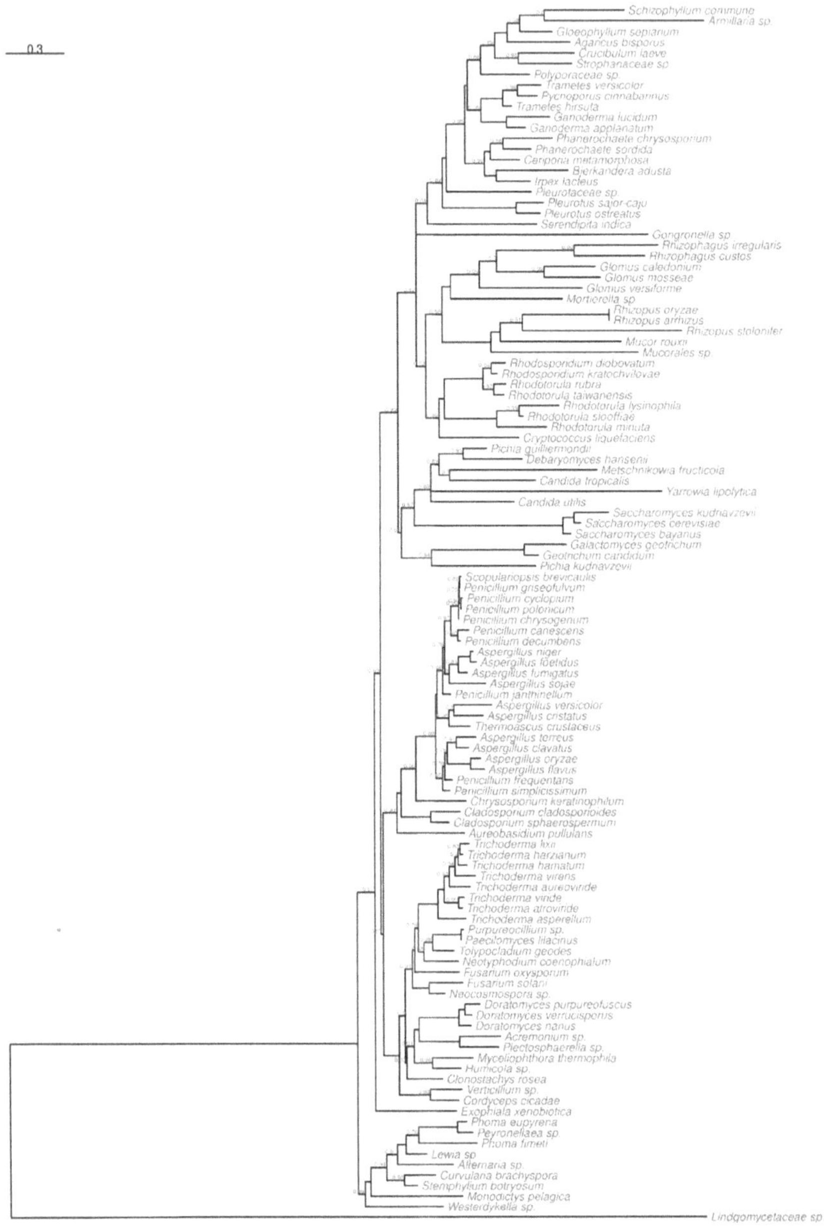

Figure 12.2 Phylogenetic relationships among fungi used in bioremediation. We show the Maximum Likelihood tree for the data set of the internal transcribed spacer (ITS) region (ITS1, ITS2 and 5.8S rDNA) of the nuclear ribosomal DNA (only species with ITS sequence length > 270 bp were used). Numbers in the tree correspond to *P*-values of the approximate likelihood ratios (SH-test; only numbers higher than 0.5 are shown). The scale on the bottom left represents genetic distances in substitutions per nucleotide. The colors of the labels correspond to phylum of the species. *Red*, Basidiomycota; *blue*, Mucoromycota; *green*, Ascomycota.

Every year, millions of tons of organic agrochemicals (OACs) and heavy metals have been added to agricultural fields worldwide through the liberation of manures, biosolids, and wastewater, resulting in contamination and elevated environmental risks. Several studies highlighted the ability of fungi to transform and degrade these recalcitrant pollutants. Ligninolytic fungi are promising organisms due their unique set of extracellular enzymes which degrade lignin; they can transform recalcitrant compounds and remove diverse organic contaminants, such as phenolic and nonphenolic compounds, pesticides, industrial precursors and products, pharmaceuticals, and steroid hormones (Spina et al., 2018).

It has been observed that *Trametes versicolor* fungus degrades the compounds of the nonphenolic type Trichloroethyl chloroformate (TrOC) while the removal of hydrophilic TrOC was negligible, highlighting that the cell culture showed considerably higher degradation of the major compounds, due to the production of intracellular enzymes of mycelium (Nguyen et al., 2014).

Wu et al. (2015) provided some insight into how to choose fungi in OAC degradation. They investigated the genome of more than 40 edible and medicinal species of fungi to evaluate the variability among different nutritional modes of saprotrophic fungi, brown-rot fungi, white-rot fungi, symbiotic fungi, and soft rot fungi; a relation between nutrition mode and the enzymes for lignocellulose degradation was observed. Furthermore, they found that symbiotic fungi have a smaller number of enzymes than saprotrophic fungi, and soft rot fungi and white-rot fungi have a greater number than brown-rot fungi.

The fungal model, *Saccharomyces cerevisiae*, has been used to clean water and treats soil containing heavy metals. It was found that its production of sulfur compounds comes from the expression of glutathione (GSH) pathways under As (III) exposure (Ledezma-Villanueva, Adame-Rodríguez, & Aréchiga-Carvajal, 2018). Two versions of a gene (PtMT2b) coding for a metallothionein to bioremediate Cd were used to transform the wild-type *S. cerevisiae* (WT) (De Oliveira, Ullah, Dunwell, & Tibbett, 2020).

AMF promote a better soil aggregation (soil structure) and stimulate plant defense responses against abiotic stresses (e.g., heavy metal contamination), inducing natural bioremediation processes. AMF mitigate cadmium (Cd) toxicity, using a common physical barrier between mycorrhizal fungi and plant as protection; however, this mutualism goes beyond as a physical barrier, promoting significant changes in plant metabolism to increase its defense to several metals' toxicity, including Cd. The AMF release substances such as metallothioneins, siderophores, glomalin, oxalic acid, malic acid, and succinic acid with chelating function which reduce the Cd availability in the soil, mitigating the Cd toxicity, and enhance its uptake by other organisms present on the ground, for example, plants and bacteria (Dar et al., 2015, 2016; Vilela, 2021).

High concentrations of the toxic element, Cr (VI) ion, are hazardous to the fauna and flora, particularly in aquatic environments. Untreated and pretreated fungal biomass of *Rhizopus* sp. was used to remove this element in aqueous solutions (Espinoza-Sánchez, Arévalo-Niño, Quintero-Zapata, Castro-González, & Almaguer-Cantú, 2019). The potential of *Trichoderma lixii*, isolated from a heavy metal-polluted electroplating wastewater, to remove Cu^{2+} was evaluated; *T. lixii* removes cooper (Cu) efficiently between a 5 and 8 pH range, and in the presence of other cooccurring contaminant such as metabolic inhibitors, anions, and heavy metals (Kumar & Dwivedi, 2021).

Ramoutar, Mohammed, and Ramsubhag (2019) found a fungus isolated from a natural indigenous consortium coming from petroleum in West Indies, with potential to degrade hydrocarbons, diesel, naphthalene, pyrene, and asphaltene. Most isolates belonged to the major genera of *Fusarium* and *Aspergillus*, while few of them belonged to the genera *Curvularia*, *Penicillium*, and *Mucorales*. Similarly, degradation of crude oil was studied using the isolated fungi from the oily sludge-contaminated soil. Strains of genera, *Trichoderma* sp. and *Aspergillus* sp., were identified (Khandelwal et al., 2021). This study suggested that fungi isolated from the soil, particularly *Trichoderma atroviride* and *Aspergillus sydowii*, have potential to degrade aliphatic hydrocarbons in the environment. On the other hand, Mucorales such as *Mucor rouxii* (Fig. 12.3) have been described as a fungus with capacity of biosorption of heavy metals (Mullen, Wolf, Beveridge, & Bailey, 1992; Yan & Viraraghavan, 2008).

Fungi have been used in bioremediation of sites containing acidic radioactive waste. *Rhodotorula taiwanensis* grows under extreme conditions and in the presence of high concentrations of chromium (Cr) and mercury (Hg) and can even form biofilms onto acidic radioactive waste sites (Tkavc et al., 2018). Srivastava et al. (2011) isolated and examined 15 fungal strains from agricultural soils contaminated with arsenic (As). All these fungal strains were cultivated on medium supplemented with sodium arsenate, and 30 days after cultivation under laboratory conditions, radial growth of these strains was determined. Only five fungal strains survived and were found resistant, highlighting species *Rhizopus* sp., *Neocosmospora* sp., *Trichoderma* sp., *Penicillium* sp., and *Aspergillus* sp.

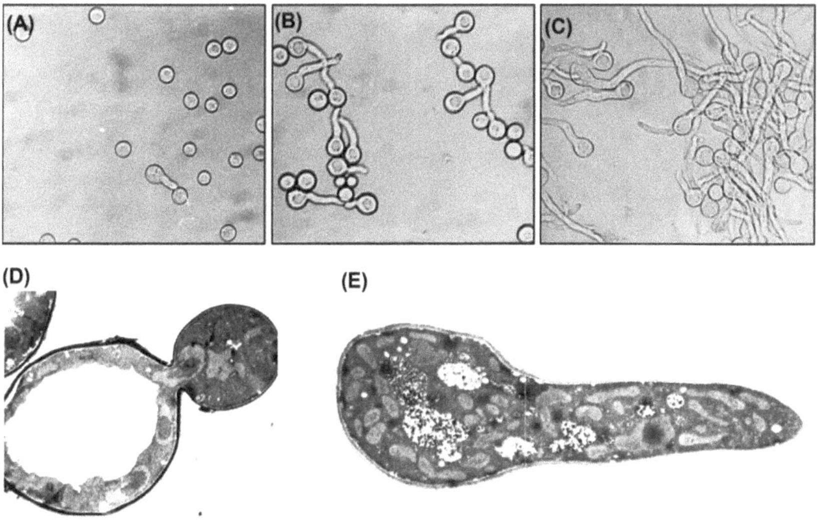

Figure 12.3 The Mucoromycota fungus *Mucor rouxii* growes in liquid YPG medium. *M. rouxii* is a fungus involved in the biosorption of heavy metals. (A) Spores. (B) Early stage of pores germination. (C) Late stage of spores germination and mycelial growth. (D) and (E) Photographs of the spore germination under TEM. *TEM*, Transmission electron microscopy.

The HCH is a persistent organic contaminant around the world, there is a global concern because of its potential toxic effects on ecosystems and humans. Ceci et al. (2019) investigated the bioremediation potential of a saprotrophic fungus, *Penicillium griseofulvum*, isolated from soils with high concentrations of isomers of HCH. They found that *P. griseofulvum* was able to grow in the presence of HCH and in stressful nutritional conditions and at different concentrations of sugar in the medium. In addition, the activation of certain metabolic pathways in response to oxidative stress was studied by the microarray technique. Similarly, same metabolic pathways were activated in responses to a mixture of HCH isomers and toxic metals (e.g., vanadium, lead) in solid and liquid media, an indication of a tolerance/resistance of these fungus to xenobiotics (Ceci et al., 2018).

Marine fungi, such as *Penicillium polonicum, Penicillium chrysogenum,* and *Penicillium cyclopium*, played an important role in the biodegradation of hydrocarbons from petroleum in marine environments (Maamar et al., 2020). The aerobic yeast *Yarrowia lipolytica* (Fig. 12.4) has the capacity to biodegrade crude oil, fatty acids, oils, n-alkenes, glycerol, etc. (Ferreira, Coelho, & Miguez da Rocha-Leao, 2012; Hassanshahian, Tebyanian, & Cappello, 2012; Martins, Ferreira, Azevedo, & Coelho, 2012). Moreover,

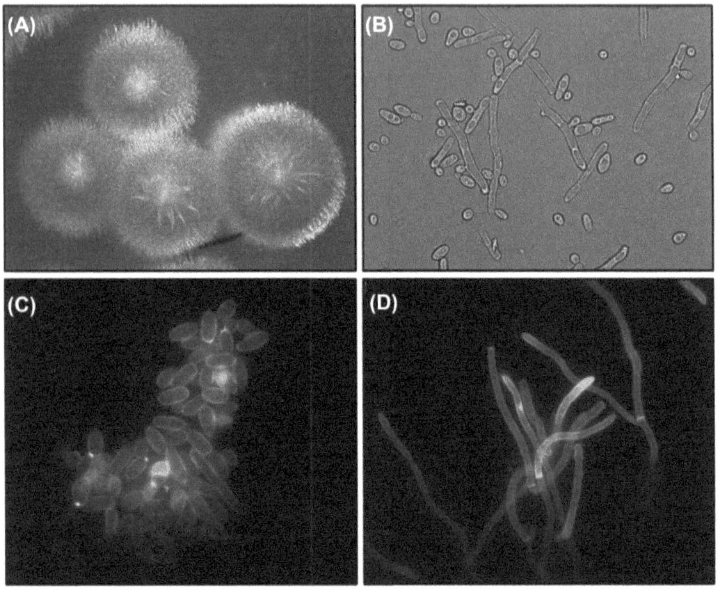

Figure 12.4 The Ascomycota model fungus *Yarrowia lipolytica*. (A) Colonies of *Y. lipolytica* in PDA solid medium. (B) *Y. lipolytica* growing in PDA liquid medium. (C) Yeast growth of *Y. lipolytica* in YNB liquid medium supplemented with $(NH_4)_2SO_4$. (D) Mycelial growth (multicellularity) of *Y. lipolytica* in YNB liquid medium supplemented with *N*-acetylglucosamine (GlcNAc).

this yeast has been used successfully as a model for the study of cellular processes of eukaryotes organisms such as dimorphism or multicellularity (Martínez-Soto & Ruiz-Herrera, 2013; Martínez-Soto, Velez-Haro, & León-Ramírez, 2020; reviewed by Martínez-Soto, Ortiz-Castellanos, Robledo-Briones, & León-Ramírez, 2020), an epigenetically regulated phenomenon (Martínez-Soto, González-Prieto, & Ruiz-Herrera, 2015). It has been described that *Y. lipolytica* and other fungi sense nutrients and stress conditions through the evolutionarily conserved mitogen-activated protein kinase (MAPK) pathways (Martínez-Soto & Ruiz-Herrera, 2017). Therefore these pathways are involved in the activation and induction of the genetic response of *Y. lipolytica* to biodegradation of contaminants.

12.7 Fungal interactions with microorganisms or superior organisms for bioremediation

In nature, fungi live interacting with other organisms such as plants, bacteria, archaea, and animals (Fig. 12.5). This symbiosis, in several cases,

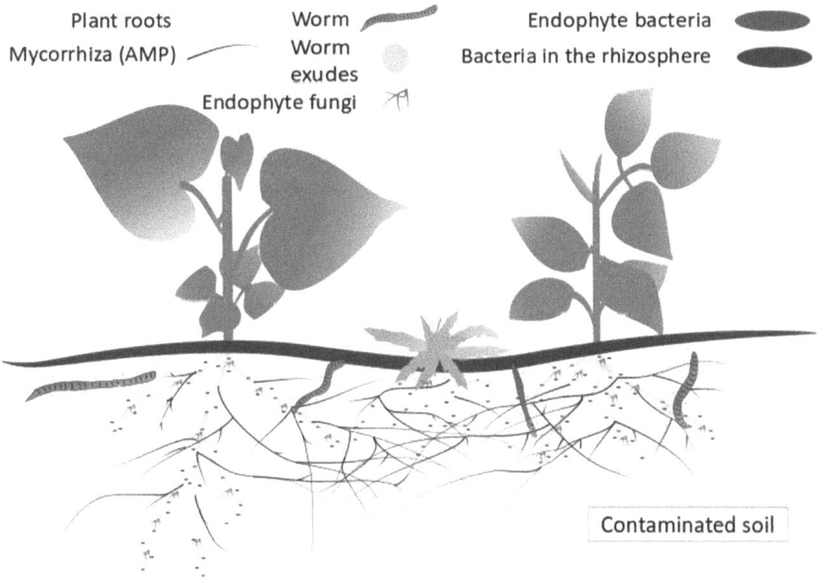

<u>Endophytes (fungi and bacteria):</u>
- Regulation of plant metabolism.
- Production of antibiotics, secondary metabolites.
- Enhance plant defense responses.
- Increase phosphate uptake.
- Increase plant biomass.
- Phytohormone production.
- Quenching the quorum sensing (QS) of phytopathogens by bacteria.
- Bacteria enhance the release of metals into the atmosphere in volatile forms (phytovolatilization).
- Regulation of plant metabolism.

<u>Worms:</u>
- Provide soil structure.
- Absorption of contaminants.
- Increase water infiltration and soil aeration.
- Contribute to breakdown organic pollutants.
- Enhance metal uptake by plants.
- Induce enzymes for pesticide detoxification.

<u>Mycorrhiza (AMP):</u>
- Secretion of enzymes and chelating agents.
- Absorption of contaminants in cell walls and spores.
- Storage of contaminants in vacuoles, vesicles, and/ or arbuscules (bioaccumulation).
- Phosphate uptake.

<u>Bacteria in the rhizosphere</u>:
- Secretion of enzymes and chelating agents and organic acids.
- Absorption of contaminants.
- Transformation, degradation, and mineralization of contaminants.
- Enhance hydrocarbons, other contaminants, and nutrients availability for plants.

<u>Plant roots</u>:
- Secretion of enzymes and chelating agents.
- Absorption of contaminants (phytoextraction).
- Storage of contaminants in vacuoles (bioaccumulation).
- Translocation of contaminants for transport and release by Phytovolatilization.
- Stabilization of metals (Phytostabilization).

Figure 12.5 Contribution of fungi with roots of plants, worms, and bacteria in the rhizosphere along phytoremediation process. The use of a consortium of fungi, bacteria, worms, and plants for bioremediation promises to improve the bioremediation of contaminants in soil and probably in other substrates.

enhances fungi tolerance to pollutant concentrations as well as improves the processes of bioremediation in combination with plants, and in consortia with bacteria and/or another fungus (Deshmukh et al., 2016; Krishnamurthy & Naik, 2017).

The symbiosis of fungi and plants can be negative, causing plant diseases, neutral, or positive, bringing mutual benefits for both. The interest of positive plant–fungi interactions, or mutualism in bioremediation, is because fungi enhance the tolerance, accumulation, absorption, and performance of plants in contaminated environments, including the reduction of toxicity of contaminants for plants. The use of fungi in phytoremediation can improve the efficiency of plants processes removing a variety of contaminants under several conditions, reducing their time and costs (Abdel-Shafy & Mansour, 2018; Deshmukh et al., 2016; Ijaz, Imran, & Anwar ul Haq, 2016; Kulikova & Perminova, 2021; Rajtor & Piotrowska-Seget, 2016). In Table 12.1 we summarize representative examples of fungi symbiosis with plants, bacteria, and other fungi, interactions which looks promising to apply in bioremediation of heavy metals, polycyclic aromatic hydrocarbons (PAHs), arsenic (As), and dyes. Rajtor and Piotrowska-Seget (2016) compilate the studies with plants assisted by fungus in phytoremediation of PAHs (Chen et al., 2015; Krishnamurthy & Naik, 2017).

AMF and endophyte fungi are mostly used in phytoremediation due the intimacy of their interactions with plants; the first one interacts with the cells of roots and the second lives in practically all plant tissue and organs and is in contact with plant cells; both do not cause diseases in their host plants. However, endophytes could turn pathogenic according to changes in their environment into the plants (Feng et al., 2017; Ijaz et al., 2016; Lenoir, Sahraoui, & Lounes-Hadj, 2016). AMF interchange nutrients with plant roots and solubilize phosphate compounds in the rhizosphere, enhancing the plants' uptake of phosphate. Also, they can secrete enzymes which help to transform highly toxic compound to another less harmful for plants (e.g., metals ions) or compounds as siderophores, which help with the absorption of contaminants by fungus, reducing their levels in the rhizosphere (Cabral, Soares, Giachini, & Siqueira, 2015; Ijaz et al., 2016; Kumar et al., 2019; Rajtor & Piotrowska-Seget, 2016). Therefore some endophytes are known for triggering the production of phytohormones in plants, as well the production of similar compounds, enhancing the plant response to biotic and abiotic stresses.

Mycoremediation, or bioremediation using fungi, implies in most cases the use of more than one species or consortia of microorganisms for the removal of contaminants in soil, water, and other substrates. The use of consortia in bioremediation has been extensively explored with bacteria and with fungi, including the presence of bacteria. These consortia assist phytoremediation by

Table 12.1 Representative fungi interaction examples with plants and other organisms, as bacteria, fungus, and worms in bioremediation processes.

Division	Taxon	Consortium/association	Effect	Contaminant	References
Plants					
Mucoromycota	*Glomus mosseae*	*Cannabis sativa*	Enhance the accumulation and increase plant biomass	Cr and Ni	Citterio et al. (2005)
Ascomycota	Several taxa *Alternaria* sp. and *Fusarium* sp., are the most representative	*Polygonum aviculare*	Plants enhance the fungus processes	Petroleum hydrocarbons	Mohsenzadeh et al. (2010)
Ascomycota	*Neotyphodium coenophialum*, *Neotyphodium uncinatum*	*Festuca arundinacea*, *Festuca pratensis*	Enhance the accumulation and increase plant biomass	Cd	Soleimani et al. (2010)
Ascomycota	*Fusarium oxysporum*	*Sedum alfredii*	Enhance the accumulation and increase plant biomass	Cd and Zn	Zhang et al. (2012)
Mucoromycota	*Glomus caledonium*, *G. mosseae*	*S. alfredii*, *Lolium perenne*	Enhance the absorption and stabilization	Cd	Hu, Wu, and Wu (2013)
Mucoromycota	*G. mosseae*, *Glomus intraradices*	*Pteris vittata*	Enhance the accumulation and increase plant biomass	As	Leung, Leung, Ye, Cheung, and Yung (2013)
Mucoromycota	*Rhizophagus custos*	*Daucus carota*	Enhance the biosorption	PAHs	Aranda et al. (2013)

(*Continued*)

Table 12.1 (Continued)

Division	Taxon	Consortium/association	Effect	Contaminant	References
Plants					
Ascomycota	*Penicillium janthinellum*	*Solanum lycopersicum*	Enhance the tolerance, reducing toxicity	Cd	Khan, Waqas, and Hussain (2014)
Ascomycota	*Galactomyces geotrichum*	*Glycine max*	Enhance the resistant	Zn	Waqas, Khan, and Kang (2014)
Basidiomycota	*Piriformospora indica*	*Nicotiana tabacum*	Enhance the tolerance	Cd	Hui, Liu, Gao, and Lou (2015)
Mucoromycota	*Glomus versiforme*	*Solanum nigrum*	Enhance the accumulation and increase plant biomass	Cd	Liu et al. (2015)
Ascomycota	*Neotyphodium* sp.	*F. arundinacea, L. perenne*	Enhance the accumulation and increase plant biomass	Zn	Zamani, Sabzalian, Khoshgoftarmanesh, and Afyuni (2015)
Ascomycota	35 different taxa. Dominant: *Plectosphaerella* sp., *Cladosporium* sp., *Verticillium* sp., *Phoma* sp., *Peyronellaea* sp., *Alternaria* sp.	*Dysphania ambrosioides*	Enhance the tolerance	Pb, Zn, and Cd	Li et al. (2016)
Mucoromycota, Basidiomycota	*Rhizophagus irregularis, Bjerkandera adusta, Mortierella* sp.	*S. lycopersicum*	Alleviate oxidative stress in plants	Cu, As, Pb, and Zn	Fuentes, Almonacid, and Ocampo (2016)

Mucoromycota	*Rhizophagus intraradices*	*Robinia pseudoacacia, Trifolium pratense, Medicago sativa, L. perenne*	Enhance the accumulation and increase plant biomass	Pb	Yang, Liang, and Han (2016)
Ascomycota	*Purpureocillium* sp.	*Kandelia candel*	Enhance the capacity of retention	Cu	Gong, Liu, Liao, Song, and Zhang (2017)
Ascomycota	*Lewia* sp.	*F. arundinacea*	Alleviate oxidative stress in plants and increase plant biomass	Hydrocarbons	Mendarte-Alquisira, Gutiérrez-Rojas, and González-Márquez (2017)
Ascomycota	*Gaeumannomyces cylindrosporus*	*Zea mays*	Enhance the tolerance	Pb	Ban et al. (2017)
Ascomycota	*Tricoderma* sp.	*Helianthus annuus*	Enhance the accumulation on plant tissue	As and Pb	Govarthanan, Mythili, Selvankumar, Kamala-Kannan, and Kim (2018)
Mucoromycota, Basidiomycota	*R. irregularis, Serendipita indica*	*Ocimum basilicum*	Increase plant biomass	Cu and Pb	Sabra et al. (2018)
Ascomycota	*Aureobasidium pullulans*	*Cucumis sativus*	Enhance the tolerance	Cd and Pb	Ali et al. (2019)
Basidiomycota	*Crucibulum leave*	*Z. mays*	Enhance the accumulation on plant tissue	PAHs	García-Sánchez, Košnář, Mercl, Aranda, and Tlustoš (2018)
Ascomycota	*Humicola* sp.	*Bacopa monnieri*	Reduce the accumulation on plant tissue	As	Tripathi et al. (2020)

(*Continued*)

Table 12.1 (Continued)

Division	Taxon	Consortium/association	Effect	Contaminant	References
Plants					
Ascomycota	*Aspergillus* sp., *Curvularia geniculata*, *Lindgomycetaceae* sp., *Westerdykella* sp.	*Aeschynomene fluminensis*, *Z. mays*	Enhance the bioaccumulation. Increase plant biomass	Hg	Pietro-Souza et al. (2020)
Plant and bacteria					
Ascomycota	*P. janthinellum*, *Aspergillus carneus*, *Aspergillus terreus*	*Cyperus laxus*, *Bacillus cereus*, *Pseudomonas* sp., *Gordonia rubripertincta*, *Kocuria rosea*, *Arthrobacter oxydans*, *Bacillus subtilis* (two strains), *Micrococcus luteus* (two unidentified strains)	Enhance the accumulation on plant tissue	Hydrocarbons	Escalante-Espinosa, Gallegos-Martínez, Favela-Torres, and Gutiérrez-Rojas (2005)
Plant and worms					
Mucoromycota	*G. intraradices*, *G. mosseae*	*Z. mays*, *Lumbricus rubellus*	Alleviation of toxicity	Cd	Aghababaei et al. (2014)
Mucoromycota	*G. caledonium*	*F. arundinacea*, *Eisenia foetida*	Enhance the degradation	PAHs	Lu and Lu (2015)
Fungi					
Mucoromycota	Several taxa	*Mucor circinelloides*, *Actinomucor* sp., *Mortierella* sp.	Enhance the degradation and reduce the time of remotion	Zn, Pb, Mn	Cui, Zhang, Yang, and Sun (2017)
Ascomycota	*Aspergillus* sp.	*Aspergillus flavus*, *Aspergillus fumigatus*	Enhance the biosorption	Cr and Cd	Talukdar et al. (2020)

Ascomycota	Several taxa	*Curvularia brachyspora*, *Penicillium chrysogenum*, *Scopulariopsis brevicaulis*, *Cladosporium sphaerospermum*, *Alternaria alternate*, *Stemphylium botryosum*	Enhance the degradation	Petroleum hydrocarbons	Hamad et al. (2021)
Bacteria					
Ascomycota	*A. flavus*	*Pseudomonas aeruginosa*	Enhance the degradation	Industrial dyes	Mawad et al. (2020)
Plant and bacteria					
Mucoromycota	*Glomus* sp., *Acaulospora* sp., *Scutellospora* sp.	*Pennisetum glaucum*, *Sorghum bicolor*, *Streptomyces* sp., *Azotobacter* sp., *Pseudomonas* sp., *Paenibacillus* sp.	Enhance the biosorption	Fe	Mishra et al. (2016)

plants. Therefore the use of microorganisms, together with plants in bioremediation, brings new possibilities to develop new technologies to implement in contaminate sites with a mixture of hazardous waste, allowing the best management of these contaminants (Cabral et al., 2015; Ortega, Torres-Mendoza, & Cubilla-Rios, 2020). The mechanism of fungi to neutralize the pollution can be improved or can help to overcome disadvantages than other organisms that deal under high concentration of contamination (Cabral et al., 2015; Ijaz et al., 2016; Kumar et al., 2019; Lenoir et al., 2016; Rajtor & Piotrowska-Seget, 2016).

A more recent approach is the combination of fungi, plants, and worms to improve the soil conditions to enhance the processes involved in bioremediation of plants and fungi (Fig. 12.5). The worms keep a constant interchange of gases from soil to the atmosphere, promoting a better microenvironment not only for the enzymes participating in oxidoreduction reactions but also for the growth of certain aerobic microorganisms that can contribute to the neutralization of contaminants (Aghababaei, Raiesi, & Hosseinpur, 2014; Lu & Lu, 2015).

Fungi and bacteria secrete exudates allowing the transformation, degradation, and absorption of contaminants in the rhizosphere. Both enhance the availability of nutrients for plants and trigger the secretion of phytohormones, promoting plant growth and in consequence plant biomass. Endophytic fungi and bacteria enhance tolerance to high concentrations of contaminants in the soil for plants. Mycorrhiza facilitates the availability of nutrients, such as phosphate, and also transforms contaminants and absorbs some of them in their cell walls, including the storage of several compounds into vacuoles. Worms can absorb molecules containing metals by themselves and improves the aeration of soils, as well the filtration of water, and, additionally, enhance detoxification of pesticides in soil. Fungi symbiotic interaction with plants, bacteria, and worms is an alternative to improve bioremediation processes using living organisms in soil.

12.8 Bioremediation mechanisms developed by fungi

The use of fungi in bioremediation or mycoremediation has been applied to manage the pollution of a diversity of compounds in soil and wastewater. Fungi count with a wide range of enzymes, developing several mechanisms to degrade and modify the contamination in their environments. The exposure to a diverse of contaminants, such as heavy metals, hydrocarbons, dyes, pesticides, herbicides, antibiotics, PAHs, and other compounds, let them activate metabolic passive and active processes to mobilize, immobilize, and bioaccumulate, including the biosorption and biotransformation of those pollutants in their cells or in the sites they were living (Chen et al., 2015; Khatoon et al., 2021).

Metabolic passive process did not involve the participation of enzymes to neutralize the contaminants, as the mechanism of absorption, which allows fungi to retain compounds in their cell walls. In contrast, biosorption, a metabolic active process, involves the participation of enzymes in the neutralization, transformation, and accumulation of the pollutants to introduce them into the cell for its storage. The enzymes produced by fungi can immobilize or mobilize pollutants in the contaminated subtracts, such as soil and wastewater, preventing the leaching and spreading of pollutants from the origin site to other sites. In addition, these enzymes can transform the pollutant compounds to less harmful compounds for themselves or another organism, such as plants and bacteria, allowing these organisms the use of those compounds as substrates for other metabolic pathways, as well as its uptake for storage or degradation. A summary of general mechanisms and processes developed by fungi in response to contaminated environments is provided in Table 12.2. And, the main family enzymes produced by fungi in response to these environments are briefly described in Table 12.3.

Table 12.2 General mechanisms and processes developed by fungus in the presence of contaminated environments.

Metabolism	Mechanism	Process
Metabolic passive process	Biosorption	Physicochemical and reversible, process. Involved precipitation, ion exchange, adsorption, reduction and chelation. The functional groups of cell surface promote the uptake of toxic pollutants from inactive biological sources.
Metabolic active process	Mobilization	Reduction and/or oxidation processes for the mobilization of metals, metalloids, and organometallic compounds by activating microbes. • Chelation: low-molecular-weight iron-chelating molecules are capable to bind with additional metals (Cr, Ga, Mg, Mn). • Leaching: production of low-molecular-weight organic acids that provide protons and metal-complexing organic acid anions after breakdown.

(Continued)

Table 12.2 (Continued)

Metabolism	Mechanism	Process
		• Methylation: assimilation of methyl groups that are enzymatically transferred to a metal and form different metalloids. • Alkalinization: alkyl group transformation from one molecule to other as alkyl carbonation, a free radical, and a carbanion form.
Metabolic active process	Immobilization	The microbes can reduce mobility of contaminants and modify its physical or chemical characteristics by the restriction of physical interaction among the toxic contaminants.
Metabolic active process	Bioaccumulation	The uptake of toxic pollutants only by living cells of microorganisms. The contaminant is actively transported into the cell across the cell membrane, where it is accumulated intracellularly. Bioaccumulation depends on structural and biochemical properties, genetic and physiological adaptations, environmental modification of contaminants, and the contaminant availability and its toxicity.
Metabolic active process	Biotransformation	The microorganisms modify the surrounding environment through catalysis, oxidation, methylation, and demethylation. • Bioprecipitation: the process when organic carbon mineralizes into carbon dioxide and oxygen reduces into water enhances pH and alkalinity of the microbial cell environment and with the formation of excess bicarbonate the precipitation of metal ions occur such as metal hydroxides Me(OH)x or metal carbonate Me2(CO3)x. • Biological oxidation/reduction: microorganisms, with the help of various enzymes, catalyze the reduction of heavy metals such as As, Cr, Fe, Mn, Mo, Se, and U.

Source: Modified from Khatoon, H., Rai, J.P.N., & Jillani, A. (2021). Chapter 7: Role of fungi in bioremediation of contaminated soil. In: V. Kumar Sharma, M.P. Shah, S. Parmar, A. Kumar (Eds.), *Fungi bio-prospects in sustainable agriculture, environment and nanotechnology*. Academic Press, pp. 121–156. Available from https://doi.org/10.1016/B978-0-12-821925-6.00007-1.

Table 12.3 Enzyme families produced by fungi in bioremediation processes in the presences of contaminated environments.

Family	Enzyme	Mode of action
Oxidoreductases	• Oxygenase • monooxygenase • dioxygenases • laccases	Transfer oxygen atoms from diatomic oxygen
Peroxidases	• Lignin peroxidases • Manganese peroxidase	Transfer electrons to hydrogen peroxide
Hydrolases	• Lipases • Cellulases • Carboxylesterase • Phosphotriesterase • Haloalkane dehalogenase	Catalyzes the hydrolysis of a chemical bond
Lyases	• Haloalkane dihydrochloride	Act on the glycosidic linkage producing an unsaturated bond between C4 and C5 of the uronic acid residue at the nonreducing end by an elimination mechanism.

Source: Based in Bhandahari and Sharma (2021).

12.9 Fungal genes and enzymes involved in bioremediation

Fungi are versatile organisms with a wide range of enzymes which help them to adapt to changing environments. They count with an arsenal of enzymes, which include PPOs, manganese peroxidases, laccases, amylases, proteases, lipases, and ligninolytic enzymes for the transformation and degradation of several hazardous wastes in contaminate environments (Table 12.3). These enzymes have the potential to be used in the removal of contaminants compounds such as heavy metals, antibiotics, dyes, PHAs, surfactants, and others (Bhardahari & Sharma, 2021; Deshmukh et al., 2016; Kumar et al., 2019).

12.10 Conclusion and perspectives

Currently, the world is facing huge adversity caused by contamination of soil, air, and water, due to the toxic elements generated daily. This situation is aggravated rapidly by the industrialization, as well as rapid growth of urbanization. In the last decades the interest of restoration to contaminated areas has been growing to international level. However, the complete restoration of contaminated areas involves a global quite costly challenge that requires a significant effort. Due to this, different alternatives have emerged to recover the ecological balance, for example, green remediation techniques that are based in the use of microorganisms such as fungi. The fungi are versatile organisms that have a wide range of adaptability because they have a large arsenal of enzymes which help them to adapt to changing environments, turning them a great tool for bioremediation of contaminated environments. Therefore it is essential to continue investigating more about the metabolomic mechanisms that allow them to secrete this wide range of extracellular enzymes, which will allow future improvements in current bioremediation techniques as well as the development of new techniques.

Acknowledgments

Thanks are given to Prof. Dr. Rubén Salcedo-Hernández from the University of Guanajuato for donation of photographs of M. rouxii. We also thank Mayela Fernanda Salazar-Chavez from CINVESTAV Irapuato for the comments and suggestions on this work.

References

Abdel-Shafy, H. I., & Mansour, M. S. M. (2018). Phytoremediation for the elimination of metals, pesticides, PAHs, and other pollutants from wastewater and soil. In V. Kumar, M. Kumar, & R. Prasad (Eds.), *Phytobiont and ecosystem restitution*. Singapore: Springer. Available from http://doi.org/10.1007/978-981-13-1187-1_5.

Aghababaei, F., Raiesi, F., & Hosseinpur, A. (2014). The combined effects of earthworms and arbuscular mycorrhizal fungi on microbial biomass and enzyme activities in a calcareous soil spiked with cadmium. *Applied Soil Ecology*, *75*, 33–42. Available from https://doi.org/10.1016/j.apsoil.2013.10.006.

Agrawal, P. K., Upadhyay, P., Shrivastava, R., Sharma, S., & Garlapati, V. K. (2021). Evaluation of the ability of endophytic fungi from *Cupressus torulosa* to decolorize synthetic textile dyes. *Journal of Hazardous, Toxic & Radioactive Waste*, *25*(1), 06020005.

Albert, L. (2004). Contaminación ambiental. Origen, clases, fuentes y efectos. *Revista de la facultad pública y nutrición*, 228–231.

Ali, A., Bilal, S., Abdul, L. K., Fazal, M., Al-Harrasi, A., & Lee, I. (2019). Endophytic *Aureobasidium pullulans* BSS6 assisted developments in phytoremediation potentials of *Cucumis sativus* under Cd and Pb stress. *Journal of Plant Interactions*, *14*(1), 303–313. Available from https://doi.org/10.1080/17429145.2019.1633428.

Al-Zaban, M. I., Mahmoud, M. A., AlHarbi, M. A., & Bahatheq, A. M. (2020). Bioremediation of crude oil by rhizosphere fungal isolates in the presence of silver nanoparticles. *International Journal of Environmental Research and Public Health, 17*(18), 6564. Available from https://doi.org/10.3390/ijerph17186564.

Ani, E., Adekunle, A. A., Kadiri, A. B., & Njoku, K. L. (2021). Rhizoremediation of hydrocarbon contaminated soil using *Luffa aegyptiaca* (Mill) and associated fungi. *International Journal of Phytoremediation, 25*, 1–13. Available from https://doi.org/10.1080/15226514.2021.1901852.

Aranda, E., Scervino, J. M., Godoy, P., Reina, R., Ocampo, J. A., Wittich, R. M., & García-Romera, I. (2013). Role of arbuscular mycorrhizal fungus *Rhizophagus custos* in the dissipation of PAHs under root-organ culture conditions. *Environmental Pollution (Barking, Essex: 1987), 181*, 182–189. Available from https://doi.org/10.1016/j.envpol.2013.06.034.

Ban, Y., Xu, Z., Yang, Y., Zhang, H., Chen, H., & Tang, M. (2017). Effect of dark septate endophytic fungus *Gaeumannomyces cylindrosporus* on plant prowth, photosynthesis and Pb tolerance of maize (*Zea mays L.*). *Pedosphere, 27*(2), 283–292. Available from https://doi.org/10.1016/S1002-0160(17)60316-3.

Barnes, N. M., Khodse, V. B., Lotlikar, N. P., Meena, R. M., & Damare, S. R. (2018). Bioremediation potential of hydrocarbon-utilizing fungi from select marine niches of India. *3 Biotech, 8*(1), 21. Available from https://doi.org/10.1007/s13205-017-1043-8.

Bhandari, S., Kumar Poudel, D., Marahatha, R., Dawadi, S., Khadayat, K., Phuyal, S., Shrestha, S., Gaire, S., Basnet, K., Khadka, U., & Parajuli, N. (2021). Microbial enzymes used in bioremediation. *Journal of Chemistry*, 8849512. Available from https://doi.org/10.1155/2021/8849512.

Bhat, R. A., Singh, D. V., Qadri, H., Dar, G. H., Dervash, M. A., Bhat, S. A., ... Yousaf, B. (2021). Vulnerability of municipal solid waste: An emerging threat to aquatic ecosystems. *Chemosphere, 287*(2022), 132223. Available from https://doi.org/10.1016/j.chemosphere.2021.132223.

Bokade, P., Purohit, H. J., & Bajaj, A. (2021). Myco-remediation of chlorinated pesticides: Insights into fungal metabolic system. *Indian Journal of Microbiology*. Available from https://doi.org/10.1007/s12088-021-00940-8.

Cabral, L., Soares, C. R., Giachini, A. J., & Siqueira, J. O. (2015). Arbuscular mycorrhizal fungi in phytoremediation of contaminated areas by trace elements: Mechanisms and major benefits of their applications. *World Journal of Microbiology and Biotechnology, 31* (11), 1655–1664. Available from https://doi.org/10.1007/s11274-015-1918-y.

Cavero-Zegarra, D. E. & Pari-Quispe, D. (2018). *Biorremediación* (1st ed.). Perú. http://repositorio.upsc.edu.pe/handle/UPSC/4475.

Ceci, A., Pinzari, F., Riccardi, C., Maggi, O., Pierro, L., Petrangeli, P. M., ... Persiani, A. M. (2018). Metabolic synergies in the biotransformation of organic and metallic toxic compounds by a saprotrophic soil fungus. *Applied Microbiology and Biotechnology, 102*(2), 1019–1033. Available from https://doi.org/10.1007/s00253-017-8614-9.

Ceci, A., Pinzari, F., Russo, F., Persiani, A. M., & Gadd, G. M. (2019). Roles of saprotrophic fungi in biodegradation or transformation of organic and inorganic pollutants in co-contaminated sites. *Applied Microbiology and Biotechnology, 103*(1), 53–68. Available from https://doi.org/10.1007/s00253-018-9451-1.

Chen, M., Xu, P., Zeng, G., Yang, C., Huang, D., & Zhang, J. (2015). Bioremediation of soils contaminated with polycyclic aromatic hydrocarbons, petroleum, pesticides, chlorophenols and heavy metals by composting: Applications, microbes and future research needs. *Biotechnology Advances, 1*(33), 745–755. Available from https://doi.org/10.1016/j.biotechadv.2015.05.003.

Citterio, S., Prato, N., Fumagalli, P., Aina, R., Massa, N., Santagostino, A., ... Berta, G. (2005). The arbuscular mycorrhizal fungus *Glomus mosseae* induces growth and metal accumulation changes in *Cannabis sativa L. Chemosphere, 59*(1), 21–29. Available from https://doi.org/10.1016/j.chemosphere.2004.10.009.

Cota-Ruíz, K., Núñez-Gastelúm, J. A., Delgado-Ríos, M., & Martínez-Martínez, A. (2019). Biorremediación: Actualidad de conceptos y aplicaciones. *Biotecnia, 21*(1), 37–44.

Cui, Z., Zhang, X., Yang, H., & Sun, L. (2017). Bioremediation of heavy metal pollution utilizing composite microbial agent of *Mucor circinelloides*, *Actinomucor* sp. and *Mortierella* sp. *Journal of Environmental Chemical Engineering, 5*(4), 3616–3621. Available from https://doi.org/10.1016/j.jece.2017.07.021.

da Silva, T. H., Silva, D. A. S., Thomazini, A., Schaefer, C. E. G. R., & Rosa, L. H. (2019). Antarctic permafrost: An unexplored fungal microhabitat at the edge of life. In L. Rosa (Ed.), *Fungi of antarctica*. Cham: Springer. Available from http://doi.org/10.1007/978-3-030-18367-7_7.

Dao, A. T. N., Vonck, J., Janssens, T. K. S., Dang, H. T. C., Brouwer, A., & de Boer, T. E. (2019). Screening white-rot fungi for bioremediation potential of 2,3,7,8-tetrachlorodibenzo-p-dioxin. *Industrial Crops and Products, 128*, 153–161. Available from https://doi.org/10.1016/j.indcrop.2018.10.059.

Dar, G. H., Bandh, S. A., Kamili, A. N., Nazir, R., & Bhat, R. A. (2013). Comparative analysis of different types of bacterial colonies from the soils of Yusmarg Forest, Kashmir valley India. *Ecologia Balkanica, 5*(1), 31–35.

Dar, G. H., Kamili, A. N., Chishti, M. Z., Dar, S. A., Tantry, T. A., & Ahmad, F. (2016). Characterization of *Aeromonas sobria* isolated from fish Rohu (*Labeo rohita*) collected from polluted pond. *Journal of Bacteriology & Parasitology*, 7(3), 1–5. Available from https://doi.org/10.4172/2155-9597.1000273.

Dar, G. H., Kamili, A. N., Dar, S. A., Chishti, M. Z., & Ahmad, F. (2015). Detection and characterization of potentially pathogenic *Aeromonas sobria* isolated from fish *Hypophthalmichthys molitrix* (Cypriniformes: Cyprinidae). *Microbial Pathogenesis*. Available from http://doi.org/10.1016/j.micpath.2015.10.017.

Dar, G. H., Bandh, S. A., & Kamili, A. N. (2011). Bacteriological analysis of soil from yusmarg health resort of Kashmir Valley. *International Journal of Current Research, 3*(10), 53–56.

Dar, G. H., Kamili, A. N., Nazir, R., Bandh, S. A., & Bhat, R. A. (2012). A preliminary study of colony forming units of bacteria from the soils of Yusmarg Forest, Kashmir Valley, India. *International Journal of Current Research, 4*(12), 467–472.

de Lima, D. P., dos Santos, E., dos, A., Marques, M. R., Giannesi, G. C., Beatriz, A., ... dos, S. (2018). Fungal bioremediation of pollutant aromatic amines. *Current Opinion in Green and Sustainable Chemistry, 11*, 34–44. Available from https://doi.org/10.1016/j.cogsc.2018.03.012.

De Oliveira, V. H., Ullah, I., Dunwell, J. M., & Tibbett, M. (2020). Bioremediation potential of Cd by transgenic yeast expressing a metallothionein gene from *Populus trichocarpa*. *Ecotoxicology and Environmental Safety, 202*, 110917. Available from https://doi.org/10.1016/j.ecoenv.2020.110917.

Deshmukh, R., Khardenavis, A. A., & Purohit, H. J. (2016). Diverse metabolic capacities of fungi for bioremediation. *Indian Journal of Pathology & Microbiology, 56*(3), 247–264. Available from https://doi.org/10.1007/s12088-016-0584-6.

Domínguez-Gual, M. C. (2015). La contaminación ambiental, un tema con compromiso social. *Producción Limpia, 10*(1), 9–21.

Encinas Malagón, M. D. (2011). *Medio ambiente y contaminación. Principios básicos* (1st ed.). p. 119.

Erguven, G. O. (2018). Comparison of some soil fungi in bioremediation of herbicide acetochlor under agitated culture media. *Bulletin of Environmental Contamination and Toxicology, 100*(4), 570–575. Available from https://doi.org/10.1007/s00128-018-2280-1.

Escalante-Espinosa, E., Gallegos-Martínez, M. E., Favela-Torres, E., & Gutiérrez-Rojas, M. (2005). Improvement of the hydrocarbon phytoremediation rate by *Cyperus laxus Lam.* inoculated with a microbial consortium in a model system. *Chemosphere*, *59*(3), 405–413. Available from https://doi.org/10.1016/j.chemosphere.2004.10.034.

Espinoza-Sánchez, M. A., Arévalo-Niño, K., Quintero-Zapata, I., Castro-González, I., & Almaguer-Cantú, V. (2019). Cr (VI) adsorption from aqueous solution by fungal bioremediation based using *Rhizopus* sp. *Journal of Environmental Management*, *251*, 109595. Available from https://doi.org/10.1016/j.jenvman.2019.109595.

Feng, N. X., Yu, J., Zhao, H. M., Cheng, Y. T., Mo, C. H., Cai, Q. Y., ... Wong, M. H. (2017). Efficient phytoremediation of organic contaminants in soils using plant-endophyte partnerships. *The Science of the Total Environment*, *583*, 352–368. Available from https://10.1016/j.scitotenv.2017.01.075.

Ferreira, T. F., Coelho, M. A. Z., & Miguez da Rocha-Leao, M. H. (2012). Factors influencing crude oil biodegradation by *Yarrowia lipolytica*. *Brazilian Archives of Biology and Technology*, *55*, 785–791. Available from https://doi.org/10.1590/S1516-89132012000500019.

Fuentes, A., Almonacid, L., Ocampo, J. A., et al. (2016). Synergistic interactions between a saprophytic fungal consortium and *Rhizophagus irregularis* alleviate oxidative stress in plants grown in heavy metal contaminated soil. *Plant and Soil*, *407*, 355–366. Available from https://doi.org/10.1007/s11104-016-2893-2.

García-Sánchez, M., Košnář, Z., Mercl, F., Aranda, E., & Tlustoš, P. (2018). A comparative study to evaluate natural attenuation, mycoaugmentation, phytoremediation, and microbial-assisted phytoremediation strategies for the bioremediation of an aged PAH-polluted soil. *Ecotoxicology and Environmental Safety*, *147*, 165–174. Available from https://doi.org/10.1016/j.ecoenv.2017.08.012.

Garzón, J. M., Rodríguez-Miranda, J. P., & Hernández-Gómez, C. (2017). Aporte de la biorremediación para solucionar problemas de contaminación y su relación con el desarrollo sostenible. *Salus University*, *19*(2), 309–318. Available from https://doi.org/10.22267/rus.171902.93.

Gibson & Sayler. (1992). Scientific foundations of bioremediation: Current status and future needs. *American Academy of Microbiology*, Washington, DC.

Gong, B., Liu, G., Liao, R., Song, J., & Zhang, H. (2017). Endophytic fungus Purpureocillium sp. A5 protect mangrove plant *Kandelia candel* under copper stress. *Brazilian Journal of Microbiology*, *48*(3), 530–536. Available from https://doi.org/10.1016/j.bjm.2016.10.027.

González-Rojas, E. H. (2011). Concepto y estrategias de biorremediación. *Inge UAN*, *1* (1), 20–29.

Govarthanan, M., Mythili, R., Selvankumar, T., Kamala-Kannan, S., & Kim, H. (2018). Myco-phytoremediation of arsenic- and lead-contaminated soils by *Helianthus annuus* and wood rot fungi, Trichoderma sp. isolated from decayed wood. *Ecotoxicology and Environmental Safety*, *151*, 279–284. Available from https://doi.org/10.1016/j.ecoenv.2018.01.020.

Hai, F. I., Modin, O., Yamamoto, K., Fukushi, K., Nakajima, F., & Nghiem, L. D. (2012). Pesticide removal by a mixed culture of bacteria and white-rot fungi. *Journal of the Taiwan Institute of Chemical Engineers*, *43*, 459–462.

Hamad, A. A., Moubasher, H. A., Moustafa, Y. M., Nermen, H., Mohamed, E. H., & Abd-el rhim. (2021). Petroleum hydrocarbon bioremediation using native fungal isolates and consortia. *Scientific World Journal*, *2021*, 13. Available from https://doi.org/10.1155/2021/6641533.

Harms, H., Wick, L. Y., & Schlosser, D. (2017). The fungal community in organically polluted systems. In J. Dighton, & J. F. White (Eds.), *Mycology* (pp. 459–470). CRC Press, CRC Press Taylor & Francis Group 6000 Broken Sound Parkway NW, Suite 300 Boca Raton, FL 33487-2742.

Hassanshahian, M., Tebyanian, H., & Cappello, S. (2012). Isolation and characterization of two crude oil-degrading yeast strains, *Yarrowia lipolytica* PG-20 and PG-32, from the Persian Gulf. *Marine Pollution Bulletin*, *64*, 1386–1391. Available from https://doi.org/10.1016/j.marpolbul.2012.04.020.

Hu, J., Wu, S., Wu, F., et al. (2013). Arbuscular mycorrhizal fungi enhance both absorption and stabilization of Cd by Alfred stonecrop (*Sedum alfredii* Hance) and perennial ryegrass (*Lolium perenne* L.) in a Cd-contaminated acidic soil. *Chemosphere*, *93*(7), 1359–1365. Available from https://doi.org/10.1016/j.chemosphere.2013.07.089.

Hu, K., Torán, J., López-García, E., Barbieri, M. V., Postigo, C., de Alda, M. L., ... Blánquez, P. (2020). Fungal bioremediation of diuron-contaminated waters: Evaluation of its degradation and the effect of amendable factors on its removal in a trickle-bed reactor under non-sterile conditions. *The Science of the Total Environment*, *743*, 140628. Available from https://doi.org/10.1016/j.scitotenv.2020.140628.

Hui, F., Liu, J., Gao, Q., & Lou, B. (2015). *Piriformospora indica* confers cadmium tolerance in *Nicotiana tabacum*. *Journal of Environmental Sciences (China)*, *37*, 184–191. Available from https://doi.org/10.1016/j.jes.2015.06.005.

Hyde, K. D., Jeewon, R., Chen, Y. J., et al. (2020). The numbers of fungi: Is the descriptive curve flattening? *Fungal Divers*, *103*, 219–271. Available from https://doi.org/10.1007/s13225-020-00458-2.

Ijaz, A., Imran, A., Anwar ul Haq, M., et al. (2016). Phytoremediation: Recent advances in plant-endophytic synergistic interactions. *Plant and Soil*, *405*, 179–195. Available from https://doi.org/10.1007/s11104-015-2606-2.

Ipeaiyeda, A. R., Adenipekun, C. O., & Oluwole, O. (2020). Bioremediation potential of *Ganoderma lucidum* (Curt:Fr) P. Karsten to remove toxic metals from abandoned battery slag dumpsite soil and immobilisation of metal absorbed fungi in bricks. *Cogent. Environmental Sciences: An International Journal of Environmental Physiology and Toxicology*, *6*, 1. Available from https://doi.org/10.1080/23311843.2020.1847400.

Javaid, M. K., Ashiq, M., & Tahir, M. (2016). Potential of biological agents in decontamination of agricultural soil. *Scientifica (Cairo)*, *2016*, 1598325. Available from https://doi.org/10.1155/2016/1598325.

Kameshwar, A. K. S., & Qin, W. (2019). Systematic review of publicly available non-Dikarya fungal proteomes for understanding their plant biomass-degrading and bioremediation potentials. *Bioresources and Bioprocessing*, *6*, 30. Available from https://doi.org/10.1186/s40643-019-0264-6.

Khan, A. L., Waqas, M., Hussain, J., et al. (2014). Fungal endophyte *Penicillium janthinellum* LK5 can reduce cadmium toxicity in *Solanum lycopersicum* (Sitiens and Rhe). *Biology and Fertility of Soils*, *50*, 75–85. Available from https://doi.org/10.1007/s00374-013-0833-3.

Khandelwal, A., Singh, S. B., Sharma, A., Nain, L., Varghese, E., & Singh, N. (2021). Effect of surfactant on degradation of *Aspergillus* sp. and *Trichoderma* sp. mediated crude oil. *International Journal of Environmental Analytical Chemistry*. Available from https://doi.org/10.1080/03067319.2021.1879800.

Khatoon, H., Rai, J. P. N., & Jillani, A. (2021). Chapter 7: Role of fungi in bioremediation of contaminated soil. In V. Kumar Sharma, M. P. Shah, S. Parmar, & A. Kumar (Eds.), *Fungi bio-prospects in sustainable agriculture* (pp. 121–156). *Academic Press*, Environment and Nanotechnology, doi:10.1016/B978-0-12-821925-6.00007-1.

Krishnamurthy, Y. L., & Naik, B. S. (2017). Endophytic fungi bioremediation. In D. Maheshwari, & K. Annapurna (Eds.), *Endophytes: Crop productivity and protection* (16). Cham: Springer. Sustainable Development and Biodiversity. Available from https://doi.org/10.1007/978-3-319-66544-3_3.

Kulikova, N. A., & Perminova, I. V. (2021). Interactions between humic substances and microorganisms and their implications for nature-like bioremediation technologies. *Molecules (Basel, Switzerland)*, *26*(9), 2706. Available from https://doi.org/10.3390/molecules26092706.

Kumar, A., Chaturvedi, A., Yadav, K., et al. (2019). Fungal phytoremediation of heavy metal-contaminated resources: Current scenario and future prospects. In A. Yadav, S. Singh, S. Mishra, & A. Gupta (Eds.), *Recent advancement in white biotechnology through fungi*. Cham: Springer. Fungal Biology. Available from https://doi.org/10.1007/978-3-030-25506-0_18.

Kumar, V., & Dwivedi, S. K. (2021). Bioremediation mechanism and potential of copper by actively growing fungus *Trichoderma lixii* CR700 isolated from electroplating wastewater. *Journal of Environmental Management, 1*(277), 111370. Available from https://doi.org/10.1016/j.jenvman.2020.111370.

Ledezma-Villanueva, A., Adame-Rodríguez, J. M., & Aréchiga-Carvajal, E. T. (2018). Transcriptomics as a first choice gate for fungal biodegradation processes description. In R. Prasad, & E. Aranda (Eds.), *Approaches in bioremediation* (pp. 29–42). Cham: Springer. Nanotechnology in the Life Sciences. Available from https://doi.org/10.1007/978-3-030-02369-0_2.

Lenoir, I., Sahraoui, A., & Lounes-Hadj, F. J. (2016). Arbuscular mycorrhizal fungal-assisted phytoremediation of soil contaminated with persistent organic pollutants: A review. *European Journal of Soil Science, 67*(5), 624–640. Available from https://doi.org/10.1111/ejss.12375.

Leung, H. M., Leung, A. O., Ye, Z. H., Cheung, K. C., & Yung, K. K. (2013). Mixed arbuscular mycorrhizal (AM) fungal application to improve growth and arsenic accumulation of *Pteris vittata* (As hyperaccumulator) grown in as-contaminated soil. *Chemosphere, 92*(10), 1367–1374. Available from https://doi.org/10.1016/j.chemosphere.2013.04.093.

Li, X., Li, W., Chu, L., White, J. F., Jr, Xiong, Z., & Li, H. (2016). Diversity and heavy metal tolerance of endophytic fungi from *Dysphania ambrosioides*, a hyperaccumulator from Pb–Zn contaminated soils. *Journal of Plant Interactions, 11*(1), 186–192. Available from https://doi.org/10.1080/17429145.2016.1266043.

Liu, H., Yuan, M., Tan, S., Yang, X., Lan, Z., Jiang, Q., ... Jing, Y. (2015). Enhancement of arbuscular mycorrhizal fungus (*Glomus versiforme*) on the growth and Cd uptake by Cd-hyperaccumulator *Solanum nigrum*. *Applied Soil Ecology, 89*, 44–49. Available from https://doi.org/10.1016/j.apsoil.2015.01.006.

Lu, Y. F., & Lu, M. (2015). Remediation of PAH-contaminated soil by the combination of tall fescue, arbuscular mycorrhizal fungus and epigeic earthworms. *Journal of Hazardous Materials, 285*(21), 535–554. Available from https://doi.org/10.1016/j.jhazmat.2014.07.021.

Maamar, A., Lucchesi, M. E., Debaets, S., Nguyen van Long, N., Quemener, M., Coton, E., ... Matallah-Boutiba, A. (2020). Highlighting the crude oil bioremediation potential of marine fungi isolated from the port of Oran (Algeria). *Diversity, 12*(5), 196. Available from https://doi.org/10.3390/d12050196.

Martínez-Soto, D., & Ruiz-Herrera, J. (2017). Functional analysis of the MAPK pathways in fungi. *Revista Iberoamericana de Micologia: Organo de la Asociacion Espanola de Especialistas en Micologia, 34*, 192–202. Available from https://doi.org/10.1016/j.riam.2017.02.006.

Martínez-Soto, D., & Ruiz-Herrera, J. (2013). Transcriptomic analysis of the dimorphic transition of *Ustilago maydis* induced in vitro by a change in pH. *Fungal Genetics and Biology: FG & B, 58*, 116–125. Available from https://doi.org/10.1016/j.fgb.2013.08.011.

Martínez-Soto, D., González-Prieto, J. M., & Ruiz-Herrera, J. (2015). Transcriptomic analysis of the GCN5 gene reveals mechanisms of the epigenetic regulation of virulence and morphogenesis in *Ustilago maydis*. *FEMS Yeast Research, 15*(6), fov055. Available from https://doi.org/10.1093/femsyr/fov055.

Martínez-Soto, D., Ortiz-Castellanos, L., Robledo-Briones, M., & León-Ramírez, C. G. (2020). Molecular mechanisms involved in the multicellular growth of *Ustilaginomycetes*. *Microorganisms, 8*, 1072. Available from https://doi.org/10.3390/microorganisms8071072.

Martínez-Soto, D., Velez-Haro, J. M., León-Ramírez, C. G., et al. (2020). Multicellular growth of the Basidiomycota phytopathogen fungus *Sporisorium reilianum* induced by acid conditions. *Folia Microbiologica, 65*, 511–521. Available from https://doi.org/10.1007/s12223-019-00755-7.

Martins, F. F., Ferreira, T. F., Azevedo, D. A., & Coelho, M. A. Z. (2012). Evaluation of crude oil degradation by *Yarrowia lipolytica*. *Chemical Engineering Transactions, 27*, 223–228. Available from https://doi.org/10.3303/CET1227038.

Mawad, A. M. M., Hesham, A. E., Yousef, N. M. H., Shoreit, A. A. M., Gathergood, N., & Gupta, V. K. (2020). Role of bacterial-fungal consortium for enhancement in the degradation of industrial dyes. *Current Genomics, 21*(4), 283–294. Available from https://doi.org/10.2174/1389202921999200505082901.

Mendarte-Alquisira, C., Gutiérrez-Rojas, M., González-Márquez, H., et al. (2017). Improved growth and control of oxidative stress in plants of *Festuca arundinacea* exposed to hydrocarbons by the endophytic fungus *Lewia* sp. *Plant and Soil, 411*, 347–358. Available from https://doi.org/10.1007/s11104-016-3035-6.

Mishra, V., Gupta, A., Kaur, P., Singh, S., Singh, N., Gehlot, P., & Singh, J. (2016). Synergistic effects of arbuscular mycorrhizal fungi and plant growth promoting rhizobacteria in bioremediation of iron contaminated soils. *International Journal of Phytoremediation, 18*, 697–703. Available from https://doi.org/10.1080/15226514.2015.1131231.

Mohsenzadeh, F., Nasseri, S., Mesdaghinia, A., Nabizadeh, R., Zafari, D., Khodakaramian, G., & Chehregani, A. (2010). Phytoremediation of petroleum-polluted soils: Application of *Polygonum aviculare* and its root-associated (penetrated) fungal strains for bioremediation of petroleum-polluted soils. *Ecotoxicology and Environmental Safety, 73*(4), 613–619. Available from https://doi.org/10.1016/j.ecoenv.2009.08.020.

Mullen, M. D., Wolf, D. C., Beveridge, T. J., & Bailey, G. W. (1992). Sorption of heavy metals by the soil fungi *Aspergillus niger* and *Mucor rouxii*. *Soil Biology and Biochemistry, 24*(2), 129–135. Available from https://doi.org/10.1016/0038-0717(92)90268-3.

Nguyen, L. N., Hai, F. I., Yang, S., Kang, J., Leusc, F. D. L., Roddick, P. W. E., & Nghiem, L. D. (2014). Removal of pharmaceuticals, steroid hormones, phytoestrogens, UV-filters, industrial chemicals and pesticides by *Trametes versicolor*. Role of biosorption and biodegradation. *International Biodeterioration and Biodegradation, 88*, 169–175. Available from https://doi.org/10.1016/j.ibiod.2013.12.017.

Nikolaivits, E., Dimarogona, M., Karagiannaki, I., Chalima, A., Fishman, A., & Topakas, E. (2018). Versatile fungal polyphenol oxidase with chlorophenol bioremediation potential: Characterization and protein engineering. *Applied and Environmental Microbiology, 84*(23). Available from https://doi.org/10.1128/AEM.01628-18, e01628-18.

Nyanhongo, G. S., Gomes, J., Gübitz, G. M., Zvauya, R., Read, J., & Steiner, W. (2002). Decolorization of textile dyes by laccases from a newly isolated strain of *Trametes modesta*. *Water Research, 36*(6), 1449–1456. Available from https://doi.org/10.1016/s0043-1354(01)00365-7.

Ocampo-Hernández, C. E. (2021). El potencial de la biorremediación. *Herreriana II, 2*, 30–33. Available from https://doi.org/10.29057/h.v2i2.6285.

Ortega, H. E., Torres-Mendoza, D., & Cubilla-Rios, L. (2020). Patents on endophytic fungi for agriculture and bio- and phytoremediation applications. *Microorganisms, 8*(8), 1237. Available from https://doi.org/10.3390/microorganisms8081237.

Ostrem-Loss, E. M., & Jae-Hyuk, Y. (2018). Bioremediation and microbial metabolism of benzo(a) pyrene. *Molecular Microbiology, 109*(4), 433–444. Available from https://doi.org/10.1111/mmi.14062.

Pietro-Souza, W., de Campos Pereira, F., Mello, I. S., Stachack, F. F. F., Terezo, A. J., Cunha, C. N. D., … Soares, M. A. (2020). Mercury resistance and bioremediation mediated by endophytic fungi. *Chemosphere, 240*, 124874. Available from https://doi.org/10.1016/j.chemosphere.2019.124874.

Pointing, S. B. (2001). Feasibility of bioremediation by white-rot fungi. *Applied Microbiology and Biotechnology*, *57*(1–2), 20–33. Available from https://doi.org/10.1007/s002530100745.

Prigione, V., Trocini, B., Spina, F., Poli, A., Romanisio, D., Giovando, S., & Varese, G. C. (2018). Fungi from industrial tannins: Potential application in biotransformation and bioremediation of tannery wastewaters. *Applied Microbiology and Biotechnology*, *102* (9), 4203–4216. Available from https://doi.org/10.1007/s00253-018-8876-x.

Quintella, C. M., Mata, A. M. T., & Lima, L. C. P. (2019). Overview of bioremediation with technology assessment and emphasis on fungal bioremediation of oil contaminated soils. *Journal of Environmental Management*, *241*, 156–166. Available from https://doi.org/10.1016/j.jenvman.2019.04.019.

Rajtor, M., & Piotrowska-Seget, Z. (2016). Prospects for arbuscular mycorrhizal fungi (AMF) to assist in phytoremediation of soil hydrocarbon contaminants. *Chemosphere*, *162*, 105–116. Available from https://doi.org/10.1016/j.chemosphere.2016.07.071.

Ramoutar, S., Mohammed, A., & Ramsubhag, A. (2019). Laboratory-scale bioremediation potential of single and consortia fungal isolates from two natural hydrocarbon seepages in Trinidad, West Indies. *Bioremediation Journal*, *23*, 131–141. Available from https://doi.org/10.1080/10889868.2019.1640181.

Rhodes, C. J. (2014). Mycoremediation (bioremediation with fungi) – growing mushrooms to clean the earth. *Chem Speciation Bioavailability*, *26*, 196–198. Available from https://doi.org/10.3184/095422914X14047407349335.

Russo, F., Ceci, A., Maggi, O., Siciliano, A., Guida, M., Papini, M. P., ... Persiani, A. M. (2019). Understanding fungal potential in the mitigation of contaminated areas in the Czech Republic: Tolerance, biotransformation of hexachlorocyclohexane (HCH) and oxidative stress analysis. *Environmental Science and Pollution Research International*, *24*, 24445–24461. Available from https://doi.org/10.1007/s11356-019-05679-w.

Russo, F., Ceci, A., Pinzari, F., Siciliano, A., Guida, M., Malusà, E., ... Persiani, A. M. (2019). Bioremediation of Dichlorodiphenyltrichloroethane (DDT)-contaminated agricultural soils: Potential of two autochthonous saprotrophic fungal strains. *Applied and Environmental Microbiology*, *85*(21), e01720-19. Available from https://doi.org/10.1128/AEM.01720-19.

Sabra, M., Aboulnasr, A., Franken, P., Perreca, E., Wright, L. P., & Camehl, I. (2018). Beneficial root endophytic fungi increase growth and quality parameters of sweet basil in heavy metal contaminated soil. *Frontiers in Plant Science*, *9*, 1726. Available from https://doi.org/10.3389/fpls.2018.01726.

Sabuda, M. C., Rosenfeld, C. E., DeJournett, T. D., Schroeder, K., Wuolo-Journey, K., & Santelli, C. M. (2020). Fungal bioremediation of selenium-contaminated industrial and municipal wastewaters. *Frontiers in Microbiology*, *11*, 2105. Available from https://doi.org/10.3389/fmicb.2020.02105.

Shukla, K. P., Singh, N. K., & Sharma, S. (2010). Bioremediation: Developments, current practices, and perspectives. *Genetic Engineering & Biotechnology*, *2010*, GEBJ-3.

Silar, P., Dairou, J., Cocaign, A., Busi, F., Rodrigues-Lima, F., & Dupret, J. M. (2011). Fungi as a promising tool for bioremediation of soils contaminated with aromatic amines, a major class of pollutants. *Nature Reviews. Microbiology*, *9*(6), 477. Available from https://doi.org/10.1038/nrmicro2519-c1.

Soldi, E., Casey, C., Murphy, B. R., & Hodkinson, T. R. (2020). Fungal endophytes for grass based bioremediation: An endophytic consortium isolated from *Agrostis stolonifera* stimulates the growth of *Festuca arundinacea* in lead contaminated soil. *Journal of Fungi (Basel)*, *6*(4), 254. Available from https://doi.org/10.3390/jof6040254.

Soleimani, M., Hajabbasi, M. A., Afyuni, M., Mirlohi, A., Borggaard, O. K., & Holm, P. E. (2010). Effect of endophytic fungi on cadmium tolerance and bioaccumulation by *Festuca arundinacea* and *Festuca pratensis*. *International Journal of Phytoremediation*, *12*(6), 535–549. Available from https://doi.org/10.1080/15226510903353187.

Spina, F., Cecchi, G., Landinez-Torres, A., Pecoraro, L., Russo, F., Wu, B., ... Persiani, A. M. (2018). Fungi as a toolbox for sustainable bioremediation of pesticides in soil and water. *Plant Biosystems*, *152*. Available from https://doi.org/10.1080/11263504.2018.1445130.

Srivastava, P. K., Vaish, A., Dwivedi, S., Chakrabarty, D., Singh, N., & Tripathi, R. D. (2011). Biological removal of arsenic pollution by soil fungi. *The Science of the Total Environment*, *409* (12), 2430–2442. Available from https://doi.org/10.1016/j.scitotenv.2011.03.002.

Talukdar, D., Jasrotia, T., Sharma, R., Jaglan, S., Kumar, R., Vats, R., ... Umar, A. (2020). Evaluation of novel indigenous fungal consortium for enhanced bioremediation of heavy metals from contaminated sites. *Environmental Technology & Innovation*, *20*, 101050. Available from https://doi.org/10.1016/j.eti.2020.101050.

Tkavc, R., Matrosova, V. Y., Grichenko, O. E., Gostincar, C., Volpe, R. P., Klimenkova, P., ... Daly, M. J. (2018). Prospects for fungal bioremediation of acidic radioactive waste sites: Characterization and genome sequence of *Rhodotorula taiwanensis* MD1149. *Frontiers in Microbiology*, *8*, 2528. Available from https://doi.org/10.3389/fmicb.2017.02528.

Tomer, A., Singh, R., Singh, S. K., et al. (2021). Role of fungi in bioremediation and environmental sustainability. In R. Prasad, S. C. Nayak, R. N. Kharwar, & N. K. Dubey (Eds.), *Mycoremediation and environmental sustainability. Fungal biology*. Cham: Springer. Available from http://doi.org/10.1007/978-3-030-54422-5_8.

Tripathi, P., Khare, P., Barnawal, D., Shanker, K., Srivastava, P. K., Tripathi, R. D., & Kalra, A. (2020). Bioremediation of arsenic by soil methylating fungi: Role of humicola sp. strain 2WS1 in amelioration of arsenic phytotoxicity in *Bacopa monnieri L. The Science of the Total Environment*, *716*, 136758. Available from https://doi.org/10.1016/j.scitotenv.2020.136758.

Vasudevan, V., Gayathri, K. V., & Krishnan, M. E. G. (2018). Bioremediation of a pentacyclic PAH, Dibenz(a,h)Anthracene- a long road to trip with bacteria, fungi, autotrophic eukaryotes and surprises. *Chemosphere*, *202*, 387–399. Available from https://doi.org/10.1016/j.chemosphere.2018.03.074.

Vidali, M. (2001). Bioremediation. An overview. *Pure and Applied Chemistry*, *73*(7), 1167–1172.

Vilela, L. A. F. (2021). Chapter 17 – Molecular and cellular changes of arbuscular mycorrhizal fungi-plant interaction in cadmium contamination. In M. Hasanuzzaman, & M. N. Vara Prasad (Eds.), *Handbook of bioremediation* (pp. 277–283). *Academic Press*. ISBN 9780128193822. Available from https://doi.org/10.1016/B978-0-12-819382-2.00017-X.

Waqas, M., Khan, A. L., Kang, S. M., et al. (2014). Phytohormone-producing fungal endophytes and hardwood-derived biochar interact to ameliorate heavy metal stress in soybeans. *Biology and Fertility of Soils*, *50*, 1155–1167. Available from https://doi.org/10.1007/s00374-014-0937-4.

Wu, B., Zhang, X. L., Cui, B. K., & Dai, Y. C. (2015). Comparative genomic analysis of edible (medicinal) fungi reveals different ecological habitats. *Mycosystema*, *34*(4), 742–760. Available from https://manu40.magtech.com.cn/Jwxb/EN/10.13346/j.mycosystema.150037.

Yan, G., & Viraraghavan, T. (2008). Mechanism of biosorption of heavy metals by *Mucor rouxii*. *Engineering in Life Sciences*, *8*(4), 363–371. Available from https://doi.org/10.1002/elsc.200820237.

Yang, Y., Liang, Y., Han, X., et al. (2016). The roles of arbuscular mycorrhizal fungi (AMF) in phytoremediation and tree-herb interactions in Pb contaminated soil. *Scientific Reports*, *6*, 20469. Available from https://doi.org/10.1038/srep20469.

Zamani, N., Sabzalian, M. R., Khoshgoftarmanesh, A., & Afyuni, M. (2015). Neotyphodium endophyte changes phytoextraction of zinc in *Festuca arundinacea* and *Lolium perenne*. *International Journal of Phytoremediation*, *17*, 456–463. Available from https://doi.org/10.1080/15226514.2014.922919.

Zapana-Huarache, S. V., Romero-Sánchez, C. K., Dueñas Gonza, A. P., Torres-Huaco, F. D., & Lazarte Rivera, A. M. (2020). Design and testing of a cost-efficient bioremediation system for tannery effluents using native chromium-resistant filamentous fungi. *International Journal of Environmental Science and Technology*, *17*, 3825–3834. Available from https://doi.org/10.1007/s13762-020-02726-9.

Zhang, S., Fan, F., & Meng, F. (2020). Seasonality and community separation of fungi in a municipal wastewater treatment plant. *Applied and Environmental Microbiology*, *86*(18), e00991-20. Available from https://doi.org/10.1128/AEM.00991-20.

Zhang, X., Lin, L., Chen, M., Zhu, Z., Yang, W., Chen, B., ... An, Q. (2012). A non-pathogenic *Fusarium oxysporum* strain enhances phytoextraction of heavy metals by the hyperaccumulator *Sedum alfredii* hance. *Journal of Hazardous Materials*, *229–230*, 361–370. Available from https://doi.org/10.1016/j.jhazmat.2012.06.013.

Zhu, Z., Song, Q., & Dong, F. (2018). Taxonomy characterization and plumbum bioremediation of novel fungi. *Journal of Basic Microbiology*, *58*(4), 368–376. Available from https://doi.org/10.1002/jobm.201700469.

CHAPTER 13

Biosurfactants for the recovery and remediation of oil and petroleum waste

Vandana Singh
Department of Microbiology, School of Allied Health Science, Sharda University, Greater Noida, Uttar Pradesh, India

13.1 Introduction

Biosurfactants are a structurally diverse group of surface-active molecules synthesized by microorganism with diverse chemical properties and molecular size. These molecules reduce surface and interfacial tension in both aqueous solutions and hydrocarbon mixture, which make them a potential candidate to be used in oil recovery and bioremediation of petroleum. Biosurfactants are amphiphilic compounds mostly produced or excreted extracellularly by microbial cell and contain both hydrophobic and hydrophilic moieties that reduce surface tension and interfacial tension between individual molecules at the surface and interface, respectively. Due to emulsification properties of biosurfactant, it may often categorize with emulsifiers (Galabova et al., 2014).

In emerging petroleum industry, the biosurfactants plays an important role. Biotechnology based petroleum industry mainly emphasizes on implementation of biological processes for the production, transformation, and refining of petroleum for making valued by-products and reducing its waste as well as pollutants (Bhat et al., 2021; Silva et al., 2014). The metabolic adaptability of the microorganisms and their fundamental ability to facilitate transformation of these multifaceted petroleum raw materials in extreme circumstances such as various ranges of temperature, high salinity, different pH values along with pressure, and hydrophobicity contribute to the development of these compounds (Dar, Bandh, & Kamili, 2011). Many researchers have explored numerous applications of biosurfactant in oil industry and found it more promising (Silva et al., 2014).

Microbial Consortium and Biotransformation for Pollution Decontamination
DOI: https://doi.org/10.1016/B978-0-323-91893-0.00015-8

The potential of biosurfactants has been explored and well proved by many researchers as multifunctional materials and have lots of applications in different industries (Vigneshwaran, Vasantharaj, Jerold, & Sivasubramanian, 2016) other than petroleum, such as medical, cosmetics, food, and pharmaceuticals (Fig. 13.1). Biosurfactants are generally produced by microbes as surface-active compounds to improve or boost the bioaccessibility of hydrophobic immiscible and mostly unreachable substrates required by them for enhancing their survival under low-humid or extreme conditions. The synthesis of biosurfactant generally necessitates the presence of hydrocarbons or miscible oil substrates as a carbon source (Dar, Kamili, Nazir, Bandh, & Bhat, 2012; Dar, Bandh, Kamili, Nazir, & Bhat, 2013). The use of waste products for the synthesis process makes them more economical and eco-friendly. Today, biologically produced surfactants are more in trend, especially in oil and petroleum industries for extraction, recovery, and refining processes than their chemically synthesized counterparts. In petroleum industry, biosurfactants are also explored for the cleaning of oil tanks, oil pipelines, oil sludges, and oil spillage at aquatic resources (Nikolova & Gutierrez, 2021). This chapter mainly emphasizes

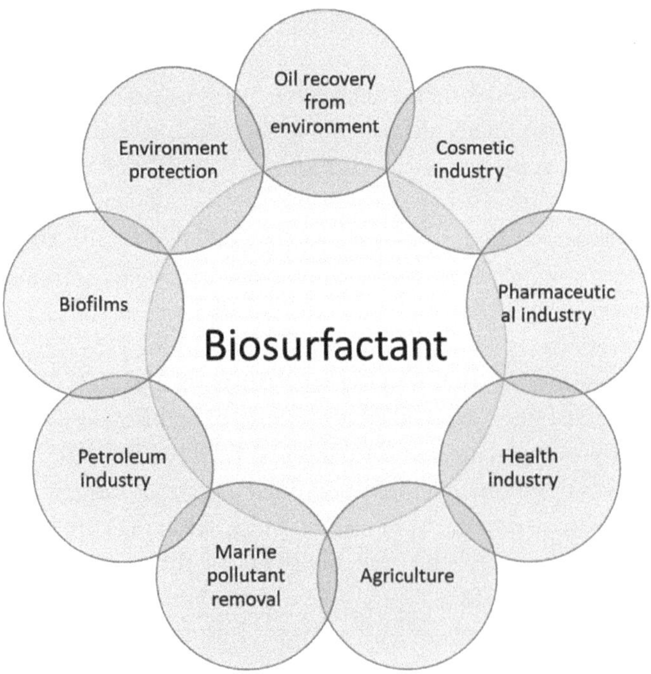

Figure 13.1 Applications of biosurfactants.

on the types of biosurfactant and its applications in bioremediation of hydrocarbon and petroleum oil.

13.1.1 Classification of biosurfactants and their microbial origin

Biosurfactants are mainly categorized on the basis of their origin and chemical composition. The basic structure of biosurfactants composed of a hydrophilic moiety along which may consist of anion or cation of peptide or amino acid, di- or polysaccharide and hydrophobic moiety of saturated or unsaturated fatty acids. As microbial surfactants are complex molecules and may cover wide range of chemical types, broadly, the biosurfactants are classified into two classes Table 13.1 (Galabova et al., 2014):

A. Molecules with low molecular weight

1. Glycolipids
2. Lipopeptides and lipoproteins
3. Phospholipids

B. Polymers with high molecular weight

4. Polymeric and particulate surfactants

1. Glycolipids: The most commonly isolated and stated biosurfactants are belonging to the group of glycolipids. It's a combination of carbohydrates with long chain of hydroxyl aliphatic group. The glycolipid can be further categorized as rhamnolipids, sophorolipids, and trehalolipids.
   a. Rhamnolipids: The biosurfactants having chemical composition rhamnose (glycolipid) was first time isolated or synthesized from *Pseudomonas aeruginosa*. The basic chemical structure is given in

Table 13.1 Major classes of microbial biosurfactants.

Type of biosurfactants	Examples of microbial flora producing biosurfactant
Rhamnolipids	*Rhodococcus* spp., *Pseudomonas* spp., *Arthrobacter* spp.
Trehalose lipids	*Rhodococcus* spp., *Mycobacterium* spp., *Nocardia* spp., *Corynebacterium* spp., *Arthrobacter* spp.
Sophorolopids	*Torulopsis bombicola, Candida* spp., yeast
Lipopeptides and lipoproteins	*Thiobacillus* spp., *Aspergillus* spp., *Arthrobacter* spp., *Bacillus* spp.
Polymeric and particulate surfactants	*Acinetobacter* spp., *Candida* spp., *Schizonella* spp.
Fatty acids and phospholipids	*Thiobacillus* spp., *Aspergillus* spp., *Arthrobacter* spp., *Bacillus* spp.
Natural lipids	*Corynebacterium* spp.

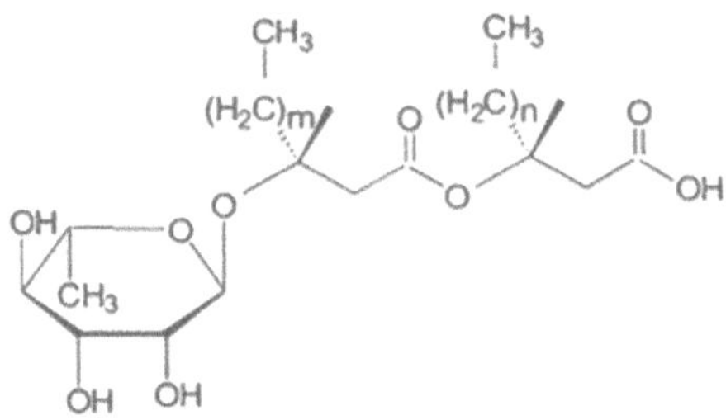

Figure 13.2 Basic structure of rhamnolipids (monomer).

Fig. 13.2. This type of biosurfactant composed of a hydrophilic rhamnose head and a tail of 3-(hydroxy-alkanoyloxy) alkanoic acid (HAA) fatty acid, which can vary in length (Astuti et al., 2019). The phenomenon of employing biosurfactants to dissolve the hydrophobic chemical compounds is not new. Numerous microbes use this biosurfactant phenomenon for accessing unavailable carbon sources provided in the medium. Besides mono-rhamnolipids, *P. aeruginosa* is also known for producing di-rhamnolipids, which composed of two hydrophilic rhamnose heads. In *Pseudomonas*, the genes, which are accountable for making the mono-rhamnolipids, are known as Rhl-A and Rhl-B. The pair of these genes activate the enzyme which form subunits of rhamnosyl-transferase-1 and are systematized in an operon which increases its proficiency. The gene Rhl-A mainly involves in creating tail of fatty acid, whereas Rhl-B attaches the tail to a rhamnose molecule. For the detection and confirmation of these rhamno-biosurfactants, thin-layered chromatography and mass spectrometry were always preferred (Han et al., 2014).

b. Trehalolipids: These are also known as "cord factor." These type of biosurfactants are produced by *Actinomycetes* spp. Various types of microbial trehalolipid biosurfactants have been reported by researchers such as di-saccharide trehalose linked at C6 and C6 to oxalic acid is mainly associated with *Mycobacterium, Nocardia and Corynebacterium* spp. The cord factor from different organisms is different in size and the basic structure of it is shown in Fig. 13.3 (Franzetti et al., 2010; Vigneshwaran et al., 2016).

c. Sophorolopids: They are generally reported to be produced by strains of yeast *Torulopsis* which consists of two beta-1,2-linked glucose units. The hydroxyl group is generally acylated and to its

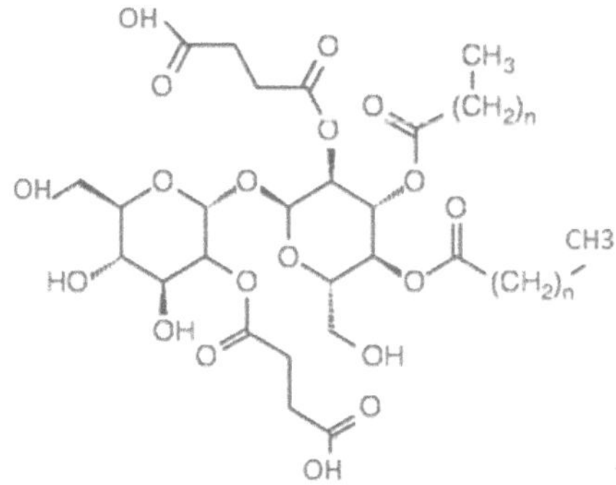

Figure 13.3 Structure of trehalolipids.

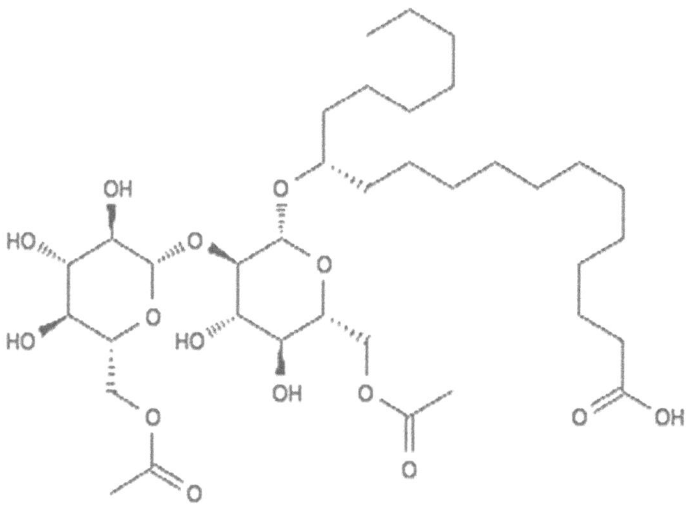

Figure 13.4 Basic structure of sophorolipids.

reducing end lipid portion attached through the glycosidic linkage. Researchers also have been reported the high quantity production of sophorolipids by *Torulopsis bombicola*. The basic structure is shown in Fig. 13.4 (Roy, 2018).

2. Lipopeptides and lipoproteins: Several bacterial and yeast strains are reported for producing large quantity of fatty acid and phospholipid surfactants during their growth on n-alkanes. Phospholipids are the major components of microbial membrane. Several bacterial and fungal strains have been explored for the production of phospholipid biosurfactant. Few examples are *Thiobacillus thiooxidans*, *Aspergillus* spp., *Arthrobacter* spp., and *P. aeruginosa*. These compounds have potent

surfactant properties and are able to produce optically clear microemulsions of alkanes in water. Cyclic lipopeptide produced by *Bacillus subtilis* is the most effective biosurfactant discovered so far, it is even superior to sodium lauryl sulfate. The structural example of the same is shown in Fig. 13.5 (Roy, 2018; Shoeb et al., 2013).

3. Polymeric and particulate surfactants: The best studied polymeric biosurfactants are emulsan, liposan, mannoprotein, and other polysacc—protein complexes. In the same concern *Acinetobacter* spp. (*RAG-1*) has been explored and it produces biosurfactant called emulsan (Fig. 13.6). Emulsan consists of heteropolysaccharide backbone containing a repeating trisaccharide of *N*-acetyl-D-galactosamine, *N*-acetylgalactosamine uronic acid, and unidentified *N*-acetyl amino sugar. Emulsan is a very effective emulsifying agent for hydrocarbons in water even at low concentration. It is one of the most powerful emulsion stabilizers known till date. Particulate biosurfactants is produced by some microbes from extracellular membrane vesicles which help in partition of hydrocarbon from oil—water molecule to form microemulsion which plays an important role in alkane uptake. *Acinetobacter* spp. *Aeromonas* spp., *Serratia* spp., *Bacillus brevis*, and *Acinetobacter calcoaceticus* are some of the strains of microbes studied for the production of particulate biosurfactant (Shoeb et al., 2013).

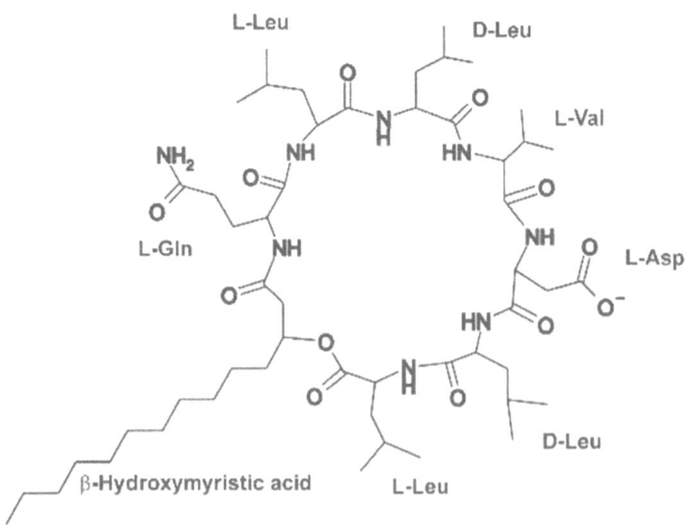

Figure 13.5 Structural example of lipoprotein biosurfactant.

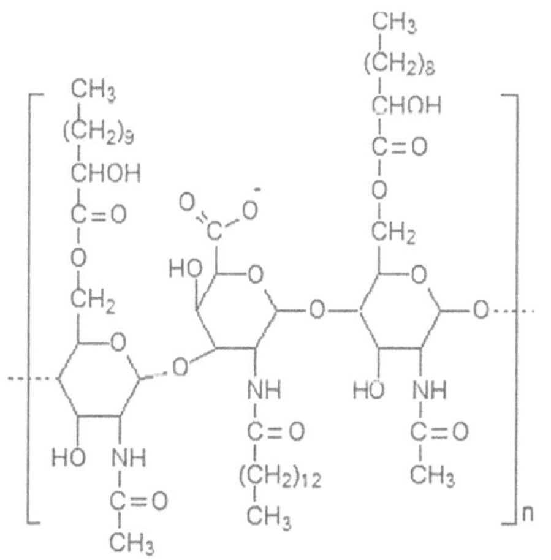

Figure 13.6 Structural example of polymeric and particulate surfactants (emulsan).

These biosurfactant molecules have numerous significance and advantages over their chemical counterparts as they are economical, compatible to environment, biodegradable, of negligible toxicity, heat stable, biocompatible, and of less maintenance in terms of extreme conditions, such as salinity, acidity, and pH (Md, 2012). These properties of biologically produced biosurfactants make them eligible candidate to be used in oil industry.

13.1.2 Properties of biosurfactants to be used in pollutant remediation

1. Biosurfactants are eco-friendly, effective, nontoxic, biodegradable, and cost-effective product (Md, 2012; Calvo et al., 2009).
2. These have property of emulsifier for various substances such as hydrocarbon and are highly active in diverse pH and temperature range.
3. These regulate hydrophobic elements into microbial cells and help in biodegradation.
4. Organic substances of these can be easily synthesized from bacterial strain with industrial waste and these are safe, renewed, and natural.
5. These have high synergetic action when applied with synthetic surfactants.

13.2 Biosurfactants in petroleum industries

Petroleum industry is supposed to be originated from microbiomes and their organic matters, which further deposited over the years to form plankton. This may not get oxidized by oxidation process and keep amassed at the ocean's bottom and sheltered by other remains or sediments. The interaction between sediments and organic matter, along with appropriate thermochemical conditions, is the major requirement for initiation of procedural chain vital for the formation of petroleum. Crude oil typically involves three diverse mechanisms and stages: liquid, solid, and gas phase. After collection of crude oil, the petroleum engineering uses numerous separation mechanics and machineries, such as prospecting, drilling, and seismic activities to separate these all phases from each another (De Almeida et al., 2016).

The crude oil recovery mainly possesses three processes called primary, secondary, and tertiary (shown in Fig. 13.7). The primary recovery is principally based on reservoir and its natural distinctive energy used to dislocate oil from the porous rocks, while in secondary recovery process, injection of water and/or gas is used or involved for enhancing the oil displacement, agility, and productivity from the wells. Substantial quantity of crude oil (up to 50%) is often unrecoverable by these two conventional oil recovery procedures. The leftover oil is trapped in reservoirs again. These leftovers are then recovered by tertiary recovery procedure. The

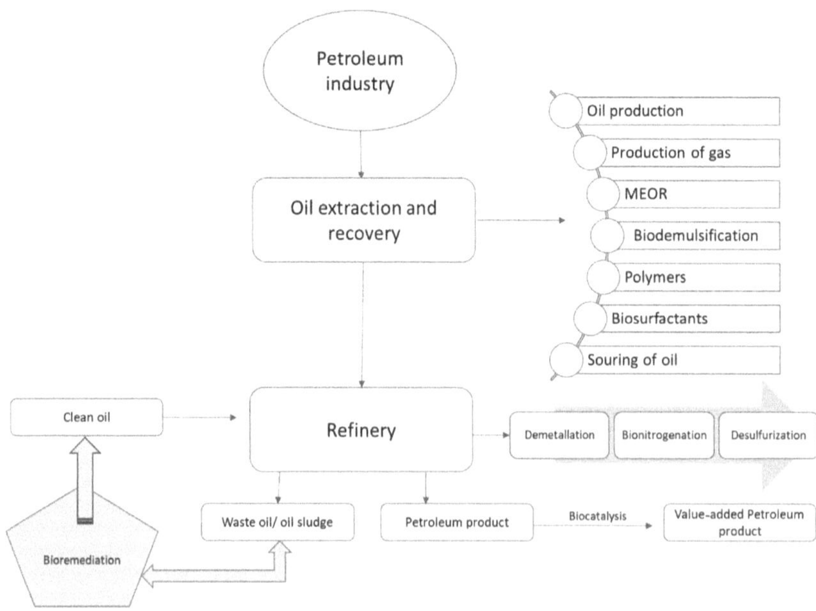

Figure 13.7 Processing steps of petroleum industry.

tertiary process is used for enhancing oil production and recovery further and which may recover significant amount of oil (remaining after performing conventional methods) (De Almeida et al., 2016; El-Sheshtawy et al., 2017).

The refining of petroleum is normally done by applying physicochemical procedures such as chemical-based catalysis and distillation with high pressures and temperatures, which make them ready to be used in market and have defined specifications, such as diesel, petrochemical, and gasoline. The practice of bioprocesses such as biodemetallation, biodesulfurization, biodenitrogenation, and biotransformation in petroleum industry has increased their application associated with refining. These technologies may upgrade the quality and quantity of fuels and also help in producing fine chemical compounds, decreasing souring process occurred during production, and enhancing supplementing approaches such as biodegradation, microbial enhanced oil recovery (MEOR) and bioremediation (Khire, 2010). By considering biotechnology to be better in use than the other conventional methods for oil production, the biosurfactants have been observed as most promising because these can be applied to almost in all phases of the oil production chain (Perfumo, Rancich, & Banat, 2010). It has been well stated that biosurfactant has played an important role in enhanced recovery of crude oil from their exhausted reservoirs, increased and stationary petroleum production, along with refining processes. Further it's also important to keep the environmental safe and help in remediation and disposal practices (Nikolova & Gutierrez, 2021).

13.2.1 Characteristics of biosurfactants to be used in petroleum industry

1. Biosurfactants are amphipathic complexes having hydrophobic and hydrophilic moieties that favorably create a barrier at the edge between two different phases such as gas–liquid, liquid–solid, solid–gas, and with liquids of diverse polarities (oil/water and water/oil) and hydrogen bonding. These molecules reduce the surface and interfacial tension, conferring many properties such as detergency, emulsifying, foaming, and dispersing, making them versatile chemical process. Petroleum industry mostly employs petrochemical-based synthetic surfactants as mobilizing agents in their activities. Hence, in search of sustainable methodology in terms of natural, environment-friendly, and biodegradable compounds, biosurfactants came to the depiction (Mazaheri Assadi & Tabatabaee, 2010).

2. As we know, biosurfactants are principally produced by microbes grow on water immiscible substrates and that is may be one of the reasons that they have access to hydrophobic substrates, such as oil and hydrocarbons. Their low level of toxicity and ability of biodegradability make them useful for oil and petroleum industries. But at the same time, the high cost of production is a main constrain for the market growth of biosurfactants. If it can be produced from economical renewable agricultural resources and waste products, then the cost will significantly decrease (Silva et al., 2014).
3. As per resent research and increased environmental awareness, these green synthesized surfactants have been a choice for the replacement of chemical surfactants (Akhundova & Atakishiyeva, 2015).
4. Researchers have patented the properties of biosurfactant in the various processes of petroleum industry such as wetting, separation at different phases, emulsification, solubilization, foaming, de-emulsification, corrosion inhibition, and for reducing viscosity of heavy crude oils (De Almeida et al., 2016).
5. The biosurfactants are also explored to provide ease in forming composition and techniques for the transportation and combustion of extremely viscous hydrocarbon emulsions (in water) and stabilized bioemulsifier emulsions of hydrocarbon (in water) (Ireland, 2010).
6. Researchers have also patented recovery of crude petroleum oil from their reservoir by incorporating MEOR method and also as bioemulsifier for the stabilization of hydrocarbons (Geetha, Banat, & Joshi, 2018).
7. Biosurfactants were also explored as cleaning agent for contaminated oil tankers, retrieval of oil from sludge oil-storing containers, and also in carriage of heavy crude oil (Cameotra & Singh, 2008).

13.2.2 Oil waste treatment using biosurfactants

In oil or petroleum industries, at every step and operation of oil from extraction of crude oil to recovery and biodegradability, biosurfactant plays an important role. Various processes of petroleum industry are extraction and production of oil, transportation, refining, and preparation of commercial products' waste management. Biosurfactants have the ability to deal with every step, along with bioremediation of oil and oil pollution (Table 13.2) (Calvo et al., 2009; De Almeida et al., 2016).

Table 13.2 The usage and application of biosurfactants in different stages of petroleum/oil production processes in petroleum/oil industries.

Petroleum/oil industries Stages	Claim
Oil extraction	• Modification of reservoir wettability • Reduction of oil viscosity • Mud drilling
Oil transportation/oil pipes	• Reduction of oil viscosity • Stabilization of oil emulsion • Deposition of asphalt/paraffin
Oil container/tank/pipe cleanup	• Reduction of oil viscosity • Dispersion of oil hydrocarbon • Emulsification of oily sludge
Recovery/treatment of oil waste	• Mobilization of oil • Solubilization of oil

13.2.3 Mechanism for recovery and removal of oil

The mechanisms behind enhanced recovery and removal of petroleum/oil by the use of biosurfactants are either through mobilization, solubilization, emulsification or by enhancing the surface area for contact of hydrocarbons (De Almeida et al., 2016). Solubilization property of surfactants has the ability to intensify the solubility of oily or hydrophobic components in liquid medium. As a result, a substantial increase in solubilization forms micelles and causes segregation of the hydrocarbon from hydrophobic portion of the micelles. In this type of progression, increased amount of biosurfactants (in terms of quantity) are essentially required, as solubility of hydrocarbon exclusively depends on the concentration of biosurfactant. The mechanism, mobilization is principally working on two segments: dispersion and displacement (Shi et al., 2019). Dispersion is a process which depends on the concentration of biosurfactant along with interfacial tension. In this process, hydrocarbons are dispersed into the aqueous phases by the formation of emulsions and consequently it is allied to the both interfacial tension and concentration of surfactant. Whereas displacement occurs when droplets of hydrocarbon are coming out from the absorbent or porous medium because of the reduction process in interfacial tension. The entrapped hydrocarbon endures displacement process and when sufficient reduction of interfacial tension occurs between the oil and aqueous phases by disabling the capillary forces which may cause the formation of saturated residuals. Therefore the displacements are mainly associated with interfacial tension which may occur between oil and aqueous solution (Nikolova & Gutierrez, 2021).

13.2.4 Extraction of crude oil by the use of biosurfactants

The oil production approaches are mainly divided into three recovery processes: primary, secondary, and tertiary. By using primary recovery process, the preliminary oil is mined under normal pressure and often recovers about 20% of the original oil from the resource (Dhanarajan et al., 2018). Primary recovery is always followed by the secondary recovery processes due to less amount of oil yields from reservoir and the secondary recovery process uses gas or water injections for recovery. By the secondary recovery process, the yield of oil can be recovered up to 50% (Bachmann, Johnson, & Edyvean, 2014). The tertiary or improved oil recovery process comprises thermal- and chemical-based techniques. Thermal processes mainly performed by using high temperature, hot water, steam, and combustible gas injection, which may increase the temperature of the gases and oil present in their respective reservoirs and enable their flow to the extraction wells. On the other hand, the chemical processes involve the inoculation of hydrocarbon solvents, gas, surfactants, or their combinations for the extraction of residual oil, by dropping the interfacial tension between water and oil (Bachmann et al., 2014). These techniques are very expensive and even not environment-friendly, so for the same, cost and eco-friendly substitutes are required (Perfumo, Banat, & Marchant, 2018).

For overcoming the abovementioned problems, MEOR can be explored as solution. Many researchers explored MEOR as tertiary recovery process for oil recovery, in which microbial agents and their metabolic products are employed for enhanced recovery of residual oil. This method is less expensive than their chemical-based counterparts, principally when microbes are employed to produce adequate products such as biosurfactants and polymers with the help of economical substrates and raw ingredients (Khire, 2010). The key principle of employing biosurfactants for petroleum and oil recovery is its main focus on improving mobility of hydrocarbon and therefore it enhances the recovery of crude oil from their respective reservoirs (El-Sheshtawy et al., 2017).

There are three main strategies for biosurfactants used in MEOR:

1. Ex situ production by using bioreactors for continuous or batch culture followed by sufficient injection of biosurfactant into the reservoir

accompanied by the water flood (also known as ex situ MEOR) (Geetha et al., 2018). As we know, the synthesis or production of biosurfactant is reliant on the composition of medium along with other environmental conditions and controlled setting, these parameters are also important for ex situ production of these surface-active agents produced by the exogenous mixed culture of microbes budding in in situ or supplemented in inoculated flood waters comprehending hydrophobic substrate. The enhanced amount of energy and carbon sources also promotes the synthesis of these surface-active agents called biosurfactant (Bachmann et al., 2014).

2. Biosurfactant-producing microbial flora is added in oil reservoirs to allow in situ dispersal. These microbes play an important part in the surface interactions at interphases between water and oil. It has been stated by many researchers that at the water/oil interphase, the formed emulsions are proportional to the total microbial biomass produced; hence, increased amount of emulsion is seen with higher measurable values of biomass (Geetha et al., 2018).
3. Nutrients: Addition of vital elements (nutritional compounds) into the reservoir is very much required for the stimulation of required microbial growth and production of biosurfactant. Microbes grow exponentially under favorable environments and also produce required metabolic products (biosurfactant) along with gases, which increases the mobilization of residual oil from wells (Khire, 2010).

The biosurfactant-based approaches increase the yields of petroleum from their depleted reservoir by reducing interfacial tension and oil-rock surface along with passageway forces, which may obstruct oil mobility through the pores of rock. These microbial surfactants also break down the oil film from reservoir and form stable water–oil emulsion which maximizes the oil extraction. With lots of significance, the MEOR however also has some drawbacks which may comprise augmented corrosive reaction in nonresistant apparatus because of the air arrayed in aerobic MEOR or by inoculation of high nutrient medium/additives through downhole piping (Khire, 2010).

13.2.5 Biosurfactants for the transportation of crude oil

Due to high viscosity, crude oil often has an ability to form sludge deposition in the transportation pipes. The crude oil is regularly transported through tubes–lines from the reservoir or extraction unit to shipping tanks and then to depots and refineries. During such transportation, heavy or sometime extra-heavy crude forms sludges in the inner lining of these pipelines which may cause obstructions in transportation. The high grade of viscosity of crude oil is due to the presence of high level of asphaltene and paraffin contents. The presence of asphaltenes may lead to the formation of precipitate under acidic conditions and if ferric ions are also present then it may form asphaltene mud which would obstruct the flow of crude oil in the pipeline. On the other hand, the paraffin may decrease the liquidity of oil as it has cyclic hydrocarbons in high amount that can get solidify and deposit at room temperature which may result as blockage in pipeline and so in transportation. To solve these problems an eco-friendly option is biosurfactant, which may work as emulsifier and may facilitate the oil mobility (Cameotra & Singh, 2008; Mazaheri Assadi & Tabatabaee, 2010).

13.2.6 Cleaning of oil storage vessels for oil recovery

Oil-storing containers and vessels are required to be clean and maintained on regular basis. The oil sludge that forms at the bottom and on the walls of these tanks is also extremely viscid and very difficult to be removed by using conventional methods (pumping). These sludge materials are often required to be cleaned up either manually or by using hot/steam water or by applying chemical-based solvents, which are hazardous, both time and labor exhaustive, expensive, and usually result in the production of large amounts of waste material for disposal (Cameotra & Singh, 2008). The cleaning of oil-based sludges from the storage tanks by using surfactants was introduced in 1981, but the use of biologically synthesized surfactants for the same was applied in 1991 at the Kuwait Oil Company. It was also reported that by using biosurfactant, about 91% of the hydrocarbon was recovered from the sludge (De Almeida et al., 2016).

By using the same principle, many oil-based industries are working in the arena of oil remediation and diffusion. Biosurfactants are replacing the counter chemical-based surfactants as biologically synthesized have better biodegradability, inducibility, and bioremediation ability (Akhundova & Atakishiyeva, 2015).

13.3 Biosurfactants for oil waste treatment and bioremediation

The considerable amount of oil-based sludge generates during extraction, transportation, storage, and refining processes. The discarding of such remainders has been always a foremost problem confronted by fuel industries. These sludges of oil are a compound of emulsion which may consist of hydrocarbon (petroleum), water, some solid particles, and heavy metals (Ireland, 2010; Zhang et al., 2012).

Worldwide, numerous approaches and technologies have been opted by petroleum industry for the management of these oil-generated wastes. Several chemical, physical, and physiochemical methods have been explored, for instance, dewatering, incineration, solvent extraction, pyrolysis, stabilization, applying hot water, steam or chemical surfactant, and, above all, biodegradation. Various in vitro studies and pilot studies have proved that the biodegradation-based biological approaches are considered to be more apt because of their high selectivity, enhanced recovery, and low toxicity (Fenibo et al., 2019; Zhang et al., 2012).

Petroleum industry unavoidably generates large volumes of oily wastewater which has become an urgent challenge for most oilfields and petroleum companies, focusing attention toward its efficient treatment techniques. The separation procedures, such as ultrafiltration, centrifugation, decantation, flocculation, and flotation, are some examples of physiochemical methods which are efficiently employed for segregation of water—oil combinations. Some time while performing these procedures chemical-based coagulants have also been used for enhancing the efficiency (Shi et al., 2019). In the same procedure, biosurfactants can act as biologically produced coagulants and increase the efficiency of these procedures (Maikudi Usman et al., 2016).

13.4 Biosurfactants as demulsifying agents

While refining petroleum, at every stage, oilfield emulsion may form and is a major problem because it often forms amphiphilic molecules such as dammar-like segment which may contain asphaltenes and naphthenic acids and sometime fine artefacts (solid state) such as wax, scales, and clay crystals. These emulsions can be broken by de-emulsification process which may include disruption of the thermodynamic environments at the edge, primarily to disrupt the stable surfaces amid the internal and bulk phases. Consequently, de-emulsification is an important process and need to be performed before downstream oil processing. As de-emulsification is a difficult process, it is generally supported by physiochemical methods which may include chemical treatment, heat treatment, centrifugation, and electrical treatment. The chemical-based de-emulsifier may have environmental issues in terms of disposal, so here biosurfactants can replace the chemical substitute and also have potential to solve the problem of emulsion. De-emulsifiers originated from microbial origin usually have good biodegradability and low toxicity which cannot be matched by chemical-based counterparts. It has been explored by many researchers and stated that biologically synthesized emulsifier (biosurfactant) can be easily recovered and removed at the completion of the de-emulsification process (Silva et al., 2018). Glycolipids, phospholipids, glycoproteins, and polysaccharides are some microbiologically produced metabolites which are capable of removing emulsification from water–oil interface (Perfumo et al., 2010).

13.5 Bioremediation of oil waste and spilling

Crude oil is composed of multiple hydrocarbon and nonhydrocarbon metals and species, each having their different aqueous solubility. It is highly hydrophobic, so when it is mixed with water phase, it floats on the water surface as it has lower density in comparison to water. Oil spill in water bodies during transportation, extraction, and other process or sometimes by ships may affect the marine life in different ways (Karlapudi et al., 2021). It causes sublethal damage or kills fish larvae and eggs and increases mortality, fouling gill structures, habitat destruction, etc. The polycyclic aromatic hydrocarbons are mainly responsible for toxicity in oil (Dar, Kamili, Dar, Chishti, & Ahmad, 2015; Dar et al., 2016). A small concentration of polycyclic aromatic hydrocarbons leads to death, malformed morphological structure, stunted development, circulatory failure, and low appetite in the early stages of fish (Silva et al., 2018). It is because crude oil affects them at their gene level and mainly to the genes that maintain the water amount, ion, and morphogenesis of specific organs or tissues. Dispersed oil becomes available for biodegradation through hydrocarbon-degrading bacteria as dispersed oil is mainly found as a fine droplet of neutral buoyant with higher surface-to-volume ratio in comparison to nondispersed oil (Galabova et al., 2014).

Many researchers have explored marine oil spill bioremediation studies with oil-degrading bacterial strains such as *Brevibacillus* and *Ochrobactrum* without/with rhamnolipids. It was also observed that the efficiency of oil removal by bacterial alliance alone was lower by 6% when it was compared with the combination of biosurfactant and bacterial alliance. It was also noted that biodegradation of alkanes was enhanced in the presence of rhamnolipid (Müller et al., 2012; Zhang et al., 2012). Researchers also explored biosurfactant as bioemulsifier and also aid in oil spill response with positive results (Phulpoto et al., 2021). An exopolysaccharide bioemulsifier produced by strain *A. calcoaceticus* wax explored for biodegradation of crude oil in seawater and as a result, total oil biodegradation was enhanced by twofold (Karlapudi et al., 2021; Patowary et al., 2018).

13.6 Biodegradation of diesel by biosurfactants

Diesel is one of the hydrophobic pollutants inclined to bind with the particles of soil and water in such a manner that it is difficult for the microbe to access it for degradation. Generally, chemical-based surfactants are in use for remediation of these pollutants (Ireland, 2010). At the same term, the biosurfactants have also been explored and found to be useful for removal of these oils (petrol or diesel) in the case of oil spills or contaminations. Biosurfactants are well proficient for reduction of surface tension and enhancement of the solubility of diesel (Bento et al., 2005). Many researchers have explored rhamnolipid for removal and degradation of diesel from water systems. Some researchers employed rhamnolipid along with surfactin for the degradation of diesel and stated better method for removal or degradation of hydrocarbon. *Achromobacter* spp. was also used for the synthesis of rhamnolipid and saponin surfactants which exponentially increased the biodegradation of diesel and also altered the surface of the cell. Similarly, *Rhodococcus* (hydrocarbon-oxidizing strain) was examined for the degradation of petroleum and was found to be efficient (Ganesh & Lin, 2009). Emulsification of oil, hydrocarbons, and petroleum products was evaluated by use of *Pseudomonas* sp., and as a result, hexadecane exhibited that microbial surfactant efficiently emulsified these hydrophobic substrates. Based on these properties, biosurfactant synthesized by *Pseudomonas* sp. was also examined for the treatment of oil-spilled or polluted sites at several environmental circumstances (Karlapudi et al., 2021; Szulc et al., 2014).

13.7 Bioremediation of metal-contaminated sites by biosurfactants

Heavy metals are generally nonbiodegradable and highly toxic and may cause several environmental problems. Due to the excessive use of heavy metals in industrial use, these are widely distributed as environmental contaminant and may be the part of water and soil pollution. The remediation techniques generally used for the removal of these contaminants (metal) from soil are either removal from the site or dumping at the site (Zouboulis et al., 2004). Removal procedures may include soil extraction and phytoremediation, whereas, in dumping at site requires solidification and stabilization process for decreasing their availability. The removal of heavy metals from soil matrix is a difficult task and use of chemical surfactants as resolution for the same may increase chemical level at soil and water. Sluggish release and desorption from soil to water phase confirm lasting contaminant reason and may create hindrance in remediation. For the removal of these metals from the medium, microbial biosurfactant is a better choice. Biosurfactants have the properties of stabilizing water–oil emulsions and increasing hydrocarbon solubility along with biodegradation rate (Sarubbo et al., 2015). The removal efficiency of biosurfactant is measured on the basis of physical and chemical properties (Pacwa-Płociniczak et al., 2011). The removal efficiency is in fact determined by the physical and chemical properties of the biosurfactants and not by their effects on the cell surface properties and microbial metabolism. The mechanism of these interactions between metals and biosurfactants is not exactly known, but based on assumption, it may be because of their requirement of metabolic activity. Hence, biosurfactants are adequate to be used as bioremediation agent. It is safer than their counterparts (chemical-based surfactants) to living organisms.

13.8 Conclusion

This chapter is mainly emphasized on the recovery of oil from oil waste and its bioremediation by the use of biosurfactants, naturally produced by several microbial floras. Biosurfactants can be very useful at any stage and form of petroleum or oil recovery. It is concluded that advances in the petroleum industry may require an effective and eco-friendly mechanism to solve all types of problems related to oil biotechnology. Biosurfactant is increasingly acquainting credit and appreciation in the oil industry and also in aquatic oil spilling problems due to their effective action and replacement to their chemical counterparts. These complexes are not only supporting the processes of petroleum industry but are improving the quality, which make them essential compounds in fuel biotechnology. The one major benefit of biosurfactants is their biodegradability property which potentially reduces the harmful impacts on environment. Biosurfactants have all the potential to be explored as environmental control or in bioremediation. As biosurfactants are a complex organic molecule with specific functional groups and often specific in their action, such as detoxification of specific pollutants and de-emulsification of industrial emulsion, these can be efficiently used for handling industrial emulsions, control of oil spills, biodegradation and detoxification of industrial effluents, and bioremediation of oil-contaminated soil. This chapter also comprises potential of biosurfactants in protecting environment from oil pollutants, along with their (biosurfactants) biological and physicochemical properties.

References

Akhundova, E., & Atakishiyeva, Y. (2015). Phytoremediation for green energy. *Phytoremediation for Green Energy*. Available from https://doi.org/10.1007/978-94-007-7887-0.

Astuti, D. I., et al. (2019). Screening and characterization of biosurfactant produced by *Pseudoxanthomonas* sp. G3 and its applicability for enhanced oil recovery. *Journal of Petroleum Exploration and Production Technology*, *9*(3), 2279–2289. Available from https://doi.org/10.1007/s13202-019-0619-8.

Bachmann, R. T., Johnson, A. C., & Edyvean, R. G. J. (2014). Biotechnology in the petroleum industry: An overview. *International Biodeterioration and Biodegradation*, *86*, 225–237. Available from https://doi.org/10.1016/j.ibiod.2013.09.011.

Bento, F. M., et al. (2005). Diversity of biosurfactant producing microorganisms isolated from soils contaminated with diesel oil. *Microbiological Research*, *160*(3), 249–255. Available from https://doi.org/10.1016/j.micres.2004.08.005.

Bhat, R. A., Singh, D. V., Qadri, H., Dar, G. H., Dervash, M. A., Bhat, S. A., ... Yousaf, B. (2021). Vulnerability of municipal solid waste: An emerging threat to aquatic ecosystems. *Chemosphere*, *287*(2022), 132223. Available from https://doi.org/10.1016/j.chemosphere.2021.132223.

Calvo, C., et al. (2009). Application of bioemulsifiers in soil oil bioremediation processes. Future prospects. *Science of the Total Environment*, *407*(12), 3634–3640. Available from https://doi.org/10.1016/j.scitotenv.2008.07.008.

Cameotra, S. S., & Singh, P. (2008). Bioremediation of oil sludge using crude biosurfactants. *International Biodeterioration and Biodegradation*, *62*(3), 274–280. Available from https://doi.org/10.1016/j.ibiod.2007.11.009.

Dar, G. H., Bandh, S. A., & Kamili, A. N. (2011). Bacteriological analysis of soil from Yusmarg Health Resort of Kashmir Valley. *International Journal of Current Research*, *3*(10), 53–56.

Dar, G. H., Bandh, S. A., Kamili, A. N., Nazir, R., & Bhat, R. A. (2013). Comparative analysis of different types of bacterial colonies from the soils of Yusmarg Forest, Kashmir Valley India. *Ecologia Balkanica*, *5*(1), 31–35.

Dar, G. H., Kamili, A. N., Chishti, M. Z., Dar, S. A., Tantry, T. A., & Ahmad, F. (2016). Characterization of *Aeromonas sobria* isolated from Fish Rohu (*Labeo rohita*) collected from polluted pond. *Journal of Bacteriology & Parasitology*, 7(3), 1–5. Available from https://doi.org/10.4172/2155-9597.1000273.

Dar, G. H., Kamili, A. N., Dar, S. A., Chishti, M. Z., & Ahmad, F. (2015). Detection and characterization of potentially pathogenic *Aeromonas sobria* isolated from fish *Hypophthalmichthys molitrix* (Cypriniformes: Cyprinidae). *Microbial Pathogenesis* (2015). Available from https://doi.org/10.1016/j.micpath.2015.10.017.

Dar, G. H., Kamili, A. N., Nazir, R., Bandh, S. A., & Bhat, R. A. (2012). A preliminary study of colony forming units of bacteria from the soils of Yusmarg Forest, Kashmir Valley, India. *International Journal of Current Research*, *4*(12), 467–472.

De Almeida, D. G., et al. (2016). Biosurfactants: Promising molecules for petroleum biotechnology advances. *Frontiers in Microbiology*, 7(OCT), 1–14. Available from https://doi.org/10.3389/fmicb.2016.01718.

Dhanarajan, G. et al. (2018). 'Performance evaluation of biosurfactant stabilized microbubbles in enhanced oil recovery', *bioRxiv*. doi: 10.1101/504431.

El-Sheshtawy, H. S., et al. (2017). Enhancement the bioremediation of crude oil by nanoparticle and biosurfactants. *Egyptian Journal of Chemistry*, *60*(5), 835–848. Available from https://doi.org/10.21608/EJCHEM.2017.1321.1082.

Fenibo, E. O., et al. (2019). Microbial surfactants: The next generation multifunctional biomolecules for applications in the petroleum industry and its associated environmental remediation. *Microorganisms*, 7(11), 1–29. Available from https://doi.org/10.3390/microorganisms7110581.

Franzetti, A., et al. (2010). Production and applications of trehalose lipid biosurfactants. *European Journal of Lipid Science and Technology*, *112*(6), 617–627. Available from https://doi.org/10.1002/ejlt.200900162.

Galabova, D., et al. (2014). *Role of microbial surface-active compounds in environmental protection. The role of colloidal systems in environmental protection.* Elsevier B.V. Available from https://doi.org/10.1016/B978-0-444-63283-8.00003-X.

Ganesh, A., & Lin, J. (2009). Diesel degradation and biosurfactant production by Gram-positive isolates. *African Journal of Biotechnology*, *8*(21), 5847–5854. Available from https://doi.org/10.5897/ajb09.811.

Geetha, S. J., Banat, I. M., & Joshi, S. J. (2018). Biosurfactants: Production and potential applications in microbial enhanced oil recovery (MEOR). *Biocatalysis and Agricultural Biotechnology*, *14*, 23–32. Available from https://doi.org/10.1016/j.bcab.2018.01.010.

Han, L., et al. (2014). Engineering the biosynthesis of novel rhamnolipids in *Escherichia coli* for enhanced oil recovery. *Journal of Applied Microbiology*, 139–150. Available from https://doi.org/10.1111/jam.12515.

Ireland, N. (2010). *Handbook of Hydrocarbon and Lipid Microbiology*. Available from https://doi.org/10.1007/978-3-540-77587-4.

Karlapudi, A. P., et al. (2021). Role of biosurfactants in bioremediation of oil pollution - A review. *Petroleum*, 7(2), 230. Available from https://doi.org/10.1016/j.petlm.2021.01.007.

Khire, J. M. (2010). Bacterial biosurfactants, and their role in microbial enhanced oil recovery (MEOR). *Advances in Experimental Medicine and Biology*, *672*, 146–157. Available from https://doi.org/10.1007/978-1-4419-5979-9_11.

Maikudi Usman, M., et al. (2016). Application of biosurfactants in environmental biotechnology; remediation of oil and heavy metal. *AIMS Bioengineering*, *3*(3), 289–304. Available from https://doi.org/10.3934/bioeng.2016.3.289.

Mazaheri Assadi, M., & Tabatabaee, M. S. (2010). Biosurfactants and their use in upgrading petroleum vacuum distillation residue: A review. *International Journal of Environmental Research*, *4*(4), 549–572. Available from https://doi.org/10.22059/ijer.2010.242.

Md, F. (2012). Biosurfactant: Production and application. *Journal of Petroleum & Environmental Biotechnology*, *3*(4). Available from https://doi.org/10.4172/2157-7463.1000124.

Müller, M. M., et al. (2012). Rhamnolipids-next generation surfactants? *Journal of Biotechnology*, *162*(4), 366–380. Available from https://doi.org/10.1016/j.jbiotec.2012.05.022.

Nikolova, C., & Gutierrez, T. (2021). Biosurfactants and their applications in the oil and gas industry: Current state of knowledge and future perspectives. *Frontiers in Bioengineering and Biotechnology*, *9*. Available from https://doi.org/10.3389/fbioe.2021.626639.

Pacwa-Płociniczak, M., et al. (2011). Environmental applications of biosurfactants: Recent advances. *International Journal of Molecular Sciences*, *12*(1), 633–654. Available from https://doi.org/10.3390/ijms12010633.

Patowary, R., et al. (2018). Application of biosurfactant for enhancement of bioremediation process of crude oil contaminated soil. *International Biodeterioration and Biodegradation*, *129*, 50–60. Available from https://doi.org/10.1016/j.ibiod.2018.01.004.

Perfumo, A., Banat, I. M., & Marchant, R. (2018). Going green and cold: Biosurfactants from low-temperature environments to biotechnology applications. *Trends in Biotechnology*, *36*(3), 277–289. Available from https://doi.org/10.1016/j.tibtech.2017.10.016.

Perfumo, A., Rancich, I., & Banat, I. M. (2010). Possibilities and challenges for biosurfactants use in petroleum industry. *Advances in Experimental Medicine and Biology*, *672*, 135–145. Available from https://doi.org/10.1007/978-1-4419-5979-9_10.

Phulpoto, I. A., et al. (2021). Effect of natural microbiome and culturable biosurfactants-producing bacterial consortia of freshwater lake on petroleum-hydrocarbon degradation. *Science of the Total Environment*, *751*, 141720. Available from https://doi.org/10.1016/j.scitotenv.2020.141720.

Roy, A. (2018). A review on the biosurfactants: Properties, types and its applications. *Journal of Fundamentals of Renewable Energy and Applications*, *8*(1). Available from https://doi.org/10.4172/2090-4541.1000248.

Sarubbo, L. A., et al. (2015). Some aspects of heavy metals contamination remediation and role of biosurfactants. *Chemistry and Ecology*, *31*(8), 707–723. Available from https://doi.org/10.1080/02757540.2015.1095293.

Shi, K., et al. (2019). Mechanism of degrading petroleum hydrocarbons by compound marine petroleum-degrading bacteria: Surface adsorption, cell uptake, and biodegradation. *Energy and Fuels*, *33*(11), 11373–11379. Available from https://doi.org/10.1021/acs.energyfuels.9b02306.

Shoeb, E., et al. (2013). Part-I: Natural and applied sciences classification and industrial applications of biosurfactants. *Academic Research International*, *4*(3). Available at:http://www.savap.org.pk%5Cnhttp://www.journals.savap.org.pk.

Silva, E. J., et al. (2018). Recovery of contaminated marine environments by biosurfactant-enhanced bioremediation. *Colloids and Surfaces B: Biointerfaces*, *172*, 127–135. Available from https://doi.org/10.1016/j.colsurfb.2018.08.034.

Silva, R. D. C. F. S., et al. (2014). Applications of biosurfactants in the petroleum industry and the remediation of oil spills. *International Journal of Molecular Sciences*, *15*(7), 12523–12542. Available from https://doi.org/10.3390/ijms150712523.

Szulc, A., et al. (2014). The influence of bioaugmentation and biosurfactant addition on bioremediation efficiency of diesel-oil contaminated soil: Feasibility during field studies. *Journal of Environmental Management*, *132*, 121–128. Available from https://doi.org/10.1016/j.jenvman.2013.11.006.

Vigneshwaran, C., Vasantharaj, M., Jerold, K., & Sivasubramanian, V. (2016). A review on biosurfactants and its environmental applications. *IOSR Journal of Environmental Science, Toxicology and Food Technology*, *10*(8), 152–160. Available from https://doi.org/10.9790/2402-100802152160.

Zhang, X., et al. (2012). Isolation and identification of biosurfactant producing and crude oil degrading *Pseudomonas aeruginosa* strains. *Chemical Engineering Journal*, *209*, 138–146. Available from https://doi.org/10.1016/j.cej.2012.07.110.

Zouboulis, A. I., et al. (2004). Erratum: The use of biosurfactants in flotation: Application for the removal of metal ions (Minerals Engineering (2003) 16 (1231-1236)). *Minerals Engineering*, *17*(1), 105. Available from https://doi.org/10.1016/j.mineng.2003.11.017.

CHAPTER 14

Biofilm: a doable microbial continuum for the treatment of wastewater

Sumit Kumar Dubey and Arun Kumar Kashyap
Department of Biotechnology, Govt. E Raghavendra Rao Postgraduate Science College, Bilaspur, Chhattisgarh, India

14.1 Introduction

Water availability to humans is decreasing day by day due to the gradual increase in population and industrialization. Urbanization attracts population which leads to water pollution in existing water reservoirs. The quality of water systems is reduced because of the household and industrial waste effluents discharged into it without prior treatment. The agricultural pesticides, chemical fertilizers, landfills, waste disposals, unprocessed sewage, along with human and animal excretory products also contribute to reduce groundwater quality. Water pollution causes biodiversity reduction and water-related infections on living organisms of the ecosystem. WHO reported that approximately 40% of deaths happened due to the polluted water across the world (Kantawanichkul, Kladprasert, & Brix, 2009).

The major source of wastewater is often household, agriculture, and industrial effluents (Bhat et al., 2021; Olutiola, 2010), which comprise nutrients (i.e., nitrogen, phosphorous, and ammonia); inorganic metals (i.e., mercury, lead, cadmium, and nickel); fats and oils; proteins, carbohydrates; variety of microorganisms (i.e., bacteria, viruses, protozoa, and helminthes); and other synthetic derivatives. Wastewater constituents vary as per the source and contain harmful substances that are the root cause of various diseases and abnormalities to the living organisms. The level of harmful effects on living organisms may vary as per their metabolic pathway and cellular system.

The biological, chemical, and physical methods are most often used in wastewater treatment. These methods are associated with different degrees of effectiveness, advantages, and disadvantages. Some methods are

Microbial Consortium and Biotransformation for Pollution Decontamination
DOI: https://doi.org/10.1016/B978-0-323-91893-0.00013-4

environmentally and economically suitable. However, biological wastewater treatment method has attracted researchers because of cost-effective operation, easy handling, and relatively less injurious to ecosystem. On the basis of biomass constituents, biological wastewater treatment system is divided into dispersed growth and attached growth (Shama & Iffat, 2016). Dispersed growth occurs when cells are suspended in broth without any attachment. Activated sludge systems and extended aeration system are most common examples of the dispersed growth. In attached growth system, the cells are attached to inert support medium and form a layer of biofilm, that is, trickling filters (TFs), membrane bioreactors (MBRs), and rotating biological contactor (RBC).

Biofilm-assisted wastewater treatment technology is well developed. In biofilm-assisted wastewater treatment, the solid media provides attachment surfaces to microbes to form biofilms inside suspended growth reactors. An increase in the microbial concentration is positively proportional to contaminant degradation rate in the biofilms system. Biofilm-assisted wastewater treatment uses biodegradation, bioaccumulation, biosorption, and biomineralization to improve waste degradation efficiency (Pal, Sarkar, & Dasgupta, 2010). The biofilm breaks different nutrients (i.e., phosphorous and nitrogen derivatives and carbon compounds) and traps the pathogens from the wastewater. However, biofilm-assisted wastewater treatment could be integrated with other relevant wastewater treatment technologies to improve the quality of treated water.

Biofilm-assisted wastewater treatment system provides several advantages such as operational resilience, systematic and compact arrangements, minimum hydraulic retention time (HRT), flexible to altered environmental conditions, higher biomass retention time, raised active biomass concentration, enhanced ability to degrade recalcitrant wastewater contaminants, consistent and sluggish microbial growth rate, and minimum sludge generation rate.

A variety of bacterial species produce biofilms that include Gram-positive bacterial species such as *Bacillus* sp., *Listeria* sp., *Staphylococcus* sp., and *Lactobacillus* sp. (Danhorn & Fuqua, 2007). Gram-negative bacterial species are *Escherichia coli* and *Pseudomonas aeruginosa* (Danhorn & Fuqua, 2007). Bacterial communities such as cyanobacteria, nitrogen-fixing bacteria (i.e., *Rhizobium leguminosarum*), entophytes (e.g., *Pseudomonas putida*), and relevant plant symbionts have also been reported for biofilm production (Danhorn & Fuqua, 2007; Rossi & De Philippis, 2015). Apart from bacteria, some archaea (Chimileski, Franklin, & Papke, 2014) and eukaryotic organisms such as fungi (e.g., *Cryptococcus laurentii*) and microalgae (Joubert, Wolfaardt, & Botha, 2006) have also been mentioned in literatures for biofilm production. Another study revealed that diatoms (also referred to as microalgae) can colonize in various water bodies such as lakes, ponds, and marines, as biofilms (Aslam, Cresswell-Maynard, Thomas, & Underwood, 2012; Van, Underwood, Serôdio, & Paterson, 2014). Microbial cells involved in pollutant removal from wastewater are tabulated in Table 14.1. The biofilms are most often considered as the

Table 14.1 Microbial cells involved in pollutant removal from wastewater.

Microbial cell	Target pollutant	Efficiency	References
Pseudomonas sp. *Rhodococcus* sp.	Chlorophenol	99.9% reduction	Puhakka et al. (1995)
*Sphingomonas xenophaga*QYY	Bromoamine acid	50% COD reduction	Qu et al. (2009)
Arthrobacter viscosus	Cr (III) and Cr(VI)	14 mg reduction per gram of zeolite (initial 100 mg)	Silva, Figueiredo, Quintelas, Isabel, and Nevesb (2008)
Proteobacteria *Bacteroidetes*, *Nitrospirae* *Cyanobacteria* *Actinobacteria*	COD Ammonia	47% and 84.90% COD and ammonia reduction by biofilm on ceramsite and sponge, respectively	Zhang, Li, Yu, Xu, and Wu (2016)
Geobacter metallireducens + bacteria present in the wastewater	COD	80% COD reduction by biofilm on graphite	Liu, Ramnarayanan, and Logan (2004)
Desulfovibrio desulfuricans	Sulfate	99% sulfate reduction by biofilm on activated carbon	Zhao et al. (2008)

COD, chemical oxygen demand.

general mode of microbial growth. One example of biofilm is the oral streptococci that adapted for their sessile growth in surface of teeth. Presently, biofilm-based moving bed reactors biologically active filters, fluidized-bed reactors, RBCs, and TFs are frequently used in the industrial settings (Lewandowski & Boltz, 2011).

This chapter will describe the brief mechanism of wastewater treatment via biofilm-producing microbes, types and techniques, factors affecting, problems, current research status, and prospects of the biofilm-assisted wastewater treatment technology.

14.2 Mechanism of biofilm formation

System enclosed microbial cells generated extracellular polymeric substances (EPSs) that form biofilm. The development of microbial biofilms has five steps (Fig. 14.1) (Kjelleberg, Marshall, & Givskov, 2007; Webb, 2007):

1. Initially, the microbial cells are attached over the surface of solid support materials.
2. Irreversible attachment has occurred by the production of microbial-mediated EPSs. A polyhydroxyl group of EPSs helps microbial cells to attach over solid surface via hydrogen bonding.
3. Replication of microbial cells forms monolayer microcolonies that fixed on solid surface and generate primary layer of biofilm.
4. Maturation of primary layer of biofilm into a thick three-dimensional arrangement for the interaction to adjacent environment for specific task.
5. Dispersion of biofilm can be active or passive and can be done by quorum sensing (QS) and cell-to-cell signaling-based interaction among microbial cells.

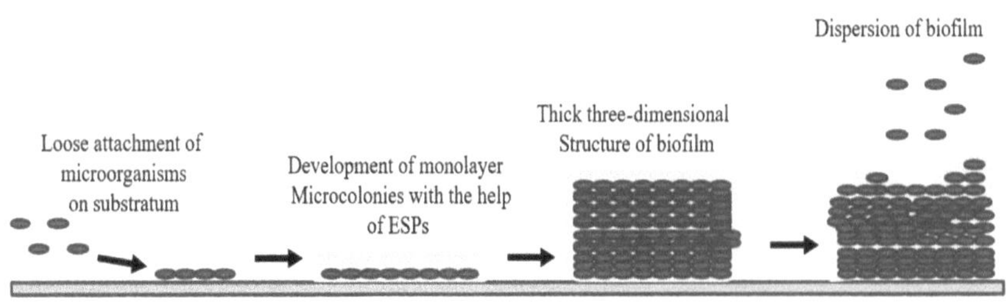

Figure 14.1 Steps in biofilm formation.

14.2.1 Three major events of microbial extracellular biofilm formation

14.2.1.1 Adherence on surfaces

Biofilm-producing microbes secreted extracellular matrix (Flemming & Wingender, 2010) that consisted of EPSs, carbohydrate-binding proteins at the surface area (Branda, Chu, Kearns, Losick, & Kolter, 2006; Tielker et al., 2005). Pili, flagella (Cegelski et al., 2009; Zogaj, Nimtz, Rohde, Bokranz, & Romling, 2001), extracellular DNA (eDNA) (Vilain, Pretorius, Theron, & Brozel, 2009; Whitchurch, Tolker-Nielsen, Ragas, & Mattick, 2002), and other adhesive fibers act as stabilizing agent for the three-dimensional structure of biofilms. Nutrients are entrapped in the matrix for further metabolic consumptions and water molecules are held together by hydrogen bonding along with hydrophilic polysaccharides (Conrad et al., 2003). The composition of EPS is modified by the enzymes that secreted by bacteria during altered nutrition (Sauer et al., 2004) which helps biofilm architecture to withstand the specific environmental condition. Thus the structure of the matrix is getting hydrated, compact, and higher tensile strength which favor close proximity of bacterial cells and assist in cell-to-cell interactions and DNA exchange between them (Flemming & Wingender, 2010; Koo, Xiao, Klein, & Jeon, 2010). Biofilms are also reported to protect the microbial biomass from dehydration, oxidizing agents, and radiation (Jefferson, Goldmann, & Pier, 2005; Walters, Roe, Bugnicourt, Franklin, & Stewart, 2003).

14.2.1.2 Maturation

Bacterial adherence to the solid surface triggers gene expression modification, regulating factors associated with sessility, that is, formation of EPS (Beloin, Roux, & Ghigo, 2008; Bhomkar, Materi, Semenchenko, & Wishart, 2010; Morici et al., 2007). Cellulose has been reported to be an important constituent of biofilms (*E. coli*) (Bokranz, Wang, Tschape, & Romling, 2005; Romling, 2002; Zogaj et al., 2001). Literature revealed that polyglucosamine and colanic acid are important toward the formation of biofilm architecture (Agladze, Wang, & Romeo, 2005; Wang, Preston, & Romeo, 2004). *P. aeruginosa* has also been exhibited biofilm formation that relies on environmental conditions (Harmsen, Yang, Pamp, & Tolker-Nielsen, 2010). *Pel* and *Psl* have been reported for primary EPS constituents (Ma, Jackson, Landry, Parsek, & Wozniak, 2006; Matsukawa & Greenberg, 2004; Vasseur, Vallet-Gely, Soscia, Genin, & Filloux, 2005). However, *Pel* and *Psl* have been reported in human body which trigger the biofilm formation in epithelial cells in cystic fibrosis patients and these are associated with *P. aeruginosa* (Govan & Deretic, 1996; Ma et al., 2006; Starkey et al., 2009). CdrA has been reported for adhesin secretion during biofilm maturation and triggered by the 3,5-cyclic diguanylic acid, binds to *Psl*, and thus stabilizes biofilm structure (Borlee et al., 2010). *P. aeruginosa*-derived EPS contains alginate which is associated with antibiotic resistance. In the case of human diseases, it defends host immune system (Hentzer et al., 2001; Leid et al., 2005). Studies have revealed that diguanylate cyclase MucR promotes alginate synthesis when c-di-GMP is present in higher concentration (Hay, Remminghorst, & Rehm, 2009). eDNA also promotes cell-to-cell links and biofilm stabilization during maturation (Whitchurch et al., 2002; Yang et al., 2007). Growing biofilms are more sensitive to DNase in contrast to mature biofilms, which indicates the positive role of eDNA during biofilm maturation process (Whitchurch et al., 2002). Mohamed, Huang, Nallapareddy, Teng, and Murray (2004) revealed that a mutant with Atn^{-} autolysin showed 30% decrease in biofilm development and maturation rate. Another supporting study was done by Guiton et al. (2009) which showed that Atn^{+} has an important role in the sequential DNA regulation at the time of biofilm development.

14.2.1.3 Dispersion

When biofilm matures, bustling occurred for active exchanges to maintain biofilm structure and favorable environmental conditions for inhabitant bacterial communities. Biofilms are dispersed due to the alterations of nutrient availability, oxygen fluctuations, toxic secondary metabolites, stress-inducing condition, and shear forces (Hong, Lee, & Wood, 2010; Karatan & Watnick, 2009; Rowe, Withers, & Swift, 2010; Sauer et al., 2004). Iron, carbon, and nitrogen sources have been reported for biofilm dispersion over the solid surface (Rowe et al., 2010). Biofilms are also generated by dead cells that serve to the dispersion of *P. aeruginosa*-based biofilm development (Webb et al., 2003). Bacteria that disperse are able to restart the biofilm initiation and development at suitable environmental conditions. A study conducted on *Bacillus subtilis* revealed that biofilm disassembly mediated by D-amino acids such as D-leucine and D-tyrosine (Kolodkin-Gal et al., 2010). D-amino acids are formed during the stationary phase of biofilm development (Lam et al., 2009). D-amino acids interfere with adhesive fibers anchoring efficiency that results in dissociation of biofilm adhesive fiber and loss of adherence but not affect the bacterial growth (Kolodkin-Gal et al., 2010).

14.3 Biofilm-producing microbes

Bacteria, cyanobacteria, nitrogen-fixing bacteria, entophytes, and plant symbionts have been observed for biofilm production (Danhorn & Fuqua, 2007; Rossi & De Philippis, 2015). Some archaea have also been reported for biofilm production (Chimileski et al., 2014). Eukaryotic organisms such as fungi and microalgae (Joubert et al., 2006) have been characterized for biofilm generation. Diatoms have been recorded to colonize as biofilms in water bodies such as lakes and marines (Aslam et al., 2012; Van et al., 2014). Biofilms from autochthonous microorganisms have also been reported to be helpful in wastewater treatment (Radojević, Grujić, & Ranković, 2019). List of microbial cells involved in pollutant removal from wastewater is given in Table 14.1.

14.3.1 Why do microbial cells grow as biofilm?

Defense and colonization encourage microbial cells to make biofilm. Defense includes stress response of microbial cells against external environment. Biofilm-mediated defenses are resistant to shear forces, pH changes, oxygen scavengers, antibiotics, disinfecting agents, and phagocytosis (Donlan & Costerton, 2002; Lewis, 2001). Certain microbial species showed EPS-mediated defense mechanism against pathogens prior immune system (Shirtliff, Mader, & Camper, 2002), for example, stimuli such as iron deficiency and osmotic stress encourage the expression of genes in *Staphylococci* and *Enterococci* that encode EPS synthesis (Baldassarri et al., 2001; Deighton & Borland, 1993). The stress response regulators Dgk (*Streptococcus mutans*), rB (*Staphylococcus aureus*), and RpoS (*E. coli*) promote biofilm development (Adams & McLean, 1999; Yoshida & Kuramitsu, 2002). However, some literatures questioned about the function of *rB* and *RpoS* in biofilm initiation and development (Corona-Izquierdo & Membrillo-Hernandez, 2002; Kies, Otto, Vuong, & Götz, 2001). Moreover, RpoH of *P. aeruginosa* is associated with stress (Whiteley et al., 2001). The stress is triggered by the environmental condition and genetic background of the microbial cells. Colony of microbial cells tends to be attached to the solid surface or host that is mediated by biofilm formation. Bacteria, for example, *Pseudomonas*, *Vibrio cholerae*, *E. coli*, and *Staphylococci* can produce EPS. Glucose and other utilizable carbon sources also significantly induce EPS synthesis (Ammendolia, Di Rosa, Montanaro, Arciola, & Baldassarri, 1999; Jefferson, Pier, Goldmann, & Pier, 2004; O'Toole, Kaplan, & Kolter, 2000).

14.4 Types of biofilm system for wastewater treatment

The biomass needs a support medium to grow and generate a biofilm in the attached growth systems. Attachment efficiency depends on the composition of the media, types of cells and their interactions, and nature of surface molecules (Boks, Norde, van der Mei, & Busscher, 2008). The support material can be natural (i.e., rocks, gravels, and sand) or synthetic (i.e., rubber and plastic). The organic matter and nutrients consistently flow over biofilms to feed them. Biofilm needs fresh substrates for increased microbial activity. For removal of pollutant with high efficiency, currently microbial biofilm rectors are used extensively for wastewater treatment (Abdelfattah, Hossain, & Cheng, 2020). Many reactors have been developed for the treatment of wastewater, including TF rector, rotating biological rectors, membrane biofilm reactor, and passive aeration simultaneous nitrification and denitrification biofilm reactor (Fig. 14.2) which are capable of removing carbon, nitrogen, sulfur, and oxidized contaminants such as chlorate and bromate (Abdelfattah et al., 2020).

14.4.1 Trickling filters

TFs are well-characterized wastewater treatment technology. TF vessel is filled with solid inert materials such as gravel, rock, coke, ceramic, or plastic. The wastewater is sprinkled over TF and the contaminants of the wastewater trickle via TF solid inert media. External mechanical or gravitational force can be used to the consistent flow of wastewater in TF system. Biological slime (precursor of biofilm) grows on the solid inert media and the microbes on mature biofilm absorb the dissolved organic compounds for their growth. Some metals can be adsorbed or another relevant interaction is made to remove contaminant of wastewater (Pal et al., 2010). TFs are appropriate for MSMEs (micro, small, and medium enterprises) with an advantage of high filter-loading efficiency, effortless process, autocleaning system, and proficient ammonia removal. Major disadvantages are huge sludge discharge and a high rate of clogging (Donlan & Costerton, 2002).

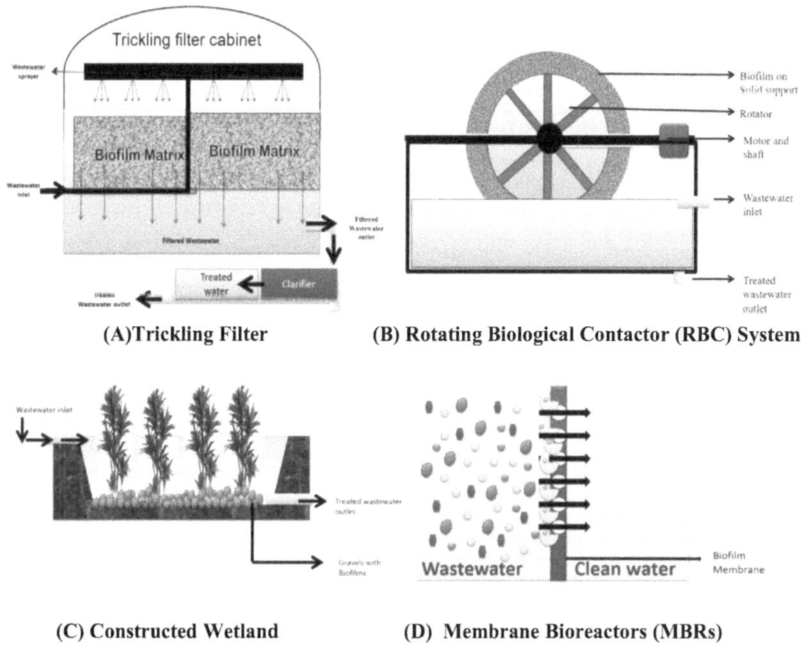

Figure 14.2 Different types of biofilm systems for wastewater treatment: (A) trickling filter, (B) RBC system, (C) constructed wetland, and (D) membrane bioreactors (MBRs). *RBC*, rotating biological contactor.

14.4.2 Rotating biological contactor system

The RBC is a competent purification system for the removal of contaminants from wastewater effluents from a variety of industries such as food, beverages, refinery, and paper processing. In addition, it can also be applicable in municipal wastewater and landfill leachate liquid treatment. The RBC comprises solid media as plastic derivatives such as polyethylene, polyvinyl chloride, and polystyrene. Solid media has to be slowly rotated to ensure proper contact between biofilm and wastewater so that microbes can interact with biodegradable contaminants of wastewater. When wastewater comes across the solid media, naturally occurring or any specific microbes (in the case of predetermined microbial consortium for specific purpose) absorb oxygen from the rotating system and get nutrients from wastewater and then biofilm (layer of microbes) formation initiated. Excess biomass can be removed by shearing and gravitational force. The RBC system has some advantages such as less cycle needed, minimum sludge generation, small space required, easy operational setup, stable, and significant biological oxygen demand (BOD) reduction.

14.4.3 Constructed wetland system

Constructed wetlands (CWs) are known for its well-engineered attached growth systems. The CWs' microbial growth system consists of solid medium such as sand, soil, and gravel as beds which comprise appropriate microbial flora. These microbial floras play an important role to remove contaminants from the surface water and groundwater system. CWs can be classified into subsurface flow and surface flow. The water runs over the surface in surface flow system (SFS), while water runs below the surface in subsurface flow system (SbFS). The odor can be significantly minimized by using SbFS. The contact between bacteria, substrate, and wastewater seems to be lower in SFS than that in SbFS. Due to this, enhanced treatment efficiency can be obtained with SbFS (Kadlec & Reddy, 2001). SbFS is also grouped into horizontal and vertical subsurface flow on the basis of flow path. Horizontal and vertical subsurface flow is proficient in eliminating recalcitrant contaminants and pathogens present in the wastewater. The combination of vertical and horizontal flows referred to as hybrid CWs and believed to be a suitable for least water loss during the treatment process (Molle, Prost-Boucle, & Lienard, 2008).

CWs' water purification system comprises a series of physical (e.g., adsorption, sedimentation, and filtration); chemical (e.g., chemical precipitation); and biological (i.e., microbial activities) processes. The waste elimination efficacy of CWs system depends on HRT, temperature, substrate constituents, types of solid media, and efficiency of microorganisms (Sehar et al., 2015). In CW system, macrophytes (aquatic plant) have an essential role in nutrient removal from wastewater and accelerate wastewater purification process by enhancing the biochemical reactions in rhizosphere. The merits of CWs system are lower operational cost, smaller amount of energy requirement, less sludge production, and eco-friendly (Shalaby, Altalhy, & Mosallam, 2008).

14.4.4 Membrane bioreactors

Membrane bioreactors (MBRs) are used for municipal and industrial effluent treatment. MBRs are outfitted with microfiltration or ultrafiltration membrane-based contaminant removal. The membrane can separate wastewater contaminants based on the relative size or electrical charge as per the system design. MBRs are classified into gravity-driven vacuum systems and pressure-driven cartridge systems. The vacuum system has hollow fiber membranes inside the reactors, while the pressure-driven system has an external cartridge fitted outside the reactor. An initial step in MBR system includes microbial assisted bioprocess that takes place in membrane region where pollutants are degraded. Filtration is carried out by a series of submerged membrane unit. Each membrane unit is called modules and chain of multiple modules is referred to as membrane unit. Air is introduced through diffusers into MBR system to facilitate mixing and to maintain aerobic oxygenated environment. The major advantages of MBRs system are low sludge generation, significant BOD and chemical oxygen demand (COD) removal efficiencies, decreased footprints, and less final finishing requisite for disinfection and clarification. But, the major obstacles are membrane fouling, membrane clogging, and complex process that increase the operational costs of the MBR system (Judd, 2008).

14.5 Factors affecting biofilm-based wastewater treatment

The major factors that affect biofilm-based wastewater treatment are nutrients, pH, temperature, surface topography, velocity, turbulence, hydrodynamics, EPSs, eDNA, and divalent cations.

14.5.1 Effects of nutrients, pH, and temperature

Biofilm formations depend upon nutrient conditions. Nutrient-rich environment encourages the transformation of microbial cells into biofilm (a layer of microbial cells). The nutrient depletion results in detachment of biofilm from solid surfaces. The microbial biofilms get nutrients from concentrated trace organics via extracellular polymer, waste products, and pooled biological resources via enzymatic breakdown of substrate or waste.

Biofilm is affected by the pH change because biofilms are pH dependent for better nutrient uptake and metabolic process for proper growth and development of microbial cells. Microbes regulate the active protein synthesis linked with various cellular metabolic activities in response changes in pH. The optimum pH for microbial exopolysaccharide secretion rate differs from species to species. A study suggests that most of the bacterial cells adopt pH 7.0 for biofilm development (Oliveira, Melo, Oliveira, & Salgueiro, 1994).

Biofilm formation by microbes is sensitive to temperature fluctuation. Every microbe has optimum temperature for healthy growth and reproduction. The biofilm formation and microbial metabolic activities are dependent on enzymatic activity and it is greatly affected by the variation in temperature and pH during the biofilm development. The optimum temperature for bacterial and fungal growth is about 40°C ± 03 and 27°C ± 03, respectively (Ells & Hansen, 2006).

14.5.2 Surface topography

Surface topography influences microbial adherence to the surface of solid support media. At the first step of colonization, the surface topography promotes the adhesion of microbial cells to substrates by providing greater surface area for attachment. Surface roughness decreases the shear force on microbial cells and liquids at high flow rates, for example, water supply pipes in industrial settings. Chemical modification of solid surface affects microbial attachment. Charge, hydrophobicity, and elasticity play important role in microbial attachment (Prakash, Veeregowda, & Krishnappa, 2003).

14.5.3 Velocity, turbulence, and hydrodynamics

The surface area with less effective turbulent flow is recognized as the boundary layer and also flow velocity is insufficient to remove biofilms from solid surface. The area outside of boundary layer has higher level of turbulent flow which influences the adherence of microbial cells to the solid surface. The thickness and length of the boundary layer is relying on the flow velocity. When the boundary layer reduced cells are experienced higher turbulence level under the high flow velocities of water. Biofilms formation, structure, mass, thickness, metabolic activities, and rate of EPS production are influenced by the hydrodynamics.

14.5.4 Gene regulation and quorum sensing

Literature claimed that gene regulation (up- and downregulation) is associated with primary attachment of cells with substrate. The biofilm formation by *P. aeruginosa* is regulated by approximately 22% of upregulation gene and 16% of downregulation gene (Steyn, Oosthuizen, MacDonald, Theron, & Brözel, 2001). The rpoS, algD, algU, and polyphosphokinase (PPK) gene synthesis are included in upregulated gene during the biofilm generation in *P. aeruginosa* (Prakash, Veeregowda, & Krishnappa, 2003).

Biofilms of *S. aureus* are upregulated by the enzyme-encoding genes (i.e., phosphoglycerate mutase, triphosphate, and alcohol dehydrogenase) that involved in glycolysis (aerobic) or fermentation (often anaerobic) (Becker, Hufnagle, Peters, & Herrmann, 2001).

QS, a type of cell-to-cell signaling in bacterial communities, has also been played an important role in the attachment and detachment of microbial cell to biofilms. Biofilm development is mediated by certain chemical signals along with EPS matrix. QS uses a transcriptional activator protein signaling molecules to promote the expression of desired genes that result in changes in chemical behavior. The accumulation of enough protein signaling molecule such as small autoinducer signaling molecule promotes intercellular communication that is associated with morphological differentiation, gene expression, and the development of responses in bacterial cells (Xiong & Liu, 2010).

14.5.5 Production of extracellular polymeric substances

EPSs consist of complex mixture of high-molecular-weight polymeric substance that is secreted by microbial cells. EPSs are synthesized from microbial assisted hydrolysis of organic matters of wastewater or other nutritional sources. EPSs are the matrix of polymers, that is, polysaccharides, proteins, glycoprotein's, nucleic acids (DNA and RNA), phospholipids, and humic acids (Flemming & Wingender, 2010). EPSs are hydrated as they retain high amounts of water molecules by hydrogen bond. EPSs are reported to promote the gel-like network to stay bacteria together in hydrophobic interactions of biofilms. EPSs protect bacteria from lethal environmental settings, help them to adhere to solid surface, and facilitate nutrient uptake by bacterial cells (Zhao et al., 2013). The rate of biofilm production efficiency of microbes differs and amount of EPS production is directly proportional to the age of biofilms (O'Toole, 2011).

14.5.6 Extracellular DNA

eDNA or naked DNA promotes the bacterial cells to produce EPSs that are somewhat chromosomal DNA with respect to its primary sequence (Bockelmann et al., 2006). eDNA initiates the adhesion of bacterial cells, aggregation, and primary biofilm generation. It strengthens biofilms that help microbial cells to be protected from physical stress and antibiotics (Das, Sehar, & Manefield, 2013). eDNA can also be used for the engineering of biofilms for specific applications such as bioremediation and fuel production by bioelectrochemical bioreactors.

14.5.7 Divalent cations

Presence of divalent cations such as Ca^{2+} are significant in the initiation of microbial attachment to the clumps of activated sludge granules and biofilms to the negative surface charge sites on extracellular polymers (Kerchove & Elimelech, 2008). Enough calcium present in terrestrial and aquatic environments that mediate biofilm-associated microbial growth on the ecosystem. Divalent cations associated with eDNA as chelates help to modify the properties of bacterial cell surface that help the resistance of biofilms against surfactants and antimicrobial agents (Mulcahy, Charron-Mazenod, & Lewenza, 2008). Divalent cations are responsible for thickness, density, and stability of biofilm (Sarkisova, Patrauchan, Berglund, Nivens, & Franklin, 2005). Calcium involved in biofilm virulence, cell signaling, and regulation of extracellular secretions during biofilm formation (Das et al., 2014).

14.6 Wastewater pollutants remediated by biofilms

The quality of industrial and household wastewaters is distinguished by COD, BOD, total suspended solids, concentration of heavy metals, pH, and turbidity of wastewater. These parameters vary as per the sources such as household, textile, pharmaceutical, food processing, tannery, and other relevant manufacturing industrial sectors. In contrast to municipal sewage, industrial effluents consisted of higher organic load, an extreme physico-chemical condition (i.e., high acid or alkaline pH and raised salinity), and higher concentration of heavy metals or specific by-product from certain industrial effluents. As discussed earlier that microbial biofilm can resist harsh environmental condition which makes them suitable for wider range of complex wastewater treatment. Microbial biofilms are being used in removal of organic and inorganic pollutants, oil spills, heavy metals (i.e., cadmium, copper, uranium, and chromium), pesticides, xenobiotics, and water contaminates (Dar, Kamili, Dar, Chishti, & Ahmad, 2015; Dar et al., 2016; Diels, De Smet, Hooyberghs, & Corbisier, 1999; Valls & de Lorenzo, 2002). Heavy metal remediation mediated by phosphatase enzyme which facilitates metal precipitation in both aerobic (presence of oxygen) and anaerobic (absence of oxygen) conditions (Macaskie et al., 1997).

Biofilm formation in contaminated groundwater acts as barrier to reduce the flow of pollutants and minimize its spread.

Persistent organic pollutants such as polycyclic aromatic hydrocarbons have long half-life that found in air, water, and less often in soils that are considered as harmful to ecosystem quoted by the United States Environmental Protection Agency (Chen et al., 2015; Dar, Kamili, Nazir, Bandh, & Bhat, 2012; Dar, Bandh, Kamili, Nazir, & Bhat, 2013; Shukla, Mangwani, Rao, & Mangwani, 2014; Smith & Gadd, 2000). Bacterial biofilms have been reported for persistent organic pollutants (POP) degradation and can be used in bioremediation]. Bacteria, that is, *Pseudomonas*, *Arthrobacter*, *Rhodococcus*, and *Bacillus* spp. have been studied for POP degradation (Chen et al., 2015). Biofilm of sulfate-reducing bacteria has been studied for the heavy metal remediation from contaminated water (Muyzer & Stams, 2008). Recent studies revealed that some bacterial strains are able to form electrochemically active biofilms (Erable, Duţeanu, & Ghangrekar, 2009) and it can be applicable for the reduction of heavy metals from groundwater and soil system (Dar, Bandh, & Kamili, 2011; Li, Zhang, & Lei, 2008).

14.7 Research paradigm on biofilm

Literatures published over last decades showed the variety of research paradigms in terms of microbial biofilm, molecular biology of cell surface interaction and adhesion, mechanism, and application. There are two aspects of biofilm research paradigm (1) to utilize biofilm for industrial application and (2) to inhibit harmful biofilm development inside the living cells and ecosystem. Many research have focused on the molecular biology of biofilm development; genes, molecules, promoter involved in biofilm formation because these informations are much contribute to fulfill both aspects mentioned earlier. As the chapter deals with the application of biofilm for the wastewater treatment, so here only relevant research paradigms are mentioned.

Arciola, Campoccia, Ravaioli, and Montanaro (2015) explored adhesin in polysaccharide intercellular biofilm in terms of structural and regulatory characteristics. Mayer et al. (1999) pointed the intermolecular interactions model for *P. aeruginosa*-derived biofilms and found that electrostatic interactions and hydrogen bonds are prevailing forces between constituent molecules of the biofilm. Reichhardt, Wong, Passos da Silva, Wozniak, and Parsek (2018) found that CdrA interfaces in *P. aeruginosa* biofilm defend proteolysis and catalyze cellular packing. Rodesney et al. (2017) studied *P. aeruginosa* biofilm and found that mechanosensing of shear promotes levels of the cyclic-di-GMP signal to start biofilm formation over the solid support. Fritts, LaSarre, Stoner, Posto, and McKinlay (2017) examined *Rhodopseudomonas palustris* biofilm for various photoheterotrophic conditions and also rhizobiales-specific unipolar polysaccharide mediated adhesin contribution to biofilm formation. Keogh et al. (2018) claimed that metabolism in *Enterococcus faecalis* biofilm is powered by the extracellular electron transfer.

Devaraj, Justice, Bakaletz, and Goodman (2015) examined uropathogenic *E. coli* (UPEC) and revealed that DNABII proteins have a crucial role in biofilm generation in UPEC which protect it from unfavorable environmental condition and antibiotics. Jin and Riedel-Kruse (2018) analyzed pDawn-Ag43 in *E. coli* to initiated high-resolution patterned biofilm development. Glass and Riedel-Kruse (2018) experimented on synthetic bacterial cells to explore cell-to-cell adhesion for programing multicellular morphologies and significant patterns. Floyd et al. (2015) divulge that adhesive fiber stratification in UPEC biofilm has showed oxygen-mediated control in type-1 pili of the cell. Stewart and Franklin (2008) studied the bacterial biofilm possessing wider range of physiological heterogeneity. Huseby et al. (2010) worked on *staphylococcal* biofilms and revealed that β-toxin promotes the nucleoprotein matrix in biofilm.

More, Yan, Tyagi, and Surampalli (2015) studied the production biopolymers by microbial consortium and explored its applications in industrial and municipal wastewater treatment. Subramanian, Yan, Tyagi, and Surampalli (2010) worked on isolation, molecular identification, EPS characterization, and performance for sludge sedimentation and dewatering via EPS-producing bacterial strains which isolated from the sludge of municipal wastewater. Sulzacova, Trtilek, and Rataj (2015) examined the efficiency of microalgal biofilm for the significant phosphorus removal during tertiary stage of wastewater treatment using modified biofilm photobioreactor. Christenson (2011) examined biofilm production by algal cells and also attempted wastewater treatment generated from biofuel by-products. The biofilm producing cynobacteria from biomass and lipid contents were characterised by Bruno et al. (2012).

Boelee, Janssen et al. (2014) and Boelee, Temmink, Janssen, Buisman, and Wijffels (2014) conducted study on symbiotic association of microalgae and bacterial cells with reference to treatment of synthetic municipal wastewater by biofilm-based reactors. Boelee, Janssen et al. (2014) and

Boelee, Temmink et al. (2014) revealed the effect of harvesting during biomass generation and nutrient elimination from the phototrophic biofilm reactors for effluent treatment and polishing. Kesaano and Sims (2014) explored wastewater treatment technology based on the algal biofilm. Boelee, Temmink, Janssen, Buisman, and Wijffels (2011) examined municipal wastewater effluent for microalgal biofilm-assisted nitrogen and phosphorus removal from it. Li, Lin, Podola, and Melkonian (2015) studied on consistent zinc removal from industrial effluent by using microalgal biofilm. Yang et al. (2007) studied to reveal that DNA release during biofilm development in *P. aeruginosa* is affected by iron.

Roeselers, Van Loosdrecht, and Muyzer (2008) studied phototrophic biofilms for their potential applications. Larsen et al. (2007) explored most abundant amyloid mediated adhesins in natural biofilms. Laue et al. (2006) examined that alginate and levan showed significant involvement during biofilm formation in *Pseudomonas syringae*. Ma et al. (2006) analyzed conditional *Psl* variant in *P. aeruginosa* to divulge its functions toward *Psl* polysaccharide during adhesion and also keep biofilm structure stable. Rice et al. (2007) worked on biofilm development in *S. aureus* and observed that cidA murein hydrolase regulator was associated with DNA release to initiate biofilm generation on bacterial species.

14.8 Conclusion

The biofilm-based system is being explored for the treatment of wastewater. Due to the ease and simplicity of operation, this technique can be beneficial than other known techniques. Many research works have been done to explore the biofilm for sustainable wastewater treatment. However, the systematic literature series on biofilm-mediated wastewater treatment techniques are still need to be reviewed for the better utilization of microbial biofilm for efficient pollutant removal from wastewater. The molecular biology-based tools for biofilm development can contribute to the better understating to optimize and develop biofilm for customized industrial and household wastewater treatment technologies to make it potential and cost-effective. The application of the technology is also limited in industrial scale due to lack of information on sustainability, system performance, and reliability. Further research is needed to be done to make the system more efficient and high performing even at industrial scale.

References

Abdelfattah, A., Hossain, M. I., & Cheng, L. (2020). High-strength wastewater treatment using microbial biofilm reactor: a critical review. *World Journal of Microbiology and Biotechnology*, *36*, 75. Available from https://doi.org/10.1007/s11274-020-02853-y.

Adams, J. L., & McLean, R. J. (1999). Impact of rpoS deletion on *Escherichia coli* biofilms. *Applied and Environmental Microbiology*, *65*, 4285–4287.

Agladze, K., Wang, X., & Romeo, T. (2005). Spatial periodicity of *Escherichia coli* K-12 biofilm microstructure initiates during a reversible, polar attachment phase of development and requires the polysaccharide adhesin PGA. *Journal of Bacteriology*, *187*, 8237–8246.

Ammendolia, M. G., Di Rosa, R., Montanaro, L., Arciola, C. R., & Baldassarri, L. (1999). Slime production and expression of the slime-associated antigen by *Staphylococcal* clinical isolates. *Journal of Clinical Microbiology*, *37*, 3235–3238.

Arciola, C. R., Campoccia, D., Ravaioli, S., & Montanaro, L. (2015). Polysaccharide intercellular adhesin in biofilm: Structural and regulatory aspects. *Frontiers in Cellular and Infection Microbiology*, *5*, 7.

Aslam, S. N., Cresswell-Maynard, T., Thomas, D. N., & Underwood, G. J. (2012). Production and characterization of the intra- and extracellular carbohydrates and polymeric substances (EPS) of three sea-ice diatom species, and evidence for a cryoprotective role for EPS. *Journal of Phycology*, *48*(6), 1494–1509. Available from https://doi.org/10.1111/jpy.12004. Available from 27009999.

Baldassarri, L., Cecchini, R., Bertuccini, L., Ammendolia, M. G., Iosi, F., Arciola, C. R., ... Creti, R. (2001). *Enterococcus spp.* produces slime and survives in rat peritoneal macrophages. *Medical Microbiology and Immunology*, *190*, 113–120.

Becker, P., Hufnagle, W., Peters, G., & Herrmann, M. (2001). Detection of different gene expression in biofilm-forming vs planktonic populations of *Staphylococcus aureus* using micro-representational-difference analysis. *Applied Environmental Microbiology*, *67*, 2958–2965.

Beloin, C., Roux, A., & Ghigo, J. M. (2008). *Escherichia coli* biofilms. *Current Topics in Microbiology and Immunology*, *322*, 249–289.

Bhat, R. A., Singh, D. V., Qadri, H., Dar, G. H., Dervash, M. A., Bhat, S. A., ... Yousaf, B. (2021). Vulnerability of municipal solid waste: An emerging threat to aquatic ecosystems. *Chemosphere*, *287*(2022), 132223. Available from https://doi.org/10.1016/j.chemosphere.2021.132223.

Bhomkar, P., Materi, W., Semenchenko, V., & Wishart, D. S. (2010). Transcriptional response of *E. coli* upon FimH-mediated fimbrial adhesion. *Gene Regulation and Systems Biology*, *4*, 1–17.

Bockelmann, U., Janke, A., Kuhn, R., Neu, T. R., Wecke, J., Lawrence, J. R., & Szewzyk, U. (2006). Bacterial extracellular DNA forming a defined network-like structure. *FEMS Microbiology Letters*, *262*, 31–38.

Boelee, N. C., Janssen, M., Temmink, H., Taparaviciute, L., Khiewwijit, R., Janoska, A., ... Wijffels, R. H. (2014). The effect of harvesting on biomass production and nutrient removal in phototrophic biofilm reactors for effluent polishing. *Journal of Applied Phycology*, *26*(3), 1439–1452.

Boelee, N. C., Temmink, H., Janssen, M., Buisman, C. J. N., & Wijffels, R. H. (2011). Nitrogen and phosphorus removal from municipal wastewater effluent using microalgal biofilms. *Water Research*, *45*(18), 5925–5933.

Boelee, N. C., Temmink, H., Janssen, M., Buisman, C. J. N., & Wijffels, R. H. (2014). Balancing the organic load and light supply in symbiotic microalgal-bacterial biofilm reactors treating synthetic municipal wastewater. *Ecological Engineering*, *64*, 213–221.

Bokranz, W., Wang, X., Tschape, H., & Romling, U. (2005). Expression of cellulose and curli fimbriae by *Escherichia coli* isolated from the gastrointestinal tract. *Journal of Medical Microbiology*, *54*, 1171–1182.

Boks, N. P., Norde, W., van der Mei, H. C., & Busscher, H. J. (2008). Forces involved in bacterial adhesion to hydrophilic and hydrophobic surfaces. *Microbiology (Reading, England)*, *154*(10), 3122–3133.

Borlee, B. R., Goldman, A. D., Murakami, K., Samudrala, R., Wozniak, D. J., & Parsek, M. R. (2010). *Pseudomonas aeruginosa* uses a cyclic-di-GMP-regulated adhesin to reinforce the biofilm extracellular matrix. *Molecular Microbiology*, *75*, 827–842.

Branda, S. S., Chu, F., Kearns, D. B., Losick, R., & Kolter, R. (2006). A major protein component of the *Bacillus subtilis* biofilm matrix. *Molecular Microbiology*, *59*, 1229–1238.

Bruno, L., Di Pippo, F., Antonaroli, S., Gismondi, A., Valentini, C., & Albertano, P. (2012). Characterization of biofilm-forming cyanobacteria for biomass and lipid production. *Journal of Applied Microbiology*, *113*(5), 1052–1064.

Cegelski, L., Pinkner, J. S., Hammer, N. D., Cusumano, C. K., Hung, C. S., Chorell, E., ... Almqvist, F. (2009). Small-molecule inhibitors target *Escherichia coli* amyloid biogenesis and biofilm formation. *Nature Chemical Biology*, *5*, 913–919.

Chen, M., Xu, P., Zeng, G., Yang, C., Huang, D., & Zhang, J. (2015). Bioremediation of soils contaminated with polycyclic aromatic hydrocarbons, petroleum, pesticides, chlorophenols and heavy metals by composting: Applications, microbes and future research needs. *Biotechnology Advances*, *33*, 745–755.

Chimileski, S., Franklin, M. J., & Papke, R. T. (2014). Biofilms formed by the archaeon *Haloferax volcanii* exhibit cellular differentiation and social motility, and facilitate horizontal gene transfer. *BMC Biology*, *12*, 65. Available from https://doi.org/10.1186/s12915-014-0065-5, ISSN 1741–7007. PMC 4180959. Available from 25124934.

Conrad, A., Suutari, M. K., Keinanen, M. M., Cadoret, A., Faure, P., Mansuy-Huault, L., & Block, J. C. (2003). Fatty acids of lipid fractions in extracellular polymeric substances of activated sludge flocs. *Lipids*, *38*, 1093–1105.

Corona-Izquierdo, F. P., & Membrillo-Hernandez, J. (2002). A mutation in rpoS enhances biofilm formation in *Escherichia coli* during exponential phase of growth. *FEMS Microbiology Letters*, *211*, 105–110.

Danhorn, T., & Fuqua, C. (2007). Biofilm formation by plant-associated bacteria. *Annual Review of Microbiology*, *61*, 401–422. Available from https://doi.org/10.1146/annurev.micro.61.080706.093316. Available from 17506679.

Dar, G. H., Bandh, S. A., Kamili, A. N., Nazir, R., & Bhat, R. A. (2013). Comparative analysis of different types of bacterial colonies from the soils of Yusmarg Forest, Kashmir Valley India. *Ecologia Balkanica*, *5*(1), 31–35.

Dar, G. H., Kamili, A. N., Chishti, M. Z., Dar, S. A., Tantry, T. A., & Ahmad, F. (2016). Characterization of *Aeromonas sobria* isolated from Fish Rohu (*Labeo rohita*) collected from polluted pond. *Journal of Bacteriology & Parasitology*, 7(3), 1–5. Available from https://doi.org/10.4172/2155-9597.1000273.

Dar, G. H., Kamili, A. N., Dar, S. A., Chishti, M. Z., & Ahmad, F. (2015). Detection and characterization of potentially pathogenic *Aeromonas sobria* isolated from fish *Hypophthalmichthys molitrix* (Cypriniformes: Cyprinidae). *Microbial Pathogenesis* (2015). Available from https://doi.org/10.1016/j.micpath.2015.10.017.

Dar, G. H., Bandh, S. A., & Kamili, A. N. (2011). Bacteriological analysis of soil from Yusmarg Health Resort of Kashmir Valley. *International Journal of Current Research*, *3* (10), 53–56.

Dar, G. H., Kamili, A. N., Nazir, R., Bandh, S. A., & Bhat, R. A. (2012). A preliminary study of colony forming units of bacteria from the soils of Yusmarg Forest, Kashmir Valley, India. *International Journal of Current Research*, *4*(12), 467–472.

Das, T., Sehar, S., Koop, L., Wong, Y. K., Ahmed, S., Siddiqui, K. S., & Manefield, M. (2014). Influence of calcium in extracellular DNA mediated bacterial aggregation and biofilm formation. *PLoS One*, *9*(3). Available from https://doi.org/10.1371/journal.pone.0091935.

Das, T., Sehar, S., & Manefield, M. (2013). The roles of extracellular DNA in the structural integrity of extracellular polymeric substance and bacterial biofilm development. *Environmental Microbiology Reports*, *5*(6), 778–786.

Deighton, M., & Borland, R. (1993). Regulation of slime production in *Staphylococcus epidermidis* by iron limitation. *Infection and Immunity*, *61*, 4473–4479.

Devaraj, A., Justice, S. S., Bakaletz, L. O., & Goodman, S. D. (2015). DNABII proteins play a central role in UPEC biofilm structure. *Molecular Microbiology*, *96*, 1119–1135.

Diels, L., De Smet, M., Hooyberghs, L., & Corbisier, P. (1999). Heavy metals bioremediation of soil. *Molecular Biotechnology*, *12*, 149–158.

Donlan, R. M., & Costerton, J. W. (2002). Biofilms: Survival mechanisms of clinically relevant microorganisms. *Clinical Microbiology Reviews*, *15*, 167–193.

Ells, T. C., & Hansen, L. T. (2006). Strain and growth temperature influence *Listeria* spp. attachment to intact and cut cabbage. *International Journal of Food Microbiology*, *111* (1), 34–42.

Erable, B., Duţeanu, N. M., Ghangrekar, M. M., et al. (2009). Application of electroactive biofilms. *Biofouling*, *26*, 57–71.

Flemming, H. C., & Wingender, J. (2010). The biofilm matrix. *Nature Reviews of Microbiology*, *8*, 623–633.

Floyd, K. A., Moore, J. L., Eberly, A. R., Good, J. A., Shaffer, C. L., Zaver, H., ... Hadjifrangiskou, M. (2015). Adhesive fiber stratification in uropathogenic *Escherichia coli* biofilms unveils oxygen -mediated control of type 1 pili. *PLoS Pathogens*, 11.

Fritts, R. K., LaSarre, B., Stoner, A. M., Posto, A. L., & McKinlay, J. B. (2017). A Rhizobiales-specific unipolar polysaccharide adhesin contributes to *Rhodopseudomonas* palustris biofilm formation across diverse photoheterotrophic conditions. *Applied and Environmental Microbiology*, *83*, 03035 -03016.

Glass, D. S., & Riedel-Kruse, I. H. (2018). A synthetic bacterial cell-cell adhesion toolbox for programming multicellular morphologies and patterns. *Cell*, *174*, 64–658.

Govan, J. R., & Deretic, V. (1996). Microbial pathogenesis in cystic fibrosis: Mucoid *Pseudomonas aeruginosa* and *Burkholderia cepacia*. *Microbiological Reviews*, *60*, 539–574.

Guiton, P. S., Hung, C. S., Kline, K. A., Roth, R., Kau, A. L., Hayes, E., ... Hultgren, S. J. (2009). Contribution of autolysin and Sortase A during *Enterococcus faecalis* DNA-dependent biofilm development. *Infection and Immunity*, 77, 3626–3638.

Harmsen, M., Yang, L., Pamp, S. J., & Tolker-Nielsen, T. (2010). An update on *Pseudomonas aeruginosa* biofilm formation, tolerance, and dispersal. *FEMS Immunology and Medical Microbiology*, *59*, 253–268.

Hay, I. D., Remminghorst, U., & Rehm, B. H. (2009). MucR, a novel membrane-associated regulator of alginate biosynthesis in *Pseudomonas aeruginosa*. *Applied and Environmental Microbiology*, *75*, 1110–1120.

Hentzer, M., Teitzel, G. M., Balzer, G. J., Heydorn, A., Molin, S., Givskov, M., & Parsek, M. R. (2001). Alginate overproduction affects *Pseudomonas aeruginosa* biofilm structure and function. *Journal of Bacteriology*, *183*, 5395–5401.

Hong, S. H., Lee, J., & Wood, T. K. (2010). Engineering global regulator Hha of *Escherichia coli* to control biofilm dispersal. *Microbial Biotechnology*, *3*, 717–728.

Huseby, M. J., Kruse, A. C., Digre, J., Kohler, P. L., Vocke, J. A., Mann, E. E., ... Earhart, C. A. (2010). Beta toxin catalyzes formation of nucleoprotein matrix in *staphylococcal* biofilms. *Proceedings of the National Academy of Sciences of the United States of America*, *107*, 14407–14412.

Jefferson, K. K., Goldmann, D. A., & Pier, G. B. (2005). Use of confocal microscopy to analyze the rate of vancomycin penetration through *Staphylococcus aureus* biofilms. *Antimicrobial Agents and Chemotherapy*, *49*, 2467–2473.

Jefferson, K. K., Pier, D. B., Goldmann, D. A., & Pier, G. B. (2004). The teicoplanin-associated locus regulator (TcaR) and the intercellular adhesin locus regulator (IcaR) are transcriptional inhibitors of the ica locus in *Staphylococcus aureus*. *Journal of Bacteriology*, *186*, 2449–2456.

Jin, X., & Riedel-Kruse, I. H. (2018). High-resolution patterned biofilm deposition using pDawn-Ag43. *Journal of Visualized Experiments*, *140*, 58625.

Joubert, L. M., Wolfaardt, G. M., & Botha, A. (2006). Microbial exopolymers link predator and prey in a model yeast biofilm system. *Microbial Ecology*, *52*(2), 187–197. Available from https://doi.org/10.1007/s00248-006-9063-7. Available from 16897306, S2CID 20431229.

Judd, S. (2008). The status of membrane bioreactor technology. *Trends in Biotechnology*, *26* (2), 109–116.

Kadlec, R. H., & Reddy, K. R. (2001). Temperature effect in treatment wetlands. *Water Environment Research*, *73*, 543–557.

Kantawanichkul, S., Kladprasert, S., & Brix, H. (2009). Treatment of high-strength wastewater in tropical vertical flow constructed wetlands planted with *Typha angustifolia* and *Cyperus involucratus*. *Ecological Engineering*, *35*(2), 238–247.

Karatan, E., & Watnick, P. (2009). Signals, regulatory networks, and materials that build and break bacterial biofilms. *Microbiology and Molecular Biology Reviews: MMBR*, *73*, 310–347.

Keogh, D., Lam, L. N., Doyle, L. E., Matysik, A., Pavagadhi, S., Umashankar, S., ... Kline, K. A. (2018). Extracellular electron transfer powers *Enterococcus faecalis* biofilm metabolism. *Molecular Biosystems*, 9. Available from https://doi.org/10.1128/mBio.00626-17.

Kerchove, A. J., & Elimelech, M. (2008). Calcium and magnesium cations enhance the adhesion of motile and nonmotile *Pseudomonas aeruginosa* on alginate films. *Langmuir: the ACS Journal of Surfaces and Colloids*, *24*, 3392–3399.

Kies, S., Otto, M., Vuong, C., & Götz, F. (2001). Identification of the sigB operon in *Staphylococcus epidermidis*: construction and characterization of a sigB deletion mutant. *Infection and Immunity*, *69*, 7933–7936.

Kjelleberg, S., Marshall, K. C., & Givskov, M. (2007). The biofilm mode of life. In S. Kjelleberg, & M. Givskov (Eds.), *The Biofilm Mode of Life: Mechanisms and Adaptations* (pp. 5–21). Norfolk, UK: Horizon Bioscience.

Kolodkin-Gal, I., Romero, D., Cao, S., Clardy, J., Kolter, R., & Losick, R. (2010). D-amino acids trigger biofilm disassembly. *Science (New York, N.Y.)*, *328*, 627–629.

Koo, H., Xiao, J., Klein, M. I., & Jeon, J. G. (2010). Exopolysaccharides produced by *Streptococcus* mutans glucosyltransferases modulate the establishment of microcolonies within multispecies biofilms. *Journal of Bacteriology*, *192*, 3024–3032.

Lam, H., Oh, D. C., Cava, F., Takacs, C. N., Clardy, J., de Pedro, M. A., & Waldor, M. K. (2009). D-amino acids govern stationary phase cell wall remodeling in bacteria. *Science (New York, N.Y.)*, *325*, 1552–1555.

Larsen, P., Nielsen, J. L., Dueholm, M. S., Wetzel, R., Otzen, D., & Nielsen, P. H. (2007). Amyloid adhesins are abundant in natural biofilms. *Environmental Microbiology*, *9* (12), 3077–3090. Available from https://doi.org/10.1111/j.1462-2920.2007.01418.x.

Laue, H., Schenk, A., Hongqiao, L., Lambertsen, L., Neu, T. R., Molin, S., & Ullrich, M. (2006). Contribution of alginate and levan to biofilm formation by *Pseudomonas syringae*. *Microbiology (Reading, England)*, *152*, 2909–2918.

Leid, J. G., Willson, C. J., Shirtliff, M. E., Hassett, D. J., Parsek, M. R., & Jeffers, A. K. (2005). The exopolysaccharide alginate protects *Pseudomonas aeruginosa* biofilm bacteria from IFN-g-mediated macrophage killing. *Journal of Immunology*, *175*, 7512–7518.

Lewandowski, Z., & Boltz, J. P. (2011). Biofilms in water and wastewater treatment. In P. Wilderer (Ed.), *Treatise on Water Science* (4, pp. 529–570). Oxford: Academic Press.

Lewis, K. (2001). Riddle of biofilm resistance. *Antimicrobial Agents and Chemotherapy*, *45*, 999–1007.

Li, T., Lin, G. Y., Podola, B., & Melkonian, M. (2015). Continuous removal of zinc from wastewater and mine dump leachate by a microalgal biofilm PSBR. *Journal of Hazardous Materials*, *297*, 112–118.

Li, Z., Zhang, X., & Lei, L. (2008). Electricity production during the treatment of real electroplating wastewater containing Cr6 + using microbial fuel cell. *Process Biochemistry*, *43*, 1352–1358.

Liu, H., Ramnarayanan, R., & Logan, B. E. (2004). Production of electricity during wastewater treatment using a single chamber microbial fuel cell. *Environmental Science & Technology*, *38*, 2281–2285.

Ma, L., Jackson, K. D., Landry, R. M., Parsek, M. R., & Wozniak, D. J. (2006). Analysis of Pseudomonas aeruginosa conditional Psl variant reveals roles for the Psl polysaccharide in adhesion and maintaining biofilm structure postattachment. *Journal of Bacteriology*, *188*, 8213–8221.

Macaskie, L. E., Yong, P., Doyle, T. C., Roig, M. G., Diaz, M., & Manzano, T. (1997). Bioremediation of uranium-bearing wastewater: biochemical and chemical factors influencing bioprocess application. *Biotechnology and Bioengineering*, *53*, 100–109.

Matsukawa, M., & Greenberg, E. P. (2004). Putative exopolysaccharide synthesis genes influence *Pseudomonas aeruginosa* biofilm development. *Journal of Bacteriology*, *186*, 4449–4456.

Mayer, C., Moritz, R., Kirschner, C., Borchard, W., Maibaum, R., Wingender, J., & Flemming, H.-C. (1999). The role of intermolecular interactions: studies on model systems for bacterial biofilms. *International Journal of Biological Macromolecules*, *26*(1), 3–16.

Mohamed, J. A., Huang, W., Nallapareddy, S. R., Teng, F., & Murray, B. E. (2004). Influence of origin of isolates, especially endocarditis isolates, and various genes on biofilm formation by *Enterococcus faecalis*. *Infection and Immunity*, *72*, 3658–3663.

Molle, P., Prost-Boucle, S., & Lienard, A. (2008). Potential for total nitrogen removal by combining vertical flow and horizontal flow constructed wetlands: a full-scale experiment study. *Ecological Engineering*, *34*, 23–29.

More, T. T., Yan, S., Tyagi, R. D., & Surampalli, R. Y. (2015). Biopolymers production by mixed culture and their applications in water and wastewater treatment. *Water Environment Research: A Research Publication of the Water Environment Federation*, *87*(6), 533–546.

Morici, L. A., Carterson, A. J., Wagner, V. E., Frisk, A., Schurr, J. R., Honer zu Bentrup, K., … Schurr, M. J. (2007). *Pseudomonas aeruginosa* AlgR represses the Rhl quorum-sensing system in a biofilm-specific manner. *Journal of Bacteriology*, *189*, 7752–7764.

Mulcahy, H., Charron-Mazenod, L., & Lewenza, S. (2008). Extracellular DNA chelates cations and induces antibiotic resistance in *Pseudomonas aeruginosa* biofilms. *PLoS Pathogens*, *4*(11). Available from https://doi.org/10.1371/journal.ppat.1000213.

Muyzer, G., & Stams, A. J. (2008). The ecology and biotechnology of sulphate-reducing bacteria. *Nature Reviews. Microbiology*, *6*, 441–454.

O'Toole, G., Kaplan, H. B., & Kolter, R. (2000). Biofilm formation as microbial development. *Annual Review of Microbiology*, *54*, 49–79.

O'Toole, G. A. (2011). Microtiter dish biofilm formation assay. *Journal of Visualized Experiments*, *47*, 1–2.

Oliveira, R., Melo, L., Oliveira, A., & Salgueiro, R. (1994). Polysaccharide production and biofilm formation by *Pseudomonas fluorescens*: effects of pH and surface material. *Colloids and Surfaces B: Biointerfaces*, *2*(1), 41–46.

Olutiola, P. O. (2010). Relationship between bacterial density and chemical composition of a tropical sewage oxidation pond. *African Journal of Environmental Science and Technology*, *4*(9), 595–602.

Pal, S., Sarkar, U., & Dasgupta, D. (2010). Dynamic simulation of secondary treatment processes using trickling filters in a sewage treatment works in Howrah, West Bengal, India. *Desalination*, *253*(1), 135–140.

Prakash, B., Veeregowda, B. M., & Krishnappa, G. (2003). Biofilms. A survival strategy of bacteria. *Current Science*, *85*(9), 1299–1307.

Puhakka, J. A., Melin, E. S., Järvinen, K. T., Koro, P. M., Jukka, A., Hartikainen, P., ... Ferguson, J. F. (1995). Fluidized-bed biofilms for chlorophenol mineralization. *Water Science and Technology: A Journal of the International Association on Water Pollution Research*, *31*, 227–235. Available from https://doi.org/10.1016/0273-1223(95)00170-R.

Qu, Y. Y., Zhou, J. T., Wang, J., Xing, L. L., Jiang, N., Gou, M., & Salah, U. M. (2009). Population dynamics in bioaugmented membrane bioreactor for treatment of bromoamine acid wastewater. *Bioresource Technology*, *100*, 244–248.

Radojević, I. D., Grujić, S. M., Ranković, B. R., et al. (2019). Single-species biofilms from autochthonous microorganisms: Biotechnological potential in automotive wastewater treatment. *International Journal of Environmental Science and Technology.*, *16*, 6189–6198. Available from https://doi.org/10.1007/s13762-019-02265-y.

Reichhardt, C., Wong, C., Passos da Silva, D., Wozniak, D. J., & Parsek, M. R. (2018). CdrA interactions within the *Pseudomonas aeruginosa* biofilm matrix safeguard it from proteolysis and promote cellular packing. *American Society for Microbiology*. Available from https://doi.org/10.1128/mBio.01376-18.

Rice, K. C., Mann, E. E., Endres, J. L., Weiss, E. C., Cassat, J. E., Smeltzer, M. S., & Bayles, K. W. (2007). The cidA murein hydrolase regulator contributes to DNA release and biofilm development in Staphylococcus aureus. *Proceedings of the National Academy of Sciences of the United States of America*, *104*, 8113–8118.

Rodesney, C. A., Roman, B., Dhamani, N., Cooley, B. J., Katira, P., Touhami, A., & Gordon, V. D. (2017). Mechanosensing of shear by *Pseudomonas aeruginosa* leads to increased levels of the cyclic -di -GMP signal initiating biofilm development. *Proceedings of the National Academy of Sciences of the United States of America*, *114*, 5906–5911.

Roeselers, G., Van Loosdrecht, M. C. M., & Muyzer, G. (2008). Phototrophic biofilms and their potential applications. *Journal of Applied Phycology*, *20*(3), 227–235.

Romling, U. (2002). Molecular biology of cellulose production in bacteria. *Research in Microbiology*, *153*, 205–212.

Rossi, F., & De Philippis, R. (2015). Role of cyanobacterial exopolysaccharides in phototrophic biofilms and in complex microbial mats. *Life (Chicago, Ill.: 1978)*, *5*(2), 12181238. Available from https://doi.org/10.3390/life5021218, PMC 4500136. PMID. Available from 25837843.

Rowe, M. C., Withers, H. L., & Swift, S. (2010). Uropathogenic *Escherichia coli* forms biofilm aggregates under iron restriction that disperse upon the supply of iron. *FEMS Microbiology Letters*, *307*, 102–109.

Sarkisova, S., Patrauchan, M. A., Berglund, D., Nivens, D. E., & Franklin, M. J. (2005). Calcium-induced virulence factors associated with the extracellular matrix of mucoid *Pseudomonas aeruginosa* biofilms. *Journal of Bacteriology*, *187*, 4327–4337.

Sauer, K., Cullen, M. C., Rickard, A. H., Zeef, L. A., Davies, D. G., & Gilbert, P. (2004). Characterization of nutrient-induced dispersion in *Pseudomonas aeruginosa* PAO1 biofilm. *Journal of Bacteriology*, *186*, 7312–7326.

Sehar, S., Sumera, S., Naeem, S., Perveen, I., Ali, N., & Ahmed, S. (2015). A comparative study of macrophytes influence on wastewater treatment through subsurface flow hybrid constructed wetland. *Ecological Engineering*, *81*, 62–69.

Shalaby, I. M., Altalhy, A. D., & Mosallam, H. A. (2008). Preliminary field study of a model plant for sewage water treatment using gravel bed hydroponics method. *World Applied Science Journal*, *4*(2), 238–243.

Shama, S., & Iffat, N. (2016). Role of the biofilms in wastewater treatment. In D. Dhanasekaran, & N. Thajuddin (Eds.), *Microbial biofilms - Importance and applications*. IntechOpen. Available from 10.5772/63499. Available from. Available from https://www.intechopen.com/books/microbial-biofilms-importance-and-applications/role-of-the-biofilms-in-wastewater-treatment.
Shirtliff, M. E., Mader, J. T., & Camper, A. K. (2002). Molecular interactions in biofilms. *Chemistry & Biology*, *9*, 859–871.
Shukla, S. K., Mangwani, N., Rao, T. S., & Mangwani, N. (2014). Biofilm-mediated bioremediation of polycyclic aromatic hydrocarbons. In S. Das (Ed.), *Microbial Biodegradation and Bioremediation* (pp. 203–232). Oxford: Elsevier.
Silva, B., Figueiredo, H., Quintelas, C., Isabel, C., & Nevesb, T. T. (2008). Zeolites as supports for the biorecovery of hexavalent and trivalent chromium. *Microporous Mesoporous Mater*, *116*, 555–560. Available from https://doi.org/10.1016/j.micromeso.2008.05.015.
Smith, W. L., & Gadd, G. M. (2000). Reduction and precipitation of chromate by mixed culture sulphate reducing bacterial biofilms. *Journal of Applied Microbiology*, *88*, 983–991.
Starkey, M., Hickman, J. H., Ma, L., Zhang, N., De Long, S., Hinz, A., ... Starner, T. D. (2009). *Pseudomonas aeruginosa* rugose small-colony variants have adaptations that likely promote persistence in the cystic fibrosis lung. *Journal of Bacteriology*, *191*, 3492–3503.
Stewart, P. S., & Franklin, M. J. (2008). Physiological heterogeneity in biofilms. *Nature Reviews. Microbiology*, *6*, 199–210.
Steyn, B., Oosthuizen, M. C., MacDonald, R., Theron, J., & Brözel, V. S. (2001). The use of glass wool as an attachment surface for studying phenotypic changes in *Pseudomonas aeruginosa* biofilms by two dimensional gel electrophoresis. *Proteomics*, *1*(7), 871–879.
Subramanian, S. B., Yan, S., Tyagi, R. D., & Surampalli, R. Y. (2010). Extracellular polymeric substances (EPS) producing bacterial strains of municipal wastewater sludge: Isolation, molecular identification, EPS characterization and performance for sludge settling and dewatering. *Water Research*, *44*, 2253–2266.
Sulzacova, K., Trtilek, M., & Rataj, T. (2015). Phosphorus removal using a microalgal biofilm in a new biofilm photobioreactor for tertiary wastewater treatment. *Water Research*, *71*, 55–63.
Tielker, D., Hacker, S., Loris, R., Strathmann, M., Wingender, J., Wilhelm, S., ... Jaeger, K. E. (2005). *Pseudomonas aeruginosa* lectin LecB is located in the outer membrane and is involved in biofilm formation. *Microbiology (Reading, England)*, *151*, 1313–1323.
Valls, M., & de Lorenzo, V. (2002). Exploiting the genetic and biochemical capacities of bacteria for the remediation of heavy metal pollution. *FEMS Microbiology Reviews*, *26*, 327–338.
Van, C. C., Underwood, G. C., Serôdio, J., & Paterson, D. M. (2014). Ecology of intertidal microbial biofilms: Mechanisms, patterns and future research needs. *Journal of Sea Research*, *92*, 2–5. Available from https://doi.org/10.1016/j.seares.2014.07.003.
Vasseur, P., Vallet-Gely, I., Soscia, C., Genin, S., & Filloux, A. (2005). The pel genes of the *Pseudomonas aeruginosa* PAK strain are involved at early and late stages of biofilm formation. *Microbiology (Reading, England)*, *151*, 985–997.
Vilain, S., Pretorius, J. M., Theron, J., & Brozel, V. S. (2009). DNA as an adhesin: *Bacillus cereus* requires extracellular DNA to form biofilms. *Applied and Environmental Microbiology*, *75*, 2861–2868.
Walters, M. C., Roe, F., Bugnicourt, A., Franklin, M. J., & Stewart, P. S. (2003). Contributions of antibiotic penetration, oxygen limitation, and low metabolic activity to tolerance of *Pseudomonas aeruginosa* biofilms to ciprofloxacin and tobramycin. *Antimicrobial Agents and Chemotherapy*, *47*, 317–323.

Wang, X., Preston, J. F., 3rd, & Romeo, T. (2004). The pgaABCD locus of *Escherichia coli* promotes the synthesis of a polysaccharide adhesin required for biofilm formation. *Journal of Bacteriology, 186*, 2724–2734.

Webb, J. S., Thompson, L. S., James, S., Charlton, T., Tolker- Nielsen, T., Koch, B., ... Kjelleberg, S. (2003). Cell death in *Pseudomonas aeruginosa* biofilm development. *Journal of Bacteriology, 185*, 4585–4592.

Webb, J. S. (2007). Differentiation and dispersal in biofilms. In S. Kjelleberg, & M. Givskov (Eds.), *The Biofilm Mode of Life: Mechanisms and Adaptations* (7, pp. 167–174). Norfolk, UK: Horizon Bioscience.

Whitchurch, C. B., Tolker-Nielsen, T., Ragas, P. C., & Mattick, J. S. (2002). Extracellular DNA required for bacterial biofilm formation. *Science (New York, N.Y.), 295*, 1487.

Whiteley, M., Bangera, M. G., Bumgarner, R. E., Parsek, M. R., Teitzel, G. M., Lory, S., & Greenberg, E. P. (2001). Gene expression in *Pseudomonas aeruginosa* biofilms. *Nature, 413*, 860–864.

Xiong, Y., & Liu, Y. (2010). Biological control of microbial attachment: A promising alternative for mitigating membrane biofouling. *Applied Environmental Microbiology, 86*, 825–837.

Yang, L., Barken, K. B., Skindersoe, M. E., Christensen, A. B., Givsko, M., & Tolker-Nielsen, T. (2007). Effects of iron on DNA release and biofilm development by *Pseudomonas aeruginosa. Microbiology (Reading, England), 153*, 1318–1328.

Yoshida, A., & Kuramitsu, H. K. (2002). Multiple *Streptococcus* mutans genes are involved in biofilm formation. *Applied and Environmental Microbiology, 68*, 6283–6291.

Zhang, X., Li, J., Yu, Y., Xu, R., & Wu, Z. (2016). Biofilm characteristics in natural ventilation trickling filters(NVTFs) for municipal wastewater treatment: Comparison of three kinds of biofilm carriers. *Biochemical Engineering Journal, 106*, 87–96. Available from https://doi.org/10.1016/j.bej.2015.11.009.

Zhao, F., Rahunen, N., Varcoe, J. R., Chandra, A., Avignone-Rossa, C., Thumser, A. E., & Slade, R. C. T. (2008). Activated carbon cloth as anode for sulfate removal in amicrobial fuel cell. *Environmental Science & Technology, 42*, 4971–4976.

Zhao, K., Tseng, B. S., Beckerman, B., Jin, F., Gibiansky, M. L., Harrison, J. J., ... Wong, G. C. L. (2013). Psl trails guide exploration and microcolony formation in *Pseudomonas aeruginosa* biofilms. *Nature, 497*, 388–391.

Zogaj, X., Nimtz, M., Rohde, M., Bokranz, W., & Romling, U. (2001).) The multicellular morphotypes of *Salmonella typhimurium* and *Escherichia coli* produce cellulose as the second component of the extracellular matrix. *Molecular Microbiology, 39*, 1452–1463.

CHAPTER 15

Biotechnology: the sustainable tool for effective treatment of wastewater

Muhammad Usman, Shahid Adeel, Zunaira Amjad, Tanveer Hussain Bokhari, Nadia Akram and Sapna Anam
Department of Chemistry, Government College University Faisalabad, Faisalabad, Pakistan

15.1 Introduction

Dyes from natural sources such as mollusks, insects, and plants were in use till the late 19th century. It was the first time, after 1856 when a synthetic dye called "mauveine or aniline purple" was discovered by W. H. Perkin in his attempt to make antimalarial drug quinine. After that, the synthetic dye, mauveine, started to be manufactured at a large industrial scale (Robinson, McMullan, Marchant, & Nigam, 2001). Nowadays, the production rate of commercial dyes is high. According to an estimate, about 7×10^5 tons of dyes are produced annually and 100,000 dyes exist already. Several industries are regarded as one of the potent consumers of dyes, especially textile industry, which also uses water in excess. Approximately, 2.8×10^5 tons of dyes annually are discharged through water by the textile industry (Jin, Liu, Xu, & Tao, 2007). The concentration of unbound dyes in textile effluents is reported as 0.6−0.8 g dye per liter (Frank, Hermanutz, & Opperman, 1994). Approximately, $1 - 1.5 \times 10^5 8$ tons of dyes have become part of water systems (Rai et al., 2005). The use of synthetic dyes has enhanced the production of xenobiotics that are carcinogenic and are enlisted as highly toxic compounds by the United States Environment Protection Agency (Wang & Hu, 2008). Industries such as textile, leather, food, cosmetics, and pharmaceuticals are legally and ethically bound to shed their effluents toward the environment after proper treatment (Rai et al., 2005). It is stated that azo dyes, as well as anthraquinone dyes, are making the largest part of the wastewater pollution. The azo dyes have structural diversity due to the presence of many substituents, including chloro (−Cl), methyl ($-CH_3$),

Microbial Consortium and Biotransformation for Pollution Decontamination
DOI: https://doi.org/10.1016/B978-0-323-91893-0.00007-9

nitro ($-NO_2$), amino ($-NH_2$), hydroxyl ($-OH$), and carboxyl ($-COOH$). The triphenylmethane, polycyclic, and anthraquinone groups are extensively used in the textile and printing process (Saratale, Saratale, Chang, & Govindwar, 2011). The presence of mentioned dyes in the effluent is affecting the concentration of dissolved oxygen, the rate of photosynthesis, and the equilibrium of the ecosystem (Van der Zee & Villaverde, 2005). The complex aromatic structure and xenobiotic nature of dyes are responsible for the less effective treatment of effluents. Even the end products of dye degradation are also, sometimes, toxic and carcinogenic. A variety of methods ranging from the physical to chemical decolorization are there but are neither cost and energy nor time effective (Forgacs, Cserhati, & Oros, 2004; Ranganathan, Karunagaran, & Sharma, 2007). It, therefore, seems right to use selected microorganisms for the degradation process on account of their sustainable and green nature. These processes can also completely degrade the dyes without disturbing the aquatic ecosystem (Sarayu & Sandhya, 2012). Many microorganisms, including bacteria, fungi, yeasts, algae, and actinomycetes, have the potential to decolorize the azo dyes (Saratale et al., 2011). That's why for biodegradation to come into play, at an industrial scale, it is crucial to select microbes that can withstand diverse pH, temperature, and high dye toxicity (Kaushik & Malik, 2009). Thus the need of the hour is to study the degradation ability of microbes for industrial wastewater to design an industrial biodegradation process. The choice of any biological system to deal with the removal of the dyes needs not only to have the capacity to handle various dyes but also to have a remarkable power to resist the toxicity of the dyes, their end products, surfactants, detergents, salts, extreme condition of pH, and temperature simultaneously (Anjaneyulu, Chary, & Raj, 2005; Chen, 2006). The best microbial system to degrade the azo dyes is bacterial cultures as some of the bacterial systems have the capacity to grow faster in all the extreme conditions of pH, temperature, salinity, and presence or absence of oxygen (Solís, Solís, Pérez, Manjarrez, & Flores, 2012).

15.2 Classification of biodegradation processes

There are different methods for degradation of dyes based on the mode of action but biodegradation has proved it to be more sustainable. It is done by employing various microorganisms. Some of them are discussed in the following sections.

15.2.1 Bacterial biodegradation

The available literature has reported bacteria as efficient biodecolorizers. The degradation through bacterial strains follows eco-friendly and cost-effective methodology such as microbial decolorization that works by cleaving the dye into fragments (Khehra, Saini, Sharma, Chadha, & Chimni, 2005; McMullan et al., 2001). A complete breakdown of dyes into carbon dioxide, biomass, and other inorganic compounds by bacteria has been reported (Kolekar et al., 2012).

15.2.2 Algal biodegradation

Algal biodegradation is relatively less reported microbial degradation. Algal biodegradation is considered effective in shallow stable ponds as it cleaves the azo linkage and even degrades the aromatic amines. The limitations of algal biodegradation include lesser adsorption and lesser degradation as compared to bacterial degradation. The algal adsorption is factor-dependent. The examples include (35.62 mg dye per gram biomass) for Supranol Red 3BW, (44.98 mg dye per gram biomass) for Lanaset Red 2 G, and (43.17 mg dye per gram biomass) for Levafix Navy Blue EBNA by *Chlorella vulgaris* (Lim, Chu, & Phang, 2010; Priya, Uma, Ahamed, Subramanian, & Prabaharan, 2011).

15.2.3 Fungal biodegradation

Some species of wood-rotting fungi, for example, *Phanerochaete chrysosporium* and *Trametes* sp. are reported to degrade not only dyes but also lignin. The sludge formation in fungal biodegradation is reported rarely, thus fungal biodegradation is considered a slow and tough process. The fungal enzymes responsible for lignin degradation are nonspecific in nature. Lignin peroxidase, manganese peroxidase, and laccase are extracellular fungal enzymes that regulate oxidation reactions, for example, nonphenolics are degraded by lignin peroxidases and laccase are responsible for the degradation of phenolics (McMullan et al., 2001). Since its early discovery as a dye degrader in 1983 white rot fungi is deeply investigated and fungal biodegradation is found to be able to degrade any sort of chemical dyes (Glenn & Gold, 1983). However, fungal biodegradation is affected by the diversity and capability of fungal species. Aromatic structures are degraded by secondary metabolism in the case of limiting nutrients (e.g., C, N, and S) (Kirk & Farrell, 1987). The factors essential for fungal cultivation and propagation include the addition of energy substrates, cofactor availability, and pH of the medium. The stable fungal bioreactors for degradation of synthetic dyes are successful but the use of fungal degradation for dyes in industrial effluents is not much successful as the impurities of water cause irregular biomass growth and untrustworthy enzyme production (Mielgo, Moreira, Feijoo, & Lema, 2001; Stolz, 2001). Another hurdle is massive wastewater volumes as they affect the crucial steps of the process such as extraction and purification of dyes (Nigam, Armour, Banat, Singh, & Marchant, 2000). Lignin peroxidase requires acidic pH of (4.5−5.0) for proper working that, if managed, can harm the other beneficial microbes, for example, bacteria. Aromatics present in wastewater also interfere with the process (Stolz, 2001).

15.3 Factors affecting biodegradation process: an overview

A lot of factors are there which can affect the degradation process; however, the followings have major contributions to affect the rate of this process.

- concentration of biomass
- molecular structure, physiochemical properties, and concentration of substrate
- treatment conditions, including pH, moisture, temperature, metals, dissolved oxygen, nitrates, and carbon and energy resources Uddin et al. (2007)

15.3.1 Effects of dye concentration

The effect of the concentration of dye on biodecolorization is a well-researched topic. The higher the concentration of dyes the lower will be the biodegradation because the toxic degradation products inhibit microbial growth (Kalyani, Patil, Jadhav, & Govindwar, 2008). The degradation of Scarlet R (50 mg/L) by *Micrococcus glutamicus* is reported to be 100% in an incubation period of 20 h that decreased to 18% in an incubation period of 72 h with an increase in concentration to 250 mg/L (Saratale, Saratale, Kalyani, Chang, & Govindwar, 2009). The degradation of Malachite Green by *Lasiodiplodia* sp. is found to decrease to 89.18% from 96.9% by an increase in dye concentration from 50 to 500 mg/L (Arunprasath, Sudalai, Meenatchi, Jeyavishnu, & Arumugam, 2019).

15.3.2 Effects of molecular structure

The molecular structure and concentration of pollutants highly impact the biodegradation treatment (Khan, Bhawana, & Fulekar, 2013).

15.3.3 Effects of pH

The efficiency of biodegradation is affected by the pH of the medium. The rate of degradation has been reported to be high at optimum pH and decreased on moving either side of the scale (Chaudhari, Tapase, Markad, & Kodam, 2013; Saratale et al., 2009; Yoo, Libra, & Adrian, 2001). The pH affects the rate-controlling step of the process as it affects the movement of dye molecules through cell membranes (Kodam, Soojhawon, Lokhande, & Gawai, 2005). The optimum pH for algal and fungal biodegradation lies in the neutral or acidic region. While for bacterial biodegradation, it falls in the basic or neutral region. The optimum pH for the bacterial function is reported to be in the range of 6.0 to 10.0 (Kolekar & Kodam, 2012; Kolekar et al., 2008). The decolorization of 27 diverse dyes, belonging to mono-azo, diazo, phthalocyanine, and triphenylmethane classes, has been reported by utilizing 21 diverse basidiomycetes run best at optimum pH 3.0–5.0 (Nozaki et al., 2008). The biodegradation of methyl red by a yeast, *Galactomyces geotrichum*, has found to be completed within the pH range of 3.0–5.0 (Jadhav, Kalme, & Govindwar, 2008). The decolorization of Indigo Carmine, Remazol Brilliant Blue R, Bromophenol Blue, and Methyl Red by a white rot fungi *Lentinus polychrous* has been found to be optimum at a pH of 9.0, 3.0, 4.0, and 4.0–5.0, respectively (Sarnthima, Khammuang, & Svasti, 2009). *Trametes* sp., a fungus, degrades the Remazol Brilliant Blue R well at optimum pH 4.5 (Yang, Zhao, Liu, Zheng, & Qian, 2009). The reported optimum pH for the biodecolorization of Reactive Black 5 by *Enterobacter* sp. EC3 is 7.0 as bacterial strain shows maximum growth at this pH (Wang et al., 2009).

15.3.4 Effects of temperature

Temperature is one of the limiting factors in the biodegradation process as it affects the microbial growth. Degradation potential is found to be directly proportional to the temperature within the optimum temperature range (Nachiyar & Rajkumar, 2003). Azoreductase enzymes play a significant rule in the degradation process. After optimum temperature, these entities undergo denaturation and thus degradation process stops (Saratale et al., 2009). *Lysinibacillus* sp. KMK-A is reported to degrade reactive M2R dye and chromate at an optimum temperature range of 20°C–50°C (Khan et al., 2013).

The optimum temperature for fungal growth is reported as 25°C–35°C (Fu & Viraraghavan, 2001). Cotton Blue is decolorized by *Penicillium ochrochloron* MTCC 517 at an optimum temperature of 25°C (Shedbalkar, Dhanve, & Jadhav, 2008). The decolorization of Methyl Red by a yeast *G. geotrichum* has been studied at different temperatures and 30°C was reported as optimum temperature for the process (Jadhav et al., 2008). The optimum temperature for the decolorization of Reactive Black 5 is 37°C. While 42°C temperature has a negative effect on the same process, this results in denaturation of the involved enzymes (Wang et al., 2009).

15.3.5 Effects of nitrogen content

The nitrogen concentration affects the microbial growth thus biodecolorization. The concentration of fungal enzymes is affected by the nitrogen content both positively and negatively (Kaushik & Malik, 2009). A limited nitrogen concentration is favorable for the azo dye's decolorization and degradation (Hu, 1998). The nitrogen addition to a medium containing nitrogen nutrient inhibits the Congo Red decolorization (Tatarko & Bumpus, 1998).

15.3.6 Effects of impurities

The wastewater especially textile effluents host a variety of acids, bases, salts, heavy metals, and other impurities (Khalid, Arshad, & Crowley, 2008). Microbial decolorization is affected by the high salt concentrations (up to 10 g/L NaCl or Na_2SO_4). However, low salt concentrations do not affect most of the microbes. The high salt concentrations cause cell plasmolysis. *Gracilibacillus* sp. GTY strain is among some bacteria that can work at high salt concentrations even of 15% (w/v) of NaCl without affecting efficiency (Uddin et al., 2007). Decolorization of Reactive Black 5, Direct Red 81, Acid Red 88, and Disperse Orange 3 by *Shewanella putrefaciens* in the presence of different concentrations of NaCl showed no decolorization above 60 g/L NaCl (Khalid et al., 2008). Heavy metals affect the biodegradation of dyes by affecting microbial growth. It has been reported that the presence of the metals in the medium causes no effect on the degradation of organic material in a biochemical process (Boopathy, 2000).

15.3.7 Effects of agitation

There is a contradiction for the effect of shaking on the decolorization of dyes. Higher aeration and better distribution of nutrients are reported to enhance the decolorization of dyes (Kaushik & Malik, 2009). On the other hand, decolorization of Reactive Red 2 by *Pseudomonas* sp. *SUK1* is found to be 96% in 6 h under static conditions and 24 h under agitated conditions (Kalyani, Telke, Dhanve, & Jadhav, 2009). Static conditions supported the biodegradation of Reactive Red 120 and Direct Red 81 by *Aspergillus niger* (Husseiny, 2008). Navy Blue HER found to be degraded to 100% by *Trichosporon beigelii* NCIM-3326 when not agitated and 30% under agitation. It is reported that the static condition favors the growth of the microbes (Saratale, Saratale, Chang, & Govindwar, 2009).

15.3.8 Effects of aerobatic conditions

According to the available literature, oxygen availability or aerobic conditions do not favor dye decolorization or degradation, as dyes resist microbial attack under these conditions (Forgacs et al., 2004).

The use of aerobic and facultative anaerobic microorganisms for biodegradation has been reported in the literature (Moosvi, Keharia, & Madamwar, 2005; Nigam, Banat, Singh, & Marchant, 1996). These cultures developed aerobically and perform only under anaerobic or anoxic treatment conditions (Kolekar et al., 2012). *Pseudomonas luteola, Aeromonas hydrophila, Bacillus subtilis, Pseudomonas* sp., *Proteus mirabilis, Bacillus fusiformis*, and *Alishewanella* sp. have been reported to decolorize dyes both under hypoxic and static hypoxic conditions in the pure form (Chang et al., 2001; Kolekar & Kodam, 2012; Kolekar et al., 2008). The degradation efficiency up to 75% of different dyes reported to be achieved under hypoxic conditions as compared to the 30% by agitation (Chaudhari et al., 2013; Kolekar et al., 2008). The anaerobic step of this treatment reduces the azo dyes to aromatic amines that are degraded in the next aerobic treatment. The aerobic microbes are also reported to cleave the azo bond by producing azoreductase enzyme under oxygen availability (Nachiyar & Rajakumar, 2005). However, the aerobic decolorization of dyes by bacterial strains is less reported as they are unable to use dyes as an electron acceptor and require carbon input (Stolz, 2001). *Pseudomonas aeruginosa* is reported to decolorize textile dyes by using glucose input in the presence of oxygen (Nachiyar & Rajkumar, 2003). The requirement of organic input in forms of yeast, peptone, or carbohydrates is essential for not only mixed but also for pure bacterial cultures (Chen, Wu, Liou, & Hwang, 2003; Khehra et al., 2005).

The literature suggests that oxygen availability or aerobic conditions are necessary to help microbes grow and anaerobic conditions to degrade the azo bond.

The next is an extensive explanation of each of the abovementioned biodegradation and biodecolorization processes.

15.4 Bacterial biodegradation and biodecolorization

Bacteria can work to remove the dyes from the textile effluent efficiently. More specifically, bacteria and their enzymes performed well against the azo dyes, which are a substantial part of the industrial wastewater. The *Pseudomonas*, *Bacillus*, *Rhodococcus* were found to decolorize wastewater (Telke, Kalyani, Jadhav, & Govindwar, 2008). Bacteria can do degradation under the aerobic, anaerobic, acidic, and basic conditions. These bacteria were found in the soil contaminated with dyes containing industrial effluents (Dawkar, Jadhav, Jadhav, & Govindwar, 2008; Dhanve, Shedbalkar, & Jadhav, 2008; Kalyani et al., 2009).

15.4.1 Mechanism of bacterial dye degradation

The followings are the mechanisms of bacterial degradation processes. The site of these mechanisms may be extracellular and intracellular depending on the type of bacterial strain. Irrespective of the nature of the mechanism, bacterial degradation of azo dyes involves the symmetrical breakdown of the azo group ($-N=N$).
1. mechanism that involves bacterial enzymatic tool/direct method
2. mechanism that involves redox mediator/indirect method
3. mechanism that involves biogenic reductants

15.4.2 Mechanism that involves enzymatic tool

The following are some enzymes that are efficiently playing their role in the dye decolorization.

15.4.2.1 Decolorization through azoreductase

Anaerobic bacteria possess extracellular azoreductases that are sensitive to oxygen. For example, two azoreductases were purified one is Orange I azoreductase from *Pseudomonas K24* and the other one is Orange II azoreductase from *Pseudomonas KF46* (Fig. 15.1). The later one acts with a hydroxyl group at the ortho-position of the aromatic ring while the former reductase needs it on the para-position for its action (Zimmermann, Gasser, Kulla, & Leisinger, 1984).

Clostridium and *Eubacterium* are the names of some bacterial strains that decolorize the sulfonated azo dyes present in their growing mediums.

The anaerobic azoreductases are flavoprotiens. These reductases find in the extracellular and intracellular spaces of the cell membrane. These enzymes also need the electron donors for the reduction of the azo-bonds, for example,

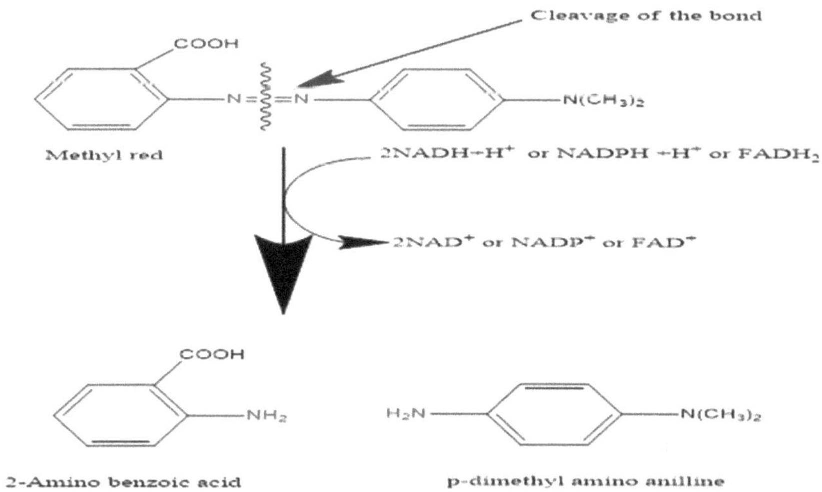

Figure 15.1 Degradation of methyl red by azoreductases.

NADH, NADPH, and FADH (Rafii, Franklin, & Cerniglia, 1990; Russ, Rau, & Stolz, 2000).

Cofactors, which are located in the cytoplasm, act as the hydrogen donors in the reduction process (Russ et al., 2000). The extracts, starving and lysed cells break azo-bonds with a high reduction rate. The reason is, in these cells, the cofactors are released extracellularly and the availability of the cofactors makes the azoreductases to work efficiently. The cofactors are FADH2, FMNH2, NADH, and NADPH (Dubin & Wright, 1975; Mechsner & Wuhrmann, 1982; Roxon, Ryan, & Wright, 1967; Walker, 1970; Wuhrmann, Mechsner, & Kappeler, 1980). However, the intracellular function of the azoreductase is uncertain in the detoxification of the azo dyes, because the azo dyes have high molecular weight and are complex in the structure that makes it difficult to get access to the cytoplasm.

15.4.2.2 Decolorization by nonspecific reductases

Some nonspecific reductases also play a significant role in the detoxification of the azo dyes (Parshetti, Kalme, Saratale, & Govindwar, 2006). There are two nonspecific reductases, the way of action of which is still unknown, as given next:

1. **NADH-DCIP reductase:** These enzymes use NADH as an electron donor to reduce the substrate. The complete reduction results in the disappearance of the blue color of the DCIP substrate (Fig. 15.2).
2. **MG reductase:** One of the nonspecific reductases, that play a primary role in the biodegradation is called Malachite Green reductase (Fig. 15.3).

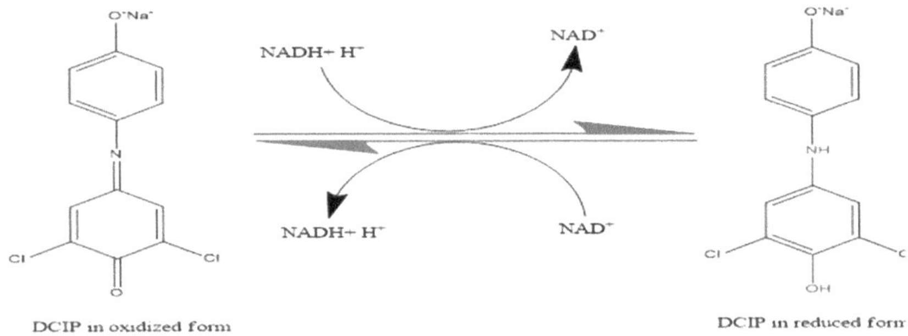

Figure 15.2 Reduction reaction of DCIP substrate by NADH-DCIP reductase.

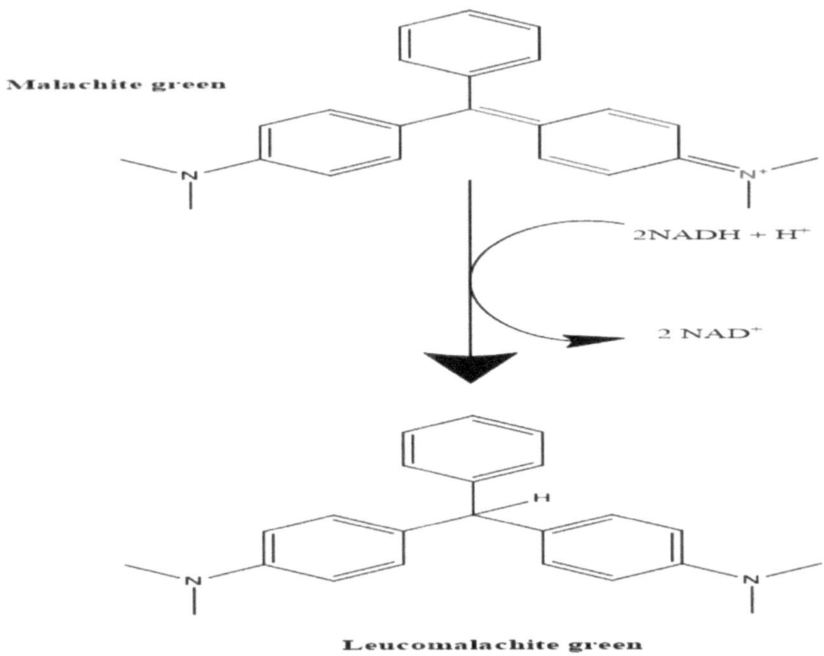

Figure 15.3 Reduction of Malachite Green by MG reductase enzyme.

MG reductase enzymes also use the NADH as an electron donor, in the reduction process of Malachite Green (Tien, Kirk, Bull, & Fee, 1986).

15.4.2.3 Decolorization by lignin peroxidase

These enzymes are N-glycosylated proteins and

during action on peroxide. These enzymes belong to the family of oxidoreductases. Lignin peroxidase has the heme group at the active site and molecular weight in the range of 38–47 kDa (Tien & Kirk, 1983). They participate in the oxidation reaction of the compounds such as lignin either by abstracting an electron to form reactive and unstable radicals or to break down the aromatic rings in the structure (Kersten, Tien, Kalyanaraman, & Kirk, 1985; Umezawa & Higuchi, 1987).

15.4.2.4 Decolorization through laccases

Their molecular weight exists in the range of 60 to 390 kDa (Claus, 2003). These act as an electron acceptor in the oxidation of the compounds such as substituted phenolic and nonphenolic and degrade the azo dyes, and produce nonaromatic amines (Claus, 2004; Givaudan et al., 1993; Solano & Sanchez-Amat, 1999). *Azospirillum lipoferum* was the first bacterium in which the first prokaryotic laccase was reported (Stauffer, 1994).

15.4.2.5 Decolorization by polyphenol oxidase (tyrosinase)

Streptomyces glaucescens, *Streptomyces antibioticus*, *Bacillus licheniformis*, *Bacillus natto*, and *Bacillus sphaericus* are the names of some bacteria in which polyphenol oxidases are found. These enzymes catalyzed the oxidation of many colors and phenolics compounds to quinones in the presence of the molecular oxygen (Tamboli, Telke, Dawkar, Jadhav, & Govindwar, 2011).

15.4.2.6 Veratryl alcohol oxidase

The vet is a member of the family oxidoreductases. These take part in the parallel reaction (Levine, 1991) (Table 15.1).

15.4.3 Mechanism that involves redox mediator/electron shuttle

The redox mediator also called electron shuttle plays a notable role in the azo dye decolorization process. The electron shuttle transfer the electrons from the bacterial cell to the extracellular spaces. During this activity, the redox mediator continuously changes from oxidized form into reduced form. The anthraquinone-2-sulfonate is a synthetic electron carrier that enhances the rate of the decolorization of the azo dye molecules (Pereira, Pereira, Alves, & Pereira, 2014). The natural humic acids are regarded as an enriched source of quinone groups that act as redox mediators. Several carbonaceous materials such as carbon nanotube, activated carbon, and

Table 15.1 List of enzymes for degradation of textile effluents.

Enzyme name	Bacterial species	Mechanism of enzyme	Product of dye degradation
Veratryl alcohol oxidase	*Comamonas* sp. UVS	Oxidative cleavage.	The desulfonation of unknown product produced naphthalene-1,2,5-triol (Jadhav, Dawkar, Tamboli, & Govindwar, 2009)
Azoreductase	*Exiguobacterium* sp. RD3	Azo bond cleavage by azoreductase (Dhanve et al., 2008; Dhanve, Kalyani, Phugare, & Jadhav, 2009).	1. *N, N*-dimethyl benzene1,4 diamine and 2-amino benzoic acid 2. Diamino naphthyl sulfonic acid and benzidine
Laccase	*Bacillus* sp. ADR (Telke et al. 2009)	Asymmetric cleavage of an azo bond (Telke, 2009).	1. p-hydroperoxy dimethylamino benzene 2. p-phenyldizine sulfonic acid and p-hydroxy *N, N*′-dimethylamino benzene
Lignin peroxidase	*Bacillus laterosporus*	First intermediate cyclo-2, 5-ene-1-oneproduce the sulfate ion, *N, N* dimethylamine, and a nitrogen molecule. Condensation of 2 molecules of the first intermediate	p-isopropanal-2,5-cyclohexa-dienone (Gomare, Jadhav, & Govindwar, 2008)

(*Continued*)

Table 15.1 (Continued)

Enzyme name	Bacterial species	Mechanism of enzyme	Product of dye degradation
		produced 1, 1′-dicyclohex-2, 5-ene-4-one. α-cleavage of this cyclohexene produced 4-substituted hexanoic acid which produced 4-cyclohexenone which further produced lactone cation on the removal of a methyl group. Elimination of iminium moiety from lactone cation produced the final product.	

carbon black are also found useful for the dye wastewater treatment (Pandey, Singh, & Iyengar, 2007).

15.4.4 Mechanism that involves biogenic reductants

The end products of the metabolism reactions such as sulfide and ferrous ions have an effective role in the decolorization of the azo dyes in a pure chemical reaction (Kargi, 2002). The different inducers such as $CaCO_3$, o-toluidine, veratrole, vanillin, and indole of oxidative enzymes increase the rate of detoxification of azo dyes (Dawkar et al., 2008). Sodium acetate, sodium format, sodium succinate, sodium citrate, and sodium pyruvate are electron donors that enhance the reduction rate of the detoxification of the azo dyes by *Bacillus* sp. ADR.

15.4.5 Factors affecting bacterial decolorization

The following are the physical/chemical parameters that influence the bacterial degradation process of dyes.

15.4.5.1 The pH of the medium

An optimal pH range for the decolorization of the azo dyes is 6.0–10. A combination of *G. geotrichum* and *Bacillus* sp. is reported to be inert toward the pH change during the decolorization of the Brilliant Blue G.

15.4.5.2 Effects of temperature change

Temperature is always a key factor that influences the growth of cells and the function of enzymes of microorganisms. For the degradation process, industries require bacterial strains that can tolerate temperature variations. For this purpose, the azoreductase enzyme of *B acillus badius* is the best choice for high-temperature tolerance. Similarly, the thermophilic bacterial strains are the best choice for the temperature variation.

15.4.5.3 Effects of dissolved oxygen

The decolorization process slows down in the presence of oxygen because the oxygen molecules and the azo dyes compete for the electrons.

15.4.5.4 Effects of agitation

It is reported that agitation makes the fair distribution of the nutrients and finally improve the decolorization rate.

15.4.5.5 Effects of saline condition

These conditions affect microbial flora growth and ultimately slow down azo dye removal process. Dilution is the only way to reduce salinity before wastewater treatment. It will result in a large affluent mass that will make the treatment process more difficult. Many bacterial strains work well to remove the azo dyes under the high salinity condition, for example, *Shewanella* sp. AS96 (Saratale et al., 2009).

15.4.5.6 Effects of carbon/nitrogen supplement

It is stated that the carbon and nitrogen sources impose a positive effect on the azo dye decolorization. For example, glucose, acetate, glycerin, sucrose, salicylate, and citrate molecules make the detoxification process more operative (Ruiz-Arias et al., 2010). On the other hand, at high concentrations, these supplements cause negative effects by making the medium more acidic (Parshetti et al., 2011).

15.4.5.7 Effects of the dye's structure

Azo dyes have a broad range of structures. Low-molecular-weight dyes, with simple structures, are easily decolorized by the bacterial cultures. Mono-azo dyes have a more decolorization rate than tri-azo and diazo dyes.

15.4.5.8 Effects of the dye's concentration

There is an inverse relationship between the percentage of decolorization and dye concentration (Saratale et al., 2011).

15.4.5.9 Effects of aeration supply

The proper aeration provides the bacterial cell with the necessary oxygen. Azoreductases reduced their efficiency in the presence of oxygen. During the reduction of the azo dye molecules, many simple aromatic compounds are formed, the degradation of which occurs through the ring-opening and hydroxylation. These processes need the presence of the oxygen to complete. So, it is good to proceed with the degradation process first under the anaerobic condition and then under aerobic treatment (Saratale et al., 2011; Solís et al., 2012).

Example: Here are some dyes to which degradation is reported by bacterial strains.

1. **Crystal Violet:** The degradation of this dye is reported by some bacterial strains, including *Agrobacterium radiobacter, Rhizobium radiobacter, Sphingomonas paucimobilis, Stenotrophomonas maltophila* LK-24, *Pseudomonas mendocina* MCM B-402, *Enterobacter cloacae,* and *B. subtilis* IFO 13719 (An et al., 2002; Cheriaa & Bakhrouf, 2009; Kim et al., 2002; Sani & Banerjee, 1999; Sarnaik & Kanekar, 1999; Yatome, Ogawa, & Matsui, 1991).
2. **Brilliant Green:** *S. paucimobilis, Citrobacter* sp. strain KCTC 18061P, and *Kurthia* sp. are involved in the degradation of Brilliant Green dye (Dönmez & Aksu, 2002; Mohan, Rao, Prasad, & Karthikeyan, 2002).

15.5 Fungal biodegradation and biodecolorization

Mycoremediation is a general term for fungal decontamination of the environment, while "myco" means "fungus."

Characteristics:

1. Fungus being the microorganism has the high cell-to-surface area ratio. The advantage associated with this feature is that these organisms can make the greatest contact with the environment.

2. Extracellular nature of the fungal enzymes makes them enable to survive in the intense toxic environment.
3. As compared to the single cell, the mycelial growth is faster.

15.5.1 Mechanism of mycoremediation

Fungal biodegradation/mycoremediation is believed to opt one of following three pathways for the biodegradation and biodecolorization of synthetic dyes:

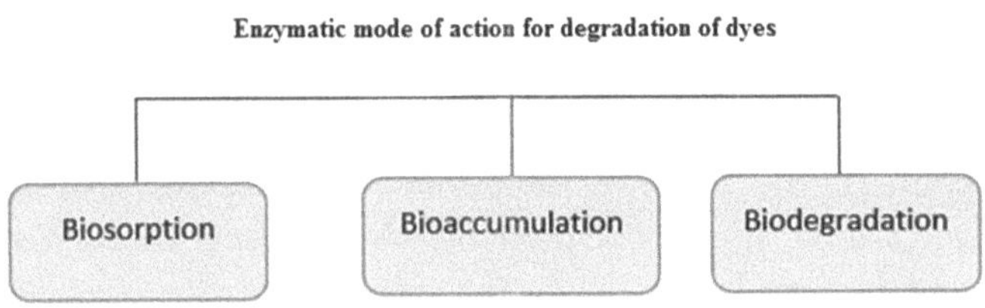

15.5.2 Biosorption

The fungal cell wall has binding functional groups such as carboxyl, amino, thiol, and phosphates that absorb the dye molecules on the surface of the cell (Fan, Yang, Gao, & Yuan, 2012; Tan, Li, Lin, Lu, & Chen, 2011).

Features: The following features are described for the biosorption process:

1. It is a passive process.
2. The biosorption is proceeded by the dead/living fungi (Srinivasan & Viraraghavan, 2010).
3. Adsorption of the dye is the primary step.
4. It is a reversible process.
5. There is no need to maintain the nutrients of the medium.
6. The rate of the process is fast and needs only a few hours to be completed.
7. This process is not affected by the toxicity of the dyes.

15.5.2.1 Factors affecting biosorption

The following are the parameters that cause influence on the biosorption process (Srinivasan & Viraraghavan, 2010):

1. the temperature of the solution
2. the concentration of the dye in the medium

3. types of the dye being used in the process
4. the pH of the medium
5. presence of another chemical with the dyes in the wastewater
6. ionic strength
7. ionic state of dyes

15.5.3 Bioaccumulation

Bioaccumulation is the mechanism through which dyes are removed from the wastewater by accumulation in the cytoplasm of the microorganism where different metabolism processes occur (Chojnacka, 2010). Bioaccumulation is achieved in two different ways (Das, Charumathi, & Das, 2011; Dias, Lucas, Sampaio, Peres, & Bezerra, 2010; Taskin & Erdal, 2010):

1. using single-cell fungi
2. mycelial fungi such as *A. niger*

Features: The followings are the features of the bioaccumulation process.

1. It is an active process.
2. The biomass is alive for the initiation of the process.
3. Absorption of the dyes is the principle step of the process.
4. This process is partially reversible.
5. There is a need to maintain the nutrients of the medium.
6. This process is called a double-stage process (elaborated under basic principle).
7. The rate of the process is slow.
8. This process also depends on the energy as well as on metabolism.
9. The contaminants have the strongest effect on this process.
10. Cellular growth is the basic requirement of the process.

15.5.4 Biodegradation

In this process, enzymes are used to degrade the dye molecules.

Features: The following are the attributes of the biodegradation process:

1. It is an active process.
2. The biomass, in the process, is alive.
3. The enzymes are the basic elements in the process to degrade the dye molecules.

4. The site for the degradation depends on the process selected for the degradation. For example, if the enzymes are secreted extracellularly by the fungi then the degradation is performed in vitro, otherwise the molecules are adsorbed and degraded in vivo.
5. It is also an irreversible process.
6. The preservation of the nutrition is required specifically the limitation of nitrogen is beneficial.
7. It is called the double-stage process.
8. The rate of the process is slow.
9. It is a metabolically control process.
10. The contaminants show deleterious effects on the process.
11. Cellular growth is important for the process to occur at a good pace.

15.5.5 Advantage of fungal biodegradation and biodecolorization

The following are some benefits that are associated with the use of fungi for the biodegradation (Daassi, Mechichi, Nasri, & Rodriguez-Couto, 2013; Jin & Ning, 2013; Martorell, Pajot, Rovati, & Figueroa, 2012; Moreira-Neto, Mussatto, Machado, & Milagres, 2013).

1. The secretion of the enzymes by the fungi is extracellular. This causes biodegradation without the need for the diffusion of the dye molecules into the fungi cell for the initiation of dye degradation.
2. The nature of the produced enzymes does not get affected by the presence of the pollutants because it only depends on the nutrients such as carbon and nitrogen.

Example: Following are the examples of some fungal degraded dyes:

1. **Methyl Violet:** *Cyathus bulleri* 195062 and *P. chrysosporium* are some white rot fungi that degraded the methyl violet (Bumpus & Brock, 1988; Vasdev, Kuhad, & Saxena, 1995).
2. **Victoria Blue:** *Irpex lacteus* is responsible for the degradation of *Victoria Blue* (Novotny et al., 2001).
3. **Methylene Blue, 80% degradation:** *Ganoderma lucidum* and *Pycnoporus cinnabarinus*, which are white rot fungi, are responsible for the degradation of the methylene blue to a greater extent within 20 days.
4. **Methylene Blue, 40% degradation:** While *Phaseolus coccineus*, *Stereum ostrea*, *Tulipa. suaveolen*, *Fomes fomentarius*, and *Coriolus versicolor* are effective only to 40% degradation of the dye in the same time (Jayasinghe et al., 2008).

5. **Brilliant Green:** Some fungi are responsible for the degradation of the Brilliant Green, including *Trametes versicolor, Trametes trogii, Grammothele subargentea* LPSC no. 436, *Fomes sclerodermeus, Phlebia floridensis, Phlebia fascicularia, Dichomitus squalens, Pycnoporus sanguineus, C. versicolor, Pleurotus ostreatus*, and *P. chrysosporium* (Casas, Parella, Vicent, Caminal, & Sarra, 2009; Gill, Arora, & Chander, 2002; Levin, Papinutti, & Forchiassin, 2004; Papinutti, Mouso, & Forchiassin, 2006; Pointing & Vrijmoed, 2000; Saparrat et al., 2008; Sun, Hu, Bi, & Cao, 2009).

15.6 Algal biodegradation and biodecolorization

Algal biodegradation is an important process that follows the mechanisms mentioned in the following for the biodegradation of synthetic dyes.

15.6.1 Dye-ion accumulation and biocoagulation

Algae have a high binding affinity and high surface area (Daneshvar, Kousha, Sohrabi, Khataee, & Converti, 2012). Different types of functional groups are present on the cell wall of the algae that are sulfate, carboxyl, hydroxyl, and various charge groups. The charge groups are produced by lipid components and complex heteropolysaccharides. The charge groups are responsible for the adsorption of the dyes from the wastewater. This versatile nature of the cell wall causes the biosorption process to occur by complexation and electrostatic attraction (Aksu & Tezer, 2005; Özer, Akkaya, & Turabik, 2006). So, in the first step of degradation, the dyes ions are accumulated on the surface of algae.

15.6.2 Diffusion

Another process is the diffusion of the dye's molecules into the biopolymer of the algae (Kumar, Sivanesan, & Ramamurthi, 2005).

15.6.3 Biocoagulation

Algae release some long-chain biopolymers called metabolic intermediates. These extracellular polymers have functional groups on the surface. These biopolymers, when release, accumulate the dyes that left behind in the wastewater and settle down (Mohan et al., 2002).

Example:

Malachite Green: Under the alkaline pH condition, a cationic dye, Malachite Green gets adsorbed on the surface of the algae. The alkaline condition enhances the adsorption capacity of the algae (Daneshvar, Ayazloo, Khataee, & Pourhassan, 2007; Khataee, Zarei, & Pourhassan, 2009).

15.6.4 Factors affecting algal biodegradation

Algal biodegradation is susceptible to the following factors:

15.6.4.1 pH

Algal biodegradation and biodecolorization of synthetic dyes are found to perform best under acidic or neutral pH. The efficiency of the process is affected badly on moving either side of optimum pH (Kolekar & Kodam, 2012; Kolekar et al., 2008; Yoo et al., 2001). The isoelectric point for algae species is at pH 3.0 (Kumar et al., 2005). The algal decolorization of Malachite Green by *Cosmarium* sp. showed the variation in dye removal with variation in pH value (Daneshvar et al., 2007).

15.6.4.2 Temperature

Temperature is among the limiting factors in the biodegradation process as it affects the microbial growth. Degradation potential is found to be directly proportional to the temperature within the optimum temperature range (Kolekar et al., 2008; Nachiyar & Rajkumar, 2003). The reason behind a fall in degradation after optimum temperature lies in the denaturation of an azoreductase enzyme that is responsible for the degradation process. The increase in temperature is found to favor the decolorization of Malachite Green by *Cosmarium* sp. within a temperature range of 5°C−45°C (Saratale et al., 2009).

15.6.4.3 Dye concentration

The effect of the concentration of dye on biodecolorization is a well-researched topic. However, different trends are reported. The higher the concentration of dyes the lower will be the biodegradation because the toxic degradation products inhibit microbial growth (Kalyani et al., 2008). The high initial dye concentration is found to have a positive effect on the decolorization of Malachite Green by *Cosmarium* sp., an algal species (Daneshvar et al., 2007).

15.7 Future prospectus: an importance

Followings are the elements that justify the act of biotechnology, the sustainable tool for effective treatment of wastewater in terms of energy requirement, coculture application, simple setup of bioreactors, availability characterization techniques, and toxicity level detection capability.

15.7.1 Energy requirement

Energy is the primary requirement for any process to be completed and the biodegradation process is no exception. The energy source for the biodegradation of the dyes is easily available. For example, the pollutants act as an energy source for microorganisms.(Kolekar et al., 2008).

A few bacteria are reported to get their energy from amines produced by the reduction of the azo bond (Pandey et al., 2007). These microbes are a few in number but highly desirable to solve the issues in real sense. Crystal carbon acts as a source of carbon and energy for *A. hydrophila* for the degradation of azo, anthraquinone, and triphenylmethane dyes (Ren, Guo, Zeng, & Sun, 2006). *Aspergillus flavus* and *Alternaria solani* feed on Malachite Green for carbon and energy (Deller, Macheroux, & Sollner, 2008). Navy Blue HER is degraded by *T. beigelii* NCIM-3326 and it is found that the microbe can feed on this dye (Saratale et al., 2009).

15.7.2 Coculture application

Coculture can be used to get nontoxic products to result from biodegradation. For example, bacterial decolorization of azo dyes reported the production of carcinogenic aromatic amines in the result of enzymatic action of azoreductases, while the production of such species is not reported in the case of fungal decolorization of azo dyes by oxidase action. The use of enzymes from the coculture of fungi and bacteria is reported to enhance the biodegradation of azo dyes without the production of aromatic amines (Lade, Waghmode, Kadam, & Govindwar, 2012).

15.7.3 Bioreactor for effective decolorization of textile dyes

The applicability of the biodegradation process on an industrial scale requires a proper setup. For this purpose, bioreactors are formed. The development of bioreactors is a well-reported subject that leads to the development of several types of bioreactors, including upflow anaerobic sludge blanket (UASB) reactor and activated sludge reactors. UASB is designed in the late 1970s for the treatment of industrial effluents and is also used for sewage treatment (Lettinga, Van Velsen, Hobma, De Zeeuw, & Klapwijk, 1980). The components of the USAB include gas separator, effluent distribution, and effluent withdrawal systems. The sludge bed of this anaerobic reactor hosts semimobile microbes that are responsible for the production of biogas as a result of the degradation of organic compounds (Chinwetkitvanich, Tuntoolvest, & Panswad, 2000).

15.7.4 Techniques used for the characterization

Characterization of treated effluents is necessary as they might contain toxic compounds produced in the result of biodegradation of dyes (Forss & Welander, 2009). Decolorization and degradation of dyes not only aim to disappear the color but also to eliminate the toxicity of dyes (Sun et al., 2009). The process of biodegradation would be more effective if it results in nontoxic products. Therefore it is compulsory to identify the end products.

15.7.5 Identification of intermediates

Advanced chromatographic and spectroscopic techniques are available to isolate and characterize the intermediates of dye degradation (Saeed et al., 2016).

15.7.6 Identification of degradation products

The following techniques help in the identification of degradation products in the mentioned way:

1. UV–visible spectrophotometry: This technique helps in the determination of the path of decolorization like whether it is by adsorption or by microbial cell ingestion. In the case of surface-mediated decolorization, all peaks decrease, while in the case of degradation by microbial cell ingestion, disappearance of major peak and formation of the new peak is observed (Chen et al., 2003; Saratale et al., 2009). The cells assisting surface adsorption will be colorized and the cells degrading dye by engulfing it do not show enhanced colors (Sun et al., 2009).
2. Thin-layered chromatography: Comparison of R_f values of original dye and degradation products is enough to identify the nature of degradation products.
3. For detailed information, more advanced techniques, including GC-Ms, NMR, and HPLC. These techniques are also good to provide a detailed mechanism of the degradation process.

15.7.7 Assessment of detoxification of dye degradation products

The effectiveness of biodegradation depends upon the toxicity levels of degradation products that can be assessed by comparing the phytotoxicity and microbial toxicity assays of original dye and degradation products.

A model plant is selected to treat with original dye and its degradation product for phytotoxicity assay. The evaluation of percent germination and plumule length of the sample compared with the control plant seed help in the assessment of degradation product's toxicity. The procedure is the same for microbial toxicity assay but the assessment relies on microbial cell count per millimeter. For the tests conducted using algae, the results of treated and untreated "chlorophyll a" are compared to assess the relative toxicity of dye and its degradation product.

The research conducted on the toxicity of original dyes as compared to degradation products in previous years is summarized in the table:

Model plant/ model microbe	Dye and its biodegradation products	Treatment	Results	References
Triticum aestivum and *Phaseolus mungo*	Malachite Green	Phytotoxicity assay	The plumule length and plant germination is affected more by original dye, thus the degradation products are less toxic.	Parshetti et al. (2006)
Sorghum vulgare and *P. mungo*	Reactive Red 2	Phytotoxicity assay	The degradation products are less toxic than the original dye.	Kalyani et al. (2009)
P. mungo and *S. vulgare*	Scarlet R	Phytotoxicity assay after degrading with mixed strains of *Proteus. vulgaris* and *Micrococcus glutamic*	Original dye caused 50% and 60% inhibition in plants, respectively. No, inhibition observed by degradation products.	Saratale et al. (2009)
Bacillus sp. VUS	Textile azo dye Navy Blue 2GL	Phytotoxicity assay	The original dye reduced the	Dawkar, Jadhav, Ghodake,

(Continued)

Model plant/ model microbe	Dye and its biodegradation products	Treatment	Results	References
			germination of both plants by 80% and 70% in a 7-day incubation period, while degradation products caused 2%–5% of growth reduction.	and Govindwar (2009)
Sorghum bicolor and *T. aestivum*	Methyl Red	Phytotoxicity assay of biodegraded dye by *Galactomyces geotrichum* MTCC 1360	An 88% and 72% germination inhibition is observed by the original dye in both plants, respectively, while degradation products did not show any inhibition.	Jadhav et al. (2008)
P. mungo and *S. vulgare*	Navy Blue HER	Phytotoxicity assay after treating with *Trichosporon beigelii* NCIM-3326	Degradation products are detoxified.	Saratale et al. (2009)
T. aestivum and *Ervum lens* L	Cotton Blue	Phytotoxicity assay after biodegrading with *Penicillium ochrochloron* MTCC 517	Original dye caused 30% and 20% inhibition, while degradation products caused a 10% inhibition in	Shedbalkar et al. (2008)

(*Continued*)

Model plant/ model microbe	Dye and its biodegradation products	Treatment	Results	References
			T. aestivum and no inhibition in *Ervum lens* L.	
Gloeocapsa			*pleurocapsoides, Phormidium ceylanicum, Chroococcus minutes*	FF Sky Blue, Amido Black
"Chlorophyll a" content	The dyes caused a decrease in	"Chlorophyll a" content that also depicts growth inhibition.	Parikh and Madamwar (2005)	
Escherichia coli strain JM 109	Crystal Violet	Microbial toxicity assay after decolorizing with *Shewanella* decoloration is NTOU1	Dye detoxified by microbial strain.	Chen (2006)

15.8 Conclusion

The toxicity, carcinogenicity, mutagenicity, and clastogenicity of synthetic dyes make them harmful for living beings. The health and environmental hazards of synthetic dyes alarmed the global senses for almost three decades that impacted the legislations to control dye concentration in industrial effluents. Since then, physical, biological, and chemical methods utilized to degrade the synthetic dyes from which physical and chemical methods are greater in costs and production of secondary contaminants than biological methods. The biological method is bioremediation or the use of microbes for the degradation of dyes and other synthetic compounds. Microbes act as microreactors that degrade xenobiotics (synthetic organic compounds) to less harmful or completely nontoxic final products. Biodegradation also is known for the complete degradation or mineralization of compounds to water, CO_2, and salts. The factors affecting biodegradation depend upon the type of dye and microbe used. But generally, the process is susceptible to change in pH, temperature, the concentration of dye, salt concentration in the medium, agitation, nitrogen, and oxygen content, and aeration, etc. The comparison of degradation ability of different microbes is, thus, needed to consider these factors. The ability of microbes to use dye molecules for energy and C-content makes them highly favorable dye degraders. The research activities on the removal of synthetic dyes by using bioremediation enhanced significantly in the past three decades and still have a lot to work on and explore.

References

Aksu, Z., & Tezer, S. (2005). Biosorption of reactive dyes on the green alga *Chlorella vulgaris*. *Process Biochemistry*, *40*(3–4), 1347–1361.

An, S.-Y., Min, S.-K., Cha, I.-H., Choi, Y.-L., Cho, Y.-S., Kim, C.-H., & Lee, Y.-C. (2002). Decolorization of triphenylmethane and azo dyes by *Citrobacter* sp. *Biotechnology Letters*, *24*(12), 1037–1040.

Anjaneyulu, Y., Chary, N. S., & Raj, D. S. S. (2005). Decolourization of industrial effluents—available methods and emerging technologies—A review. *Reviews in Environmental Science and Bio/Technology*, *4*(4), 245–273.

Arunprasath, T., Sudalai, S., Meenatchi, R., Jeyavishnu, K., & Arumugam, A. (2019). Biodegradation of triphenylmethane dye malachite green by a newly isolated fungus strain. *Biocatalysis and Agricultural Biotechnology*, *17*, 672–679.

Boopathy, R. (2000). Factors limiting bioremediation technologies. *Bioresource Technology*, *74*(1), 63–67.

Bumpus, J. A., & Brock, B. J. (1988). Biodegradation of crystal violet by the white rot fungus *Phanerochaete chrysosporium*. *Applied and Environmental Microbiology*, *54*(5), 1143–1150. Available from https://doi.org/10.1128/aem.54.5.1143-1150.1988.

Casas, N., Parella, T., Vicent, T., Caminal, G., & Sarra, M. (2009). Metabolites from the biodegradation of triphenylmethane dyes by *Trametes versicolor* or laccase. *Chemosphere*, *75*(10), 1344–1349. Available from https://doi.org/10.1016/j.chemosphere.2009.02.029.

Chang, J. S., Chou, C., Lin, Y. C., Lin, P. J., Ho, J. Y., & Hu, T. L. (2001). Kinetic characteristics of bacterial azo-dye decolorization by *Pseudomonas luteola*. *Water Research*, *35*(12), 2841–2850. Available from https://doi.org/10.1016/s0043-1354(00)00581-9.

Chaudhari, A. U., Tapase, S. R., Markad, V. L., & Kodam, K. M. (2013). Simultaneous decolorization of reactive Orange M2R dye and reduction of chromate by *Lysinibacillus* sp. KMK-A. *Journal of Hazardous Materials*, *262*, 580–588. Available from https://doi.org/10.1016/j.jhazmat.2013.09.006.

Chen, H. (2006). Recent advances in azo dye degrading enzyme research. *Current Protein & Peptide Science*, 7(2), 101–111. Available from https://doi.org/10.2174/138920306776359786.

Chen, K. C., Wu, J. Y., Liou, D. J., & Hwang, S. C. (2003). Decolorization of the textile dyes by newly isolated bacterial strains. *Journal of Biotechnology*, *101*(1), 57–68. Available from https://doi.org/10.1016/s0168-1656(02)00303-6.

Cheriaa, J., & Bakhrouf, A. (2009). Triphenylmethanes, malachite green and crystal violet dyes decolourisation by *Sphingomonas paucimobilis*. *Annals of Microbiology*, *59*(1), 57–61.

Chinwetkitvanich, S., Tuntoolvest, M., & Panswad, T. (2000). Anaerobic decolorization of reactive dyebath effluents by a two-stage UASB system with tapioca as a co-substrate. *Water Research, 34*(8), 2223–2232.
Chojnacka, K. (2010). Biosorption and bioaccumulation—The prospects for practical applications. *Environment International, 36*(3), 299–307.
Claus, H. (2003). Laccases and their occurrence in prokaryotes. *Archives of Microbiology, 179*(3), 145–150.
Claus, H. (2004). Laccases: Structure, reactions, distribution. *Micron (Oxford, England: 1993), 35*(1-2), 93–96.
Daassi, D., Mechichi, T., Nasri, M., & Rodriguez-Couto, S. (2013). Decolorization of the metal textile dye Lanaset Grey G by immobilized white-rot fungi. *Journal of Environmental Management, 129*, 324–332. Available from https://doi.org/10.1016/j.jenvman.2013.07.026.
Daneshvar, E., Kousha, M., Sohrabi, M. S., Khataee, A., & Converti, A. (2012). Biosorption of three acid dyes by the brown macroalga *Stoechospermum marginatum*: Isotherm, kinetic and thermodynamic studies. *Chemical Engineering Journal, 195*, 297–306.
Daneshvar, N., Ayazloo, M., Khataee, A. R., & Pourhassan, M. (2007). Biological decolorization of dye solution containing Malachite Green by microalgae *Cosmarium* sp. *Bioresource Technology, 98*(6), 1176–1182. Available from https://doi.org/10.1016/j.biortech.2006.05.025.
Das, D., Charumathi, D., & Das, N. (2011). Bioaccumulation of the synthetic dye Basic Violet 3 and heavy metals in single and binary systems by *Candida tropicalis* grown in a sugarcane bagasse extract medium: Modelling optimal conditions using response surface methodology (RSM) and inhibition kinetics. *Journal of Hazardous Materials, 186*(2-3), 1541–1552.
Dawkar, V. V., Jadhav, U. U., Ghodake, G. S., & Govindwar, S. P. (2009). Effect of inducers on the decolorization and biodegradation of textile azo dye Navy blue 2GL by *Bacillus* sp. VUS. *Biodegradation, 20*(6), 777–787. Available from https://doi.org/10.1007/s10532-009-9266-y.
Dawkar, V. V., Jadhav, U. U., Jadhav, S. U., & Govindwar, S. P. (2008). Biodegradation of disperse textile dye Brown 3REL by newly isolated *Bacillus* sp. VUS. *Journal of Applied Microbiology, 105*(1), 14–24. Available from https://doi.org/10.1111/j.1365-2672.2008.03738.x.
Deller, S., Macheroux, P., & Sollner, S. (2008). Flavin-dependent quinone reductases. *Cellular and Molecular Life Sciences: CMLS, 65*(1), 141–160. Available from https://doi.org/10.1007/s00018-007-7300-y.
Dhanve, R., Shedbalkar, U., & Jadhav, J. P. (2008). Biodegradation of diazo reactive dye Navy Blue HE2R (Reactive Blue 172) by an isolated *Exiguobacterium* sp. RD3. *Biotechnology and Bioprocess Engineering, 13*(1), 53–60.
Dhanve, R. S., Kalyani, D. C., Phugare, S. S., & Jadhav, J. P. (2009). Coordinate action of exiguobacterial oxidoreductive enzymes in biodegradation of reactive yellow 84A dye. *Biodegradation, 20*(2), 245–255. Available from https://doi.org/10.1007/s10532-008-9217-z.
Dias, A. A., Lucas, M. S., Sampaio, A., Peres, J. A., & Bezerra, R. M. (2010). *Decolorization of azo dyes by yeasts. Biodegradation of azo dyes* (pp. 183–193). Springer.
Dubin, P., & Wright, K. L. (1975). Reduction of azo food dyes in cultures of *Proteus vulgaris*. *Xenobiotica; the Fate of Foreign Compounds in Biological Systems, 5*(9), 563–571. Available from https://doi.org/10.3109/00498257509056126.
Dönmez, G., & Aksu, Z. (2002). Removal of chromium (VI) from saline wastewaters by *Dunaliella* species. *Process Biochemistry, 38*(5), 751–762.
Fan, H., Yang, J., Gao, T., & Yuan, H. (2012). Removal of a low-molecular basic dye (Azure Blue) from aqueous solutions by a native biomass of a newly isolated

Cladosporium sp.: Kinetics, equilibrium and biosorption simulation. *Journal of the Taiwan Institute of Chemical Engineers, 43*(3), 386–392.

Forgacs, E., Cserhati, T., & Oros, G. (2004). Removal of synthetic dyes from wastewaters: A review. *Environment International, 30*(7), 953–971. Available from https://doi.org/10.1016/j.envint.2004.02.001.

Forss, J., & Welander, U. (2009). Decolourization of reactive azo dyes with microorganisms growing on soft wood chips. *International Biodeterioration & Biodegradation, 63*(6), 752–758.

Frank, G., Hermanutz, F., & Opperman, W. (1994). Ozonation-an important technique to comply with new German laws for textile wastewater treatment. *Water Science and Technology, 30*(3), 255.

Fu, Y., & Viraraghavan, T. (2001). Fungal decolorization of dye wastewaters: A review. *Bioresource Technology, 79*(3), 251–262. Available from https://doi.org/10.1016/s0960-8524(01)00028-1.

Gill, P. K., Arora, D. S., & Chander, M. (2002). Biodecolourization of azo and triphenylmethane dyes by *Dichomitus squalens* and *Phlebia* spp. *Journal of Industrial Microbiology & Biotechnology, 28*(4), 201–203. Available from https://doi.org/10.1038/sj/jim/7000222.

Givaudan, A., Effosse, A., Faure, D., Potier, P., Bouillant, M.-L., & Bally, R. (1993). Polyphenol oxidase in *Azospirillum lipoferum* isolated from rice rhizosphere: Evidence for laccase activity in non-motile strains of *Azospirillum lipoferum. FEMS Microbiology Letters, 108*(2), 205–210.

Glenn, J. K., & Gold, M. H. (1983). Decolorization of several polymeric dyes by the lignin-degrading basidiomycete *Phanerochaete chrysosporium. Applied and Environmental Microbiology, 45*(6), 1741–1747.

Gomare, S. S., Jadhav, J. P., & Govindwar, S. P. (2008). Degradation of sulfonated azo dyes by the purified lignin peroxidase from *Brevibacillus laterosporus* MTCC 2298. *Biotechnology and Bioprocess Engineering, 13*(2), 136–143.

Hu, T. (1998). Degradation of azo dye RP2B by *Pseudomonas luteola. Water Science and Technology, 38*(4-5), 299–306.

Husseiny, S. M. (2008). Biodegradation of the reactive and direct dyes using Egyptian isolates. *Journal of Applied Sciences Research, 4*(6), 599–606.

Jadhav, S., Kalme, S., & Govindwar, S. (2008). Biodegradation of methyl red by *Galactomyces geotrichum* MTCC 1360. *International Biodeterioration & Biodegradation, 62*(2), 135–142.

Jadhav, U. U., Dawkar, V. V., Tamboli, D. P., & Govindwar, S. P. (2009). Purification and characterization of veratryl alcohol oxidase from *Comamonas* sp. UVS and its role in decolorization of textile dyes. *Biotechnology and Bioprocess Engineering, 14*(3), 369–376.

Jayasinghe, C., Imtiaj, A., Lee, G. W., Im, K. H., Hur, H., Lee, M. W., . . . Lee, T. S. (2008). Degradation of three aromatic dyes by white rot fungi and the production of ligninolytic enzymes. *Mycobiology, 36*(2), 114–120. Available from https://doi.org/10.4489/MYCO.2008.36.2.114.

Jin, X., & Ning, Y. (2013). Laccase production optimization by response surface methodology with *Aspergillus fumigatus* AF1 in unique inexpensive medium and decolorization of different dyes with the crude enzyme or fungal pellets. *Journal of Hazardous Materials, 262*, 870–877. Available from https://doi.org/10.1016/j.jhazmat.2013.09.024.

Jin, X. C., Liu, G. Q., Xu, Z. H., & Tao, W. Y. (2007). Decolorization of a dye industry effluent by *Aspergillus fumigatus* XC6. *Applied Microbiology and Biotechnology, 74*(1), 239–243. Available from https://doi.org/10.1007/s00253-006-0658-1.

Kalyani, D. C., Patil, P. S., Jadhav, J. P., & Govindwar, S. P. (2008). Biodegradation of reactive textile dye Red BLI by an isolated bacterium *Pseudomonas* sp. SUK1. *Bioresource Technology, 99*(11), 4635–4641. Available from https://doi.org/10.1016/j.biortech.2007.06.058.

Kalyani, D. C., Telke, A. A., Dhanve, R. S., & Jadhav, J. P. (2009). Ecofriendly biodegradation and detoxification of Reactive Red 2 textile dye by newly isolated *Pseudomonas* sp. SUK1. *Journal of Hazardous Materials*, *163*(2-3), 735–742. Available from https://doi.org/10.1016/j.jhazmat.2008.07.020.

Kargi, F. (2002). Empirical models for biological treatment of saline wastewater in rotating biodisc contactor. *Process Biochemistry*, *38*(3), 399–403.

Kaushik, P., & Malik, A. (2009). Fungal dye decolourization: Recent advances and future potential. *Environment International*, *35*(1), 127–141. Available from https://doi.org/10.1016/j.envint.2008.05.010.

Kersten, P. J., Tien, M., Kalyanaraman, B., & Kirk, T. K. (1985). *The* ligninase of *Phanerochaete chrysosporium* generates cation radicals from methoxybenzenes. *The Journal of Biological Chemistry*, *260*(5), 2609–2612.

Khalid, A., Arshad, M., & Crowley, D. E. (2008). Decolorization of azo dyes by *Shewanella* sp. under saline conditions. *Applied Microbiology and Biotechnology*, *79*(6), 1053–1059. Available from https://doi.org/10.1007/s00253-008-1498-y.

Khan, R., Bhawana, P., & Fulekar, M. (2013). Microbial decolorization and degradation of synthetic dyes: A review. *Reviews in Environmental Science and Bio/Technology*, *12*(1), 75–97.

Khataee, A. R., Zarei, M., & Pourhassan, M. (2009). Application of microalga *Chlamydomonas* sp. for biosorptive removal of a textile dye from contaminated water: Modelling by a neural network. *Environmental Technology*, *30*(14), 1615–1623. Available from https://doi.org/10.1080/09593330903370018.

Khehra, M. S., Saini, H. S., Sharma, D. K., Chadha, B. S., & Chimni, S. S. (2005). Decolorization of various azo dyes by bacterial consortium. *Dyes and Pigments*, *67*(1), 55–61.

Kim, J.-D., Yoon, J.-H., Park, Y.-H., Fusako, K., Kim, H.-T., Lee, D.-W., & Kang, K.-H. (2002). Identification of *Stenotrophomonas maltophilia* LK-24 and its degradability of crystal violet. *Journal of Microbiology and Biotechnology*, *12*(3), 437–443.

Kirk, T. K., & Farrell, R. L. (1987). Enzymatic" combustion": The microbial degradation of lignin. *Annual Reviews in Microbiology*, *41*(1), 465–501.

Kodam, K., Soojhawon, I., Lokhande, P., & Gawai, K. (2005). Microbial decolorization of reactive azo dyes under aerobic conditions. *World Journal of Microbiology and Biotechnology*, *21*(3), 367–370.

Kolekar, Y. M., & Kodam, K. M. (2012). Decolorization of textile dyes by *Alishewanella* sp. KMK6. *Applied Microbiology and Biotechnology*, *95*(2), 521–529. Available from https://doi.org/10.1007/s00253-011-3698-0.

Kolekar, Y. M., Nemade, H. N., Markad, V. L., Adav, S. S., Patole, M. S., & Kodam, K. M. (2012). Decolorization and biodegradation of azo dye, reactive blue 59 by aerobic granules. *Bioresource Technology*, *104*, 818–822. Available from https://doi.org/10.1016/j.biortech.2011.11.046.

Kolekar, Y. M., Pawar, S. P., Gawai, K. R., Lokhande, P. D., Shouche, Y. S., & Kodam, K. M. (2008). Decolorization and degradation of Disperse Blue 79 and Acid Orange 10, by *Bacillus fusiformis* KMK5 isolated from the textile dye contaminated soil. *Bioresource Technology*, *99*(18), 8999–9003. Available from https://doi.org/10.1016/j.biortech.2008.04.073.

Kumar, K. V., Sivanesan, S., & Ramamurthi, V. (2005). Adsorption of malachite green onto *Pithophora* sp., a fresh water algae: Equilibrium and kinetic modelling. *Process Biochemistry*, *40*(8), 2865–2872.

Lade, H. S., Waghmode, T. R., Kadam, A. A., & Govindwar, S. P. (2012). Enhanced biodegradation and detoxification of disperse azo dye Rubine GFL and textile industry effluent by defined fungal-bacterial consortium. *International Biodeterioration & Biodegradation*, *72*, 94–107.

Lettinga, G., Van Velsen, A., Hobma, Sd, De Zeeuw, W., & Klapwijk, A. (1980). Use of the upflow sludge blanket (USB) reactor concept for biological wastewater treatment, especially for anaerobic treatment. *Biotechnology and Bioengineering*, *22*(4), 699–734.

Levin, L., Papinutti, L., & Forchiassin, F. (2004). Evaluation of Argentinean white rot fungi for their ability to produce lignin-modifying enzymes and decolorize industrial dyes. *Bioresource Technology*, *94*(2), 169–176.

Levine, W. G. (1991). Metabolism of azo dyes: implication for detoxication and activation. *Drug Metabolism Reviews*, *23*(3-4), 253–309. Available from https://doi.org/10.3109/03602539109029761.

Lim, S.-L., Chu, W.-L., & Phang, S.-M. (2010). Use of Chlorella vulgaris for bioremediation of textile wastewater. *Bioresource Technology*, *101*(19), 7314–7322.

Martorell, M. M., Pajot, H. F., Rovati, J. I., & Figueroa, L. I. (2012). Optimization of culture medium composition for manganese peroxidase and tyrosinase production during Reactive Black 5 decolourization by the yeast *Trichosporon akiyoshidainum*. *Yeast (Chichester, England)*, *29*(3-4), 137–144. Available from https://doi.org/10.1002/yea.2896.

McMullan, G., Meehan, C., Conneely, A., Kirby, N., Robinson, T., Nigam, P., ... Smyth, W. F. (2001). Microbial decolourisation and degradation of textile dyes. *Applied Microbiology and Biotechnology*, *56*(1-2), 81–87. Available from https://doi.org/10.1007/s002530000587.

Mechsner, K., & Wuhrmann, K. (1982). Cell permeability as a rate limiting factor in the microbial reduction of sulfonated azo dyes. *European Journal of Applied Microbiology and Biotechnology*, *15*(2), 123–126.

Mielgo, I., Moreira, M. T., Feijoo, G., & Lema, J. M. (2001). A packed-bed fungal bioreactor for the continuous decolourisation of azo-dyes (Orange II). *Journal of Biotechnology*, *89*(2–3), 99–106. Available from https://doi.org/10.1016/s0168-1656(01)00319-4.

Mohan, S. V., Rao, N. C., Prasad, K. K., & Karthikeyan, J. (2002). Treatment of simulated Reactive Yellow 22 (Azo) dye effluents using *Spirogyra* species. *Waste Management*, *22*(6), 575–582.

Moosvi, S., Keharia, H., & Madamwar, D. (2005). Decolourization of textile dye Reactive Violet 5 by a newly isolated bacterial consortium RVM 11.1. *World Journal of Microbiology and Biotechnology*, *21*(5), 667–672.

Moreira-Neto, S. L., Mussatto, S. I., Machado, K. M., & Milagres, A. M. (2013). Decolorization of salt-alkaline effluent with industrial reactive dyes by laccase-producing Basidiomycetes strains. *Letters in Applied Microbiology*, *56*(4), 283–290. Available from https://doi.org/10.1111/lam.12049.

Nachiyar, C. V., & Rajkumar, G. S. (2003). Degradation of a tannery and textile dye, Navitan Fast Blue S5R by *Pseudomonas aeruginosa*. *World Journal of Microbiology and Biotechnology*, *19*(6), 609–614.

Nachiyar, C. V., & Rajakumar, G. S. (2005). Purification and characterization of an oxygen insensitive azoreductase from *Pseudomonas aeruginosa*. *Enzyme and Microbial Technology*, *36*(4), 503–509.

Nigam, P., Armour, G., Banat, I. M., Singh, D., & Marchant, R. (2000). Physical removal of textile dyes from effluents and solid-state fermentation of dye-adsorbed agricultural residues. *Bioresource Technology*, *72*(3), 219–226.

Nigam, P., Banat, I. M., Singh, D., & Marchant, R. (1996). Microbial process for the decolorization of textile effluent containing azo, diazo and reactive dyes. *Process biochemistry*, *31*(5), 435–442.

Novotny, C., Rawal, B., Bhatt, M., Patel, M., Sasek, V., & Molitoris, H. P. (2001). Capacity of *Irpex lacteus* and *Pleurotus ostreatus* for decolorization of chemically different dyes. *Journal of Biotechnology*, *89*(2-3), 113–122. Available from https://doi.org/10.1016/s0168-1656(01)00321-2.

Nozaki, K., Beh, C. H., Mizuno, M., Isobe, T., Shiroishi, M., Kanda, T., & Amano, Y. (2008). Screening and investigation of dye decolorization activities of basidiomycetes. *Journal of Bioscience and Bioengineering, 105*(1), 69–72. Available from https://doi.org/10.1263/jbb.105.69.

Özer, A., Akkaya, G., & Turabik, M. (2006). The removal of Acid Red 274 from wastewater: combined biosorption and biocoagulation with *Spirogyra rhizopus*. *Dyes and pigments, 71*(2), 83–89.

Pandey, A., Singh, P., & Iyengar, L. (2007). Bacterial decolorization and degradation of azo dyes. *International Biodeterioration & Biodegradation, 59*(2), 73–84.

Papinutti, L., Mouso, N., & Forchiassin, F. (2006). Removal and degradation of the fungicide dye malachite green from aqueous solution using the system wheat bran–*Fomes sclerodermeus*. *Enzyme and Microbial Technology, 39*(4), 848–853.

Parikh, A., & Madamwar, D. (2005). Textile dye decolorization using cyanobacteria. *Biotechnology Letters, 27*(5), 323–326.

Parshetti, G., Kalme, S., Saratale, G., & Govindwar, S. (2006). Biodegradation of Malachite Green by *Kocuria rosea* MTCC 1532. *Acta Chimica Slovenica, 53*(4).

Parshetti, G. K., Parshetti, S. G., Telke, A. A., Kalyani, D. C., Doong, R. A., & Govindwar, S. P. (2011). Biodegradation of crystal violet by *Agrobacterium radiobacter*. *Journal of Environmental Sciences (China), 23*(8), 1384–1393. Available from https://doi.org/10.1016/s1001-0742(10)60547-5.

Pereira, R., Pereira, M., Alves, M., & Pereira, L. (2014). Carbon based materials as novel redox mediators for dye wastewater biodegradation. *Applied Catalysis B: Environmental, 144*, 713–720.

Pointing, S. B., & Vrijmoed, L. (2000). Decolorization of azo and triphenylmethane dyes by *Pycnoporus sanguineus* producing laccase as the sole phenoloxidase. *World Journal of Microbiology and Biotechnology, 16*(3), 317–318.

Priya, B., Uma, L., Ahamed, A. K., Subramanian, G., & Prabaharan, D. (2011). Ability to use the diazo dye, CI Acid Black 1 as a nitrogen source by the marine cyanobacterium *Oscillatoria curviceps* BDU92191. *Bioresource Technology, 102*(14), 7218–7223.

Rafii, F., Franklin, W., & Cerniglia, C. E. (1990). Azoreductase activity of anaerobic bacteria isolated from human intestinal microflora. *Applied and Environmental Microbiology, 56*(7), 2146–2151. Available from https://doi.org/10.1128/aem.56.7.2146-2151.1990.

Rai, H. S., Bhattacharyya, M. S., Singh, J., Bansal, T., Vats, P., & Banerjee, U. (2005). Removal of dyes from the effluent of textile and dyestuff manufacturing industry: a review of emerging techniques with reference to biological treatment. *Critical Reviews in Environmental Science and Technology, 35*(3), 219–238.

Ranganathan, K., Karunagaran, K., & Sharma, D. (2007). Recycling of wastewaters of textile dyeing industries using advanced treatment technology and cost analysis—Case studies. *Resources, Conservation and Recycling, 50*(3), 306–318.

Ren, S., Guo, J., Zeng, G., & Sun, G. (2006). Decolorization of triphenylmethane, azo, and anthraquinone dyes by a newly isolated *Aeromonas hydrophila* strain. *Applied Microbiology and Biotechnology, 72*(6), 1316–1321. Available from https://doi.org/10.1007/s00253-006-0418-2.

Robinson, T., McMullan, G., Marchant, R., & Nigam, P. (2001). Remediation of dyes in textile effluent: A critical review on current treatment technologies with a proposed alternative. *Bioresource Technology, 77*(3), 247–255. Available from https://doi.org/10.1016/s0960-8524(00)00080-8.

Roxon, J. J., Ryan, A. J., & Wright, S. E. (1967). Enzymatic reduction of tartrazine by *Proteus vulgaris* from rats. *Food and Cosmetics Toxicology, 5*(5), 645–656. Available from https://doi.org/10.1016/s0015-6264(67)83216-4.

Ruiz-Arias, A., Juarez-Ramirez, C., de los Cobos-Vasconcelos, D., Ruiz-Ordaz, N., Salmeron-Alcocer, A., Ahuatzi-Chacon, D., & Galindez-Mayer, J. (2010). Aerobic biodegradation of a sulfonated phenylazonaphthol dye by a bacterial community immobilized in a multistage packed-bed BAC reactor. *Applied Biochemistry and Biotechnology*, *162*(6), 1689–1707. Available from https://doi.org/10.1007/s12010-010-8950-z.

Russ, R., Rau, J., & Stolz, A. (2000). The function of cytoplasmic flavin reductases in the reduction of azo dyes by bacteria. *Applied and Environmental Microbiology*, *66*(4), 1429–1434. Available from https://doi.org/10.1128/AEM.66.4.1429-1434.2000.

Saeed, M., Adeel, S., Ilyas, M., Shahzad, M. A., Usman, M., Haq, E.-u, & Hamayun, M. (2016). Oxidative degradation of Methyl Orange catalyzed by lab prepared nickel hydroxide in aqueous medium. *Desalination and Water Treatment*, *57*(27), 12804–12813.

Sani, R. K., & Banerjee, U. C. (1999). Decolorization of triphenylmethane dyes and textile and dye-stuff effluent by *Kurthia* sp. *Enzyme and Microbial Technology*, *24*(7), 433–437.

Saparrat, M. C., Mocchiutti, P., Liggieri, C. S., Aulicino, M. B., Caffini, N. O., Balatti, P. A., & Martínez, M. J. (2008). Ligninolytic enzyme ability and potential biotechnology applications of the white-rot fungus *Grammothele subargentea* LPSC no. 436 strain. *Process Biochemistry*, *43*(4), 368–375.

Saratale, R., Saratale, G., Kalyani, D., Chang, J.-S., & Govindwar, S. P. (2009). Enhanced decolorization and biodegradation of textile azo dye Scarlet R by using developed microbial consortium-GR. *Bioresource Technology*, *100*(9), 2493–2500.

Saratale, R. G., Saratale, G. D., Chang, J. S., & Govindwar, S. P. (2009). Decolorization and biodegradation of textile dye Navy blue HER by *Trichosporon beigelii* NCIM-3326. *Journal of Hazardous Materials*, *166*(2-3), 1421–1428. Available from https://doi.org/10.1016/j.jhazmat.2008.12.068.

Saratale, R. G., Saratale, G. D., Chang, J.-S., & Govindwar, S. (2011). Bacterial decolorization and degradation of azo dyes: A review. *Journal of the Taiwan Institute of Chemical Engineers*, *42*(1), 138–157.

Sarayu, K., & Sandhya, S. (2012). Current technologies for biological treatment of textile wastewater–A review. *Applied Biochemistry and Biotechnology*, *167*(3), 645–661. Available from https://doi.org/10.1007/s12010-012-9716-6.

Sarnaik, S., & Kanekar, P. (1999). Biodegradation of methyl violet by *Pseudomonas mendocina* MCM B-402. *Applied Microbiology and Biotechnology*, *52*(2), 251–254. Available from https://doi.org/10.1007/s002530051517.

Sarnthima, R., Khammuang, S., & Svasti, J. (2009). Extracellular ligninolytic enzymes by *Lentinus polychrous* Lév. under solid-state fermentation of potential agro-industrial wastes and their effectiveness in decolorization of synthetic dyes. *Biotechnology and Bioprocess Engineering*, *14*(4), 513–522.

Shedbalkar, U., Dhanve, R., & Jadhav, J. (2008). Biodegradation of triphenylmethane dye cotton blue by *Penicillium ochrochloron* MTCC 517. *Journal of Hazardous Materials*, *157*(2-3), 472–479. Available from https://doi.org/10.1016/j.jhazmat.2008.01.023.

Solano, F., & Sanchez-Amat, A. (1999). Note: Studies on the phylogenetic relationships of melanogenic marine bacteria: Proposal of *Marinomonas mediterranea* sp. nov. *International Journal of Systematic and Evolutionary Microbiology*, *49*(3), 1241–1246.

Solís, M., Solís, A., Pérez, H. I., Manjarrez, N., & Flores, M. (2012). Microbial decolouration of azo dyes: A review. *Process Biochemistry*, *47*(12), 1723–1748.

Srinivasan, A., & Viraraghavan, T. (2010). Decolorization of dye wastewaters by biosorbents: A review. *Journal of Environmental Management*, *91*(10), 1915–1929. Available from https://doi.org/10.1016/j.jenvman.2010.05.003.

Stauffer, C. E. (1994). , 1993. $185 00 (xv + 625 pages) ISBN 0 8247 9148 7 In J. R. Whitaker, & M. Dekker (Eds.), *Principles of enzymology for the food sciences* (2nd ed.). Elsevier.

Stolz, A. (2001). Basic and applied aspects in the microbial degradation of azo dyes. *Applied Microbiology and Biotechnology, 56*(1–2), 69–80. Available from https://doi.org/10.1007/s002530100686.

Sun, J., Hu, Y.-y, Bi, Z., & Cao, Y.-q (2009). Simultaneous decolorization of azo dye and bioelectricity generation using a microfiltration membrane air-cathode single-chamber microbial fuel cell. *Bioresource Technology, 100*(13), 3185–3192.

Tamboli, D. P., Telke, A. A., Dawkar, V. V., Jadhav, S. B., & Govindwar, S. P. (2011). Purification and characterization of bacterial aryl alcohol oxidase from *Sphingobacterium* sp. ATM and its uses in textile dye decolorization. *Biotechnology and Bioprocess Engineering, 16*(4), 661.

Tan, C.-y, Li, M., Lin, Y.-M., Lu, X.-Q., & Chen, Z.-l (2011). Biosorption of Basic Orange from aqueous solution onto dried *A. filiculoides* biomass: Equilibrium, kinetic and FTIR studies. *Desalination, 266*(1-3), 56–62.

Taskin, M., & Erdal, S. (2010). Reactive dye bioaccumulation by fungus *Aspergillus niger* isolated from the effluent of sugar fabric-contaminated soil. *Toxicology and Industrial Health, 26*(4), 239–247. Available from https://doi.org/10.1177/0748233710364967.

Tatarko, M., & Bumpus, J. A. (1998). Biodegradation of Congo red by *Phanerochaete chrysosporium. Water Research, 32*(5), 1713–1717.

Telke, A., Kalyani, D., Jadhav, J., & Govindwar, S. (2008). Kinetics and mechanism of Reactive Red 141 degradation by a bacterial isolate *Rhizobium radiobacter* MTCC 8161. *Acta Chimica Slovenica, 55*(2).

Telke, A. A. (2009). *Microbes and its enzymes as bioremediation tool for degradation of bisphenol A and textile dyes.*

Tien, M., & Kirk, T. K. (1983). Lignin-degrading enzyme from the *Hymenomycete Phanerochaete chrysosporium* Burds. *Science (New York, N.Y.), 221*(4611), 661–663. Available from https://doi.org/10.1126/science.221.4611.661.

Tien, M., Kirk, T. K., Bull, C., & Fee, J. A. (1986). Steady-state and transient-state kinetic studies on the oxidation of 3,4-dimethoxybenzyl alcohol catalyzed by the ligninase of *Phanerocheate chrysosporium* Burds. *The Journal of Biological Chemistry, 261*(4), 1687–1693.

Uddin, M. S., Zhou, J., Qu, Y., Guo, J., Wang, P., & Zhao, L. (2007). Biodecolorization of azo dye acid red B under high salinity condition. *Bulletin of Environmental Contamination and Toxicology, 79*(4), 440–444. Available from https://doi.org/10.1007/s00128-007-9260-1.

Umezawa, T., & Higuchi, T. (1987). Mechanism of aromatic ring cleavage of β-O-4 lignin substructure models by lignin peroxidase. *FEBS Letters, 218*(2), 255–260.

Van der Zee, F. P., & Villaverde, S. (2005). Combined anaerobic–aerobic treatment of azo dyes—A short review of bioreactor studies. *Water Research, 39*(8), 1425–1440.

Vasdev, K., Kuhad, R. C., & Saxena, R. K. (1995). Decolorization of triphenylmethane dyes by the bird's nest fungus *Cyathus bulleri. Current Microbiology, 30*(5), 269–272.

Walker, R. (1970). The metabolism of azo compounds: A review of the literature. *Food and Cosmetics Toxicology, 8*(6), 659–676. Available from https://doi.org/10.1016/s0015-6264(70)80455-2.

Wang, B. E., & Hu, Y. Y. (2008). Bioaccumulation vs adsorption of reactive dye by immobilized growing *Aspergillus fumigatus* beads. *Journal of Hazardous Materials, 157*(1), 1–7. Available from https://doi.org/10.1016/j.jhazmat.2007.12.069.

Wang, H., Zheng, X. W., Su, J. Q., Tian, Y., Xiong, X. J., & Zheng, T. L. (2009). Biological decolorization of the reactive dyes Reactive Black 5 by a novel isolated bacterial strain *Enterobacter* sp. EC3. *Journal of Hazardous Materials, 171*(1-3), 654–659. Available from https://doi.org/10.1016/j.jhazmat.2009.06.050.

Wuhrmann, K., Mechsner, K., & Kappeler, T. (1980). Investigation on rate—Determining factors in the microbial reduction of azo dyes. *European Journal of Applied Microbiology and Biotechnology, 9*(4), 325–338.

Yang, X. Q., Zhao, X. X., Liu, C. Y., Zheng, Y., & Qian, S. J. (2009). Decolorization of azo, triphenylmethane and anthraquinone dyes by a newly isolated Trametes sp. SQ01 and its laccase. *Process Biochemistry*, *44*(10), 1185–1189.

Yatome, C., Ogawa, T., & Matsui, M. (1991). Degradation of crystal violet by *Bacillus subtilis*. *Journal of Environmental Science & Health Part A*, *26*(1), 75–87.

Yoo, E., Libra, J., & Adrian, L. (2001). *Mechanism of* decolorization of azo dyes in anaerobic mixed culture. *Journal of Environmental Engineering*, *127*(9), 844–849.

Zimmermann, T., Gasser, F., Kulla, H. G., & Leisinger, T. (1984). Comparison of two bacterial azoreductases acquired during adaptation to growth on azo dyes. *Archives of Microbiology*, *138*(1), 37–43. Available from https://doi.org/10.1007/BF00425404.

CHAPTER 16

Microbial decontamination: economic and environmental benefits

Shahid Adeel[1], Shumaila Kiran[2], Saba Naz[2] and Fatima Batool[3]
[1]Department of Chemistry, Government College University Faisalabad, Faisalabad, Pakistan
[2]Department of Applied Chemistry, Government College University Faisalabad, Faisalabad, Pakistan
[3]Department of Botany, Division of Science and Technology, University of Education Lahore, Lahore, Pakistan

16.1 Introduction

One of the biggest manufacturing industries is the textile industry. Large quantities of chemicals are being used in this industry, making it the main contributor of chemically polluted environment. The huge usage of water and complex chemical substances in the textile industry throughout the processing of textile as well as untreated and unused remaining materials are discharged as a wastewater. This textile effluent is greatly loaded with complex chemicals, hence making the effluent great in color, pH, as well as fatal substances. Addition of this toxic effluent with fresh water bodies such as ponds, canals, streams, and rivers is causing water pollution, eventually disturbing the marine life (Bhat et al., 2021; Verma, Dash, & Bhunia, 2012). The industrial effluents are a blend of various contaminating materials of chemically complex structures which are typically significant counterparts of various kinds of colorants being employed in the scheme of dyeing textiles (Kiani, Mirzaei, Ghanbari, Feizi, & Mehdipour, 2020).

The effective remediation for the fabric industry effluent has therefore become a big ecological issue. But it's very hard to remove dyes from the textile discharge, there must be appropriate remediation methods for wastewater before it's discharged into the environment without affecting environmental health (GilPavas, Dobrosz-Gómez, & Gómez-García, 2020; Gulzar, Farooq, Kiran, Ahmad, & Hameed, 2019; Kiran, Gulzar et al., 2019; Kiran, Nosheen et al., 2019; Palamthodi et al., 2011). In wider sense, the regular exposure with wastewater containing fabric dyes is a possible health hazard which is the direct or indirect reason of several health issues

Microbial Consortium and Biotransformation for Pollution Decontamination
DOI: https://doi.org/10.1016/B978-0-323-91893-0.00004-3

that are autoimmune diseases, destruction of immune system, several problems of respiration, blood circulation, neurobehavioral disorders, destruction to central nervous system, allergic reactions, copious diarrhea, several kinds of cancerous diseases, vomiting, skin disorders, eye infections, tissue damage, some lungs related issues, etc. (Foo & Hameed, 2010; Chowdhary, Bharagava, Mishra, & Khan, 2020; Kishor et al., 2021).

These days, increasing trend toward the application of microbes for eco-friendly remediation of textile wastewater has attained significance. It is informed that some microorganisms have the ability to degrade/absorb a great number of fabric dyes from wastewater (Tripathi & Srivastava, 2011; Wu, Chen, Lin, Su, & Chung, 2020). In actual, microbes associated to different assemblies, such as algae, various bacterial, and white rot fungal strains, also possess potential to decolorize and degrade several complex dye molecules which otherwise are hard to eradicate from the wastewater (Jamee & Siddique, 2019; Berkessa, Yan, Li, Jegatheesan, & Zhang, 2020). The application of microorganisms for biodegradation is appropriate since it is versatile and has potential machinery of enzymes and dynamic metabolisms. Biotreatment is a nontoxic, cost-effective, eco-friendly, and often more efficient substitute to conventional treatment methods for the textile waste (Kiran et al., 2017; Samsami, Mohamadizaniani, Sarrafzadeh, Rene, & Firoozbahr, 2020).

16.2 Textile industry wastewater

The aromatic compounds, dyes, and pigments are some of frequently employed chemical substances that have several applications in the fabric manufacturing industry, for example, in leather tanning, dyeing, and coloring of several materials being employed in textile engineering (Ramachandran, Sundharam, Palaniyappan, & Munusamy, 2013). Numerous chemicals released from fabric manufacturing become continual ecological contaminants. Severe ecological issues sometimes give rise to such ranks which can be extremely dangerous for the human health, pet animals, livestock, and aquatic biota and show adverse effects on whole environments because of the untreated wastewater being discharged from fabric industry (Gowri et al., 2014). Around the world, 0.7 million tons of dyes formed in the fabric industry yearly, it is evaluated that approximately 2 lakh tons of these stable chemicals are discharged per annum as wastewater due to defective dyeing procedures (Ogugbue & Sawidis, 2011; Bidu, Van der Bruggen, Rwiza, & Njau, 2021). Unfortunately, mostly these chemical substances stay in the environment due to their great stability against the physical, chemical, and microbial methods of decomposition and hence are a constant danger to flora and fauna and to the environment (Manzoor & Sharma, 2020; Odubanjo, Oyetibo, & Ilori, 2021; Shindhal et al., 2021) (Fig. 16.1).

The textile industry wastewater has a large value of pH; large amount of suspended solids, nitrates, chlorides, and metals such as sodium, manganese, copper, lead, iron, chromium; as well as high chemical oxygen demand (COD) and biological oxygen demand (BOD) value (Table 16.1). The effluent has a dark-brown color as well (Pal, 2017).

The azo, anthraquinone, and phthalocyanine dyes are mostly employed which are highly toxic and have carcinogenic characteristics as well. Due to of their weak absorption ability to threads and stable chemical structure, removal of them is very tough from wastewater by the traditional treatment methods (Adeel et al., 2020; Benkhaya, M'rabet, & El Harfi, 2020). The textiles such polyamine, silk, and wool which are colored by acid dyes are reflected as the biggest category of dyes. Principally, there are three kinds of fabric dyes: anionic, cationic, and nonionic dyes. The anionic dyes generally comprise direct, reactive, and acid dyes. There is just one kind of cationic dye that is being

Figure 16.1 Textile industry effluent discharged into environment.

Table 16.1 Composition and quantity of fabric industry effluent.

Wastewater parameters	Unit	Quantity based on literature
Dye concentration	mg/L	700
Chloride	mg/L	15,867
Sulfate	mg/L	1400
Total nitrogen	mg/L	23
COD	mg/L	1781
BOD	mg/L	363
NH_4	mg/L	17
NO_3	mg/L	2
PO_4	mg/L	17
Ca	mg/L	43
Mg	mg/L	4
Na	mg/L	2900
Fe	mg/L	1.2
pH	–	10

BOD, Biological oxygen demand; *COD*, chemical oxygen demand.

employed in fabric manufacturing is basic dyes. The nonionic dyes are the disperse dyes that do not go through ionization in solution (El Harfi & El Harfi, 2017). From the most widely employed dyes present in textile manufacturing, azo dyes comprise about $\approx 70\%$ of all the dyestuff. The biotreatment of azo dyes from wastewater is greatly efficient and these are the solvable dyes. The reactive dyes which are present in wastewater discharged from the fabric manufacturing industry are highly decomposition-resistant dyes (Soylu, Gökkuş, & Özyonar, 2020) (Fig. 16.2).

16.3 Treatment of textile industrial effluent

16.3.1 Bacterial biodegradation of textile effluents

It is known that under specific parameters, several bacterial strains possess great potential to degrade, decolorize, and mineralize different dyes. Several bacterial strains capable of degrading fabric dyes have gained researchers' interest and there has been a rising curiosity toward this (Pandey et al., 2016). Numerous scientists have researched on various bacterial species for biological treatment of waters, soils, metals, and dyes which are polluted by wastewater of textile industry. The dyes which have heavy metals, carcinogenic compounds, formaldehyde, bleaching agents holding chlorine, halogen holding complexes, etc. are some major

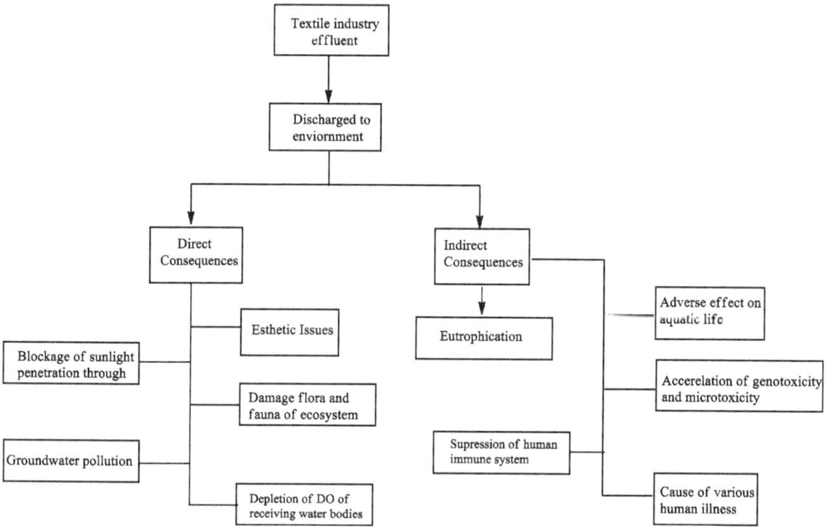

Figure 16.2 Effects of textile industry effluent on the environment and living beings.

Table 16.2 Bacterial species with azoreductase potential.

Sr. no.	Bacterial strain	References
1	*Pseudomonas balearica*	Zhuang et al. (2020)
2	*Pseudomonas aeruginosa*	Zhuang et al. (2020)
3	*Pseudomonas stutzeri*	Joshi et al. (2020)
4	*Enterococcus* spp.	Paba and Ávila (2021)
5	*Streptomyces* spp.	Kumar, Kumar, Das, Solanki, and Kapur (2020)

contaminants of effluent of textile industry (Kishor et al., 2021; Jadhav et al., 2007). Bacteria, in pure cultures, capable of degrading dye (Table 16.2) have been illustrated (Mohanty & Kumar, 2021). The microbial mixed culture tolerates more profits than the employment of pure cultures in degradation of dyes being applied in fabric manufacturing. The colorants may be attacked directly by several bacterial strains or few ones are also capable of more disintegrating the end products produced by other strains (Dar, Bandh, & Kamili, 2011; Thiruppathi et al., 2021).

The bacterial oxido-reductive enzymes that comprise azoreductases, laccase, etc. contribute to dye degradation process of bacteria particularly to remove azo dye color (Telke et al., 2011, 2015). The microbial disintegration of azo dyes includes an electron donor and the series of enzymatic steps assisted by azoreductase which catalyze the splitting of azo bonds (Barathi, Aruljothi, Karthik, & Padikasan, 2020; Sarayu & Sandhya 2010). Bacterial species with azoreductase potential are given in Table 16.2.

Thus it's possible to attain the capability of bacterial azoreductase enzymes for the catalysis of the azo bond splitting in azo molecules to develop efficient, biodegradable processes to treat textile wastewater (Mahmood et al., 2016). The bacterial disintegration of azo dye in aerobic environment is just not efficient though the collection of diverse bacterial strains has been shown to break down many azo dyes (Table 16.3). *Bacillus subtilis* and *Pseudomonas luteola* are the example of bacteria being employed for anaerobic disintegration of colorants (Horitsu et al., 1997; Gunti, Maruthiramaih, & Shivanandappa, 2021).

16.3.1.1 Microbial mechanism of azo dye decomposition

Biological treatment method of fabric engineering dyes is an effective and eco-friendly approach. The microbial degradation and decolorization mechanism of highly challenging fabric dyes which are azo dyes consist of splitting of azo bonds in azo molecules, therefore causing decolorization of azo dyes. The azo compounds first go through the reductive cleavage and after that resultant disintegration products that are aromatic amines mostly are metabolized in the presence of oxygen. Generally, dyes are decomposed to simple inorganic substances which are safe for the life of living beings (Kiran, Gulzar et al., 2019; Kiran, Nosheen et al., 2019; Yusuf, 2018).

Table 16.3 Bacterial strains decomposing synthetic dyes.

Bacterial strain	Synthetic dye	References
Bacillus subtilis	Reactive azo dyes	Horitsu et al. (1997)
Aeromonas hydrophila	Acid Orange 7, Reactive Blue 160,	Chen et al. (2003)
Escherichia coli	Ethyl Red, Methyl Red	Nakanishi et al. (2001)
Sphingomonas xenophaga	Acid Orange 7, 10	Coughlin et al. (1999)
Rhizobium radiobacter	Reactive Red 141	Coughlin et al. (1999)
Pseudomonas luteola	Reactive azo dyes	Hu et al. (1992)
Corynebacterium glutamicum	Reactive Yellow 2	Won and Yun (2008)
Enterobacter agglomerans	Methyl Red	Keharia and Madamwar (2003)
Bacillus megaterium	Acid Red	Ayman et al. (2019)

The reduced flavin nucleotides are present in microbial electron transport chain for that azo dye acts as an oxidant which causes the oxidation of flavin nucleotides. The method normally needs an extra source of carbon for the degradation and decolorization to progress at reasonable speed. Some aerobic microorganisms use the azo dyes for nitrogen and carbon as their single source, in other bacterial strains, the azo group reduction can only be catalyzed by the oxygen-holding azoreductases. The carbon is converted to carbon dioxide and methane, causing electron release. Then, the electrons, in the electron transport chain, move down to azo molecule that acts as an electron acceptor. After that reaction is happened in between azo dye and electrons, moving toward the azo bond reduction, and hence occurs decolorization (Dilek et al., 1999; Khan et al., 2021). Numerous other microorganisms such as bacterial strains, actinomycetes, algae, and fungi are widely explored for their azo dye disintegration ability (Ozdemir et al., 2008) (Fig. 16.3).

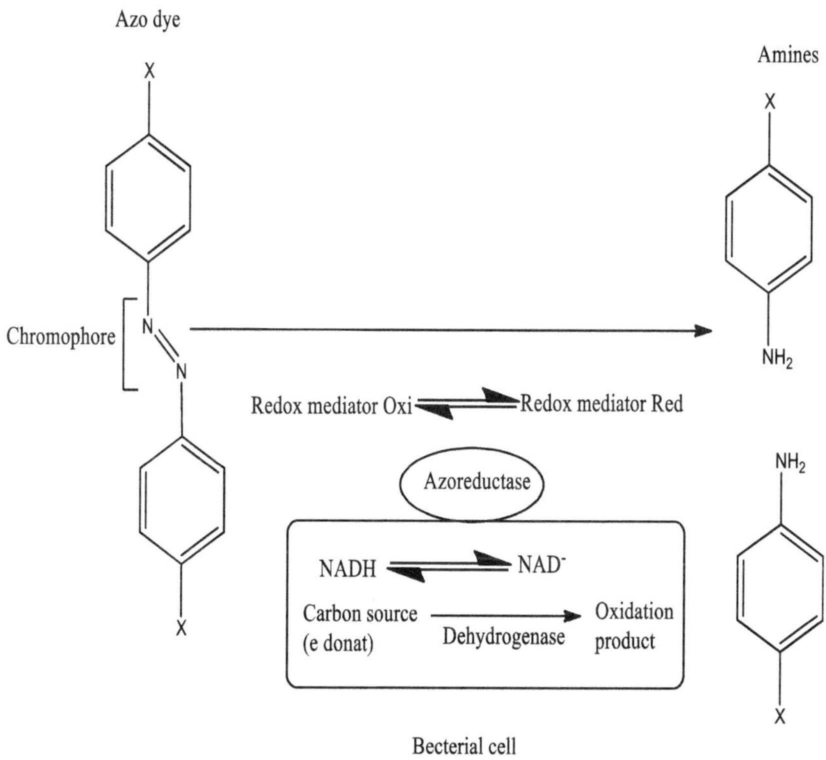

Figure 16.3 Mechanism for the bacterial decomposition of azo dye.

16.3.2 Phycoremediation: algal decomposition and decolorization of fabric dyes

Phycoremediation can be stated as the employment of microalgae or macroalgae to the removal or biological decomposition of contaminants, counting fabric dyes and xenobiotics from effluent with simultaneous biomaterial propagation. There are several methods to treat industrial effluents, contaminated water, various pigments, dyes, tanning, and bleaching agents by means of microalgae aerobically and anaerobically (Škufca et al., 2021). Though bacteria have significant role in the breakdown of ecological contaminants, microalgae are exposed to have the ability to degrade organic contaminants (Table 16.4) moreover giving oxygen for aerobic bacteria which can decompose contaminant only in the presence of oxygen. Certainly, there have been so many research works verifying that existing environmental issues such as fabric industry wastewater remediation problems can be solved using microalgae to degrade pollutants into environment-friendly substances (Abd Ellatif et al., 2021; Dar, Kamili, Nazir, Bandh, & Bhat, 2012; Dar, Bandh, Kamili, Nazir, & Bhat, 2013). The carcinogenic and mutagenic azo dyes and their intermediate complexes are the cause of several problems associated with health and are also the cause of mutagenic actions in surface water which are contaminated mainly by fabric industry wastewater. Their discharge also causes big problems into surface water, obstructs oxygen access, and also hinders light to diffuse into aquatic bodies (Alderete et al., 2021; Chung et al., 2016). The microalgae do not need carbon source, they are photoautotrophs. The use of microalgae for effluent remediation is an effective and inexpensive procedure. The development of photosynthetic algae is observable even in wastewater of fabric industry and hence they are the possible contestant for effluent remediation (Dubey et al., 2011).

Significant color deletion can be attained by algal enzymatic systems which is able to degrade azo compounds. *Oscillatoria* and *Chlorella* species have been exposed to generate aromatic amines via decomposition of azo compounds, consequently leading to the generation of simple organic molecules. Some other algae types are even capable of using azo dyes as their single nitrogen and carbon source.

More than 30 azo dyes have been shown to break down and decolorize into simple aromatic amines by *Oscillatoria* sp. and *Chlorella* sp. (Abd Ellatif et al., 2021). Few algae are present which can directly degrade azo compounds instead of changing them into aromatic compounds first (El-Naggar et al., 2021).

Table 16.4 Decomposition of synthetic dyes by algae.

Algae	Dye removed	Percentage of removal efficiency	References
Cosmarium sp.	Malachite Green	92.4	Daneshvar et al. (2005)
Green algae	Mono-azo and di-azo dyes	68	Omar (2008)
Algal biomass	Malachite Green	85	Gajare et al. (2012)
Green algae	Indigo	89.3	Elisangela et al. (2009)
Green algae	Direct Blue	79	Elisangela et al (2009)
Green algae	Remazol Brilliant Orange	75.3	Elisangela et al. (2009)
Green Algae	Crystal Violet	72.5	Elisangela et al. (2009)
Kluyveromyces maraxianus	Remazol Black-B	98	Meehan et al. (2000)
Chlorella sp.	Indigo	89.3	Cheriaa et al. (2009)
Chlorella sp.	Direct Blue	79	Cheriaa et al. (2009)
Chlorella sp.	Remazol Brilliant Orange	75.3	Cheriaa et al. (2009)
Chlorella sp.	Crystal Violet	72.5	Cheriaa et al. (2009)
Spirogyra sp.	Blue dye	78.29	Brahmbhatt and Jasrai (2016)
Spirogyra sp.	Red dye	64.21	Brahmbhatt and Jasrai (2016)
Oscillatoria sp.	Blue dye	76.48	Brahmbhatt and Jasrai (2016)
Oscillatoria sp.	Red dye	62.63	Brahmbhatt and Jasrai (2016)
Lyngbya sp.	Textile dye	73	Henciya et al. (2013)
Lyngbya lagerlerimi	Methyl Red	35.61	El-Sheekh et al. (2009)

(*Continued*)

Table 16.4 (Continued)

Algae	Dye removed	Percentage of removal efficiency	References
Nostoc linckia	Methyl Red	81.97	El-Sheekh et al. (2009)
Oscillatoria rubescens	Basic Cationic	85.80	El-Sheekh et al. (2009)
Chlorella vulgaris	Basic Fuchsin	91.20	El-Sheekh et al. (2009)
Elkatohrix viridis	G-red	90.75	El-Sheekh et al. (2009)

16.3.3 Mycoremediation: fungi's role in decomposition and decolorization of synthetic dyes

Practically, use of fungi for attaining biotreatment of contaminated soils and textile wastewater is known as mycoremediation. Several challenging substances such as polychlorinated biphenyls, petroleum hydrocarbon compounds, phenolic derived compounds, heavy metals (by biosorption), and pesticides of disintegration resistant have been informed to be broken down via mycoremediation. In this process, such dangerous chemicals are consumed by fungi as nutrient source and converted them in simple forms, causing their degradation as well (Table 16.5). The classification of fungi on the basis of their efficiency to degrade is as follows:

- fungal biosorption
- ligninolytic fungal degradation
- mycorrhizal fungal degradation (Sharif et al., 2020)

Phanerochaete chrysosporium, *Trametes versicolor*, and *Bjerkandera adusta* have been permitted for fabric dye remediation by their ligninolytic activities (Dos Santos et al., 2007). The ligninolytic fungi comprise an enzymatic scheme made up of three main fungal enzymes: manganese peroxidase (MnP), lignin peroxidase (LiP), and laccase which are enriched with greater efficiency to degrade the contaminants at a greater extent (Dao et al., 2021). Fabric manufacturing effluent comprising several dyes can be treated proficiently prior to its final release by white rot fungi which can degrade azo compounds (Pecková et al., 2021). Decomposition of Disperse Red 3B by a consortium of the fungus *Aspergillus* sp. has been given in Fig. 16.4. The fungal lignin-degrading system is found to be

Table 16.5 Dyes' decomposition proficiency (%) of various fungal species.

Fungal spp.	Dye decolorized	Proficiency of degradation (%)	References
Alternaria spp. SA4	Acid Red 151 (di-azo)	98.47	Ali et al. (2010)
Aspergillus terreus SA3	Acid Red 151 (di-azo)	55	Ali et al. (2010)
Penicillium spp. SA5	Acid Red 151 (di-azo)	65	Ali et al. (2010)
Aspergillus flavus SA2	Orange II (mono-azo)	48	Ali et al. (2010)
A. terreus SA3	Orange II (mono-azo)	44.67	Ali et al. (2010)
Alternaria spp. SA4	Orange II (mono-azo)	58.66	Asgher et al. (2008)
Penicillium spp. SA5	Orange II (mono-azo)	55.68	Asgher et al. (2008)
Pleurotus florida	Malachite Green	90	Asgher et al. (2008)

efficient in the decomposition of different industrial contaminants. The contaminant decomposition efficacy depends on following factors:

1. pollutant's nature
2. the fungal species carrying out decomposition procedures (Soundharya et al., 2016)

Several microbial enzymes such as lignin enzyme of white rot fungi are being investigated widely to research on their properties and to test their appropriateness for degradation of colorants (Del Cerro et al., 2021).

16.3.4 Decontamination of textile effluent by yeast

Unlike fungi, yeast can degrade the colorants and can stay alive even in the unfavorable environment (Ceretta, Nercessian, & Wolski, 2021). The huge captivation of heavy metals, colorants, and dyes forges the yeast a noble microorganism for biotreatment of azo compounds (Table 16.6). The yeast degrades azo dyes mainly by enzymatic degradation and adsorption or by combining both methods. There are only some research works present about yeast's degradation capability (Ali et al., 2021). *Candida krusei* and *Pseudozyma rugulosa* are the yeast strains which showed outstanding decolorization of reactive azo dyes. With contribution of azoreductase, *Saccharomyces cerevisiae* efficiently decompose Methyl Red dye at different

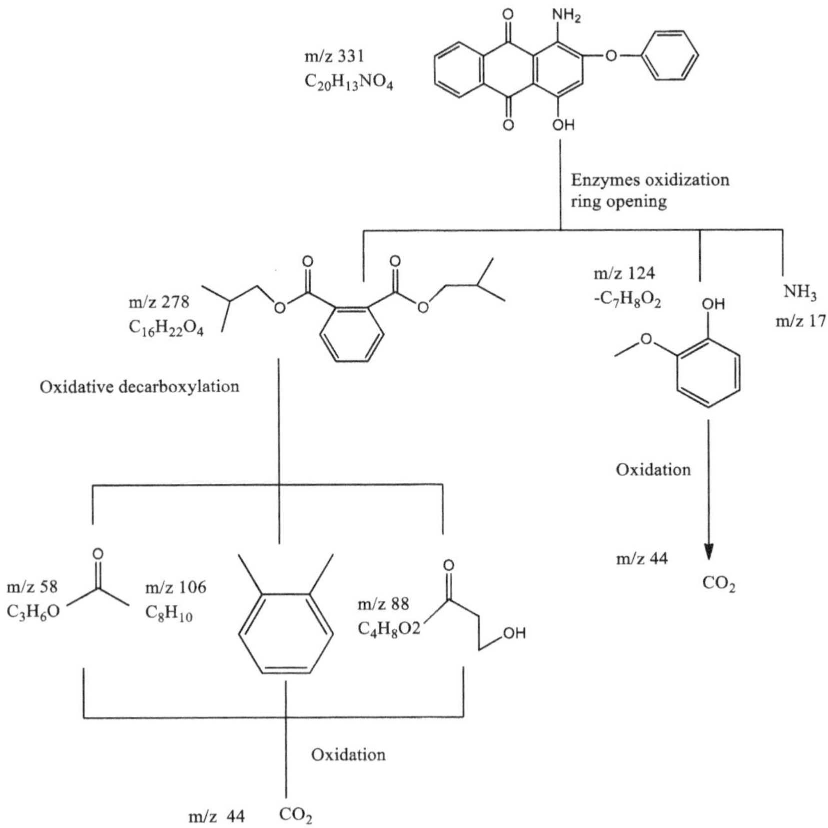

Figure 16.4 Decomposition of Disperse Red 3B by a consortium of the fungus *Aspergillus* sp. (Tang et al., 2019).

Table 16.6 Yeast (ascomycetes) involved in treatment of several dyes.

Yeast species	Dyes	References
Saccharomyces cerevisiae	Malachite Green	Jadhav et al. (2007)
Candida zeylanoides	Azo benzenesulfonates	Ramalho et al. (2006)
Candida tropicalis	Reactive Blue 19	Donmez and Asku (2002)
Candida kursei	Basic Violet 3	Charumathi and Das (2010)
Kluyveromyces marxianus	Remazol Back B	Charumathi and Das (2010)
Pseudozyma rugulosa	Reactive azo dye	Jafari et al. (2014)
Torulopsis candida	Reactive Brilliant Red	Yu and Wen (2005)

pH (Jadhav et al., 2007). There is rising attention for bioremediation of effluent compounds in textile wastewaters, for example, mostly metals, dyes, organic compounds and inorganic nutrient via biodegradation, biosorption, enzymatic mineralization, and bioaccumulation (Abioye et al., 2014). Fungi degrade azo dyes by generating extracellular enzymes such as phenol oxidases and peroxidases (Durán et al., 2002). Combination of azo dye decomposition and biodiesel production by manganese-dependent peroxidase producing oleaginous yeasts has been given in Fig. 16.5.

16.3.5 Enzymatic degradation of textile effluents

The enzymes can be taken as biochemical mean for efficient effluent remediation. Both the extracellular and intracellular microbial enzymes can be employed for the purposes of remediation. The enzymes, biological catalysts are very effectual and have capability of targeting a particular contaminant without influencing other constituents present in wastewater. (Elumalai & Saravanan, 2016). Enzymes have numerous valuable features associated to decomposition. When it comes to the composition and quantity of wastewater, the highly contaminating wastewater discharged from industries is textile effluent (Awomeso et al., 2010; Behel & Kumar, 2013). While using microbes for decomposition activities, the drawbacks which we face can be overcome simply by employing enzymes as a substitute. The enzymes are able to proficiently decompose the target contaminant (Table 16.7). Thus contaminants which have degradation resistance can simply be decomposed through enzymes. Oxidoreductases, transferases, hydrolases, and dehalogenases are few of the most significant enzymes participate in biotreatment practices. Laccase is an example of enzyme active in elimination of pollutants (Kunjadia et al., 2016). The extracellular enzymes of ligninolytic fungi, white rot fungi are significant for decomposition of dye material as well as lignin (Kunjadia et al., 2016). For degradation purposes, it is beneficial to employ white rot fungi than bacteria. Nonsubstrate nature of extracellular enzymes, for example, laccase, LiP, MnP, and Mn-independent versatile peroxidases gives white rot fungi this amazing effectiveness toward the aromatic color compounds having degradation resistance (Jin et al., 2007; Zhuo & Fan, 2021).

The microbial enzymes have several desired characteristics such as their genetic makeup can easily be manipulated; they provide highest possible yields, easily accessibility because of their absence of seasonal variations, as well as large growth of microorganisms than enzymes generated by

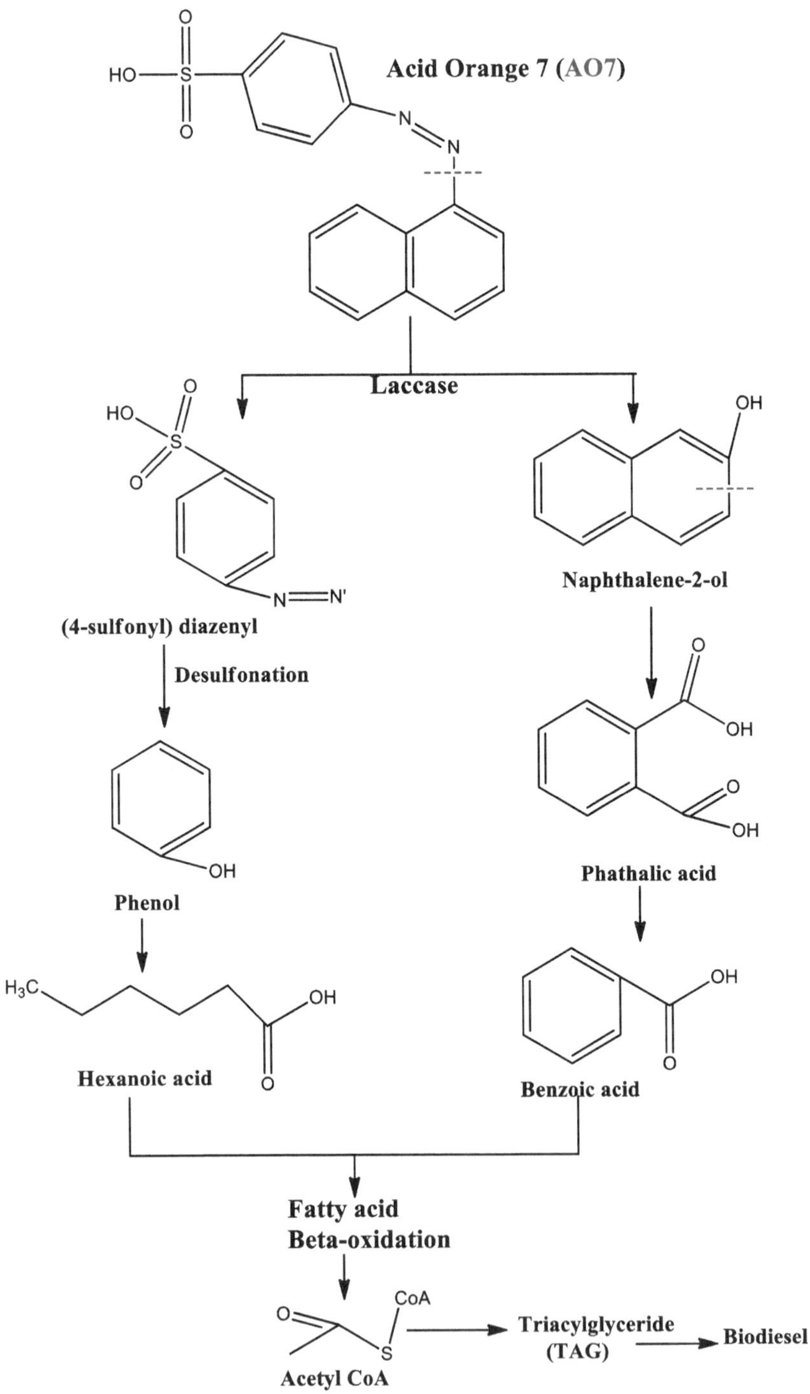

Figure 16.5 Combination of azo dye decomposition and biodiesel production by manganese-dependent peroxidase producing oleaginous yeasts (Ali et al., 2021).

Table 16.7 Some dyes being degraded by enzymatic processes.

Substrate	Enzyme and their source	References
3-(4 Dimethyl amino-1 phenylazo)benzene sulfonic acid	Laccase from *Trametes villosa*	Zille et al. (2004)
Tartrazine and ponceau	Azoreductase from green algae	Kalyuzhnyi, Yemashova, and Federovich (2006)
Reactive Yellow	Azoreductase from *Staphylococcus arlettae*	Franciscon, Zille, and Durran (2009)
Reactive Blue 19	Laccase from *Trametes versicolor*	Peralta-Zamora et al. (2003)
Reactive Blue 59	Lignin peroxidase from *Streptomyces krainskii*	Mane, Gurav, Deshmukh, and Govindwar (2008)
Remazol Brilliant Blue R	Peroxidase and laccase from *Pleurotus ostreatus*	Machado and Matheus (2006)
Azo dye	Lignin peroxidase from *Penicillium ochrochloron*	Imran et al. (2014)

animals and plants. The microbial enzymes are also significant because of their higher stability and easier production. Numerous dyes have been informed to be degraded via white rot fungi, which indicates that these entities are significant substitute for the management of industrial dye-holding wastewaters (Hasan et al., 2006; Bhandari et al., 2021; Parveen, Asgher, & Bilal, 2021). In another research, it was stated that white rot fungi decolorize the Crystal Violet dye through *N*-demethylation. Mainly reaction of azo dye degradation via bacteria involves the cleavage of azo bonds by an electron donor and an azoreductase enzyme. The fabric industrial effluent holding azo dyes can be bioremediated efficiently by emerging remediation methods which use azoreductase-generating micro-organisms. The splitting of azo dyes by reduction can be catalyzed by these azoreductase enzymes (Sarayu & Sandhya, 2010). Azo dye degradation mechanism under aerobic and anaerobic environment has been given in Fig. 16.6.

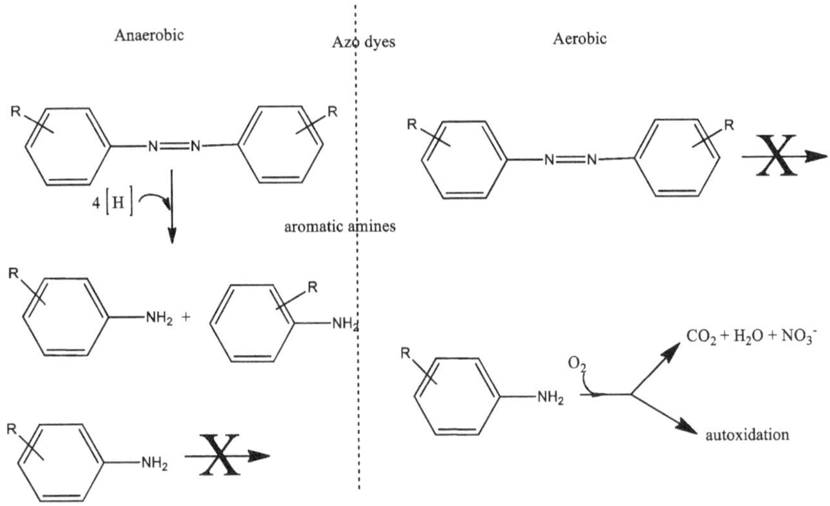

Figure 16.6 Azo dye degradation mechanism under aerobic and anaerobic environment.

Mostly significant bacterial enzymes involved in decomposition properties are azoreductases. Azoreductases cause azo bond cleavage, generating aromatic amines that are further decomposed under aerobic conditions into CO_2 and H_2O (Dafale et al., 2010).

16.4 Decontamination of textile industry effluent via biosorption

The uptake of chemicals or absorption by microbial material is called as biosorption or, more exactly, it can also be stated as "the ability of microorganisms to assemble heavy metals from effluent via physiochemical or metabolic means of absorption." Biomaterial, from different microorganisms, for example, bacteria, algae, yeast, and filamentous fungi are known to remove colorants via biosorption method (Ramachandran et al., 2013). Mainly, it happens via cell wall; however, absorption and adsorption depend on the type of biomaterial (Kiran et al., 2020; Ruscasso et al., 2021). The biomass of bacteria, algae, filamentous fungi, and yeast can be employed to remove dyes from effluent via biosorption. Lipid and heteropolysaccharide are the basic part of cell wall and have many functional groups, for example, carboxyl, hydroxyl, amino group, and other charged constituents having biosorption characteristics via the strong forces of attraction between these groups and dye molecules (Hasyimah, Furusawa, & Amirul, 2021).

Biosorption is the characteristic of particular kind of microbial biomaterial to form bond with and to amass heavy metals and wastes present in wastewater. The biomaterial has capability of performing just like an ion exchanger or chemical element which is produced from biological source. Till now, it has been broadly informed in research studies that biosorption is a flawless alternate for cleaning of metal containing wastewater (Patel, Chhaya, Rudakiya, & Joshi, 2021). The biosorption with microbes, particularly fungus, for removal of the colorants from textile effluent has attained important consideration. The synthetic dyes and dye containing wastewater degradation have been investigated by many fungi. Regarding bacterial biosorption capability, few bacterial strains such as *B. subtilis*, *Aeromonas* sp., *P. luteola*, *Staphylococcus Aureus*, and *Escherichia coli* are being employed for their biosorbent features to decolorize reactive dyes. The main benefits presented by biosorption than old remediation methods comprise cost-effectiveness, high proficiency, sludge minimization, metal recovery, regeneration of biosorbent, easy recovery of metals, etc. (Gahlout, Prajapati, Tandel, & Patel, 2021).

16.4.1 Mechanisms of biosorption

Several means are present through which microorganisms can take up metals and that's because of their complex structure. On the basis of many criteria, the biosorption processes are classified (Kiran, Gulzar et al., 2019; Kiran, Nosheen et al., 2019). Biosorption methods can be classified as follows, based on cellular metabolism:

1. metabolism-dependent biosorption
2. nonmetabolism-dependent biosorption

Moreover, the biosorption method is also categorized as cell surface sorption and extracellular/intracellular biosorption depends on the position where the metal is biosorbed from wastewater (Alluri et al., 2007).

A classic biosorption method has two phases:

1. a solid phase (biosorbent/sorbent; usually a living mass) and
2. a liquid phase (generally water), including a soluble matter to be sorbed (or a metal ion or a sorbate) (Celik, Duman, Sayin, Akar, & Akar, 2021)

There are two main mechanisms by which metals are biosorbed (Pardo, Herguedas, Barrado, & Vega, 2003; Perpetuo, Souza, & Nascimento, 2011).

1. active biosorption (metabolism-dependent)
2. passive biosorption (metabolism-independent)

Azo dye degradation mechanism under aerobic and anaerobic environment has been given in Fig. 16.7 and mechanism of heavy metal removal from wastewater has been given in Fig. 16.8. Active mechanism of biosorption needs particular conditions, for example, temperature, pH, and nutrients which are essential to balance the cellular functions. The passive biosorption methods, on the contrary, take place in living and dead cells and occur under a wide range of environmentally acceptable circumstances. The passive methods are more effective as compared to active biosorption (Papirio et al., 2017).

16.5 Environmental perspectives

The environmental problems associated with remaining dye quantity present in wastewater from fabric industries are considered a great task. The release of wastewater into aquatic bodies directly pollutes them (Zaharia, Suteu, & Muresan, 2011). It is not recognized accurately about the quantities of dyes are generated around the world nor about the quantities that are released into environment. Because the dyes are employed on commercial level, their release into environment in large amount consequently leads to severe health threat concerns. The rising effect of ecological defense on expansion of industries is subsequent of the emerging environment-friendly techniques (Forgacs, Cserhati, & Oros, 2004; Velusamy, Roy, Sundaram, & Kumar Mallick, 2021; Yang, López-Grimau, Vilaseca, & Crespi, 2020).

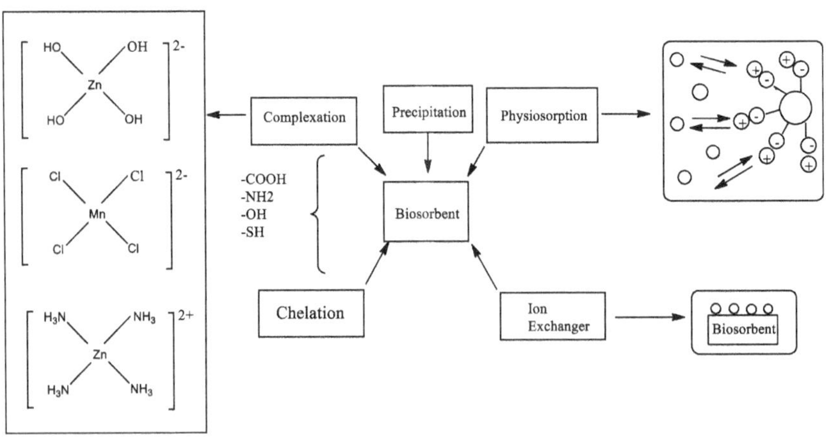

Figure 16.7 The mechanism of biosorption.

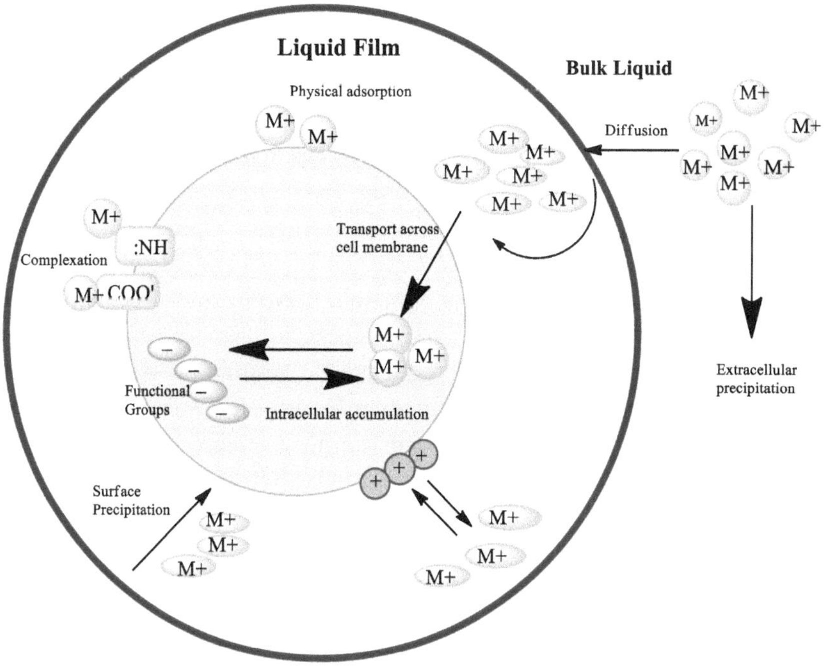

Figure 16.8 Mechanism of heavy metal removal from wastewater (Papirio et al., 2017).

It is found that most dye wastewater was not entirely decomposed by traditional effluent remedies, due to its good stability to temperature, light, chemicals, water, detergent, soap, as well as certain other parameters. The water is a vital component of life and is needed in several domestic purposes. Moreover, water is employed as a main constituent in all industries. Water is the most vital natural resource, but unfortunately it is highly contaminated because of industrialization, urbanization, increased population, and lavish lifestyle (Robinson, McMullan, Marchant, & Nigam, 2001; Zeng, Luo, Wang, & Yi, 2021).

16.5.1 Detrimental impacts to living bodies

The fabric dyes have lethal effects maybe because of the existence of aryl amine derivatives or azo dyes formed during splitting of azo bonds. The nitro-dyes are digested by nitroreductases generated by similar microbes (Umbuzeiro, Freeman, Warren, Kummrow, & Claxton, 2005). The mammalian liver enzymes catalyze the reductive splitting of azo bonds as well as the nitro-reduction of nitro groups. In both cases, if *N*-hydroxyl amines are generated, they can be the reason for DNA destruction (Arlt et al., 2002; Köktürk et al., 2021). Furthermore, the effects generated by other pollutants highly upset the transparency and quality of aquatic reservoirs such as lakes, rivers, ponds, as well as marine life (Bhatia, Sharma, Kanwar, & Singh, 2018; Wijetunga, Li, & Jian, 2010).

It was noticed that the fabric industry wastewater comprises the mixture of polluting chemicals, including dyes and organic substances (Velusamy et al., 2021). 93% of water consumption comes out polluted by dyes comprising huge quantity of heavy metals and organic complex molecules (Gupta, Suhas Nayak, Agarwal, Chaudhary, & Tyagi, 2014; Wijannarong, Aroonsrimorakot, Thavipoke, & Sangjan, 2013). Some dyes and organic substances have nonbiodegradable nature that leads to severe environmental threats. Colored effluent is harmful for water bodies, lowering the oxygenation capacity of water and threatening the aquatic life of environment and food chain (Dar, Kamili, Dar, Chishti, & Ahmad, 2015; Dar et al., 2016; Mehra, Singh, & Chadha, 2021).

16.5.2 Impacts on humans beings

It was found in previous research works that professional disclosure to few azo dyes with aromatic amines—mainly 4-aminobiphenyl, benzidine, and 2-naphthylamine enhanced the threat to bladder cancer (Chequer, Dortaz, & Oliveira, 2011; Puvaneswari, Muthukrishnan, & Gunasekaran, 2006). Such dyes are lethal and moreover comprise such chemicals which are teratogenic, carcinogenic, or mutagenic to humans (Chequer et al., 2013). It is well recognized that the carcinogenicity occurred due to the existence of azo dyes in fabric industry wastewater (Khan & Malik, 2014). In previous research, it was reported that the Malachite Green dye not just damages the reproductive systems and human immune system but also acts as a probable mutagenic and carcinogenic agent as well (Khan, Anas, & Malik, 2019). The CI Disperse Blue dye cause frameshift mutations and base pair substitutions in *Salmonella* (Umbuzeiro et al., 2005). Moreover, the cytotoxic and genotoxic influences of CI Disperse Blue dye on cells of human beings were also researched (Akarslan & Demiralay, 2015; Tsuboy et al., 2007).

16.5.3 Effects on water bodies

The color associated with fabric dyes not just the reasons for esthetic destruction of water reservoirs but also avoids the light diffusion over water (Hassan & Carr, 2018), which leads to a fall in the photosynthesis rate and dissolved oxygen scales disturbing the whole marine life (Hassan

& Carr, 2018). The fabric dyes behave as poisonous, carcinogenic, and mutagenic agents (Khatri, Nidheesh, Singh, & Kumar, 2018). It is found that azo and nitro compounds are decomposed into detrimental cancer-causing amines and dumped in water reservoirs. Such derivatives caused the water contamination and upset the marine life (Chen, 2006).

16.6 Conclusion

Fabric industry effluent needs a suitable remediation before their final discharge in environment. The remaining dyes have complex structure which is poisonous for the living beings in water and soil of the affected region. The manufacturing industries require a cost-effective and reasonable process for the treatment of their wastewaters, so when discharged into aquatic bodies, it will have less influence on the environment. The microbial and enzymatic degradation of the effluent is a feasible choice because they don't generate huge quantity of sludge, have no bad influence on the environment, and are low-cost. Hence, it can be evaluated that microbes have great ability for the treatment of wastewaters and can efficaciously be used to keep the environment healthy and clean. Thus there is a necessity to more investigation of decomposition abilities of several microorganisms.

16.7 Future perspectives

More research is required to explore and calculate the effectiveness of the treatment processes employed to treat textile industry effluent comprising complex dye combinations. Moreover, it is essential to pay attention on the comparison of the removal efficiency of every textile dye individually and in common mixtures, as this field of study remains limited. More research is needed to emphasize on metal contamination in textile industry wastewater and the process employed for their remediation.

References

Abd Ellatif, S., El-Sheekh, M. M., & Senousy, H. H. (2021). Role of microalgal ligninolytic enzymes in industrial dye decolorization. *International Journal of Phytoremediation*, *23*(1), 41–52.

Abioye, O. P., Mustapha, O. T., & Aransiola, S. A. (2014). Biological treatment of textile effluent using *Candida zeylanoides* and *Saccharomyces cerevisiae* isolated from soil. *Advanced Biology. Article ID 670394*, 1–4.

Adeel, S., Kiran, S., Ahmad, T., Habib, N., Tariq, K., & Hussaan, M. (2020). Bio-mordants in conjunction with sustainable radiation tools for modification of dyeing of

natural fibers. In: *Frontiers of textile materials: Polymers, nanomaterials, enzymes, and advanced modification techniques* (pp. 355–367). Wiley.
Akarslan, F., & Demiralay, H. (2015). Effects of textile materials harmful to human health. *Acta Physica Polonica A*, *128*(2B), 407–409.
Alderete, B. L., da Silva, J., Godoi, R., da Silva, F. R., Taffarel, S. R., da Silva, L. P., & Picada, J. N. (2021). Evaluation of toxicity and mutagenicity of a synthetic effluent containing azo dye after advanced oxidation process treatment. *Chemosphere*, *263*, 128291.
Ali, N., Hameed, A., & Ahmed, S. (2010). Role of brown-rot fungi in the bioremoval of azo dyes under different conditions. *Brazilian Journal of Microbiology*, *41*, 907–915.
Ali, S. S., Sun, J., Koutra, E., El-Zawawy, N., Elsamahy, T., & El-Shetehy, M. (2021). Construction of a novel cold-adapted oleaginous yeast consortium valued for textile azo dye wastewater processing and biorefinery. *Fuel*, *285* 119050.
Alluri, H. K., Ronda, S. R., Settalluri, V. S., Bondili, J. S., Suryanarayana, V., & Venkateshwar, P. (2007). Biosorption: An eco-friendly alternative for heavy metal removal. *African Journal of Biotechnology*, *6*, 2924–2931.
Arlt, V. M., Glatt, H., Muckel, E., Pabel, U., Sorg, B. L., Schmeiser, H. H., & Phillips, D. H. (2002). Metabolic activation of the environmental contaminant 3-nitrobenzanthrone by human acetyltransferases and sulfotransferase. *Carcinogenesis*, *23*(11), 1937–1945.
Asgher, M., Bhatti, H. N., Ashraf, M., & Legge, R. L. (2008). Recent developments in biodegradation of industrial pollutants by white rot fungi and their enzyme system. *Biodegradation*, *19*(6), 771–783.
Awomeso, J. A., Taiwo, A. M., Gbadebo, A. M., & Adenowo, J. A. (2010). Studies on the pollution of water body by textile industry effluents in Lagos, Nigeria. *Journal of Applied Sciences in Environmental Sanitation*, *5*(4), 353–359.
Ayman, M., El-Messery, S. M., Habib, E. E., Al-Rashood, S. T., Almehizia, A. A., Alkahtani, H. M., & Hassan, G. S. (2019). Targeting microbial resistance: Synthesis, antibacterial evaluation, DNA binding and modeling study of new chalcone-based dithiocarbamate derivatives. *Bioorganic Chemistry*, *85*, 282–292.
Barathi, S., Aruljothi, K. N., Karthik, C., & Padikasan, I. A. (2020). Optimization for enhanced ecofriendly decolorization and detoxification of Reactive Blue160 textile dye by *Bacillus subtilis*. *Biotechnology Reports*, *28*, e00522.
Behel, S., & Kumar, S. (2013). In situ Degradation of textile dyes using biological methods. *International Journal of Emerging Trends in Science and Technology*, 7, 26–31.
Benkhaya, S., M'rabet, S., & El Harfi, A. (2020). A review on classifications, recent synthesis and applications of textile dyes. *Inorganic Chemistry Communications*, *115*, 107891.
Berkessa, Y. W., Yan, B., Li, T., Jegatheesan, V., & Zhang, Y. (2020). Treatment of anthraquinone dye textile wastewater using anaerobic dynamic membrane bioreactor: Performance and microbial dynamics. *Chemosphere*, *238*, 124539.
Bhandari, S., Poudel, D. K., Marahatha, R., Dawadi, S., Khadayat, K., Phuyal, S., & Parajuli, N. (2021). Microbial enzymes used in bioremediation. *Journal of Chemistry*.
Bhat, R. A., Singh, D. V., Qadri, H., Dar, G. H., Dervash, M. A., Bhat, S. A., ... Yousaf, B. (2021). Vulnerability of municipal solid waste: An emerging threat to aquatic ecosystems. *Chemosphere*, *287*(2022), 132223. Available from https://doi.org/10.1016/j.chemosphere.2021.132223.
Bhatia, D., Sharma, N. R., Kanwar, R., & Singh, J. (2018). Physicochemical assessment of industrial textile effluents of Punjab (India). *Applied Water Science*, *8*(3), 1–12.
Bidu, J. M., Van der Bruggen, B., Rwiza, M. J., & Njau, K. N. (2021). Current status of textile wastewater management practices and effluent characteristics in Tanzania. *Water Science and Technology: A Journal of the International Association on Water Pollution Research*, *83*(10), 2363–2376.
Brahmbhatt, N. H., & Jasrai, R. T. (2016). Biodegradation of reactive dyes by two microalgal species. *International Journal of Science and Research*, *4*(5), 1697–1702.

Charumathi, D., & Das, N. (2010). Bioaccumulation of synthetic dyes by *Candida tropicalis* growing in sugarcane bagasse extract medium. *Advances in Biological Research, 4*, 233–240.
Chen, K. C., Wu, J. Y., Yang, W. B., & Hwang, S. C. J. (2003). Evaluation of effective diffusion coefficient and intrinsic kinetic parameters on azo dye biodegradation using PVA-immobilized cell beads. *Biotechnol Beioeng, 83(7)*, 821–832.
Cheriaa, J., Bettaieb, F., Denden, I., & Bakhrouf, A. (2009). Characterization of new algae isolated from textile wastewater plant. *Journal of Food, Agriculture and Environment*, 7(3 & 4), 700–704.
Chung, S. Y., Venkatramanan, S., Park, N., Ramkumar, T., Sujitha, S. B., & Jonathan, M. P. (2016). Evaluation of physico-chemical parameters in water and total heavy metals in sediments at Nakdong River Basin, Korea. *Environmental Earth Sciences, 75*(1), 1–12.
Celık, S., Duman, N., Sayin, F., Akar, S. T., & Akar, T. (2021). Microbial cells immobilized on natural biomatrix as a new potential ecofriendly biosorbent for the biotreatment of reactive dye contamination. *Journal of Water Process Engineering, 39*, 101731.
Ceretta, M. B., Nercessian, D., & Wolski, E. A. (2021). Current trends on role of biological treatment in integrated treatment technologies of textile wastewater. *Frontiers in Microbiology, 12*, 518.
Chen, H. (2006). Recent advances in azo dye degrading enzyme research. *Current Protein & Peptide Science*, 7, 101–111.
Chequer, F., Dortaz, D., & Oliveira, D. (2011). Azo dyes and their metabolites: Does the discharge of the azo dyes into water bodies represent human and ecological risks? In P. Hauser (Ed.), *Advances in treating textile effluents*. London: InTechOpen, 10.5772/19872 ISBN: 978-953-307-704-8.
Chequer, F. D., De Oliveira, G. R., Ferraz, E. A., Cardoso, J. C., Zanoni, M. B., & de Oliveira, D. P. (2013). Textile dyes: Dyeing process and environmental impact. In: *Eco-friendly textile dyeing and finishing* (pp. 151–176). InTechOpen, 6 (6).
Chowdhary, P., Bharagava, R. N., Mishra, S., & Khan, N. (2020). Role of industries in water scarcity and its adverse effects on environment and human health. In: *Environmental concerns and sustainable development* (pp. 235–256). Singapore: Springer.
Coughlin, M. F., Kinkle, B. K., & Bishop, P. L. (1999). Degradation of azo dyes containing aminonaphthol by *Sphingomonas* sp strain 1CX. *Journal of Industrial Microbiology and Biotechnology, 23*(4), 341–346.
Dafale, N., Wate, S., Meshram, S., & Neti, N. R. (2010). Bioremediation of wastewater containing azo dyes through sequential anaerobic–aerobic bioreactor system and its biodiversity. *Environmental Reviews, 18*, 21–36.
Daneshvar, N., Salari, D., Niaei, A., Rasoulifard, M. H., & Khataee, A. R. (2005). Immobilization of TiO_2 nanopowder on glass beads for the photocatalytic decolorization of an azo dye CI Direct Red 23. *Journal of Environmental Science and Health, Part A, 40*(8), 1605–1617.
Dao, A. T., Smits, M., Dang, H. T., Brouwer, A., & de Boer, T. E. (2021). Elucidating fungal *Rigidoporus* species FMD21 lignin-modifying enzyme genes and 2,3,7,8-tetrachlorodibenzo-p-dioxin degradation by laccase isozymes. *Enzyme and Microbial Technology, 147*, 109800.
Dar, G. H., Bandh, S. A., & Kamili, A. N. (2011). Bacteriological analysis of soil from Yusmarg Health Resort of Kashmir Valley. *International Journal of Current Research, 3*(10), 53–56.
Dar, G. H., Bandh, S. A., Kamili, A. N., Nazir, R., & Bhat, R. A. (2013). Comparative analysis of different types of bacterial colonies from the soils of Yusmarg Forest, Kashmir valley India. *Ecologia Balkanica, 5*(1), 31–35.
Dar, G. H., Kamili, A. N., Chishti, M. Z., Dar, S. A., Tantry, T. A., & Ahmad, F. (2016). Characterization of *Aeromonas sobria* isolated from fish Rohu (*Labeo rohita*) collected from polluted pond. *Journal of Bacteriology & Parasitology*, 7(3), 1–5. Available from https://doi.org/10.4172/2155-9597.1000273.

Dar, G. H., Kamili, A. N., Dar, S. A., Chishti, M. Z., & Ahmad, F. (2015). Detection and characterization of potentially pathogenic *Aeromonas sobria* isolated from fish *Hypophthalmichthys molitrix* (Cypriniformes: Cyprinidae). *Microbial Pathogenesis* (2015). Available from https://doi.org/10.1016/j.micpath.2015.10.017.

Dar, G. H., Kamili, A. N., Nazir, R., Bandh, S. A., & Bhat, R. A. (2012). A preliminary study of colony forming units of bacteria from the soils of Yusmarg Forest, Kashmir Valley, India. *International Journal of Current Research, 4*(12), 467–472.

Del Cerro, C., Erickson, E., Dong, T., Wong, A. R., Eder, E. K., Purvine, S. O., & Salvachúa, D. (2021). Intracellular pathways for lignin catabolism in white-rot fungi. *Proceedings of the National Academy of Sciences of the United States of America, 118*(9), 1–10.

Dilek, F. B., Taplamacioglu, H. M., & Tarlan, E. (1999). Colour and AOX removal from pulping effluents by algae. *Applied Microbiology and Biotechnology, 52*(4), 585–591.

Donmez, G., & Asku, Z. (2002). Removal of chromium (VI) from saline wastewater by *Dunaliella* species. *Process Biochemistry, 38*, 751–762.

Dos Santos, A. B., Cervantes, F. J., & Van Lier, J. B. (2007). Review paper on current technologies for decolourisation of textile wastewaters: Perspectives for anaerobic biotechnology. *Biores. Technol, 98*(12), 2369–2385.

Dubey, S. K., Dubey, J., Mehra, S., Tiwari, P., & Bishwas, A. J. (2011). Potential use of cyanobacterial species in bioremediation of industrial effluents. *African Journal of Biotechnology, 10*(7), 1125–1132.

Durán, N., Rosa, M. A., D'Annibale, A., & Gianfreda, L. (2002). Applications of laccases and tyrosinases (phenoloxidases) immobilized on different supports: A review. *Enzyme and Microbial Technology, 31*(7), 907–931.

El Harfi, S., & El Harfi, A. (2017). Classifications, properties and applications of textile dyes: A review. *Applied Journal of Environmental Engineering Science, 3*(3), 00000–00003.

Elisangela, F., Andrea, Z., Fabio, D. G., de Menezes, C. R., Regina, D. L., & Artur, C. P. (2009). Biodegradation of textile azo dyes by a facultative *Staphylococcus* arlettae strain VN-11 using a sequential microaerophilic/aerobic process. *International Biodeterioration and Biodegradation, 63*(3), 280–288.

El-Naggar, N. E. A., Hamouda, R. A., Saddiq, A. A., & Alkinani, M. H. (2021). Simultaneous bioremediation of cationic copper ions and anionic methyl orange azo dye by brown marine alga *Fucus vesiculosus*. *Scientific Reports, 11*(1), 1–19.

El-Sheekh, M. M., Gharieb, M. M., & Abou-El-Souod, G. W. (2009). Biodegradation of dyes by some green algae and cyanobacteria. *International Biodeterioration and Biodegradation, 63*(6), 699–704.

Elumalai, S., & Saravanan, G. K. (2016). The role of microalgae in textile dye industrial waste water recycle (phycoremediation). *International Journal of Pharma and Bio Sciences, 7*(4), 662–673.

Foo, K. Y., & Hameed, B. H. (2010). Insights into the modeling of adsorption isotherm systems. *Chemical Engineering Journal, 156*(1), 2–10.

Forgacs, E., Cserhati, T., & Oros, G. (2004). Removal of synthetic dyes from wastewaters: A review. *Environment International, 30*, 953–971.

Franciscon, E., Zille, A., & Durran, L. R. (2009). Decolorization and biodegradation of reactive sulfonated azo dyes by a newly isolated *Brevibacterium* sp. strain VN-15. *International Biodeterioration & Biodegradation, 63*, 280–288.

Gahlout, M., Prajapati, H., Tandel, N., & Patel, Y. (2021). Biosorption: An eco-friendly technology for pollutant removal. In: *Microbial rejuvenation of polluted environment* (pp. 207–227). Singapore: Springer.

Gajare, P. S. P., Meshram, P. R., Bhadane, M. A., & Tasgaonkar, R. R. (2012). Comparative evaluation studies of sustained released azithromycin microsphere prepared by using different solvent. *Inventi Rapid: Pharm Tech, 4*, 1–5.

GilPavas, E., Dobrosz-Gómez, I., & Gómez-García, M. Á. (2020). Efficient treatment for textile wastewater through sequential electrocoagulation, electrochemical oxidation and adsorption processes: Optimization and toxicity assessment. *Journal of Electroanalytical Chemistry, 878*, 114578.

Gowri, S., Gandhi, R. R., & Sundrarajan, M. (2014). Structural, optical, antibacterial and antifungal properties of zirconia nanoparticles by biobased protocol. *Journal of Materials Science & Technology, 30*(8), 782–790.

Gulzar, T., Farooq, T., Kiran, S., Ahmad, I., & Hameed, A. (2019). Green chemistry in the wet processing of textiles. In: *The impact and prospects of green chemistry for textile technology* (pp. 1–20). Woodhead Publishing.

Gunti, H., Maruthiramaih, V. V., & Shivanandappa, T. B. (2021). Bio-degradation of azo dye Acid Orange-10 by a new isolate of *Bacillus subtilis* isolated from soil sample around textile industry in South Karnataka. *Biosciences Biotechnology Research Asia, 17*(4), 707–716.

Gupta, V. K., Suhas Nayak, A., Agarwal, S., Chaudhary, M., & Tyagi, I. (2014). Removal of Ni(II) ions from water using porous carbon derived from scrap tyre. *Journal of Molecular Liquids, 190*, 215–222.

Hassan, M. M., & Carr, C. M. (2018). A critical review on recent advancements of the removal of reactive dyes from dyehouse effluent by ion-exchange adsorbents. *Chemosphere, 209*(1), 201–219.

Hasan, F., Shah, A. A., & Hameed, A. (2006). Industrial applications of microbial lipases. *Enzyme and Microbial Technology, 39*(2), 235–251.

Hasyimah, N. A. R., Furusawa, G., & Amirul, A. A. (2021). Biosorption of a dye and heavy metals using dead cells of filamentous bacterium, *Aureispira* sp. CCB-QB1. *International Journal of Environmental Science and Technology, 18*(6), 1627–1636.

Henciya, S., Shankar, A. M., & Malliga, P. (2013). Decolorization of textile dye effluent by marine cyanobacterium *Lyngbya* sp. BDU 9001 with coir pith. *International Journal of Environmental Sciences, 3*(6), 1909–1918.

Horitsu, K., Okamura, A., Ohwada, H., Kuroiwa, N., Moriyama, Y., & Fujimura, S. (1997). Increase of nicotinamide methyl-transferase activity and cell death in liver of mouse bearing Ehrlich ascites tumor under nicotinamide administration. *Cell Biology and Toxicology, 13*, 61–61.

Hu, C. A., Delauney, A. J., & Verma, D. P. (1992). A bifunctional enzyme (delta 1-pyrroline-5-carboxylate synthetase) catalyzes the first two steps in proline biosynthesis in plants. *Proceedings of the National Academy of Sciences of the United States of America, 89* (19), 9354–9358.

Imran, M., Crowley, D. E., Khalid, A., Hussain, S., Mumtaz, M. W., & Arshad, M. (2014). Microbial biotechnology for decolorization of textile wastewaters. *Reviews in Environmental Science and Bio/Technology, 9*, 32–37.

Jadhav, J. P., Parshetti, G. K., Kalme, S. D., & Govindwar, S. P. (2007). Decolourization of azo dye methyl red by *Saccharomyces cerevisiae* MTCC 463. *Chemosphere, 68*(2), 394–400.

Jafari, N., Soudi, M. R., & Kasra-Kermanshahi, R. (2014). Biodecolorization of textile azo dyes by isolated yeast from activated sludge: *Issatchenkia orientalis JKS6. Annals of Microbiology, 64*(2), 475–482.

Jamee, R., & Siddique, R. (2019). Biodegradation of synthetic dyes of textile effluent by microorganisms: An environmentally and economically sustainable approach. *European Journal of Microbiology and Immunology, 9*(4), 114–118.

Jin, X. C., Liu, G. Q., Xu, Z. H., & Tao, W. Y. (2007). Decolorization of a dye industry effluent by *Aspergillus fumigatus XC6. Applied Microbiology and Biotechnology, 74*(1), 239–243.

Joshi, A. U., Hinsu, A. T., Kotadiya, R. J., Rank, J. K., Andharia, K. N., & Kothari, R. K. (2020). Decolorization and biodegradation of textile di-azo dye Acid Blue 113 by *Pseudomonas stutzeri* AK6. *Biotech, 10*(5), 1–16.

Kalyuzhnyi, S., Yemashova, N., & Federovich, V. (2006). Bacterial degradation of azo dyes. *Water Science and Technology: A Journal of the International Association on Water Pollution Research*, *54*, 73–79.

Keharia, H., & Madamwar, D. (2003). Bioremediation concepts for treatment of dye containing wastewater: A review. *Indian Journal of Experimental Biology*, *41*, 1068–1075.

Khan, M. D., Li, D., Tabraiz, S., Shamurad, B., Scott, K., Khan, M. Z., & Yu, E. H. (2021). Integrated air cathode microbial fuel cell-aerobic bioreactor set-up for enhanced bioelectrodegradation of azo dye Acid Blue 29. *The Science of the Total Environment*, *756*, 143752.

Khan, S., Anas, M., & Malik, A. (2019). Mutagenicity and genotoxicity evaluation of textile industry wastewater using bacterial and plant bioassays. *Toxicology Reports*, *6*, 193–201.

Khan, S., & Malik, A. (2014). Environmental and health effects of textile industry wastewater. In: *Environmental deterioration and human health* (pp. 55–71). Dordrecht: Springer.

Khatri, J., Nidheesh, P. V., Singh, T. A., & Kumar, M. S. (2018). Advanced oxidation processes based on zero-valent aluminium for treating textile wastewater. *Chemical Engineering Journal*, *348*, 67–73.

Kiani, R., Mirzaei, F., Ghanbari, F., Feizi, R., & Mehdipour, F. (2020). Real textile wastewater treatment by a sulfate radicals-advanced oxidation process: Peroxydisulfate decomposition using copper oxide (CuO) supported onto activated carbon. *Journal of Water Process Engineering*, *38*, 101623.

Kiran, S., Adeel, S., Nosheen, S., Hassan, A., Usman, M., & Rafique, M. A. (2017). Recent trends in textile effluent treatments: A review. In: *Advanced materials for wastewater treatment* (pp. 29–49). Wiley.

Kiran, S., Ahmad, T., Gulzar, T., Ashraf, A., Naqvi, S. A. R., & Naz, S. (2020). Agro-waste applications for bioremediation of textile effluents. In: *Recycling from waste in fashion and textiles: A sustainable and circular economic approach* (pp. 391–421). Wiley.

Kiran, S., Gulzar, T., Iqbal, S., Habib, N., Hassan, A., & Naz, S. (2019). *Valorization of wastes for the remediation of toxicants from industrial wastewater. Integrating green chemistry and sustainable engineering* (pp. 473–525). Wiley.

Kiran, S., Nosheen, S., Abrar, S., Anjum, F., Gulzar, T., & Naz, S. (2019). Advanced approaches for remediation of textile wastewater: A comparative study. In: *Advanced functional textiles and polymers: Fabrication, processing and applications* (pp. 201–264). Wiley.

Kishor, R., Purchase, D., Saratale, G. D., Saratale, R. G., Ferreira, L. F. R., Bilal, M., & Bharagava, R. N. (2021). Ecotoxicological and health concerns of persistent coloring pollutants of textile industry wastewater and treatment approaches for environmental safety. *Journal of Environmental Chemical Engineering*, 105012.

Köktürk, M., Altindağ, F., Ozhan, G., Çalimli, M. H., & Nas, M. S. (2021). Textile dyes Maxilon blue 5G and Reactive blue 203 induce acute toxicity and DNA damage during embryonic development of *Danio rerio*. *Comparative Biochemistry and Physiology, C*, *242*, 108947.

Kunjadia, P. D., Sanghvi, G. V., Kunjadia, A. P., Mukhopadhyay, P. N., & Dave, G. S. (2016). Role of ligninolytic enzymes of white rot fungi (*Pleurotus* spp.) grown with azo dyes. *SpringerPlus*, *5*(1), 1–9.

Kumar, M., Kumar, P., Das, P., Solanki, R., & Kapur, M. K. (2020). Potential applications of extracellular enzymes from *Streptomyces* spp. in various industries. *Archives of Microbiology*, *202*, 1597–1615.

Machado, K. M. G., & Matheus, D. R. (2006). Biodegradation of Remazol brilliant blue R by ligninolytic enzymatic complex produced by *Pleurotus ostreatus*. *Brazilian Journal of Microbiology*, *37*, 468–473.

Mane, U. V., Gurav, P. N., Deshmukh, A. M., & Govindwar, S. P. (2008). Degradation of textile dye reactive navy blue RX (Reactive blue 59) by an isolated Actinomycetes *Streptomycetes Krainskii SUK 5*. *Malaysian Journal of Microbiology*, *4*, 1–5.

Mahmood, I., Imadi, S. R., Shazadi, K., Gul, A., & Hakeem, K. R. (2016). Effects of pesticides on environment. In: *Plant, soil and microbes* (pp. 253–269). Cham: Springer.
Manzoor, J., & Sharma, M. (2020). Impact of textile dyes on human health and environment. In: *Impact of textile dyes on public health and the environment* (pp. 162–169). IGI Global.
Meehan, C., Banat, I. M., McMullan, G., Nigam, P., Smyth, F., & Marchant, R. (2000). Decolorization of Remazol Black-B using a thermotolerant yeast, *Kluyveromyces marxianus* IMB3. *Environment International, 26*(1-2), 75–79.
Mehra, S., Singh, M., & Chadha, P. (2021). Adverse impact of textile dyes on the aquatic environment as well as on human beings. *Toxicology International, 28*(2), 165–176.
Mohanty, S. S., & Kumar, A. (2021). Enhanced degradation of anthraquinone dyes by microbial monoculture and developed consortium through the production of specific enzymes. *Scientific Reports, 11*(1), 1–15.
Nakanishi, K., Sakiyama, T., & Imamura, K. (2001). On the adsorption of proteins on solid surfaces, a common but very complicated phenomenon. *Journal of Bioscience and Bioengineering, 91*(3), 233.
Odubanjo, G. O., Oyetibo, G. O., & Ilori, M. O. (2021). Ecological risks of heavy metals and microbiome taxonomic profile of a freshwater stream receiving wastewater of textile industry. *Frontiers in Environmental Science, 9*, 554490.
Ogugbue, C. J., & Sawidis, T. (2011). Bioremediation and detoxification of synthetic wastewater containing triarylmethane dyes by *Aeromonas hydrophila* isolated from industrial effluent. *Biotechnology Research International, Article ID 967925*, 1–11.
Omar, H. H. (2008). Algal decolorization and degradation of monoazo and diazo dyes. *Pakistan Journal of Biological Sciences, 11*(10), 1310–1316.
Ozdemir, I., Ahrens, S., Mücklich, S., & Wielage, B. (2008). Nanocrystalline $Al-Al_2O_3p$ and SiCp composites produced by high-energy ball milling. *Journal of Materials Processing Technology, 205*(1-3), 111–118.
Paba, G. M. M., & Ávila, R. B. (2021). Enterococcus present in marine ecosystems and their potential to degrade azo dyes. In: *Coastal environments*. IntechOpen.
Papirio, S., Frunzo, L., Mattei, M. R., Ferraro, A., Race, M., D'Acunto, B., & Esposito, G. (2017). Heavy metal removal from wastewaters by biosorption: Mechanisms and modeling. In: *Sustainable heavy metal remediation* (pp. 25–63). Cham: Springer.
Pal, R. (2017) Sustainable design and business models in textile and fashion industry. In: *Sustainability in the textile industry*. (pp. 109–138). Singapore: Springer.
Palamthodi, S., Patil, D., & Patil, Y. (2011). Microbial degradation of textile industrial effluents. *African Journal of Biotechnology, 10*(59), 12657–12661.
Pandey, P., Shinde, V. N., Deopurkar, R. L., Kale, S. P., Patil, S. A., & Pant, D. (2016). Recent advances in the use of different substrates in microbial fuel cells toward wastewater treatment and simultaneous energy recovery. *Applied Energy, 168*, 706–723.
Pardo, R., Herguedas, M., Barrado, E., & Vega, M. (2003). Biosorption of cadmium, copper, lead and zinc by inactive biomass of *Pseudomonas Putida*. *Analytical and Bioanalytical Chemistry, 376*, 26–32.
Parveen, S., Asgher, M., & Bilal, M. (2021). Lignin peroxidase-based cross-linked enzyme aggregates (LiP-CLEAs) as robust biocatalytic materials for mitigation of textile dyes-contaminated aqueous solution. *Environmental Technology & Innovation, 21*, 101226.
Pecková, V., Legerská, B., Chmelová, D., Horník, M., & Ondrejovič, M. (2021). Comparison of efficiency for monoazo dye removal by different species of white-rot fungi. *International Journal of Environmental Science and Technology, 18*(1), 21–32.
Patel, Y., Chhaya, U., Rudakiya, D. M., & Joshi, S. (2021). Biological decolorization and degradation of synthetic dyes: A green step toward sustainable environment. In: *Microbial rejuvenation of polluted environment* (pp. 77–110). Singapore: Springer.

Peralta-Zamora, P., Pereira, C., Tiburtius, E., Moraes, S., Rosa, M., Minussi, R., & Duran, N. (2003). Decolorization of reactive dyes by immobilized laccase. *Applied Catalysis B: Environmental*, *42*, 131–144.

Perpetuo, E. A., Souza, C. B., & Nascimento, C. A. O. (2011). *Engineering bacteria for bioremediation. Progress in molecular and environmental bioengineering - From analysis and modeling to technology applications*. IntechOpen.

Puvaneswari, N., Muthukrishnan, J., & Gunasekaran, P. (2006). Toxicity assessment and microbial degradation of azo dyes. *Indian Journal of Experimental Biology*, *44*, 618626.

Ramachandran, P., Sundharam, R., Palaniyappan, J., & Munusamy, A. P. (2013). Potential process implicated in bioremediation of textile effluents: A review. *Advances in Applied Science Research*, *4*, 131–145.

Ramalho, P. A., Scholze, H., Cardoso, M. H., Ramalho, M. T., & Oliveira-Campos, A. M. (2002). Improved conditions for the aerobic reductive decolorisation, of azo dyes by *Candida zeylanoides*. *Enzyme and Microbial Technology*, *31*, 848–854.

Robinson, T., McMullan, G., Marchant, R., & Nigam, P. (2001). Remediation of dyes in textile effluent: A critical review on current treatment technologies with a proposed alternative. *Bioresource Technology*, *77*, 247–255.

Ruscasso, F., Bezus, B., Garmendia, G., Vero, S., Curutchet, G., Cavello, I., & Cavalitto, S. (2021). *Debaryomyces hansenii* F39A as biosorbent for textile dye removal. *Revista Argentina de Microbiologia*.

Samsami, S., Mohamadizaniani, M., Sarrafzadeh, M. H., Rene, E. R., & Firoozbahr, M. (2020). Recent advances in the treatment of dye-containing wastewater from textile industries: Overview and perspectives. *Process Safety and Environmental Protection*, *143*, 138–163.

Sarayu, K., & Sandhya, S. (2010). Aerobic biodegradation pathway for Remazol Orange by *Pseudomonas aeruginosa*. *Applied Biochemistry and Biotechnology*, *160*(4), 1241–1253.

Sharif, A., Nasreen, Z., Bashir, R., & Kalsoom, S. (2020). Microbial degradation of textile industry effluents: A review. *Pure and Applied Biology*, *9*(4), 2361–2382.

Shindhal, T., Rakholiya, P., Varjani, S., Pandey, A., Ngo, H. H., Guo, W., & Taherzadeh, M. J. (2021). A critical review on advances in the practices and perspectives for the treatment of dye industry wastewater. *Bioengineered*, *12*(1), 70–87.

Škufca, D., Kovačič, A., Prosenc, F., Bulc, T. G., Heath, D., & Heath, E. (2021). Phycoremediation of municipal wastewater: Removal of nutrients and contaminants of emerging concern. *Science of the Total Environment*, *782*(146949), 1–6.

Soundharya, K., Rashika, B., Mounica, R., & Masi, C. (2016). Enzymatic treatment of effluents from textile industries. *Journal of Chemical and Pharmaceutical Research*. *8*, 382–389.

Soylu, M., Gökkuş, Ö., & Özyonar, F. (2020). Foam separation for effective removal of disperse and reactive dyes from aqueous solutions. *Separation and Purification Technology*, *247*, 116985.

Tang, Q., Che, C., Lin, J., He, H., Zhao, W., Lv, L., & Zhao, G. (2019). Maresin1 regulates neutrophil recruitment and IL-10 expression in *Aspergillus fumigatus* keratitis. *International Immunopharmacology*, *69*, 103–108.

Telke, A. A., Ghodake, G. S., Kalyani, D. C., Dhanve, R. S., & Govindwar, S. P. (2011). Biochemical characteristics of a textile dye degrading extracellular laccase from a *Bacillus* sp. ADR. *Bioresource Technology*, *102*(2), 1752–1756.

Telke, A. A., Kadam, A. A., & Govindwar, S. P. (2015). Bacterial enzymes and their role in decolorization of azo dyes. In: *Microbial degradation of synthetic dyes in wastewaters* (pp. 149–168). Cham: Springer.

Thiruppathi, K., Rangasamy, K., Ramasamy, M., & Muthu, D. (2021). Evaluation of textile dye degrading potential of ligninolytic bacterial consortia. *Environmental Challenges*, *4*, 100078.

Tsuboy, M. S., Anjeli, J. P. F., Mantovani, M. S., Knasmiiller, S., Umbuzeiro, G. A., & Ribeiro, L. R. (2007). Genotoxic, mutagenic and cytotoxic effects of the commercial dye CI Disperse Blue 291 in the human hepatic cell line HepG2. *Toxicology In Vitro: An International Journal Published in Association with BIBRA*, *21*, 1650–1655.

Tripathi, A., & Srivastava, S. K. (2011). Ecofriendly treatment of azo dyes: biodecolorization using bacterial strains. *International Journal of Bioscience, Biochemistry and Bioinformatics*, *1*(1), 37–40.

Umbuzeiro, G. A., Freeman, H., Warren, S. H., Kummrow, F., & Claxton, L. D. (2005). Mutagenicity evaluation of the commercial product CI Disperse Blue 291 using different protocols of the Salmonella assay. *Food and Chemical Toxicology: An International Journal Published for the British Industrial Biological Research Association*, *43*(1), 49–56.

Velusamy, S., Roy, A., Sundaram, S., & Kumar Mallick, T. (2021). A review on heavy metal ions and containing dyes removal through graphene oxide-based adsorption strategies for textile wastewater treatment. *Chemical Record* (New York, N.Y.).

Verma, A. K., Dash, R. R., & Bhunia, R. (2012). A review on chemical coagulation/flocculation technologies for removal of color from textile wastewaters. *Journal of Environmental Management*, *93*, 154–168.

Wijannarong, S., Aroonsrimorakot, S., Thavipoke, P., & Sangjan, S. (2013). Removal of reactive dyes from textile dyeing industrial effluent by ozonation process. *APCBEE Procedia*, *5*, 279–282.

Wijetunga, S., Li, X. F., & Jian, C. (2010). Effect of organic load on decolourization of textile wastewater containing acid dyes in upflow anaerobic sludge blanket reactor. *Journal of Hazardous Materials*, *177*(1–3), 792–798.

Won, S. W., & Yun, Y. S. (2008). Biosorptive removal of Reactive Yellow 2 using waste biomass from lysine fermentation process. *Dyes and Pigments*, *76*(2), 502–507.

Wu, L. C., Chen, C. Y., Lin, T. K., Su, Y. Y., & Chung, Y. C. (2020). Highly efficient removal of victoria blue R and bioelectricity generation from textile wastewater using a novel combined dual microbial fuel cell system. *Chemosphere*, *258*, 127326.

Yang, X., López-Grimau, V., Vilaseca, M., & Crespi, M. (2020). Treatment of textile wastewater by CAS, MBR, and MBBR: A comparative study from technical, economic, and environmental perspectives. *Water*, *12*(5), 1306.

Yu, Z., & Wen, X. (2005). Screening and identification of yeasts for decolorizing synthetic dyes in industrial wastewater. *International Biodeterioration and Biodegradation*, *56*(2), 109–114.

Yusuf, M. (2018). *Recent trends and future prospects in bioremediation of synthetic dyes: A review. Handbook of textile effluent remediation* (pp. 247–278).

Zaharia, C., Suteu, C., & Muresan, A. (2011). Options and solutions of textile effluent decolourization using some specific physico-chemical treatment steps. *Proceedings of the 6th international conference on environmental engineering and management ICEEM'06* (pp. 121–122). Balaton Lake, Hungary.

Zeng, Z., Luo, W. G., Wang, Z., & Yi, F. C. (2021). Water pollution and its causes in the Tuojiang River Basin, China: An artificial neural network analysis. *Sustainability*, *13*(2), 792.

Zhuang, M., Sanganyado, E., Zhang, X., Xu, L., Zhu, J., Liu, W., & Song, H. (2020). Azo dye degrading bacteria tolerant to extreme conditions inhabit nearshore ecosystems: Optimization and degradation pathways. *Journal of Environmental Management*, *261*, 110222.

Zhuo, R., & Fan, F. (2021). A comprehensive insight into the application of white rot fungi and their lignocellulolytic enzymes in the removal of organic pollutants. *The Science of the Total Environment*, 146132.

Zille, A., Ramalho, P., Tzanov, A., Millward, R., Aires, V., Cardoso, M. H., ... Cavaco-Paulo, A. (2004). Enhanced decolorization of Solar brilliant red 80 textile dye by an indigenous white rot fungus *Schizophyllum commune* IBL-06. *Biotechnology Progress*, *20*, 1588–1592.

Index

Note: Page numbers followed by "*f*" and "*t*" refer to figures and tables, respectively.

C

F

G

N

O

P

U

V

W

FSC
www.fsc.org
MIX
Papier | Fördert
gute Waldnutzung
FSC® C083411

Kolibri 360 GmbH
Ferdinand-Jühlke-Straße 7
99095 Erfurt, Deutschland
produktsicherheit@kolibri360.de